P9-ECJ-463

Digital Video Transcoding for Transmission and Storage

Digital Video Transcoding for Transmission and Storage

Huifang Sun
MERL Technology Lab, Cambridge, MA, USA

Xuemin Chen
Broadman Corporation, San Diego, CA, USA

Tihao Chiang
National Chiao-Tung University, Hsinchu, Taiwan

CRC PRESS

Boca Raton London New York Washington, D.C.

Library of Congress Cataloging-in-Publication Data

Catalog record is available from the Library of Congress

Direct all inquiries to CRC Press, 2000 N.W. Corporate Blvd., Boca Raton, Florida 33431.

Visit the CRC Press Web site at www.crcpress.com

International Standard Book Number 0-8493-1694-4
Printed in the United States of America 1 2 3 4 5 6 7 8 9 0
Printed on acid-free paper

Preface

During the past two decades, there has been significant progress made in digital video processing, transmission, and storage technologies. These technologies include digital video compression, digital modulation, and digital storage. Among these technologies, digital video compression is a field in which fundamental technologies are driven by practical applications. Especially, the digital video-compression standards, developed by the Moving Pictures Expert Group (MPEG) of the International Organization for Standardization (ISO) and Video Compression Expert Group (VCEG) of the International Telecommunications Union (ITU), have enabled many successful digital-video applications.

Digital video compression plays a key role in video and multimedia industries. The compression technologies bridge the gap between the huge amount of visual data required for video, multimedia transmission, and storage and limited hardware capability, despite rapid growth in the semiconductor industry. In order to effectively employ video compression technologies, many industry standards for video coding have been developed. These are MPEG-1, MPEG-2, MPEG-4, H.261, H.263, and H.264/MPEG-4 Advanced Video Coding (AVC). Several excellent books have been published on the subject of video compression and standards. These books focus on basic video coding theory and/or the video compression standards. However, there is a need for a book that provides the explanations on the standards and at the same time explains how to convert the compressed information between standards. This is the hot topic investigated by many researchers from academia and industry; our book will meet this need and will provide some of the theories and principles of video compression and transcoding technologies.

The concepts of digital video coding are in many cases sufficiently straightforward to avoid highly theoretical mathematics. The reader will find that the primary emphasis of the book is on digital video transcoding techniques, not on the basic theories and principles of digital video coding and its standards. In fact, much of the material covered is summarized via examples of practical methods for transcoder implementation, and almost all of the video transcoding technologies introduced are related to practical applications.

This book has arisen from the authors' research results from many years in both industrial and academic societies. We would like to share our experiences and research results with colleagues in the field of video transcoding. The book takes a structured approach to transcoding technologies, starting with the basic video transcoding concepts and working gradually toward the most sophisticated transcoding systems. The practical applications of transcoders are described throughout the book. The materials are summarized from many research papers, lectures, and presentations on video compression and transcoding technologies. The text is intended to be a senior/graduate level textbook and reference book for researchers and engineers in this field. We hope that college students and engineers in the video and multimedia industry can benefit from the information in this, for the most part, self-contained text on video transcoding technologies.

The book is organized as follows. It consists of 11 chapters grouped into four parts. The first part includes two chapters, which provide the background of video coding theory, principles of video transmission, and an overview of video coding standards. The second part includes three chapters that provide the theory of video transcoding and practical problems. The third part includes three chapters and presents buffer management, packet scheduling, and encryption in the transcoding. The final part contains three chapters and describes the application of transcoding, universal multimedia access with emerging standard MPEG-21, and the end-to-end test bed. These topics occur in very recent results of research and should be welcome by the readers.

Every course has as its first lecture a sneak preview and overview of the fundamental techniques and tools useful for the following chapters. Chapter 1 provides such information. It provides fundamental concepts and theoretical basis for digital video compression techniques including fundamentals of information theory and various source coding techniques. It also covers some fundamentals for video coding such as redundancy removal using spatial, temporal, and frequency domain techniques.

In Chapter 2, the topics of digital video coding standards are discussed. First, the general principles and structures of digital video coding are presented. Then the basic tools for video coding standards are introduced. In order to improve coding efficiency and increase functionality, several enhancement tools for digital video coding standards are introduced and the basic ideas behind these tools are described. Finally, a summary of different digital video coding standards and several encoding issues are presented.

Chapter 3 provides the theoretical basis and fundamentals for the video transcoding. The general concept and possible applications for video transcoding are first introduced. Then, various transcoding architectures for specific purposes are then presented and compared. Various transcoding algorithms are developed and analyzed. Finally, an alternative to achieve transcoding, namely Fine Granularity Scalabiliy, is presented.

Chapter 4 focuses on transcoding performance optimization for various functionality. The first topic is the performance improvement for reduced spatial resolution transcoding. The second topic focuses on the temporal resolution adaptation. The third topic concentrates on the syntactical adaptation for different formats of bit streams such as MPEG-1/2/4 and H.264. Lastly, several topics such as error resilient transcoding, logo insertion, watermarking, switched pictures and picture-in-picture transcoding are covered. The complexity analysis for various transcoding architectures is also provided.

In Chapter 5, transport-level transcoding techniques are introduced. To explain transport-level transcoding, we first present the basic concept of MPEG-2 system. MPEG-2 systems have two specified stream formats: transport and program, which are specified for a set of different applications. We present the transcoding techniques, which are used to perform the conversions between the transport stream and the program stream.

In Chapter 6, we present video synchronization techniques, such as system clock recovery and time stamping for decoding and presentation. As an example, we use

MPEG-2 transport systems to illustrate the key function blocks of this video synchronization technique. In MPEG-2 systems, clock recovery is possible by transmitting time stamps called *program clock references* (PCRs) in the bit stream. The PCRs are generated at the encoder by sampling the system time clock (STC). Since the decoder's free-running system clock frequency does not exactly match the encoder's STC, the reference time is reconstructed by means of a phase-locked loop (PLL) and the received PCRs. The encoder's STC of a video program is also used to create time stamps that indicate the presentation and decoding timing of video. Methods for generating the decoding and presentation time stamps in the video encoder are discussed in this chapter. In particular, the time-stamping methods for MPEG-2 video are introduced as examples. These methods can be directly used in other compressed video.

In Chapter 7, constraints on video compression/decompression buffers and the bit rate of a compressed video bit stream have been discussed. The transmission channels impose these constraints on video encoder/transcoder buffers. First, we introduce concepts of compressed video buffers. Then conditions that prevent the video encoder/transcoder and decoder buffer overflow or underflow are derived for the channel that can transmit a variable bit rate video. Next, the discussion focuses on analyzing buffer, timing recovery, and synchronization for video transcoder. The buffering implications of the video transcoder within the transmission path have also been analyzed.

In the applications of transcoding, digital right management and security issues become more and more important. In Chapter 8, we study the basic security concepts, algorithms, and models for transporting of compressed digital video. We describe a number of important cryptographic algorithms that are used to protect high-value video content and review several configurations of conditional access systems for digital video transport systems. In particular, we outline the DES algorithm in details. We introduce four modes of operation and their applications in both block cipher and stream cipher. We also look at the area of public-key cryptosystems and examine two algorithms, the Rivest, Shamir, and Adleman encryption algorithm, based on the product of two large prime numbers, and the Diffie-Hellman key exchange algorithm, based on the discrete logarithm problem for finite field. We also review three configurations of conditional access schemes. Finally, we have introduced some basic ideas of the multi-hop encryption systems. We believe this chapter will be very useful for many researchers and engineers for developing the transcoding techniques in many applications.

Chapter 9 is devoted to the application and implementation of video transcoding. The first application is the transcoder of MPEG-2 to MPEG-4 bit streams. The MPEG-4 video coding standard has been used for video streaming and mobile terminals, but much content in video servers is encoded with MPEG-2; therefore, we need convert the MPEG-2 to MPEG-4 for users with MPEG-4 available decoder to receive the MPEG-2 encoded contents. Second, the techniques of error resilience transcoding, which are related to video transcoder for IP network, are discussed. In the applications of IP networks and wireless networks, the function of error resilience is important. How to increase the error robustness for the transcoded bit stream is introduced. Finally, the object-based transcoding technique for MPEG-4 will be presented.

In Chapter 10, we address several transcoding aspects related to the distribution of digital content. The first objective is to provide an overview of the universal multimedia access (UMA) concept. The primary function of UMA services is to provide the best quality of service or user experience by either selecting/adapting the content format to meet the playback environment, or adapting the content playback environment to accommodate the content. The second objective is to describe how the concept of UMA relates to the emerging MPEG standard, Digital Item Adaptation (DIA), which is the Part 7 of the MPEG-21 standard. The update on the standards activity in this area is presented. Finally, we address the impact that DIA will have on transcoding strategies, and we analyze some areas of future research.

In Chapter 11, we introduce the concept of real-time transport protocol and carriage of multimedia content over IP networks. The first section covers the various elements of the video streaming and transcoding system. The second section describes how content can be carried over IP networks. Details of the real-time transport protocol will be described. An implementation example of such a system is described. The last section describes some simulation results and conclusions.

Acknowledgments

We wish to acknowledge everyone who helped in the preparation of this book—in particular, the reviewers, who have made detailed comments to guide us in our final choice of content. We also wish to acknowledge Nora Konopka, the Engineering and Environmental Sciences editor at CRC Press, who helped in many aspects and made very efficient arrangements for publishing this book.

The first author would like to express his deep appreciation to his colleague, Dr. Anthony Vetro, for fruitful technical discussions related to some contents of this book and many joint publications and patents on this topic, which are reflected in this book. He would like to thank Drs. Ajay Divakaran, Fatih Porikli, Hao-Song Kong, Zafer Sahinoglu, and Jun Xin for their help in many aspects on this book. He also would like to acknowledge Drs. Richard Waters, Takashi Kan, Kent Wittenburg, and Joe Marks for their continuing support and encouragement. His acknowledgement also goes to Drs. Tokumichi Murakami and Kohtaro Asai for their friendly support and encouragement. He also would to thank Professor Wen Gao and his students Yan Lu and Lujun Yuan for their help.

Firstly, the second author wishes to thank his thesis advisor, Professor Dimitris Anastassiou, who introduced him into the field of digital video compression while it was still an emerging area. His guidance opens the door and the second author could not be here without his help. He also wishes to thank his parents, Lian-Yuan Chiang and Shing Yeh, who have been very supportive during this project. He also wishes to thank Dr. Ya-Qin Zhang for his leadership and friendship in completing this book. The second author wishes to extend special thanks to Professors Che-Ho Wei, Suh-Yin Lee, and David Lin for their help. The second author wishes to thank Professors Hsueh-Ming Hang and Chun-Jen Tsai for their support and help in this project. Professor Hang has been extremely supportive since the second author joined National Chiao Tung University in 1999. This project would not be possible without his help. The second author wishes to thank Professor Sheng-Jyh Wang for his help in this project and his continuing friendship. He would like to thank Dr. Chung-Neng Wang for his friendship and assistance in the writing of this book. He wishes to thank the group members, including Peter Chuang, Yao-Chung Lin, Chih-Hung Lee, and Hsiang-Chun Huang at the Multimedia Communications Group of the Communications and Signal Processing Laboratory at National Chiao Tung University in Taiwan. The second author wishes to thank Dr. Jun Xin, Prof. Chia-Wen Lin and Prof. Ming-Ting Sun for providing an early draft of their tutorial on transcoding. He wishes to thank Dr. Anthony Vetro for his helpful discussions on the various transcoding issues in one of the joint works. He wishes to thank Mr. Shinga Chen for his help.

The third author would like to thank Dr. Wade Wan for his review of parts of the manuscript and for his thoughtful comments. He also gratefully acknowledges Robert Eifrig, Dr. Ajay Luthra, Dr. Fan Ling, Dr. Vincent Liu, Dr. Sam Narasimhan, Dr. Krit Panusopone, Dr. Ganesh Rajan, and Dr. Limin Wang for their contributions in many joint patents, papers, and reports that are reflected in this book. He would also like to thank Professors Irving S. Reed, Tor Helleseth, Weiping Li, Ronald Crochiere, Homer H. Chen, Ya-Qin Zhang, Guirong Guo, T. K. Truong, T. C. Cheng, and Dr. Anthony Vetro for their continuing support and encouragement.

Support for the completion of the manuscript has been provided by the National Chiao-Tung University (Taiwan) and to all we are truly grateful. In particular, we truly appreciate the attentiveness that Peter Chuang has given to the preparation of the manuscript. We also wish to thank Chih-Hung Lee for his kind and timely assistance in the production phase of the project. Authors also would like to express their appreciation to many friends and colleagues of the MPEGers who provided wonderful MPEG documents and tutorial materials that are cited in many chapters of this book. It was important to be able to use many published results in the text. We would like to thank the people who made possible these important contributions.

Finally, we would like to show our great appreciation to our families for their constant help, support, and encouragement.

About the Authors

Huifang Sun is the vice president of Mitsubishi Electric Research Laboratories (MERL), and the deputy director of technology laboratory of MERL. He is a MERL Fellow and an IEEE Fellow. He graduated from Harbin Military Engineering Institute, Harbin, China, and received the Ph.D. from University of Ottawa, Canada. He was an Associate Professor at Engineering Department of Fairleigh Dickinson University before moving to Sarnoff Corporation in 1990 as a member of technical staff, and was promoted to a technology leader of Digital Video Communication later. In 1995, he joined MERL. His research interests include digital video/image compression and digital communication. He has published a textbook and more than 120 journal and conference papers. He holds 30 U.S. patents and has more pending in the area of digital video compression, processing, and communications. He received AD-HDTV Team Award in 1992 and Technical Achievement Award for optimization and specification of the Grand Alliance HDTV video compression algorithm in 1994 at Sarnoff Lab. He received the best paper award of 1992 for *IEEE Transactions on Consumer Electronics*, the best paper award of 1996 ICCE, and the best transactions paper award of 2003 *IEEE Transactions on Circuits and Systems for Video Technology.* He is now an Associate Editor for *IEEE Transactions on Circuits and Systems for Video Technology* and the Chair of Visual Processing Technical Committee of the IEEE Circuits and System Society.

Xuemin Chen is a technical director of Broadband Communications Business Group of Broadcom Corporation and an IEEE fellow. He has a Ph. D. degree in electrical engineering from University of Southern California. He had held various engineering positions such as research scientist, senior staff/manager, and senior principal scientist in American Online (Johnson & Grace), General Instrument Corporation (currently the Motorola Broadband Communication Sector), and Broadcom, respectively. He has published two graduate-level textbooks on digital communication, entitled *Error-Control Coding for Data Network* and *Transporting Compressed Digital Video.* He is an inventor of more than 70 granted or published patents worldwide in digital image/video processing and communication. He has also published over 60 research articles and contributed many book chapters in data compression and channel coding. He has served technical committees of various conferences on signal processing. He has also served as an associate editor of *IEEE Transactions on Circuit and Systems for Video Technology*, from 2000 to 2004. He actively involved in developing ISO MPEG-2 and -4 standards.His research interests include information theory, digital video compression and communication, computer architecture, VLSI system-on-a-chip, error-control coding, and data networks. His applied works concentrate on design and implementation of broadband communication system architectures, digital television systems, video transmission over cable and satellite and IP networks, and media-processor/DSP/ASIC architectures.

Tihao Chiang was born in Cha-Yi, Taiwan, Republic of China, in 1965. He received the B.S. degree in electrical engineering from the National Taiwan University, Taipei, Taiwan, in 1987, and the M.S. degree in electrical engineering from Columbia University in 1991. He received his Ph.D. degree in electrical engineering from Columbia University in 1995. In 1995, he joined David Sarnoff Research Center as a member of technical staff. Later, he was promoted as a technology leader and a program manager at Sarnoff. While at Sarnoff, he led a team of researchers and developed an optimized MPEG-2 software encoder. For his work in the encoder and MPEG-4 areas, he received two Sarnoff achievement awards and three Sarnoff team awards. Since 1992 he has actively participated in ISO's Moving Picture Experts Group (MPEG) digital video coding standardization process with particular focus on the scalability/compatibility issue. He is currently the co-editor of the part 7 on the MPEG-4 committee. He has made more than 90 contributions to the MPEG committee over the past 10 years. His main research interests are compatible/scalable video compression, stereoscopic video coding, and motion estimation. In September 1999, he joined the faculty at National Chiao-Tung University in Taiwan, R.O.C. Dr. Chiang is currently a senior member of IEEE and holder of 13 U.S. patents and 30 European and worldwide patents. He was a co-recipient of the 2001 best paper award from the IEEE *Transactions on Circuits and Systems for Video Technology.* He has published over 50 technical journal and conference papers in the field of video and signal processing. He was a guest editor for *IEEE Transactions on Circuits and Systems for Video Technology* and is the Chair of Visual Processing Technical Committee of IEEE Circuits and System Society.

Contents

1 Fundamentals of Digital Video Compression

In this chapter, we introduce the fundamentals of digital video compression. The content will include the following topics. We will discuss the concept of entropy, which pertains to the minimum number of bits necessary to carry information without any loss.

Rate distortion theory describes the fundamental bounds for compression but does not offer any specific methods for the implementation of such bounds. In practice, several lossless encoding techniques are used for compression standards. In particular, variable length codes are used to achieve compression when the more probable event is represented with shorter code and vice versa. There are several types of variable length code, but we will limit our discussion to Huffman code, Golomb code, and arithmetic code, which are commonly used in the international standards, such as MPEG, JPEG, and ITU standards.

To represent the picture in full color, the scene is typically captured and digitized as an RGB color space representation. We will review commonly used color space representations; each approach has its advantages and disadvantages for different purposes. However, such a color representation possesses significant redundancy among components, which makes it unsuitable for compression purposes. Therefore, RGB is rarely used for video compression standards, such as MPEG and ITU. For compression purposes, the removal of the spectral redundancy associated with each color component is the first step achieved with the YUV color representation. In addition to spectral redundancy, there are two other inherent redundancies, in the spatial and temporal domains. The spatial redundancy comes from the correlations between neighboring pixels within a frame. The temporal redundancy comes from the correlations between consecutive frames in a video. We will describe various techniques how such redundancies can be exploited.

To fully exploit spatial and temporal redundancy, there are practical approaches to achieve a feasible implementation. In particular, spatial redundancy is removed with spatial domain transformation into the frequency representation, such as with the discrete cosine transform (DCT), discrete wavelet transform (DWT), and subband decomposition. We will review several fast algorithms for such transformations. As for temporal redundancy removal, the typical approach used in video is block-based motion estimation and compensation. Block-based motion estimation can be implemented efficiently with existing VLSI technology. The significant advances of VLSI technology enable us to consider more complicated algorithms to achieve further compression. Thus, variable block size motion estimation and compensation is now accepted and used in state-of-the-art video standards, such as H.264/AVC.

1.1 FUNDAMENTALS OF INFORMATION THEORY

The origin of information theory can be traced back to the landmark paper "A Mathematical Theory of Communication," by Claude E. Shannon in 1948 [1-22]. The theory is now referred to as "The Mathematical Theory of Communication."

Let's consider a simple type of information — binary information source — since the binary representation of information is commonly used in data storage and computers. For example, the decimal digit is typically recorded with natural binary representation as shown in Table 1.1. We see three ways to encode the four symbols s_i in the table. The first approach is to use natural binary code to represent the information source $\boldsymbol{S}$. The advantage is its ease of interpretation and implementation of arithmetic operations, such addition and subtraction. However, natural binary representation is not the only way to store these symbols. For example, Gray code presents another method of storing the information source. The Gray code possesses only single-bit change between consecutive symbols, which is useful for achieving error protection and correction. However, both binary and Gray codes denote the symbols with an equal number of bits per symbol. Both codes are referred to as *fixed length code*.

The third approach uses code words with different lengths to represent the information source $\boldsymbol{S}$. As shown in Table 1.1, the probability for each symbol is different, so it is obvious that a more frequent symbol should be denoted with a shorter code word such that the average length is smaller. With this new code, the average number of bits used to represent each symbol is reduced from two bits to

$$\left(\frac{1}{2}\times 1+\frac{1}{4}\times 2+\frac{1}{8}\times 3+\frac{1}{8}\times 3\right)=1.75$$

The fundamental question is to determine the minimal number of bits to represent an information source with known statistics. The pursuit of the answer to such a question leads to the concept of entropy.

1.1.1 Entropy

The study of entropy starts with the understanding of the measure of information. Intuitively, the identification of a less probable symbol carries more information if such a symbol occurs. That is, an unusual event should carry more information. For example, the information on the current weather in Los Angeles has a higher probability of sunshine

TABLE 1.1
Fixed Length Code and Variable Length Code

Symbol	Probability	Natural Binary Code	Gray Code	Variable Length Code
s_1	0.5	00	00	1
s_2	0.25	01	01	01
s_3	0.125	10	11	000
s_4	0.125	11	10	001

than snow. Therefore, a snow day event will carry more information because it is less likely. The event that the sun rises in the east carries no information since it will surely occur. To implement such a concept, we define the "measure of information" as follows.

Definition 1.1 *Let the symbol s_i occur with probability p_i. The measure of information is defined by $I(s_i) = -\log p_i$ units of information. If the base of the logarithm is 2, the resulting unit of information is called a* binary unit, *or* bit.

The measure of information $I(s_i)$ is defined as $-\log p_i = \log(1/p_i)$. As p_i becomes smaller, the information it carries becomes larger. Based on the measure of information, the average of bits necessary to represent can be computed as an expected or average value of the measure of information defined as follows.

Definition 1.2 *Let the symbol s_i occur with probability p_i. The average amount of information source of symbol s_i can be computed as the entropy $H(S)$ of the source as defined by*

$$H(S) \equiv -\sum_{i=1}^{N} p_i \log p_i.$$

In his paper, Shannon proved that the defined entropy is indeed the minimal number of bits necessary to encode the information source.

Theorem 1.1 *[Source Coding Theorem] A source with entropy H can be encoded with arbitrarily small error probability at any rate as long as $R > H$.*

However, there is no unique way to achieve such a bound. Thus, it is natural to investigate how such bounds can be achieved. Before we answer that question, we want to ask another: what is the effect of the probability distribution on the entropy? We know that $\sum_{i=1}^{N} p_i = 1$ but there are infinite combinations p_i and we want to determine the maximal entropy in representing this source and under what conditions. The following theorem answers such a question without proof.

Theorem 1.2 *For a memoryless information source with **N** symbols source alphabet $\{s_i\}$, the maximum of the entropy is exactly* $\log(\mathbf{N})$. *Such maximum is achieved if and only if all the symbols are equiprobable.*

Theorem 1.1 illustrates that the entropy is maximized when the probabilities p_i satisfy the following condition: $p_1 = p_2 = \cdots = p_N = 1/N$. Obviously, the entropy is computed as an average of information

$$\begin{aligned} H(S) &\equiv -\sum_{i=1}^{N} p_i \log p_i \\ &= -\underbrace{\left[\frac{1}{N}\log\left(\frac{1}{N}\right) + \cdots + \frac{1}{N}\log\left(\frac{1}{N}\right)\right]}_{N} \\ &= \log N \end{aligned}$$

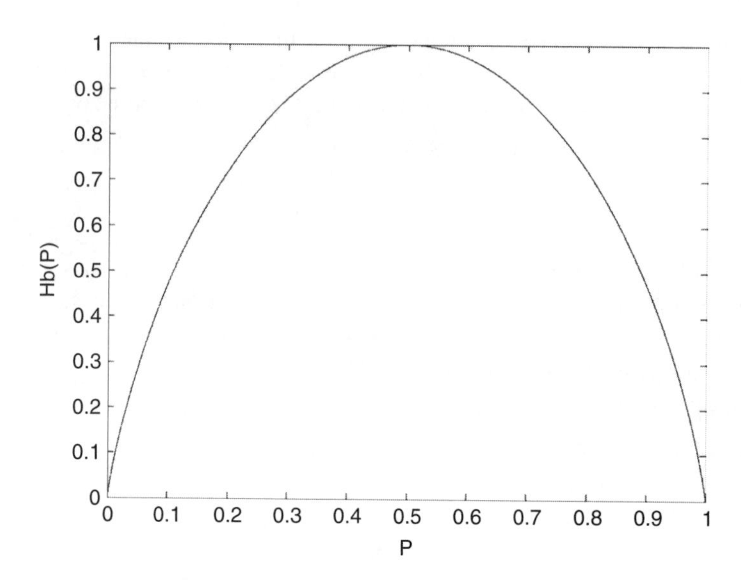

FIGURE 1.1 Entropy for the probability [1-22].

Example 1.1 *For a binary memoryless information source with probabilities* $\mathbf{p}$ *and* $1 - \mathbf{p}$*, we have the entropy* $H_b(X) = -p \log p - (1 - p)\log(1 - p)$*. This function is known as binary entropy function as shown in Figure 1.1. There are several interesting properties that can be observed in this example.*

The maximum of entropy occurs when $p = 0.5$. This means that the entropy is maximal when both symbols are equiprobable and thus the uncertainty is maximal. The minimum of entropy occurs when $p = 0$ or 1, which means that no information of the one event is certain and the other event is impossible. Such a scenario requires no information.

1.1.2 Properties of Block Codes

In order to implement a practical system, specific code words need to be assigned for each symbol. We now define the block code since both Huffman and Golomb codes belong to this category.

Definition 1.3 *A block code is a code that maps each symbol* s_i *of the information source* $\mathbf{S}$ *into a fixed sequence of the code alphabet.*

The first two codes in Table 1.2 are both block codes because each symbol s_i of the information source $\mathbf{S}$ is mapped to a fixed sequence of code alphabet. However, each sequence need not be distinct. It is obvious that indistinct code words result in confusion at the receiver side. Thus, we need to add another property for the block code, called *nonsingularity*, to remove such confusion. A nonsingular code is defined as follows.

TABLE 1.2
Nonsingular Block Codes

Symbols	Code A	Code B
s_1	0	1
s_2	10	10
s_3	110	11
s_4	1110	111

Definition 1.4 *A code is* ***nonsingular*** *if every element of the range of the source symbols maps into different code words.*

For nonsingular code, the decoder can decode each symbol if the symbols are properly separated. However, this may not be true when symbols are concatenated. For example, the sequence "111" may come from a sequence of symbols including "$s_1s_1s_1$," "s_3s_1," "s_1s_3" or "s_4." Thus, nonsingular block code does not guarantee that the symbols are decodable when multiple symbols are strung together. In information theory, such an action of concatenation is referred to as *extension*, defined as follows.

Definition 1.5 *An* $\boldsymbol{n}^{th}$ *extension of a code is a mapping from a finite length of* ***n*** *symbols to a finite length of source symbols* s_i. *A code is uniquely decodable if its* $\boldsymbol{n}^{th}$ *extension is nonsingular for all finite* ***n.***

As shown in Table 1.3, the second extension of code B has mapped both "s_3s_1" and "s_1s_3" to the same sequence, "111." The decoder cannot distinguish between these two cases, so Code ***B*** is not uniquely decodable.

In addition to unique decodability, it is desirable to have the ability to decode as soon as the full code word is received, which is referred to as *instantaneous decoding*. As shown in Table 1.4, code ***D*** is not instantaneous because the decoding of "100" from the sequence "1001" requires the fourth bit, "1," to make a decision that the first three bits represent the symbol s_3. To achieve instantaneous decoding, the following theorem states the condition as the prefix code requirements.

TABLE 1.3
Second Extension of Code B

Symbols	Code C	Symbols	Code C
s_1s_1	11	s_2s_1	101
s_1s_2	110	s_2s_2	1010
s_1s_3	111	s_2s_3	1011
s_1s_4	1111	s_2s_4	10111
s_3s_1	111	s_4s_1	1111
s_3s_2	1110	s_4s_2	11110
s_3s_3	1111	s_4s_3	11111
s_3s_4	11111	s_4s_4	111111

TABLE 1.4
Noninstantaneous Code

Symbols	Code D
s_1	1
s_2	10
s_3	100
s_4	1000

Theorem 1.3 *A code is a prefix code or instantaneous code if and only if no codeword is a prefix of any other codeword.*

1.2 VARIABLE LENGTH CODE

There are several types of variable length code. Due to its simplicity, the most commonly used code is Huffman code, which has been adopted in JPEG, MPEG, and several other standards. With given statistics, one can show that Huffman code is indeed optimal. However, the Huffman code requires a predetermined probability distribution so that it can provide an optimal code to match the statistics. If this is not the case, the code table is no longer optimal. The disadvantage can be solved using several Huffman codes for specific probability distributions or statistics.

However, there is another limitation with Huffman code, which is the need for constructing a table at both the encoder and decoder. It becomes a problem when the number of symbols is significant. Golomb code can be used to address such a problem assuming a geometrically distributed signal. The main advantage of Golomb code is its simplicity. The Golomb code does not require any code table and uses a very simple decoding method ideal for hardware implementation. It also allows easy adaptation to the statistics by selecting appropriate Golomb code. Furthermore, the Golomb code can be extended for an infinite number of source symbols. Golomb code has been used for the JPEG-LS and H.264/AVC standards.

Both Huffman and Golomb codes are block codes. Another important type of source coding technique involves non-block code. This non-block code is known as the *arithmetic* code, and has been used for several standards, such as JPEG and MPEG-4. One distinctive feature of arithmetic code is that it does not map a source symbol to a specific code word. The advantage of arithmetic code is that it does not assume any statistics for the source, as do the Huffman and Golomb codes. Instead, it estimates the statistics on the fly and uses the newly estimated statistics to improve the coding efficiency. Such a property makes it suitable to adapt to statistically nonstationary source signals, such as video and image. Another feature is the possibility of using context to classify the source signals. This allows similar improvement attained with multiple Huffman code tables without the complexity. Despite its multiple advantages, the arithmetic code suffers vulnerability to transmission noise. Error resilience is difficult to achieve for arithmetic code.

TABLE 1.5
Noninstantaneous Code

Symbols	Probability	Code word
s_1	1/2	0
s_2	3/16	11
s_3	3/16	100
s_4	2/16	101

In the following three sections, we will describe the three variable length codes typically used in video coding standards.

1.2.1 Huffman Coding

Huffman code is the optimal prefix code in terms of shortest expected length for a given distribution.

1. Sort the source symbols according to the probability.
2. Merge the two least probable symbols.
   a. Assign the top branch "0".
   b. Assign the bottom branch "1".
3. Check whether the number of symbols left is 2. If this is true, complete the code assignment. Otherwise, go to Step 1.

We will illustrate this algorithm with an example.

Example

Let us assume the information source has the symbols $\{s_1, s_2, s_3, s_4\}$ with the probabilities as shown in Table 1.5. As shown in Figure 1.2, the symbols s_3 and s_4 are combined to make a new symbol s_2'. Please note the tranposition of order from s_2 to s_3'. This is done to sort the symbols according to the magnitude of probabilities.

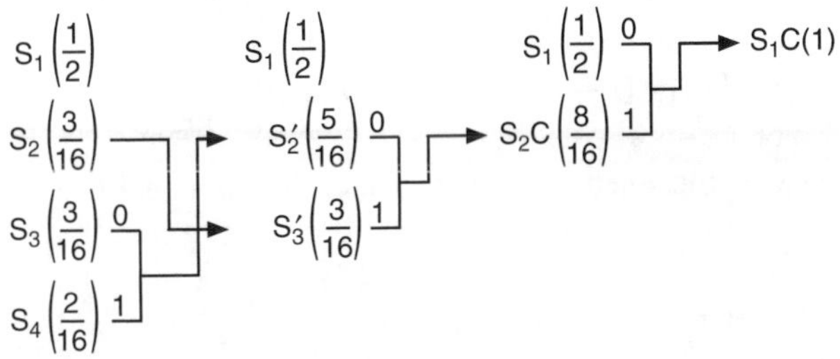

FIGURE 1.2 Huffman code construction.

TABLE 1.6
Unary Code

Symbols	Value	Probability	Code word
s_1	0	1/2	0
s_2	1	1/4	10
s_3	2	1/8	110
s_4	3	1/16	1110
⋮	⋮	⋮	111··· 0

1.2.2 Golomb Code

The Golomb code is a family of code that is designed with an assumed distribution that the probability will decrease as the value is increased [1-21]. An example of such a code is the unary code, illustrated in Table 1.6. Sometimes, we refer to the unary code as *comma code* because the termination symbol "0" acts likes a comma for the signal. A generalization of such code is to separate the code into two parts, where the first part is a unary code and the second part is a different code. The Golomb code was originally designed to code the run length of an event. In particular, it was illustrated as a run of success in a roulette game.

The main advantage of Golomb code is its ease of implementation, where no table is necessary for both the encoder and decoder. It also allows simple adaptation to the local statistics should the signals exhibit nonstationary behavior.

The Golomb code is parameterized with a nonzero positive integer $\boldsymbol{m}$**.** To represent an integer n using Golomb code with parameter m, we need to compute two numbers, namely, q and r, where q is

$$q = floor\left(\frac{n}{m}\right) \quad \text{and} \quad r = n - qm$$

Thus, we have

$$n = qm + r$$

The quotient is coded with the unary code as described in Table 1.6. The remainder r satisfies the relationship $0 \le r < m$. Thus, it takes at most $\log_2 m$ bits for representation.

It can be shown that Golomb code with parameter $\boldsymbol{m}$ is optimal for the geometrical distribution

$$P(n) = (1-p)p^{n-1} \quad \text{and} \quad m = Q\left(-\frac{1}{\log_2 p}\right)$$

The function $Q(x)$ is the smallest integer greater than or equal to x.

Example

Let us design a Golomb code for $m = 4$. Since

$$q = floor(\log_2(4)) = 2$$

The Golomb code for $m = 4$ can be expressed as in Table 1.7.

TABLE 1.7
Golomb Code with Parameter 4

Value	q	Code	r	Code	Code word
0	0	0	0	00	000
1	0	0	1	01	001
2	0	0	2	10	010
3	0	0	3	11	011
4	1	10	0	00	1000
5	1	10	1	01	1001
6	1	10	2	10	1010
7	1	10	3	11	1011

1.2.3 Arithmetic Code

The Huffman code approaches the entropy $H(S)$ for a given information source S with a set of probability distributions $\{p_i\}$. This is practically achieved with an extended alphabet, in which alphabets are concatenated to form a larger alphabet set. However, the approach becomes impractical when the $\{p_i\}$ is highly skewed and the number of alphabets is very small. In this case, the number of concatenations needed to achieve entropy may reach tens of thousand of symbols. This will make the implementation of the decoder expensive and time consuming. Furthermore, the code becomes inefficient when the probability distribution does not match the actual statistics of the signal.

The idea of arithmetic code can be considered as a mapping of a binary sequence to a number in the unit interval [0, 1). Let us consider a sequence of symbols, s_1, $s_2, s_3, s_4, \ldots, s_n$ and place a zero and decimal point before it so that we have $0.s_1, s_2$, $s_3, s_4, \ldots, s_n$." Thus, we can consider a binary sequence a real number between 0 and 1, excluding 1. A function that maps a sequence of random variables into a unit interval can be a cumulative distribution function. The arithmetic code is best illustrated with an example.

Example

Consider a binary source with an alphabet of size two {0, 1}. The set of probability is $\{\frac{3}{4}, \frac{1}{4}\}$. We would like to encode the sequence {0, 0, 1, 0}

There are a few interesting observations.

- The encoder does not generate any bit for encoding the first three symbols. This is a clear distinction between arithmetic code and the Huffman and Golomb codes, where the latter use at least one bit to represent a symbol.
- Except in Huffman code, the encoder does not require any code table to generate the bit stream.
- The less probable symbol causes the encoder to generate more bits. This matches the intuition embodied in both the Huffman and Golomb codes.

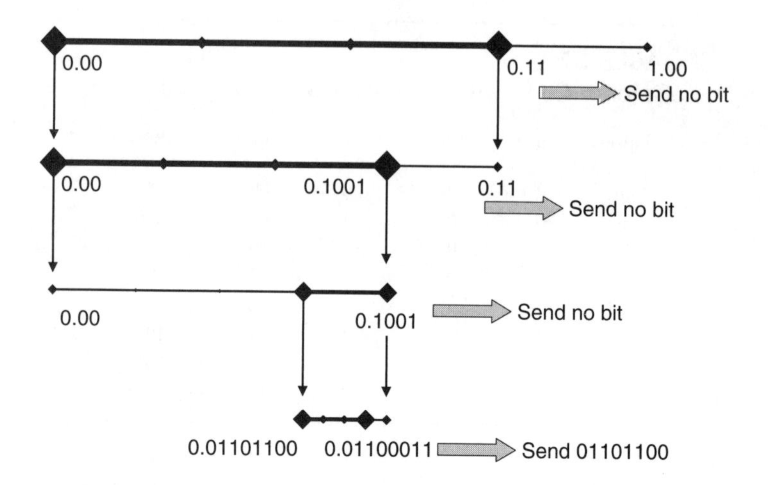

FIGURE 1.3 Arithmetic encoding of a short sequence.

- The implementation of the encoder is extremely simple. It only requires two variables storing the left and right bounds of the interval. An alternative implementation is store the left bound and the length of the interval.
- When the source is more skewed, the performance of the encoder is better; the performance gap between Huffman code and arithmetic code also widens.
- The probability model does not have to be identical for each subdivision of the interval. Thus, it is possible to update the probability model as shown in Figure 1.4. However, the encoder and decoder have to switch to the same probability model simultaneously to ensure correct decoding.

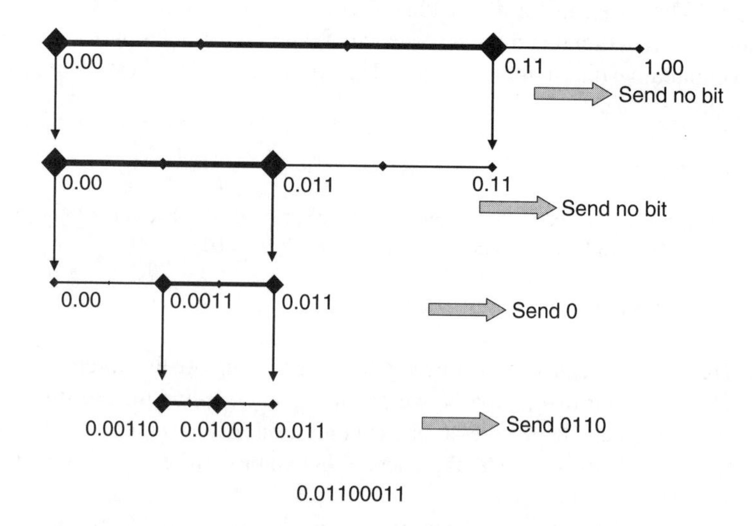

FIGURE 1.4 Arithmetic encoding of a short sequence.

1.3 Fundamentals of the Human Visual System

Color as perceived by humans can be conceptualized as a combination of the tristimuli red, green, and blue, which are called three *primary* colors. With these three primary colors, we can derive several color representations, or color spaces, using either linear or nonlinear transformations.

1.3.1 Color Space Conversion and Spectral Redundancy

The linear transformations include *YIQ*, *YUV*, and $l_1l_2l_3$ color spaces, and the nonlinear transformations include normalized RGB (Nrgb), his, and CIE spaces. The red, green, and blue components can be represented by the brightness of the object through three separate filters for each color based on the following equations.

$$R = \int_{\lambda} E(\lambda) S_R(\lambda) d\lambda$$

$$G = \int_{\lambda} E(\lambda) S_G(\lambda) d\lambda$$

$$B = \int_{\lambda} E(\lambda) S_B(\lambda) d\lambda$$

where S_R, S_G, and S_B are the color filters for the radiance $E(\lambda)$ and λ is the wavelength. The RGB color space is the most commonly used model for the television and video for consumer electronics. However, it is not frequently used for video compression because of the high correlations among each component.

Weber's law states that manipulation of colors is a linear operation, which includes either scaling or addition. Any colors can be created by the combination of the three colors and such a combination is unique. When two colors are mixed, the resultant color is the sum of the values of each color. Such a linear property enables us to perform linear transformation of color components, which can be inversed back or translated to different spaces.

The *YIQ* standard is used in American television. The transformation is defined as

$$\begin{pmatrix} Y \\ I \\ Q \end{pmatrix} = \begin{pmatrix} 0.299 & 0.587 & 0.114 \\ 0.596 & -0.274 & -0.322 \\ 0.211 & -0.253 & -0.312 \end{pmatrix} \begin{pmatrix} R \\ G \\ B \end{pmatrix}$$

where $0 \le R, G, B \le 1$. The YUV standard is used in European television and international standards for digital video. The transformation is defined as

$$\begin{pmatrix} Y \\ U \\ V \end{pmatrix} = \begin{pmatrix} 0.299 & 0.587 & 0.114 \\ -0.147 & -0.289 & 0.437 \\ 0.615 & -0.515 & -0.100 \end{pmatrix} \begin{pmatrix} R \\ G \\ B \end{pmatrix}$$

where $0 \le R, G, B \le 1$.

The normalized RGB is used to make the color representations independent of lighting changes. However, the dimensionality is reduced because when two components are determined, the third component can be determined.

$$r = \frac{R}{R+G+B}$$

$$g = \frac{G}{R+G+B}$$

$$b = \frac{B}{R+G+B}$$

It is obvious that $r + g + b = 1$. Another variation is defined as

$$Y = c_1 R + c_2 G + c_3 B$$

$$T_1 = \frac{R}{R+G+B}$$

$$T_2 = \frac{G}{R+G+B}$$

$$c_1 + c_2 + c_3 = 1.$$

The CIE color space has three primaries denoted as X, Y, and Z, which can be computed by a linear transformation from the RGB color space. The transformation matrix is defined as

$$\begin{pmatrix} X \\ Y \\ Z \end{pmatrix} = \begin{pmatrix} 0.607 & 0.174 & 0.200 \\ 0.299 & 0.587 & 0.114 \\ 0 & 0.066 & 1.116 \end{pmatrix} \begin{pmatrix} R \\ G \\ B \end{pmatrix}$$

1.4 VIDEO CODING FUNDAMENTALS

The video source consists of a sequence of frames that are temporally sampled from a three-dimensional scene at a predetermined period depending on the standards. For example, the North American standard adopts a temporal sampling frequency of 30 or 29.97 frames per second. Each frame is horizontally and vertically sampled at a progressive or interlaced sampling grid. The interlaced format is a legacy from analogue television that allows an effective temporal resolution of 60 Hz and an effective vertical resolution of 525 lines whereas the actual sampling rate is half, as shown in Figure 1.5.

The interlaced format can be considered as compression of a factor of two that fits the vast video information into a 6 MHz channel. When the analogue video information is digitized, the objective of a video coder is to compress the source as much as possible. There are three sources of redundancies including spatial, temporal, and statistical redundancies, that are available for us to exploit.

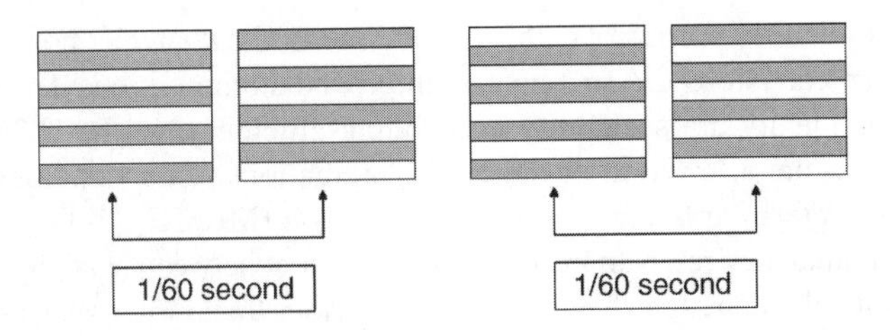

FIGURE 1.5 Interlaced formats.

1.4.1 Intrinsic Redundancy of Video Source

In the literature, video compression can be achieved with the removal of the four types of redundancy, spectral, temporal, spatial, and statistical redundancies. As shown in Figure 1.6, we list each category with approaches that can be used to remove such redundancy.

Spectral redundancy refers to the redundancy between each color component. Spectral redundancy can be removed with various color conversions, such as YUV, YIQ, and HSI. The majority of the information is compacted into the luminance components, and the chrominance components contain less visually sensitive information. Typically, the chrominance components can be subsampled horizontally and/or vertically to reduce the number of pixels without significantly degrading the visual quality. Thus, we can achieve about a factor of 2 compression ratio.

Temporal redundancy refers to the similarities between consecutive frames. Temporal redundancy can be removed with motion compensation (MC) or motion compensated temporal filtering (MCTF). The MCTF is used for a three-dimensional

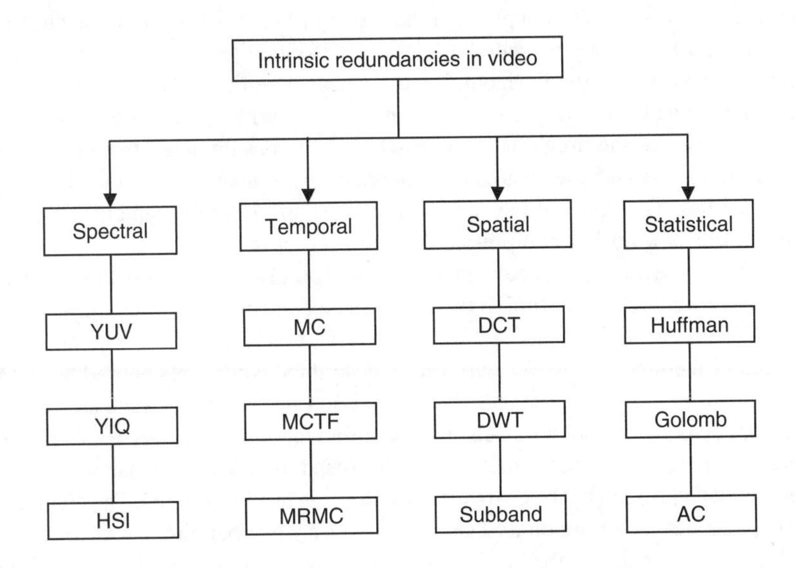

FIGURE 1.6 Intrinsic redundancies of video source.

(3-D) wavelet-based approach [1-25]. In the case of 3-D wavelet compression, the frames buffer can store $2n$ frames for improved compression. The benefits of increased complexity are scalability and coding efficiency. In the H.264 standard, multiple frames are allowed to be stored for predictions. This technique is referred to as *multiple reference frame motion compensation* (MRMC) [1-17].

Spatial redundancy refers to the correlation existent between spatially neighboring pixels. Spatial redundancy can be removed with various transforms, such as the discrete cosine transform (DCT), discrete wavelet transform (DWT), or subband decomposition. The discrete cosine transform provides very low cost implementations for energy compaction. There are fast algorithms and integer form for DCT implementation. The discrete cosine transform has been widely used in both image and video coding standards. Another tool for energy compaction is DWT. The discrete wavelet transform is mostly used for image compression and has recently been considered for video. Subband decomposition was widely used for audio coding and was used for image coding, which facilitates adaptation based on perceptual models.

Statistical redundancy is closely tied with the transform coding, in which most energy has been compacted toward the low-frequency components. The high-frequency components are more likely to become zero, and the nonzero coefficients are more likely to be located in the low-frequency area. Furthermore, the number of zero or close-to-zero coefficients is significantly larger than the number of coefficients with large amplitude. Thus, there is inherent statistical redundancy in the source that can be removed with Huffman, Golomb, or arithmetic code.

1.4.2 Temporal Redundancy

The second source of redundancy is the temporal redundancy coming from the temporal correlations because the same objects exist between frames when the sampling period is small enough such that no significant deformation occurs. Thus, a translation or more sophisticated model can describe the motion defined as a two-dimensional vector in the horizontal and vertical directions. The motion displacements are typically referred to as *motion vectors*. With the assistance of motion vectors, a block in the previous reconstructed frame is subtracted from the block in the current frame and the residual is encoded and transmitted to the decoder for reconstruction. The residual has significantly less information, which can approximately result in a further compression ratio of a factor 10.

There are several parameters that can be optimized based on the concept of motion vectors. We can optimize the performance according to the shape and size of the block, the motion model, the numerical precision of the vector, and the direction of the predictions. We will briefly describe the concepts behind these tools.

The motion models can be optimized based on various models, such as translation, affine, or perspective motion models. For the affine motion model, there are six parameters necessary to represent the motion. The overhead for transmitting the model parameters should be justified by the energy reduction of the residuals signal. The computation of the parameters is an important factor to consider when building the encoders.

The shape and size of the block can be optimized according to the physical object being predicted. The considered shapes are square, rectangle, triangle, hexagon, and polygon. In most video coding standards, the first three shapes are often used. For the rectangular block, the size can vary from 4×4 to 16×16 or larger. In the MPEG-1 video standard, only the block size of 16×16 is used. The 16×8 block size was included in the MPEG-2 video, and the 8×8 block size and the triangular mesh were included in the MPEG-4 video specifications. Smaller block sizes, such as 4×4, 4×8, 8×4, have been added in the recently unveiled H.264. The trend is obvious that smaller block sizes are used to provide a more accurate description of the model, which can lead to better coding efficiency. However, it also implies increased complexity.

The numerical precision of the motion vectors is also critical. Half-pixel precision is used in MPEG-1 and MPEG-2, while quarter-pixel precision was used in MPEG-4 and H.264 for improved performance. The increased precision of the motion vectors has several implications. The motion is more accurately described so that the coding efficiency is improved. However, the bit overhead for sending accurate motion has to outweigh the bit savings from the reduced residual energy. Furthermore, fractional pixel resolution requires more computation for searching the motion parameters. For example, the search points for quarter pixel motion vectors are four times more than the half pixel motion vectors.

The direction and number of frames stored for motion prediction are critical for its performance. As shown in Figure 1.7, the current frame can be predicted from $t-2$, $t-1$, and $t+1$. The use of the frame $t+1$ is very important because it offers information about uncovered background or new objects that cannot be found from the past frames. Frame $t-2$ offers more information at the cost of increased frame storage. Another variation of the multiple reference frames is to perform an average of the two predictions from two frames. This is typically referred to as *bidirectional* prediction. An average of two predictions was used in the MPEG standards and that has been proven to be very effective. One of the reasons for its effectiveness is that it is equivalent to fractional pixel motion compensation for translational motion. The bidirectional predicted frame is not available for future predictions in the MPEG-1/2/4 standards. Such a constraint is removed to achieve improved coding efficiency in the H.264 standards although it requires further delays and frame stores.

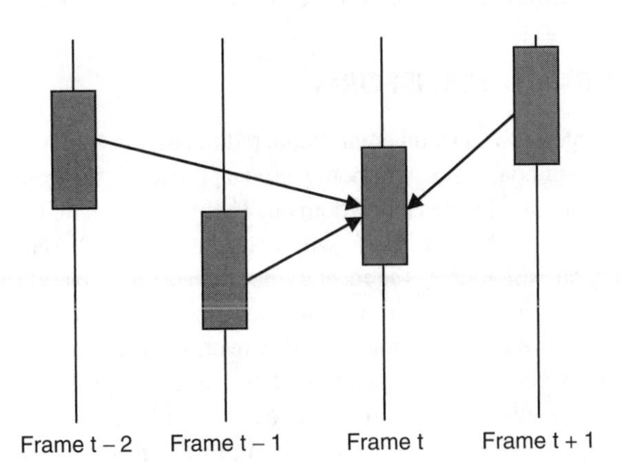

FIGURE 1.7 Motion compensation.

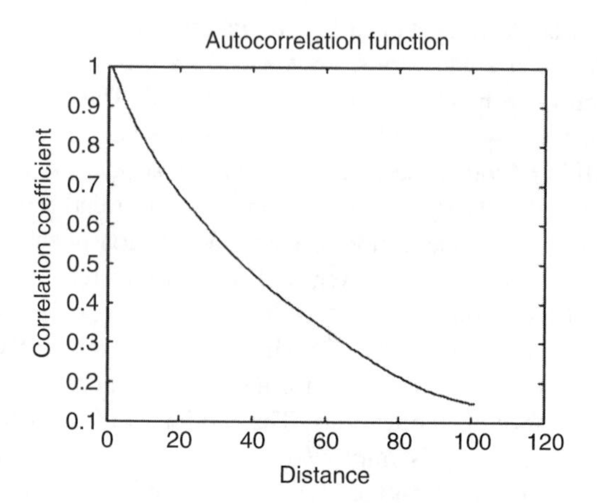

FIGURE 1.8 Autocorrelation functions of the Lena image.

1.4.3 Spatial Redundancy

The third source of redundancy is the spatial redundancy that comes from the correlations between spatially neighboring pixels. As shown in Figure 1.8, the spatial redundancy of a typical sequence can be approximated as a Markov model with using high correlations.

Typically, a compression ratio of 5 can be achieved with transformation using spatial redundancy removal. The transformations can be Discrete Cosine Transform (DCT), Discrete Since Transform (DST), and Discrete Wavelet Transform (DWT), or subband decomposition. The transformation takes advantage of the statistical redundancy to compact the energy into the low-frequency domain. The efficiency of energy compaction can be measured by the percentage of energy contained in the first few coefficients. As shown in Figure 1.9, the DCT coefficients have a Laplacian distribution that can be compressed with entropy coders easily.

1.5 BLOCK-BASED TRANSFORM

For video and image, there is an inherent spatial redundancy between neighboring pixels. To exploit such redundancy, a commonly used approach is to transform the spatial domain information into the frequency domain. There were several different transforms developed in the past. In the section, we will focus on the block-based transform coding and the next section will address the recently developed frame-based transform.

The theoretical optimal solution is the Karhunen-Loève transform, which that can achieve the most energy compaction. However, the evaluation of the Karhunen-Loève transform basis is input dependent. Thus, a less-sophisticated alternative of the transform has been sought. In this section, we will discuss various approaches to reach such a goal. In particular, we will study the Karhunen-Loève transform, discrete cosine transform, integer discrete cosine transform, subband decomposition and discrete wavelet transform.

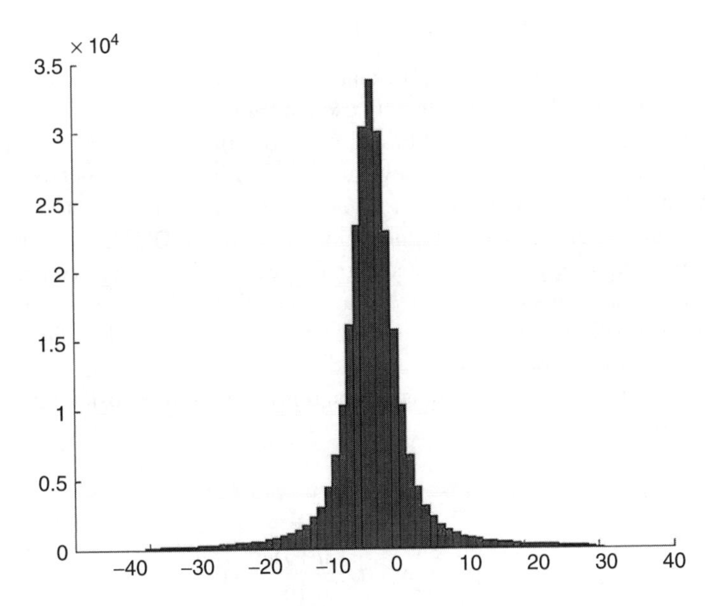

FIGURE 1.9 Histogram of DCT coefficients.

1.5.1 Karhunen-Loève transform

To fully exploit spatial redundancy, it is known that the Karhunen-Loève transform can provide the optimal solution. The rows of the discrete Karhunen-Loève transform consist of the eigenvectors for the autocorrelations matrix. The autocorrelation matrix for a stationary discrete random process $X(N)$ is given by

$$R = \begin{bmatrix} E(X_1^2) & E(X_1X_2) & \cdots & E(X_1X_n) \\ & E(X_2^2) & & \\ & & \ddots & \\ E(X_nX_1) & E(X_nX_2) & \cdots & E(X_n^2) \end{bmatrix}$$

$$= \begin{bmatrix} R_X(0) & R_X(1) & \cdots & R_X(N-1) \\ & R_X(0) & & \\ & & \ddots & \\ R_X(N-1) & R_X(N-2) & \cdots & R_X(0) \end{bmatrix}$$

It can be shown that this transform will minimize the geometric mean of the variance of the transform coefficients. Thus, it provides the best transform coding gain. However, the disadvantage of such an approach is that the source signals are typically nonstationary, so the autocorrelation function changes as time varies. Thus, the autocorrelation matrix needs to be recomputed for different information sources. Because the transform coefficient varies with time, it is necessary to transmit the coefficients to the decoder for correct decoding. The overhead or side information may not justify the additional coding gain it can achieve.

1.5.2 Discrete Cosine Transform

The discrete cosine transform (DCT) consists of a sum of cosine functions with different frequencies. The DCT is widely used in various standards, including JPEG, MPEG, and H.264. The DCT can be derived from the discrete Fourier transform (DFT), but the DCT is superior in several aspects. First, the computation of DCT does not involve any computation of complex numbers. Since it is derived from DFT, there are several known fast algorithms. Second, the DCT can perform better energy compaction for most correlated signals, such as images and video. It can be shown that the performance of DCT can approach that of KLT when we compress a Markov information source with high correlation coefficient ρ. Thus, DCT becomes the first choice for many compression standards.

The forward discrete cosine transform and inverse discrete cosine transform are defined as follows.

FDCT:

$$S_{uv} = \frac{1}{4} C_u C_v \sum_{i=0}^{7} \sum_{j=0}^{7} s_{ij} \cos \frac{(2i+1)u\pi}{16} \cos \frac{(2j+1)v\pi}{16}$$

IDCT:

$$s_{ij} = \frac{1}{4} \sum_{u=0}^{7} \sum_{v=0}^{7} C_u C_v S_{uv} \cos \frac{(2i+1)u\pi}{16} \cos \frac{(2j+1)v\pi}{16}$$

$$C_u C_v = \begin{cases} \frac{1}{\sqrt{2}} & for\, u, v = 0 \\ 1 & otherwise \end{cases}$$

1.5.3 Fast Discrete Cosine Transform Algorithms

There are fast algorithms for implementing the DCT. In the early days, the DCT can be implemented with a double-size fast Fourier transform (FFT) algorithm using complex arithmetic [1-20]. In [1-18], Chen et al. describe an efficient algorithm using only real operations for computing the DCT of a set of N points, where $N = 2^n$ and $n > 1$. It consists of alternating cosine and sine butterfly pattern with matrices to reorder the matrix elements such that the bit-reversed pattern is preserved. The Chen algorithm takes $(3N/2)(\log_2 N - 1) + 2$ real additions and $N \log_2 N$ E $3N/2 + 4$ real multiplications. This algorithm offers six times faster speed compared to the conventional approach using double-size FFT [1-18].

1.5.4 Integer Discrete Cosine Transform

In the MPEG-1/2/4 standards, the residual signal after motion compensation is compressed with 8×8 floating-point precision DCT transform. For practical applications the floating-point transform is implemented with a finite precision. There is a mismatch in the computation of DCT and IDCT at both the encoder and decoder. For static scenes, the errors in the IDCT mismatch will accumulate and result in visible artifacts. The mismatch can be removed by periodically inserting intra-coded block to stop the accumulation.

To solve the IDCT mismatch problem, a fundamental solution is to define the transform in such a way that finite precision implementation is possible. Thus, an integer DCT was defined in the recently developed H.264 standard [1-33]. There two different types of DCT for block size of 2 × 2 and 4 × 4. The forward integer DCT transform is defined as

$$\mathrm{DCT}_4 = \begin{bmatrix} 1 & 1 & 1 & 1 \\ 2 & 1 & -1 & -2 \\ 1 & -1 & -1 & 1 \\ 1 & -2 & 2 & -1 \end{bmatrix}$$

and the inverse integer DCT transform is defined as

$$\mathrm{IDCT}_4 = \begin{bmatrix} 1 & 1 & 1 & 1/2 \\ 1 & 1/2 & -1 & -1 \\ 1 & -1/2 & -1 & 1 \\ 1 & -1 & 1 & -1/2 \end{bmatrix}$$

It is obvious that such a transform can be implemented with only additions and shift without any multiplications. In the H.264/AVC standards, the quantization process is jointly optimized so that 16-bit arithmetic is possible. Using very short tables, the quantization process can be implemented without any divisions [1-16].

1.5.5 Adaptive Block Size Transform

Similar to variable block size motion compensation, the transformation can be adaptive to the local statistics with variable block size transformation. The larger block size can provide better energy compaction and detail preservation for pictures with flat area and high spatial correlation. A larger block size can represent a large area with only a few coefficients. However, the smaller block size can address picture area with high details since the picture has higher spatial correlation when the block size is reduced. It also removes the ringing artifacts that typically appear in the larger transform or discrete wavelet transform. A flexible variable block transform also enables efficient intra-picture prediction and rate-distortion optimization.

The implementation of variable block size transformation can be formulated as

$$A = T_v \cdot B \cdot T_h^T$$

The matrices A and B represent the frequency and spatial blocks of the transform. In [1-19], a combination of block sizes of 8 and 4 are proposed to achieve variable block size transformation. The block size-4 transformation matrix T_4 adopts the same transform matrix as specified in the H.264/AVC standard. The block size-8 transformation matrix T_8 can use the following matrix

$$\begin{pmatrix} 13 & 13 & 13 & 13 & 13 & 13 & 13 & 13 \\ 19 & 15 & 9 & 3 & -3 & -9 & -15 & -19 \\ 17 & 7 & -7 & -17 & -17 & -7 & 7 & 17 \\ 9 & 3 & -19 & -15 & 15 & 19 & -3 & -9 \\ 13 & -13 & -13 & 13 & 13 & -13 & -13 & 13 \\ 15 & -19 & -3 & 9 & -9 & 3 & 19 & -15 \\ 7 & -17 & 17 & -7 & -7 & 17 & -17 & 7 \\ 3 & -9 & 15 & -19 & 19 & -15 & 9 & -3 \end{pmatrix}$$

Such a size-8 transform can be implemented with an efficient fast algorithm using 36 additions, eight bit-shift operations, and ten multiplications as described in [1-19]. The improvement is about 12% in rate savings and 0.9 dB in peak signal-to-noise ratio.

1.6 FRAME-BASED TRANSFORM

In this section, several frame-based transforms are discussed. In particular, the subband and wavelet decompositions for image coding are studied. Using the frame-based wavelet transform, several image coding techniques exploit the parent-child relationships and the self-similarity property of the frequency bands at the same spatial location. In particular, we will study the well-known embedded zerotree wavelet (EZW) and set partitioning in hierarchical trees (SPIHT). Finally, generalization to the temporal directions for video coding are considered. In particular, we consider the 3-D wavelet and its variations.

1.6.1 SUBBAND DECOMPOSITION

The concept of subband coding (SBC) is to partition the original image signal into multiple frequency bands using lowpass and highpass filters as shown in Figure 1.10. The first set of filters H_0 and H_1 are referred to as the *analysis filters* while the second set of filters G_0 and G_1 are referred to as the synthesis filters. Each frequency band is separately quantized and encoded. The advantage of SBC is that the statistical

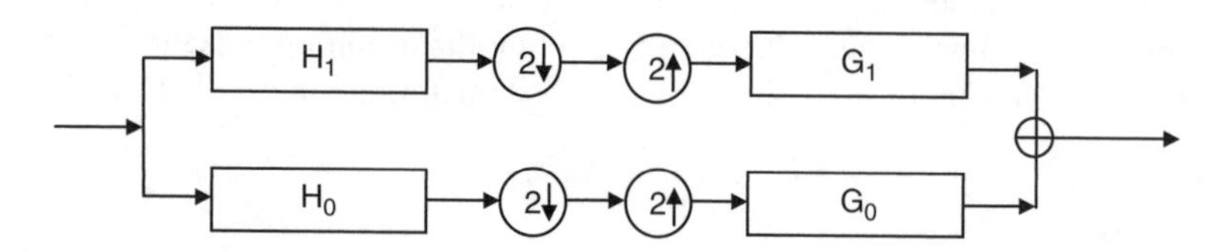

FIGURE 1.10 Filter bank for two channels.

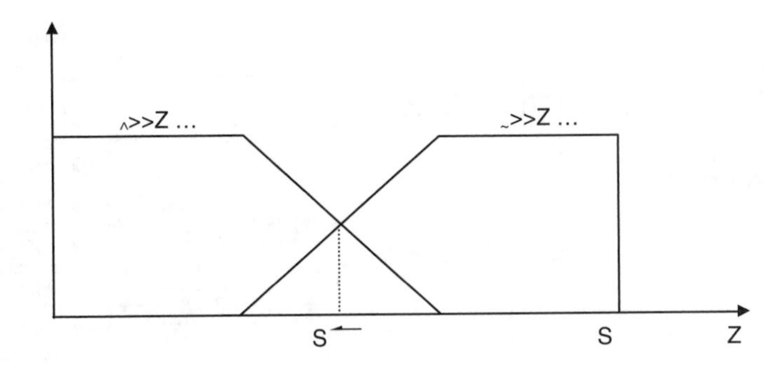

FIGURE 1.11 Quadrature mirror filters.

characteristics and redundancy of each band can be separately exploited with different quantization schemes and entropy coders. The bits can be dynamically allocated among different frequency bands and the distortion can be distributed according to human visual sensitivity.

In [1-23], Woods et al. studied the subband coding of images using quadrature mirror filters (QMF). As shown in Figure 1.11, the QMF frequency response for the one-dimension case is summed to be a constant so that perfect reconstruction at the receiver is possible. Woods et al. derived the constraints for a set of two-dimensional QMFs given a particular frequency partition. Furthermore, they demonstrate that separable one-dimensional QMFs can satisfy such constraints. There are many ways to partition the subbands. Typically, subband decomposition is performed with applications of lowpass and highpass filters in the horizontal and vertical directions of each frame. Each subband can be further decomposed into lowpass and highpass signals as illustrated in Figure 1.12. Another way to analyze the signal is to decompose the signal logarithmically. Since the human visual system is more sensitive to the distortions at the low-frequency components, it should be further analyzed with more resolution so that the encoder can perform the adaptation. The lowpass signals can be further analyzed to extract the specific frequency band as shown in Figure 1.12.

1.6.2 Discrete Wavelet Transform

In signal processing, the Fourier transform is commonly used to analyze signals in the frequency domain. The introduction of the discrete-time Fourier transform and its fast algorithm further expand the application of such an analytical tool. Fourier transform analysis can be considered as a linear expansion of a series of continuous bases made of *sine* and cosine functions. The Fourier series analysis can be also considered as using the set of periodic signals such as sines and cosines with infinite discrete frequencies as bases. The Fourier transform offers excellent resolution in the frequency domain but requires infinite time or buffer to compute because the bases expand from $-\infty$ to ∞. Such a phenomenon can be interpreted as indicating that the Fourier transform does not provide sufficient resolution in the time domain.

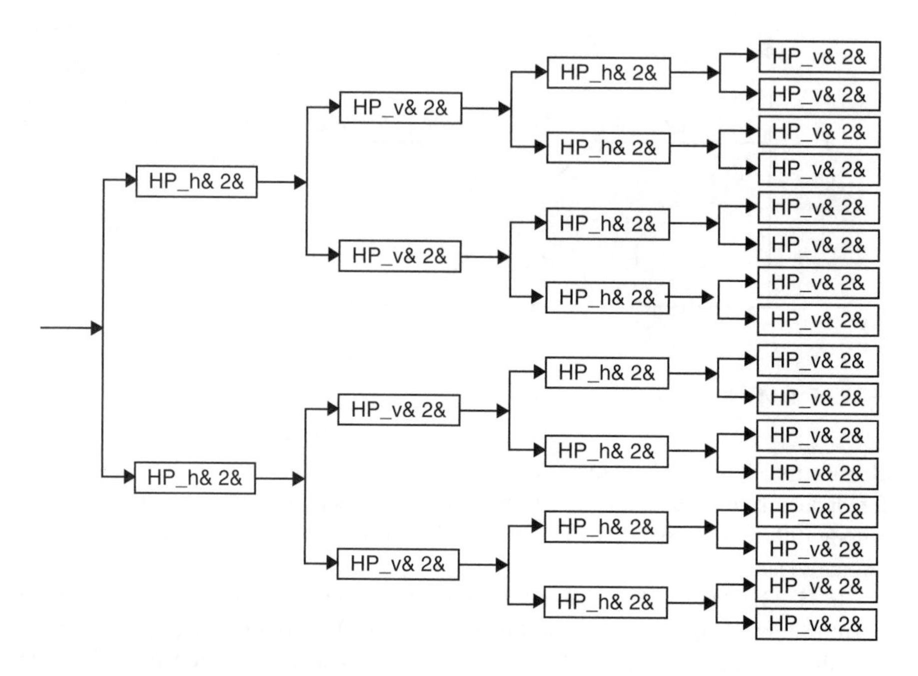

FIGURE 1.12 Equally partitioned subbands of a 2-D image.

The goal of the wavelet transform is to generalize the linear expansion concept and seek the expansion bases that can provide adequate tradeoff between time and frequency resolutions depending on the signal characteristics. The wavelet transform can use either orthonormal or biorthogonal bases or filters. The orthonormal filters conserve energy but lack the linear phase property. On the other hand, the biorthogonal filters can achieve linear phase as well as smoother approximation but cannot preserve energy, which makes it inconvenient for quantization and bit allocation.

The concept of better resolutions in the spatial or the frequency domain can be quantitatively measured as the space-frequency localization. With a measure for the spread in space and frequency of a function, one can define a region in the space-frequency plane where energy is mostly located. The wavelet filters can provide particular space-frequency localization regions or tiling with structured bases for expansion. The wavelet filters are related to each other through shifting, scaling, and/or modulations. For example, the short-time Fourier transform provides a rectangular tiling while the wavelet transform typically yields a dyadic tiling as shown in Figure 1.13 for the case of two-level analysis in the horizontal and vertical directions. The best tiling or bases is signal dependent. For example, a flat area in the picture or a lowpass signal may need better frequency resolution while a busy area in the picture may require better spatial resolution. The wavelet transform provides flexibility and tools to make such adaptations [1-29].

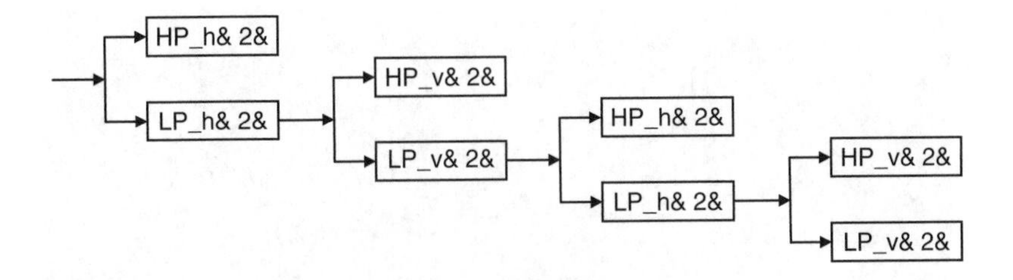

FIGURE 1.13 Two-level logarithmical partitioning of a 2-D image using the DWT.

Based on a single wavelet prototype function $\psi(t)$ with only shifting and scaling, one can construct a family of orthonormal bases $\psi_{mn}(t)$ as follows.

$$\psi_{mn}(t) = 2^{-mn}\psi\left(\frac{t-2^n}{2^m}\right)$$

In the case of dyadic tiling, the low frequency is repeatedly analyzed with the same discrete-time wavelet filter and so we are concerned whether the iterated filter is regular, i.e., such a filter converges to a continuous and differentiable function $\phi(t)$. Daubechies shows that the sufficient conditions for regularity can be obtained by placing a maximum number of zeros at $\omega = \pi$ and maintaining the orthogonality condition. Daubechies also provides a construction method to achieve regularity [1-30]. In Figure 1.14, the resultant signal is shown in Figure 1.15 with three levels of wavelet analysis.

As shown in Figure 1.14, the number of samples is identical with the original image and there are several inherent multiple spatial resolutions available in the decomposition. The multiresolution property of the wavelet transform is very desirable for progressive or error-resilient transmission.

For progressive transmission, the coarse low-frequency information can be sent first and the detail high-frequency information can be retrieved on request using additional bandwidth. The new information provides enhancement of the signal fidelity. As for the spatial resolution, the receiver can provide multiple spatial resolutions by extracting a subset of the wavelet subbands. In Figure 1.14, one can extract full, $^1/_4$, and $^1/_{16}$ resolutions of the original signal. For error-resilient transmission, the low-frequency information should be better protected depending on the transport scheme.

In a prioritized transmission scheme, the low frequency can be assigned to a channel with higher priority of less bit error rate or packet loss rate. In case of transmission errors, the receiver can still reconstruct a usable picture with less quality. This is a desirable feature, called *graceful degradation*, that is important for transmission over the Internet or terrestrial broadcasting.

1.6.3 Embedded Zerotree Wavelet

Shapiro pioneered the embedded zerotree wavelet (EZW) algorithm, which provides both progressive encoding of wavelet coefficients and excellent coding efficiency [1-26]. The wavelet coefficients are ordered according to the importance of the

FIGURE 1.14 Logarithmically partitioned wavelet trees of a 2-D image.

information. Thus, unlike most DCT-based encoding techniques, the EZW bit stream can be truncated at an arbitrary point for any given bit rate or distortion.

The EZW algorithm is based on four key elements: 1) a discrete wavelet transform, 2) zerotree description of the significant information, 3) successive-approximation quantization and 4) adaptive arithmetic coding. The EZW consists of two passes, dominant and subordinate. The dominant pass is focused on the encoding of the positions of the nonzero wavelet coefficients, which is recorded as the significance map. The subordinate pass is focused on the refinement of the nonzero coefficients using the successive approximation.

In the dominant pass, the wavelet coefficients will be compared against a predetermined threshold T to see if the magnitude is greater than T. This will result in a binary significance map. If the sign of a significant coefficient is also considered, the significant map can be encoded with three symbols: positive significant (1), negative significant (−1), and zero (0). The case of zero coefficients can be subdivided into two cases, isolated zero or zerotree. The zerotree symbol is an efficient representation that considers the self-similarity across different scales as obvious from Figure 1.14. As shown in Figure 1.15, we can observe that the wavelet coefficients

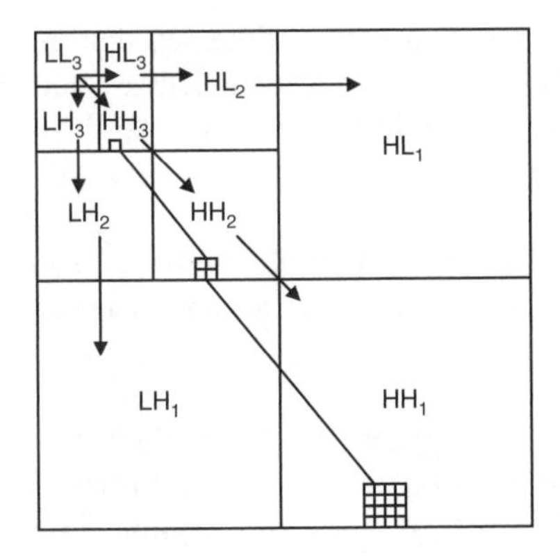

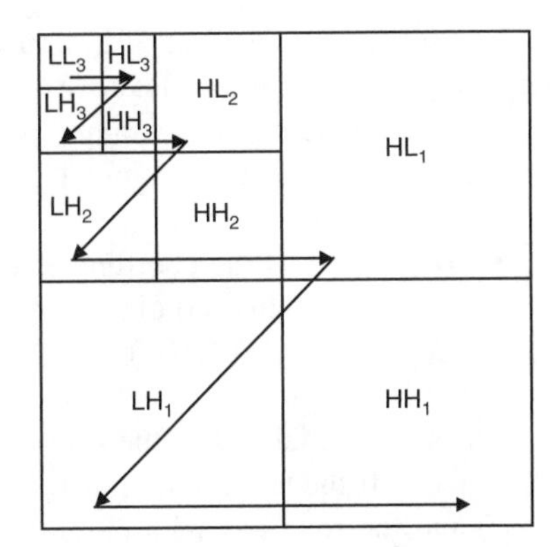

FIGURE 1.15 Parent-child relationships and scanning order of the wavelet coefficients [1-26].

that correspond to the same spatial location are similar with high correlations. These coefficients can be organized as a tree structure. Statistically, it is highly probable that the child nodes will be also zero if the parent node is zero. We use the zerotree symbol to denote a tree that is zero for the current node and has all of its descendants zero as well. In JPEG or MPEG, the end-of-block symbol has similar semantics except that it assumes that the nonzero coefficient should be low-frequency components. The significant map or trees are traversed in the scanning order as shown in Figure 1.15.

In the subordinate pass, the nonzero coefficients will be further refined with a successive-approximation quantization scheme. In the dominant pass, the encoder already identified the nonzero coefficients with the significance map. A more detailed knowledge of the magnitude can be obtained by comparing the coefficients with a sequence of thresholds T_i, where the threshold has the relationship $T_i = T_{i-1}/2$. The resultant refinement map for the threshold T_i is a binary map. The "one" symbol denotes that the coefficient is in the upper half of the interval between zero and T_{i-1}, while the "zero" symbol denotes that the coefficient will be in the lower half of the interval. If the threshold is chosen as power of 2, the refinement map corresponds to the bit planes of the wavelet coefficients. A similar bit-plane coding approach has been adopted for progressive transmission in the JPEG standard. Finally, both the significance and refinement maps are compressed with adaptive arithmetic coding technique.

1.6.4 Set Partitioning in Hierarchical Trees (SPIHT)

Based on the concept of self-similarity across wavelet scales, set partitioning in hierarchical trees (SPIHT) investigates a less complex and yet more efficient algorithm for image compression using wavelet transform. The algorithm is based on the principle of partial ordering by magnitude with a set partitioning sorting algorithm and ordered bit plane transmission [1-27]. The SPIHT algorithm consists of a sorting pass and a refinement pass. The refinement pass is similar to the refinement pass of EZW but the arithmetic encoder is not used with similar performance.

The sorting pass exploits self-similarity across scale with various sets that record the coordinates of the wavelet coefficients. The offspring of a coefficient are defined as to the coefficients of the same spatial orientation in the next finer level of the wavelet trees as shown in Figure 1.15. These sets are listed as follows:

- $\boldsymbol{O}(i, j)$: set of the coordinates of all offspring of node at coordinate (i, j)
- $\boldsymbol{D}(i, j)$: set of the coordinates of all the descendants of node at coordinate (i, j)
- $\boldsymbol{L}(i, j)$: $\boldsymbol{D}(i, j) - \boldsymbol{O}(i, j)$

The set is initialized as the set of coordinate (i, j) and $\boldsymbol{D}(i, j)$ for all of the (i, j) coefficients. If the test for significance of the set $\boldsymbol{D}(i, j)$ is true, it is partitioned into $\boldsymbol{L}(i, j)$ and the four sets with $(k, l) \in \boldsymbol{O}(i, j)$. If $\boldsymbol{L}(i, j)$ is tested as significant, it is partitioned into four sets $\boldsymbol{D}(k, l)$ with $(k, l) \in \boldsymbol{O}(i, j)$.

The sorting pass exploits self-similarity across scale with various sets that record the coordinates of the wavelet coefficients as defined above. The encoder and decoder keep track of three ordered lists, including the list of insignificant sets (LIS), the list of insignificant pixels (LIP), and the list of significant pixels (LSP). In the LIS, there are sets of either $\boldsymbol{L}(i, j)$ or $\boldsymbol{D}(i, j)$. During the sorting pass, each coefficient with its coordinate in LIP is tested for significance. If it becomes significant, the coefficient is moved to LSP. Similarly, each set in LIS is tested for significance. If the set becomes significant, the set is moved from LIS and partitioned. The new subsets with multiple elements are placed back to LIS and the single element set is placed in either the LIP or the LSP depending on its significance. In the refinement pass, all of the coefficients in the LSP are visited with similar refinement process of EZW as described in the previous section.

Generalizing SPIHT to the three-dimensional case for color video was investigated in [1-28]. The nonlinear behavior of the temporal axis in the video signal is handled with motion compensation and thus offered an MC-3-D SPIHT algorithm that performs comparably with the H.263 video standard, with additional functionality of multiresolution in the spatio-temporal domain.

1.6.5 Temporal Subband Decomposition

Ohm [1-30] proposed an early generalization of subband coding for video when he introduced the concept of a motion-compensated temporal filter for the "connected" and "unconnected" pixels as illustrated in Figure 1.16. For the connected pixels, the current frame can find corresponding pixels in the previous frame and thus a two-tap Haar filter can be used to obtain the low-frequency and high-frequency components. Previous work [1-31] has shown that longer tap filters do not bring significant coding gain. For the unconnected pixels, the original pixel scaled by $\sqrt{2}$ is the low-frequency band and the motion-compensated difference signal after being scaled by $1/\sqrt{2}$ is the high-frequency band. The inclusion of motion compensation in the 3-D subband coding greatly enhances the coding efficiency. As shown in Figure 1.17, the motion-compensated temporal filtering (MCTF) has significantly reduced the redundancy in the high-frequency subband signals.

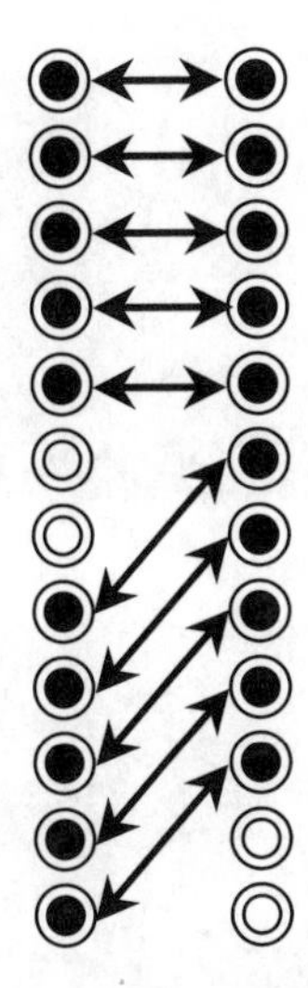

FIGURE 1.16 Connected and unconnected pixels for motion compensated temporal filtering [1-26].

In [1-24] and [1-25], Hsiang and Woods describe a spatio-temporal pyramid based on the motion-compensated temporal filtering and discrete wavelet transform to achieve both spatial and temporal scalability as shown in Figure 1.18. The proposed algorithm performs the motion estimation and MCTF to remove the temporal redundancy. The motion estimation uses a hierarchical variable size block matching algorithm with fractional pixel resolution. The temporally decomposed sequences are further analyzed for four levels using the Daubchies 9/7 analysis/synthesis filters. Thus, such a construction can provide four layers of spatial scalability, four levels of temporal scalability, and the fine granularity of SNR scalability. With such a wide range of scalability, the coding efficiency is comparable to the recently developed H.264 video specifications. In the MPEG committee, there was a recent call for proposals on scalable video coding technique; among the proposals, the interframe wavelet approach was the prime candidate for consideration.

1.7 SUMMARY

In this chapter, we briefly summarized the major results that are typically used for image and video compression standards. We first reviewed the major results in information theory that are related to standards implementation. The three major types of codes are Huffman, Golomb and arithmetic. We then reviewed the major results in color space conversion. Finally, we went over the various transforms used for image and video coding. We described DCT and its fast implementation. We then reviewed the results for subband coding and discrete wavelet transform and its generalization for video applications.

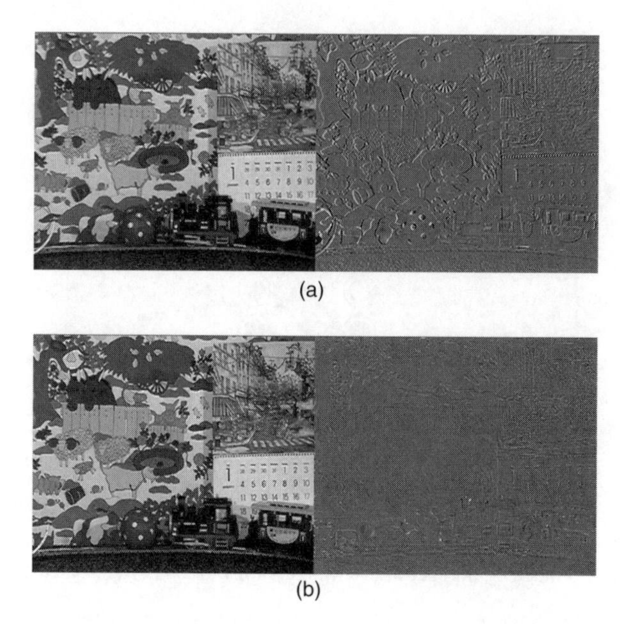

FIGURE 1.17 Temporal filtering without motion compensation (a) and with motion compensation (b) The left and right columns corresponds to the low-pass and high-pass subbands.

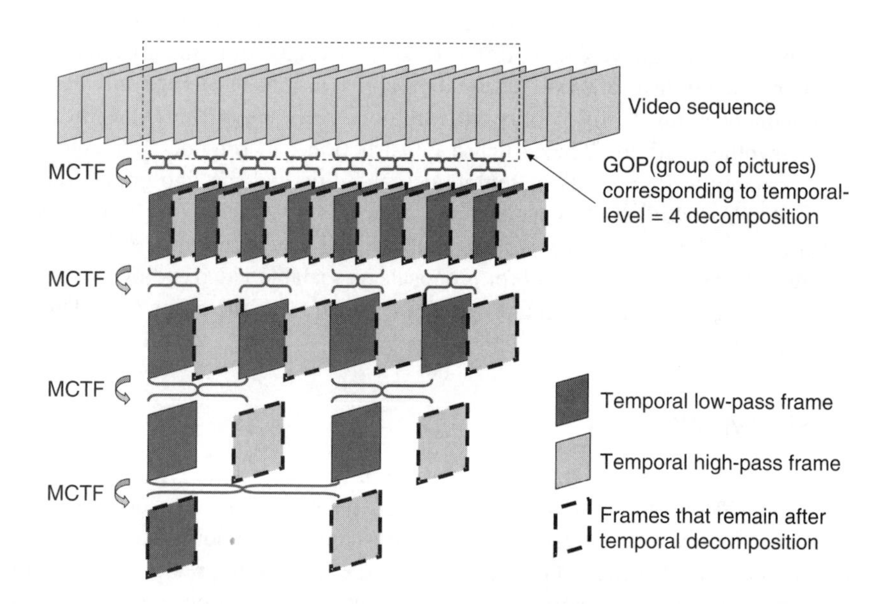

FIGURE 1.18 Spatiotemporal pyramid constructed based on motion compensated temporal filtering [1-24].

1.8 EXERCISES

1-1. Let the alphabet A consist of four symbols with probabilities as shown below.

Symbol	Probability
a_1	0.5
a_2	0.25
a_3	0.125
a_4	0.125

(a) Suppose we wish to determine an arithmetic code to represent a sequence of symbols a_3, a_2, a_2. Plot the figure for the subinterval division.
(b) Specify the binary representation of the boundaries for each interval.
(c) Specify the bits that will be transmitted for each symbol encoded?
(d) Suppose we wish to decode a bit stream "1111,1011,0" to a sequence of symbols. Illustrate the decoding interval for each bit received.
(e) Specify when a symbol can be determined without ambiguity.

1-2. Let the alphabet A consist of four symbols with its probabilities as shown below.

Symbol	Probability
a_1	0.25
a_2	0.25
a_3	0.2
a_4	0.15
a_5	0.15

(a) Construct a Huffman code using binary codeword
(b) Compute the average length using the binary codewords assignment in Part (a).
(c) Explain why it is called a prefix code.
(d) Consider a ternary code for the same alphabet. Construct a new Huffman code using ternary codewords such as {0, 1, 2}.
(e) Compute the average length using this ternary codeword assignment

1-3. Explain why you need to use different block sizes for encoding?

1-4. Compare the advantages and disadvantages of the three different source coding techniques: arithmetic code, Huffman code, and Golomb code.

REFERENCES

[1-1] K. Sayood, *Introduction to Data Compression,* Academic Press, New York, 2000.

[1-2] N. Abramson, *Information Theory and Coding,* McGraw Hill, New York, 1963.

[1-3] N.S. Jayant and P. Noll, *Digital Coding of Waveforms–Principles and Applications to Speech and Video,* Englewood Cliffs, Prentice-Hall, New Jersey, 1984.

[1-4] T. Cover and J. A. Thomas, *Elements of Information Theory,* Wiley, New York, 1991.

[1-5] ISO/IEC JTC1 IS 11172 (MPEG-1), Coding of moving picture and coding of continuous audio for digital storage media up to 1.5 Mbps, 1992.

[1-6] ISO/IEC JTC1 IS 13818 (MPEG-2), Generic coding of moving pictures and associated audio, 1994.

[1-7] ISO/IEC JTC1 IS 14386 (MPEG-4), Generic coding of moving pictures and associated audio, 2000.

[1-8] ISO/IEC JTC1 15938 (MPEG-7), Information technology—multimedia content description interface—multimedia description schemes, 2001.

[1-9] Requirements Group, MPEG-21 Overview v4, Bormans, J. Ed., ISO/IEC JTC1/SC29/WG11 N4801, Fairfax, VA, May 2002.

[1-10] ITU-T Recommendation H.261, Video codec for audiovisual services at px64 Kbit/s, March 1993.

[1-11] ITU-T Recommendation H.263, Video coding for low bit rate communication, Draft H.263, May 2, 1996.

[1-12] Joint Video Team (JVT) of ISO/IEC MPEG and ITU-T VCEG Document JVT-H036, Joint model reference encoding methods and decoding concealment methods, August 30, 2003.

[1-13] H.D. Cheng, X.H. Jiang, Y. Sun, and J. Wang, Color image segmentation: advances and prospects, *Journal of Pattern Recognition,* Volume 34, Issue 12, Pages 2259–2576, Dec. 2001.

[1-14] R. Rao, and P. Yip, *Discrete Cosine Transform: Algorithms, Advantages, Applications,* Academic Press, New York 1990.

[1-15] F. Halsall, *Multimedia Communications, Applications, Networks, Protocols and Standards,* Addison-Wesley, Reading, MA, Nov. 2000.

[1-16] H. S. Malvar, A. Hallapuro, M. Karczewicz, and L. Kerofsky, Low-Complexity transform and quantization in H.264/AVC, *IEEE Transactions on Circuits and Systems for Video Technology,* 13, 7, 598, 2003.

[1-17] M. Flierl and B. Girod, Generalized B pictures and the draft H.264/AVC video-compression standard, *IEEE Transactions on Circuits and Systems for Video Technology,* 13, 587, 2003.

[1-18] W. Chen, C.H. Smith, and S.C. Fralick, A fast computational algorithm for the discrete cosine transform, IEEE Transactions on Communications, COM-25, 1004, 1977.

[1-19] M. Wien, Variable block-size transforms for H.264/AVC, *IEEE Transactions on Circuits and Systems for Video Technology,* 13, 604, 2003.

[1-20] N. Ahmed, T. Natarjan, and K.R. Rao, "Discrete cosine transform," *IEEE Trans. Computer,* vol. C-23, no. 1, pp. 90–93, Jan. 1974.

[1-21] S.W. Golomb, "Run-Length encodings," *IEEE Transactions on Information Theory,* 12(3), July 1966.

[1-22] C.E. Shannon, A mathematical theory of communication, *Bell System Technical Journal,* 27, 379 and 623, 1948.

[1-23] J.W. Woods and S.D. OíNeil, Subband coding of images, *IEEE Transactions on ASSP,* ASSP-34, 1278, 1986.

[1-24] S.-T. Hsiang, and J.W. Woods, Embedded video coding using invertible motion compensated 3-D subband/wavelet filter bank, *Signal Processing: Image Communication* 16, 705, 2001.

[1-25] S.-J. Choi and J.W. Woods, Motion-Compensated 3-D subband coding of video, *IEEE Transactions on Image Processing,* 8, 155, 1999.

[1-26] J. Shapiro, Embedded image coding using zerotrees of wavelet coefficients, *IEEE Trans. Signal Processing,* 41, 155, 1993.

[1-27] A. Said and W.A. Peralman, A new, fast and efficient codec based on set partitioning in hierarchical trees, *IEEE Transactions on Circuits and Systems for Video Technology,* 6, 243, 1996.

[1-28] B.-J. Kim, Z. Xiong, and W.A. Pearlman, Low bit-rate scalable video coding with 3-d set partitioning in hierarchical trees (3-D SPIHT), *IEEE Trans Signal Processing,* 10, 1374, 2000.

[1-29] K. Ranchanran, M. Vetterli, and C. Herley, Wavelets, subband coding and best bases, *Proceedings of IEEE,* 84, 541, 1996.

[1-30] J.-R. Ohm, Advanced packet-video coding based on layered VQ and SBC techniques, *IEEE Transactions on Circuits and Systems for Video Technology,* 3, 208, 1993.

[1-31] D. Taubman and A. Zakhor, Multirate 3-D subband coding of video, *IEEE Transactions on Image Processing,* 3, 572, 1994.

[1-32] M. Antonini, M. Barlaud, P. Mathieu, and I. Daubechies, Image coding using wavelet transform, *IEEE Transactions on Image Processing,* 1, 205, 1992.

[1-33] A. Puri, X. Chen, and A. Luthra, Video coding using the H.264/MPEG-4 AVC video compression standard, *Signal Processing: Image Communication Journal,* 2004.

2 Digital Video Coding Standards

Many video coding standards have been developed during the past two decades. In this chapter, the general principles and structures of digital video coding are first presented. Then the basic tools for video coding standards are introduced. In order to improve coding efficiency and increase functionality, several enhancement tools for digital video coding standards have been developed and the basic ideas behind these tools are presented. Finally, the summary of different digital video coding standards and several encoding issues are described.

2.1 INTRODUCTION

During the past two decades, digital video processing, transmission, and storage techniques have made great progress. These technologies include digital video compression, digital modulation, and digital storage. Among these technologies, digital video compression is a key technology. In this area, many video coding techniques and standards have been developed to meet the increasing need for practical applications. Digital video coding standards have been developed by two major groups of standards organizations. One group is the Moving Pictures Experts Group (MPEG), of the International Organization for Standardization (ISO) and the International Electrotechnical Commission (IEC), and another is the International Telecommunications Union (ITU). The video coding standards developed by ISO/IEC include MPEG-1 [2-1], MPEG-2 [2-2], and MPEG-4 [2-3], as well as multimedia interface standard MPEG-7 [2-4] and multimedia framework standard MPEG-21 [2-5]. The standards developed by ITU include H.261 [2-6], H.262 [2-2], H.263 [2-7], and H.264/AVC [2-8]. The H.262 standard is the same as MPEG-2, which is a joint standard of MPEG and ITU. The H.264/AVC video coding standard was developed by the joint video team (JVT) of MPEG and ITU, which is also as the MPEG-4 Part 10. These standards have found many successful applications that include digital television, digital storage, DVD, streaming video on the networks, surveillance and many others. Also, with advances of computer technologies, the personal computer becomes the most flexible and most inexpensive platform for playing the compressed video stream.

Standard H.261 was completed in 1990, and it is mainly used for ISDN video conferencing.

Standard H.263 is based on the H.261 framework and completed in 1996. Standard H.263 includes more computationally intensive and efficient algorithms so as to increase the coding performance. It is used for video conferencing with analog telephone lines, desktop computers, and mobile terminals connected to the Internet.

After the baseline of H.263 was approved, more than 15 additional optional modes have been added. These optional modes provide the tools for improving coding performance for very low bit rate applications and addressing the needs of mobile video and other noisy transmission environments.

MPEG-1 was completed in 1991. The target application of MPEG-1 is digital storage media, CD-ROM, at bit rates up to 1.5 Mbps.

MPEG-2 is also referred to as H.262 and was completed in 1994. It is an extension of MPEG-1 and allows for greater input format flexibility and higher data rates for both high-definition television (HDTV) and standard definition television (SDTV). The U.S. ATSC DTV standard and European DTV standard DVB both use MPEG-2 as the source-coding standard but use different transmission systems. At the system layer, the DTV uses transport stream, which is designed for the lossy transmission environment. MPEG-2 is also used for digital video disk (DVD) data. The data format of DVD is MPEG-2 program stream, which is designed for the clear transmission environment. The DVD format can provide SDTV quality, which is much better than traditional analog VCR or digital video CD (VCD).

MPEG-4 was completed in 2000. It is the first object-based video coding standard and is designed to address the highly interactive multimedia applications. MPEG-4 can provide tools for efficient coding, object-based interactivity, object-based scalability, and error resilience. MPEG-4 provides tools not only for coding natural videos, but also for synthetic video and audio, still image, and graphics.

Standard H.264 is also referred to as *MPEG-*4 *Part 10, Advanced Video Coding.* It is a new video coding standard that was developed by the joint video team of ISO and ITU. The goal of H.264 is to provide very high coding performance, which is much higher than MPEG-2 and MPEG-4 at a large range of bit rates. In H.264, many tools have been used for improving coding performance. These tools include multiple reference frames, which can number up to 15 frames; multiple motion compensation modes with block size from 16×16 to 4×4; up to quarter-pel accuracy motion compensations; small block size transformation; in-loop deblocking filter, adaptive arithmetic entropy coding; and many other tools we will introduce later. The target applications of H.264 are broadcasting television, high-definition DVD, and digital storage.

2.2 GENERAL PRINCIPLES OF DIGITAL VIDEO CODING STANDARDS

2.2.1 Basic Principles of Video Coding Standards

As previously mentioned, several video coding standards have been developed during the past decade or so. From these standards, it can be summarized that a successful video coding standard, which is widely accepted by the industry and extensively used in many applications, should have several important features.

The most important feature is that the standard should contain a significant amount of technical merit. Such technical merits are able to solve the most difficult problems to meet the immediate needs from the industry. For example, during the development of HDTV standard, the industry was required to provide full-quality HDTV service in a single 6-MHz channel. The function of compression layer has to compress the raw data from about 1 Gbps to the data rate of approximately 19 Mbps to satisfy the 6-MHz spectrum bandwidth requirement. This goal has been achieved by using the main profile and high level of the MPEG-2 video coding standard and advanced digital modulation technique. The MPEG-2 standard uses hybrid coding techniques, which mainly combine the DCT with motion-compensated predictive coding and other techniques to provide a 50:1 compression ratio at the broadcast quality. Therefore, MPEG-2 became the basis of the standard for digital television.

The standard should consider both good coding performance and less computational complexity. The industry must especially consider the implementation issues before accepting a standard. Sometimes, the techniques adopted by standards may be a trade-off between performance and computational complexity. Some techniques have significant technical innovation but may not be adopted by the previous standard due to the implementation difficulty at that time. But it may be adopted by a later standard oran extension of a standard owing to advances in computer and VLSI technology. Many examples about this can be found in H.264/AVC.

The standard should only define the syntax and semantics of the compressed bit stream. This means that a standard defines only the decoder process. How to encode and generate the bit stream is not the normative part of a standard. In other words, the standard does not define the encoder. However, during the development of the standard, both encoder and decoder are needed to verify the syntax and semantics of the bit streams. Therefore, most standards provide the reference software encoder as an informative part of the standard. In this sense, the encoding algorithms are open for competition. Different encoding algorithms may generate the bit streams with different coding performance and different computational complexity. Only one thing is common for all kinds of encoders, the generated bit streams must be standard compliant. Therefore, the encoder manufacturers can compete on the items of coding performance and complexity. From the other side, decoder manufacturers need to produce standard-compliant decoders but can compete not only on the price but also on additional features, such as post-processing and error concealment capabilities.

The video coding standard should be independent from transmission and storage media, which is also a very important feature. The compressed video and audio bit streams can be used for different applications. The bit streams can be transmitted through broadcasting, computer networks, wireless networks, and other media. They can also be stored in the different kinds of storage. In order to achieve this goal, a system layer standard has been developed. System standards specify the packetization, multiplexing, and packet header syntax for delivering or storing the audio and video compressed bit streams in different media.

Another feature that the video coding standard should have is compatibility, is important for manufacturers. If a new standard can decode the bit streams of previous standards, it is easy to let new products be introduced to the market. For example, since the MPEG-2 decoders are able to decode the MPEG-1 bit streams, the DVD receivers, which use the MPEG-2 standard, can handle the VCDs, which contain MPEG-1-compliant bit streams. Compatibility is very important for the

DTV and DVD industry, which cannot upgrade the hardware easily. But the compatibility is not the essential condition for some applications, such as in the computer industry. Most personal computers (PCs) have big memory and high-speed CPUs, which allow PCs to use software decoders. The software decoders would easily be upgraded, or multiple decoder software can be installed in the PCs. For example, the MPEG-4 and H.264/AVC are not backward compatible with MPEG-2 and H.263, respectively.

2.2.2 General Procedure of Encoding and Decoding

The common feature of the audio/video coding standards is that the different information sources — video, audio, and data — are all converted to the digital format, which is referred to as elementary bit streams at the source coding layer. Then these elementary bit streams are mixed together to a new bit stream at the system layer with additional information, such as information for time synchronization. This new format of information, binary bit stream, was a revolutionary change in the multimedia industry, because the digitized information format, i.e., the bit stream, can be decoded by not only the traditional consumer electronics products such as television and any kind of video recorder but also the digital computers through different transmission media. As we mentioned previously, the coding standard defines only the syntax and semantics of the compressed bit stream. The compressed bit stream consists of a set of binary codewords, which are obtained by the variable length coding (VLC). Therefore, the encoding procedure is first to remove the spatial, temporal, and statistical redundancies in the video data. This decorrelation process is usually achieved by block-based discrete cosine transform coding combined with the motion-compensated predictive coding for most video coding standards. The predictive residues are converted to the bit stream according to the syntax and semantics defined by the standards with the VLC tables. In the decoder, the bit stream is decomposed to the codewords by the parser according to the syntax and semantics. Then the codewords are converted to the video data by the same VLC tables. The VLC tables represent the agreement between encoder and decoder. In other words, the encoder and decoder have the same VLC tables. The VLC tables are generated from the average statistics of a large set of training video sequences. The above descriptions are intuitively illustrated by Figure 2.1.

In the following, we introduce the basic tools and enhancement tools that are used in many video-coding standards. From these tools it is easily to understand the principles of video coding standards.

2.3 BASIC TOOLS FOR DIGITAL VIDEO CODING STANDARDS

The principal goal in the design of a video coding system is to reduce the number of bits, which is used to represent the video source, subject to some picture quality constraints. Two factors allow us to accomplish the goal of compression. One is the existing of the statistical redundancy in the video source, and another is the

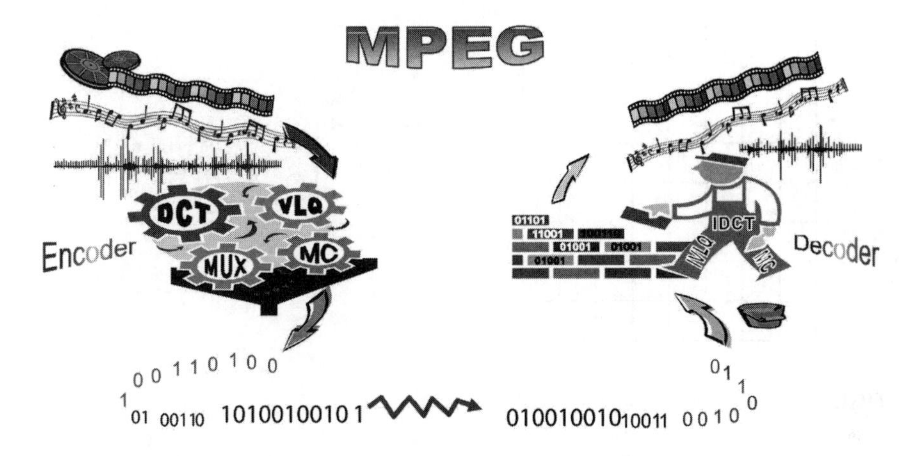

FIGURE 2.1 Illustration of audio/video coding procedure by MPEG standards.

psychophysical redundancy. The statistical redundancy includes spatial redundancy and temporal redundancy. Normally, video data are highly correlated both spatially and temporally. In most cases the results of video coding are viewed with a human observer. Therefore, the perceptual limitations of human vision can be used for video data compression. Human observers are subject to perceptual limitations in amplitude, spatial resolution, and temporal acuity. By proper design of the coding system, it is possible to discard information without affecting perception, or at least, with only minimal degradation. In summary, two factors: the statistical structure of the video data and the fidelity requirements of the end user make the compression possible. In the following subsections, we introduce the basic tools and enhancement tools, which can be used to achieve the goal for video data compression. These tools are extensively employed in the video coding standards.

2.3.1 Tools for Removing Spatial Redundancy

There are several tools that are used to remove the spatial redundancy in the video data. These tools include transformation, prediction, and adaptive quantization.

2.3.1.1 Block Transformation

In order to remove the spatial redundancy from the input video data, the video frame is first decomposed into the blocks. The blocks undergo transformation. The original data are then converted to the transform domain for each block Figure 2.2.

In the original spatial domain, all pixels are equally important, but, in the transform domain, the transformed coefficients are no longer equally important since the low-order coefficients contain more energy than the high-order coefficients, from statistical calculations for most natural videos. In other words, the original video

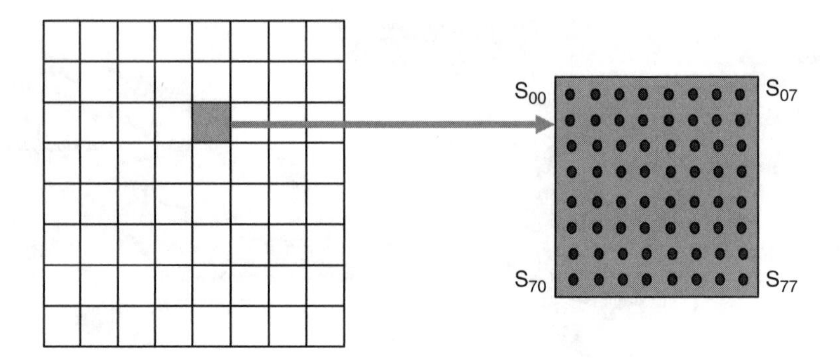

FIGURE 2.2 Partitioning into 8 × 8 blocks.

data are decorrelated by transformation. Therefore, we are able to code video more efficiently in the transform domain than in the spatial pixel domain. In the video coding standards, the block sizes used for transformation are usually 8 × 8. The discrete cosine transform is used for almost all video coding standards. Recently, the wavelet transform is used for still image coding, such as in JPEG2000 and MPEG-4 texture coding. The two-dimensional forward DCT (FDCT) and inverse DCT (IDCT) of an 8 × 8 block are defined as follows:

FDCT:

$$S_{uv} = \frac{1}{4} C_u C_v \sum_{i=0}^{7} \sum_{j=0}^{7} s_{ij} \cos\frac{(2i+1)u\pi}{16} \cos\frac{(2j+1)v\pi}{16}$$

IDCT:

$$s_{ij} = \frac{1}{4} \sum_{u=0}^{7} \sum_{v=0}^{7} C_u C_v S_{uv} \cos\frac{(2i+1)u\pi}{16} \cos\frac{(2j+1)v\pi}{16}$$

$$C_u C_v = \begin{cases} \frac{1}{\sqrt{2}} & \textit{for } u, v = 0 \\ 1 & \textit{otherwise} \end{cases} \tag{2.1}$$

The transformed coefficients are processed in zigzag order since the most energy is usually concentrated in the lower order coefficients. The zigzag ordering of elements in an 8 × 8 matrix allows for a more efficient run-length coder. Depending on the spectral distribution, the alternative scan can yield run lengths that better exploit the multitude of zero coefficients. The zigzag scan and alternative scan are shown in the Figure 2.3. The normal zigzag scan is used for MPEG-1 and as an option for MPEG-2. The alternative scan is not supported by MPEG-1 and is an

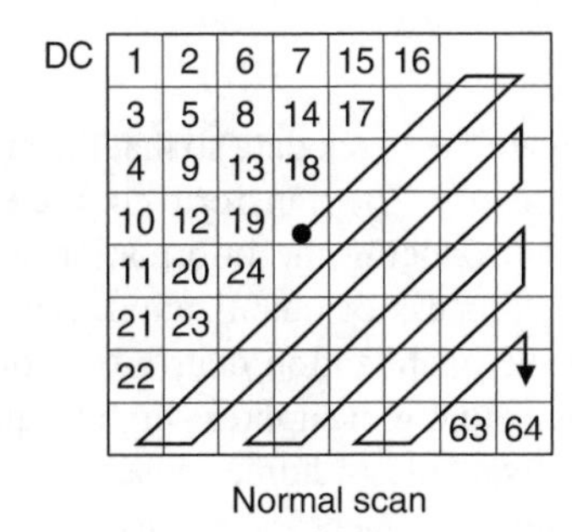

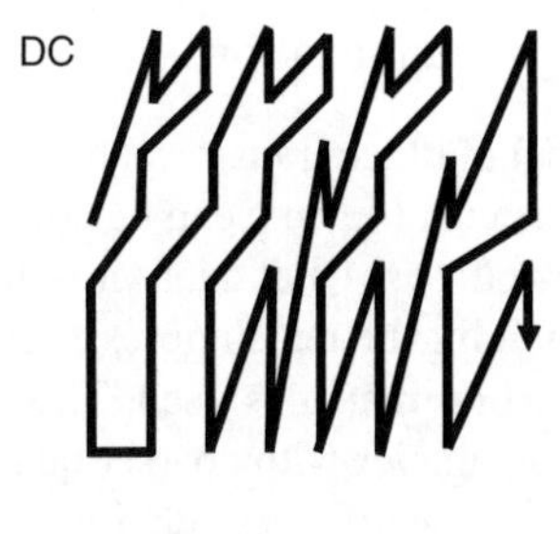

FIGURE 2.3 Normal zigzag scan order and alternative zigzag scan order.

option for MPEG-2. For frame-type DCT of interlaced video, more energy may exist at the bottom part of the block, hence the run length coding may be better with the alternative scan.

1							
2							
3							
							63
							64

As we noted, most video coding standards use an 8×8 block size and discrete cosine transform. However, in H.264/AVC, the transformation is applied to 4×4 blocks and a separable integer transform with similar properties as a 4×4 DCT is used. The transform matrix is given as

$$H = \begin{bmatrix} 1 & 1 & 1 & 1 \\ 2 & 1 & -1 & -2 \\ 1 & -1 & -1 & 1 \\ 1 & -2 & 2 & -1 \end{bmatrix}.$$

The small block size allows us to encode the pictures in a more locally adaptive way, which may reduce artifacts known as *ringing noise*. Since the inverse transform is defined by exact integer operation, the inverse-transform mismatches that exist in most other coding standards are avoided. Also, the integer transform reduces the computational complexity; it allows computation of the direct and inverse transform with just additions and minimal number of shifts, but no multiplications. It should be noted that the basic functions of the new transform in H.264/AVC do not have equal norm, which leads to an increase in the size of the quantization tables.

2.3.1.2 Quantization

The goal of block transformation is to decorrelate the block data so that the resulting transform coefficients can be coded more efficiently. The transform coefficients are then quantized. Quantization is the step for introducing the information loss, but for obtaining better compression. During the process of quantization, a weighted quantization matrix is used. The function of the quantization matrix is to quantize high frequencies with coarser quantization steps that will suppress high frequencies with no subjective degradation, thus taking advantage of human visual perception characteristics. The bits saved for coding high frequencies are used for lower frequencies to obtain better subjectively coded images. An example of coding an 8×8 block is shown in Figure 2.4.

In Figure 2.4, for the quantized DC value, QDC, is calculated as:

$$QDC(8bit) = dc//8 = 34 \tag{2.2}$$

where symbol // means integer division with rounding to the nearest integer and the half-integer values are rounded away for zero unless otherwise specified. The AC coefficients, ac(i, j), are first quantized by individual quantization factors to the value of ac ~ (i, j):

$$ac\sim(i,j) = (16*ac(i,j)//W_1(i,j) \tag{2.3}$$

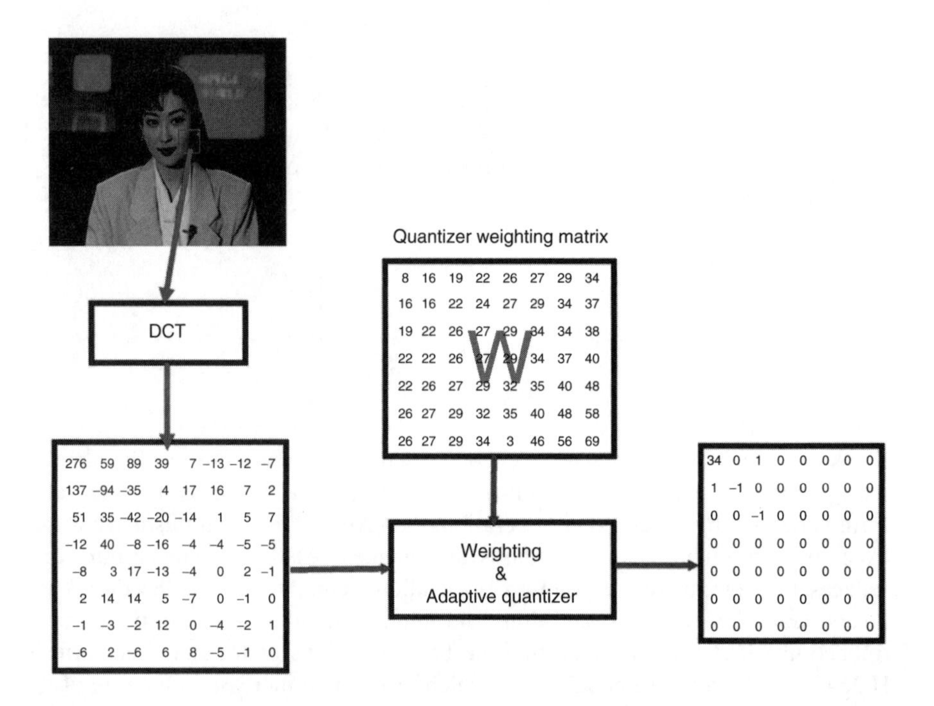

FIGURE 2.4 An example of coding an 8×8 block.

where W_I (i, j) is the element at the (i, j) position in the intra-quantizer weighting matrix shown in Figure 2.4. The quantized level QAC(i, j) is given by

$$QAC(i,j) = [ac\sim(i,j) + sign(ac\sim(i,j)*((P*mquant)//q)]/(2*mquant) \quad (2.4)$$

where mquant is the quantizer step or scale, which is derived for each macroblock by the rate control algorithm, and p = 3 and q = 4 in TM5 [2-9]. It can be seen that the resulting quantized coefficients include many zeros, which can be easily coded by the run length codes. With the zigzag order, the resulted coefficients in Figure 2.4 are lined up as 34, 0, 1, 0, −1, 1, 0, 0, 0, 0, 0, 0, −1, 0, 0, … , which is then parsed as pairs of zero-run and nonzero values: (34), (1, 1), (1, −1), (0, 1) and (6, −1). These pairs of run and value are coded by VLC. It should be noted that after quantization some information are lost, and this loss is not recoverable. However, due to the perceptual limitations of human vision, the lost information may not be perceptually noticed by a properly designed coding system.

2.3.1.3 DC and AC Prediction

After a video frame is partitioned into blocks, the transform coding is applied to each block. In such a way, the correlation between pixels within a block has been used for data compression. Since the blocks are coded independently, the correlation between blocks is not used yet. In order to remove the redundancy between blocks, the DC prediction and AC prediction are used for some video coding standards.

In MPEG-1 and MPEG-2, a video frame of the 4:2:0 format is first divided into 16 × 16 blocks for luminance and 8 × 8 for chrominance and this is called as *macroblock*. Each macroblock consists of four 8 × 8 luminance blocks and two chrominance blocks. For the luminance blocks, the order of DC prediction is shown in Figure 2.5. For the chroma block, the DC prediction is just performed along horizontal direction.

In MPEG-4 video, an adaptive DC prediction algorithm has been used to remove the redundancy between neighborhood blocks. The adaptive DC prediction involves the selection of the reference DC value from the immediately left block or the

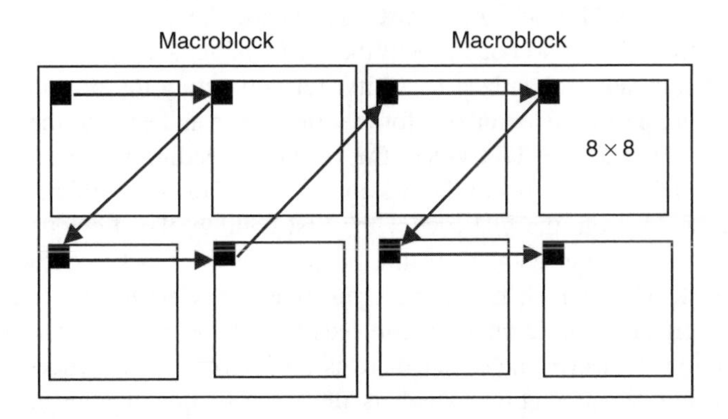

FIGURE 2.5 The DC prediction in MPEG-2 video coding.

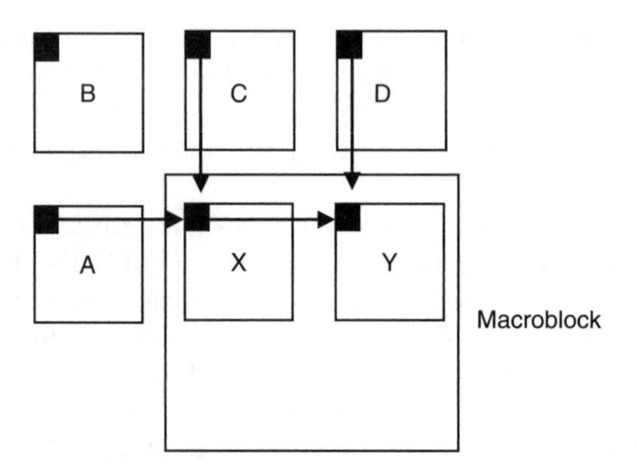

FIGURE 2.6 The DC prediction in the MPEG-4 video.

immediately above block as shown in Figure 2.6. The selection criterion is based on comparison of the horizontal and vertical DC gradients around the block to be coded. The DC prediction for chrominance is performed along the horizontal blocks only because each macroblock contains only one chrominance block for 4:2:0 format. In Figure 2.6, it is shown that the three surrounding blocks A, B, and C the current block X whose quantized DC value, QDC, is to be coded, where blocks A, B, and C are the blocks immediately left, immediately left and above, and immediately above block X, respectively. The QDC of block X, QDC_X, is predicted by either the QDC of block A, QDC_A, or the QDC of block C, QDC_C, based on the comparison of horizontal and vertical gradients as follows:

$$\begin{aligned} &\text{If } |QDC_A - QDC_B < |QDC_B - QD_C|, \\ &\text{Otherwise,} \qquad QDC_P = QDC_A \end{aligned} \tag{2.5}$$

The differential DC value is then obtained by subtracting the DC prediction, QDC_P, from QDC_X. If any of blocks A, B, or C are outside the VOP boundary, or they do not belong to an INTRA-coded block, the QDC value is assumed to take a value of 128 (if the pixel is quantized to 8 bits) for computing the prediction. The DC predictions are performed similarly for the luminance and each of the two chrominance blocks. In the MPEG-4 video, the predictive coding is not only applied to the DC coefficients, but also to the AC coefficients. For AC coefficient prediction, either coefficient from the first row or the first column of a previous coded block are used to predict the co-sited (same position in the block) coefficients in the current block. On a block basis, the same rule for selecting the best predictive direction (vertical or horizontal direction) as for the DC prediction is used for the AC prediction. However, note that there is an important difference between DC prediction and AC prediction, which is the issue of quantization scale. All DC values are quantized to the same range (8, 9, or 10 bits) for all blocks. Note also that the AC coefficients in different blocks could be quantized with different quantization scales. To address the problem of differences in the quantization of the blocks used for prediction, scaling of prediction coefficients becomes necessary. The prediction is scaled by the ratio of the current quantization step size and the quantization step size of the block used for prediction. The scaling process of quantized coefficients increases the complexity of AC prediction. Also, there may be less correlation between AC coefficients between neighborhood blocks for some video sequences. Due to above two reasons, AC prediction may be disabled sometime during the encoding. Especially, in the cases when AC coefficient prediction results in a larger range of prediction errors as compared to the original signal, it is desirable to disable AC prediction. The decision of AC prediction switched on or off is performed on macroblock basis instead of block basis to avoid the excessive overhead. The decision for switching on or off AC prediction is based on a comparison of the sum of the absolute values of all AC coefficients to be predicted in a macroblock and that of their predicted differences.

2.3.1.4 Intra Frame Coding with Directional Spatial Prediction

In the new video coding standard, H.264/AVC, a new intra frame technique based on the directional spatial prediction has been adopted. The basic idea of this technique is to predict the macroblocks to be coded as intra with the previously coded regions selected from proper spatial direction in the same frame. The merit of directional spatial prediction is able to extrapolate the edges of previously decoded parts of the current picture to the macroblocks to be coded. This can greatly improve the accuracy of the prediction and improve the coding efficiency.

For the 4×4 intra mode, in addition to DC prediction, there are total of eight prediction directions, as shown in Figure 2.7. For the 16×16 intra mode, there are four prediction modes: vertical, horizontal, DC, and plane prediction. For the technical detail, please refer to [2-8].

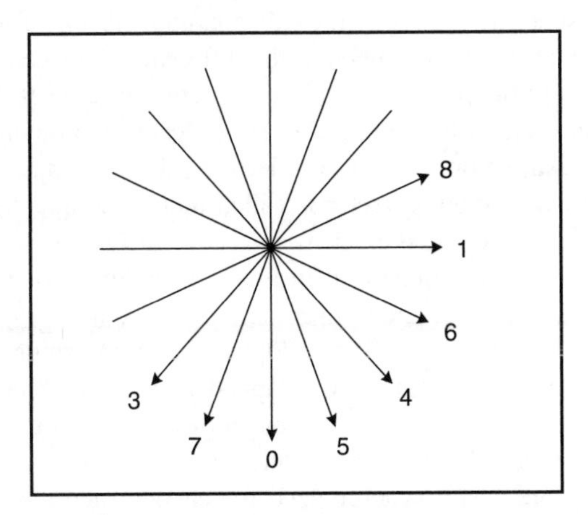

FIGURE 2.7 Eight predictive directions for intra 4×4 prediction in H.264/AVC.

2.3.2 Tools for Removing Temporal Redundancy

In this subsection, we present the basic tools for removing the temporal redundancy or interframe redundancy in the video coding standards. These tools include motion-compensated predictive coding and the structure for different frame types.

2.3.2.1 Motion-Compensated Predictive Coding

As indicated previously, a high correlation exists between successive frames in video sequences. In order to employ temporal correlation for video compression, several methods have been proposed in the past. The simplest method is to code the differences between consecutive frames by predictive coding technique, which is referred to as *frame replenishment* [2-10]. Frame replenishment can be briefly described as follows. Each pixel in a frame is classified into two classes: in the changing or in the unchanging area depending on whether the pixel difference between the current frame and previous frame at the co-located position exceeds a threshold value. The intensities and addresses of the pixels in the changing areas are coded and transmitted to the receiver for replenishment. For the pixels in the unchanging areas, nothing is coded and transmitted. Since in the frame replenishment technique only those pixels in the changing areas are coded, its coding efficiency should be higher than the techniques, which code each frame independently with the use of interframe correlation. The main problem of frame replenishment technique is that it is difficult to code the video sequences with active changes. When there are more active areas or more areas with rapid changes, more pixels need to be coded and the coding efficiency would be lower. In this case if we increase the threshold value to keep a certain number of unchanging pixels, the coding quality would be poor. In order to solve the problem with frame replenishment technique, motion-compensated predictive coding has been proposed [2-11]. In motion-compensated predictive coding, the changes between successful frames are considered to include the translation of moving objects in the image planes. In this method, the displacement vectors of objects are first estimated. The displacement vectors are called *motion vectors*. The predictive differences between frames are now defined as the differences between the pixel values in the current frame and the pixels values in the reference frame at the positions translated with motion vectors. In the encoder, the procedure to obtain motion vector is called as motion estimation. And in the decoder, the procedure to reconstruct the value be predicted with motion vector is called motion compensation. The motion-compensated predictive coding is extensively used in the video coding standards since it is more efficient than the frame replenishment technique. To avoid high overhead for coding the motion vectors and to overcome the difficulty for coding objects with arbitrary shapes, video coding standards adopt the block-matching method. The video frames are first partitioned into nonoverlapped rectangular blocks; each block is usually estimated by one motion vector, but no more than four motion vectors. In such a way, the overhead for coding motion vectors is low and the positions of rectangular blocks are easily synchronized for encoder and decoder; therefore, no side information for block address is needed. Figure 2.8 shows the principle of block-based motion estimation. For block-based motion compensation, several things have

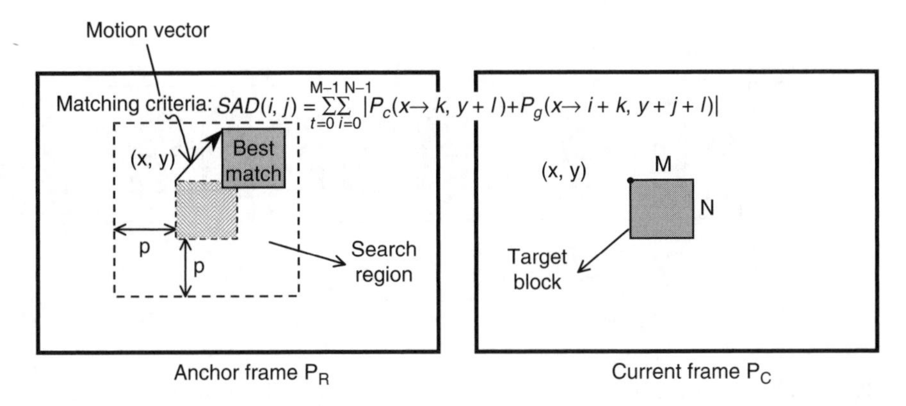

FIGURE 2.8 An example of block matching.

to be considered: the selection of block size, search window size, and matching criterion. In general, using smaller block sizes can achieve more accurate matching, but then the overhead for coding motion vectors increases. As a compromise, most video coding standards, such as MPEG-1, MPEG-2, H.261, and H.263, adopted a 16 × 16 block size for motion estimation and compensation. The advanced mode of MPEG-4 uses both 16 × 16 and 8 × 8 block size for finer estimation. The H.264/AVC has extended further for 16 × 8, 8 × 16, 8 × 4, 4 × 8, and 4 × 4 block motion estimation.

In general, the larger search window size provides more opportunities to obtain the best matching. However, the larger search window size causes a rapid increase of computational complexity. From other side, the search window size is determined by the size of correlation window and the maximum possible displacement within a frame. In addition, the search window size depends on the distance between the frame to be predicted and the reference frame, which is referred to as the *anchor frame*.

Note that in most video coding standards a translational model is used for motion estimation and compensation. Translation is a very simple model; it cannot handle complicated motion models such as rotation, zooming, occlusion, and disclosing of objects. Therefore, when using the translational model for those complicated motion cases the predictive errors are significant. In order to have good quality of coded images, both motion vectors and predictive errors are coded and transmitted. Also, in some video coding standards, more-complicated motion estimation/compensation methods are used. For example, in the MPEG-4 video standard, global motion and sprite are used to achieve high coding performance for some video sequences. We define those tools as enhancement tools and will discuss them later.

2.3.2.2 Structure for Different Frame Types

I, P, and B Frame Types

In most video coding standards, the video sequence is first divided into groups of pictures or frames (GOP) as shown in Figure 2.9. Each GOP may include three types of pictures or frames: intra-coded (I), predictive-coded (P), and bidirectionally predictive-coded (B). I-pictures are coded by intra-frame techniques only, with no

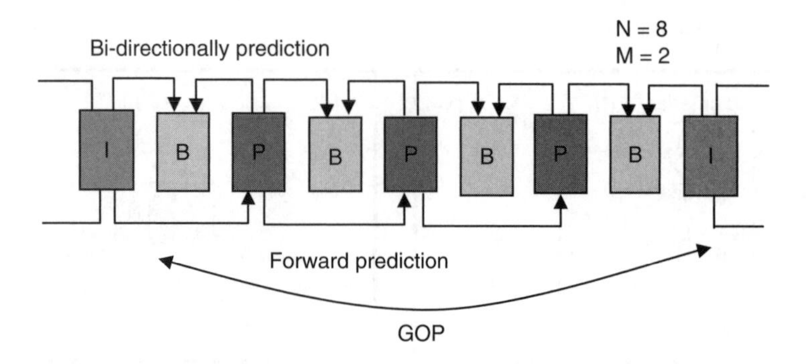

FIGURE 2.9 A group of pictures (GOP) of video sequence in display order.

need for previous information. In other words, I-pictures are self-sufficient. They are used as anchors for forward and/or backward prediction of other frames. P-pictures are coded using one-directional motion-compensated prediction from a previous anchor frame, which could be either an I- or P-picture. The distance between two nearest I-pictures is denoted by N, which is referred to as the size of GOP. The distance between two nearest anchor frames is denoted by M. The parameters N and M both are user-selectable parameters, which are selected by users during the encoding. Larger numbers of N and M will increase the coding performance but will cause error propagation or drift. Therefore, the I-picture has three functions: the first is to serve as anchor frame for P- and B-pictures, the second is to prevent the draft error propagation, and the third is as the point of random access. Usually, N is chosen from 12 to 15 and M from 1 to 3. If M is selected to be 1, this means no B-picture will be used. Regardless of the type of frame, each frame may be divided into slices; each slice consists of several macroblocks (MBs). There is no rule to decide the slice size. A slice could contain all MBs in a row of a frame or all MBs of a frame. A smaller slice size is favorable for the purpose of error resilience but decreases coding performance due to higher overhead. A macroblock contains a 16×16 luma component and spatially corresponding 8×8 chroma components. A MB has four luminance blocks and two chrominance blocks (for 4:2:0 sampling format), and the MB is also the basic unit of adaptive quantization and motion compensation. Each block contains 8×8 pixels over which the DCT operation is usually performed.

In order to exploit the temporal redundancy in the video sequence, the motion vector for each MB is estimated from two original luminance pictures using a block-matching algorithm. The criterion for the best matching between the current MB and an MB in the anchor frame is the minimum mean absolute difference. Once the motion vector for each MB is estimated, pixel values for the target MB can be predicted from the previously decoded frame. All MBs in I-frames are coded in intra mode with no motion compensation. —Macroblocks in P- and B-frames can be coded in several modes, which can be intra-coded and inter-coded, with or without motion compensation. This decision is made by mode selection. Most encoders depend on the values of predicted differences to make this decision. Within each slice, the values of motion vectors and DC values of each MB are coded using

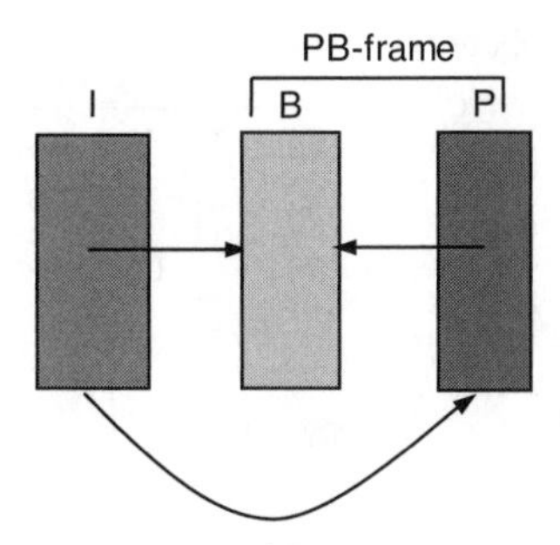

FIGURE 2.10 Prediction in PB-frame mode.

DPCM. The detailed specifications of this coding can be found in [2-1]. The structure of MPEG implies that if an error occurs within an I-frame, it will be propagated through all frames in the Group of Pictures (GOP). Similarly, an error in a P-frame will affect the related P- and B-frames, while B-frame errors will be isolated.

PB Frame Type

In H.263 and H.264, besides I-, P-, and B-pictures, there is another picture type: a PB frame is used. A PB-frame consists of two pictures, one P-picture and one B-picture, being coded as one unit as shown in Figure 2.10. Since H.261 does not have B-pictures, the concept of a B-picture comes from the MPEG video coding standards. In a PB-frame, the P-picture is predicted from the previous decoded I- or P-picture and the B-picture is bidirectionally predicted both from the previous decoded I- or P-picture and the P-picture in the PB-frame unit, which is currently being decoded. Several detailed issues have to be addressed at macroblock level in PB-frame mode, as follows.

- If a macroblock in a PB-frame is intra-coded, the P-macroblock in the PB unit is intra-coded and the B-macroblock in the PB unit is inter-coded. The motion vector of inter-coded PB-macroblocks is used for the B-macroblock only.
- A macroblock in a PB-frame contains 12 blocks for 4:2:0 format, 6 (4 luminance blocks and 2 chrominance blocks) from P-frame, and 6 from B-frame. The data for six P-blocks is transmitted first and then for the six B-blocks.
- Different parts of a B-block in a PB-frame can be predicted with different modes. For pixels where the backward vector points inside of coded P-macroblock, bidirectional prediction is used. For all other pixels, forward prediction is used.

2.3.3 Tools for Removing Statistical Redundancy, Variable Length Coding (VLC)

In the previous subsections, we presented tools used in video coding standards to decorrelate the video data by transform coding and motion-compensated predictive coding to remove spatial and temporal or interframe redundancy. In this subsection,

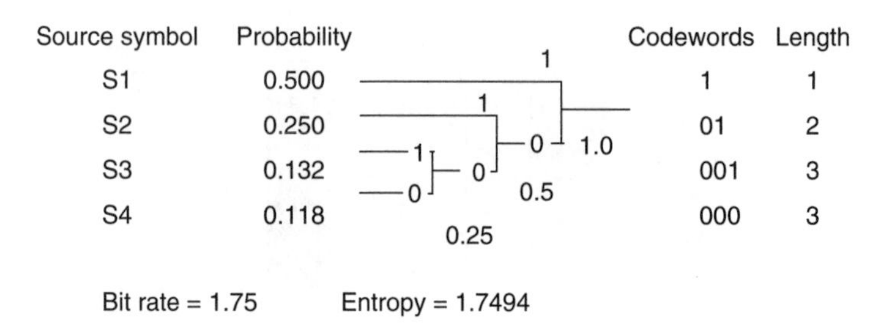

FIGURE 2.11 An example of Huffman coding.

we present two variable length coding algorithms used in video coding standards: Huffman coding and arithmetic coding. Huffman coding is a block-oriented coding technique, and arithmetic coding is a stream-oriented coding technique. Since the fundamental concepts and principles of variable length coding have been presented in Chapter 1, here we only present the implementation of variable length coding in the video coding standards.

2.3.3.1 Huffman Coding

For transmitting compressed digital video, most video coding standards use Huffman coding, which is a very popular lossless coding approach for discrete source. The basic idea of Huffman coding is to assign codewords with variable lengths to the source components according to their probabilities such that shorter codewords are used for the components with lower occurrence probability and longer codewords are used for ones with higher occurrence probability. In such a way, Huffman code results in minimum redundancy codes, which is close to the entropy of the source. The entropy is the optimal number of bits used for coding each symbol. An example of Huffman code is shown in Figure 2.11.

From the example in Figure 2.11, it can be seen that for coding a source with four symbols, we need 2 bits with fixed length code but only 1.75 bits with Huffman code, which is much closer to the entropy of the data source. As a result of Huffman coding, a set of codewords—which consists of a codebook—is created. The variable run-length coding (VLC) table in the video coding standards is an agreement between the encoder and the decoder. In video coding standards, these VLC tables are obtained by training a very large number of video sequences and in such way the occurrence probabilities for the components are guaranteed to be suitable for all video sequences. In the following, we give a real Huffman coding example used in the MPEG-2 video coding standard. An 8×8 block is transformed into 64 coefficients. The DC coefficients are coded with predictive coding. The predictive differences of the DC prediction are coded with Huffman coding. The Huffman table for dct_dc_size_luminance is shown in Table 2.1. From the table, it is noted that the small differences have larger probabilities and are coded with shorter codes, whereas the larger differences are coded with longer codes because their probabilities are smaller.

TABLE 2.1
Variable Length Codes for dct_dc_size_luminance in the MPEG-2 Standard

Variable length code	dct_dc_size_luminance
100	0
00	1
01	2
101	3
110	4
1110	5
1111 0	6
1111 10	7
1111 110	8
1111 1110	9
1111 1111 0	10
1111 1111 1	11

The problem with Huffman code is that it has to assign an output code to each symbol with an integral number of bits. This means that the shortest Huffman code is, at least, 1 bit. This would result in a longer compressed message than is theoretically possible. During the past decade, arithmetic coding has been developed to address this problem [2-12].

2.3.3.2 Arithmetic Coding

Arithmetic coding completely bypasses the method of coding an input symbol separately with a specific codeword. Instead, it encodes a stream of input symbols together with a single floating-point number. The idea behind arithmetic coding is first to define a probability line, 0 to 1, and then assign a range in this line for every symbol based on its probability in such a way that the higher the probability, the higher the range assigned to it. The output of the arithmetic coding process is a single number, which is less than 1 and greater than or equal to 0. In the following. we explain the arithmetic-coding algorithm with an example to aid understanding. To code a message "IMAGE DATA," we have a probability distribution as:

Character	Space	A	D	E	G	I	M	T
Probability	1/10	3/10	1/10	1/10	1/10	1/10	1/10	1/10

Once we get the character probabilities, we start to assign each character with a range along a probability line from 0 to 1. Along the probability line, each character is actually up to the integer number, but not including the integer number, such as for “A” which is in [0.1, 0.4).

Character	Space	A	D	E	G	I	M	T
Probability	1/10	3/10	1/10	1/10	1/10	1/10	1/10	1/10
Probability Line	0.0–0.1	0.1–0.4	0.4–0.5	0.5–0.6	0.6–0.7	0.7–0.8	0.8–0.9	0.9–1

When we encode the message of “IMAGE DATA,” the first character is “I.” In order to code the first character properly, we have to get the final output number between 0.7 and 0.8. An algorithm proposed in [2-12] is described as follows:

Set Low to 0.0
Set high to 1.0
While there are still input symbols do
Get an input symbol,
range = high − low
high = low + range • high-value (symbol)
low = low + range • low-value (symbol)
End of While
Output low

With the above algorithm we can obtain the encoding process as follows:

New character	Low value	High value
I	0.7	0.8
M	0.78	0.79
A	0.781	0.784
G	0.7828	0.7831
E	0.78295	0.78298
Space	0.782950	0.782953
D	0.7829512	0.7829515
A	0.78295123	0.78295132
T	0.782951311	0.782951320
A	0.7829513119	0.7829513146

The final low value, 0.7829513119, will be used to encode the message “IMAGE DATA” uniquely with the above algorithm. The decoding process is the revers of the encoding, as follows:

Get encoded number
Do
Find symbol whose range straddles the encoded number
Output the symbol
range = symbol low value–symbol high value
subtract symbol low value from encoded number
divide encoded number by range
until no more symbols

To perform the decoding process, we can get the following result:

Encoded Number	Output Symbol	Low	High	Range
0.7829513119	I	0.7	0.8	0.1
0.829513139	M	0.8	0.9	0.1
0.29513139	A	0.1	0.4	0.3
0.65043796666	G	0.6	0.7	0.1
0.5043796666	E	0.5	0.6	0.1
0.043796666	Space	0.0	0.1	0.1
0.43796666	D	0.4	0.5	0.1
0.3796666	A	0.1	0.4	0.3
0.932222	T	0.9	1.0	0.1
0.32222	A	0.1	0.4	0.3

2.3.3.3 Content-Based Arithmetic Encoding (CAE) for Binary Shape Coding

In the visual part of MPEG-4, the tool for coding an arbitrarily shaped video object is provided. The binary shape is coded by so-called content-based arithmetic encoding (CAE). There are two modes for binary shape coding, one is intra-mode and another is inter-mode. For intra mode, a 10-bit content is built for each pel, as illustrated in Figure 2.12a, where $C_k = 0$ for transparent pixels and $C_k = 1$ for opaque pixels. For the inter mode, temporal redundancy is exploited by using pixels from a corresponding motion-compensated binary alpha block (BAB) in the reference frame as shown in Figure 2.12b. For inter mode, a 9-bit context is built for each pixel to be coded. The context value is described by a bit pattern of 10 or 9 bits for the intra and inter mode, respectively, as follows:

$$C = \sum_k C_k \cdot 2^k \qquad (2.6)$$

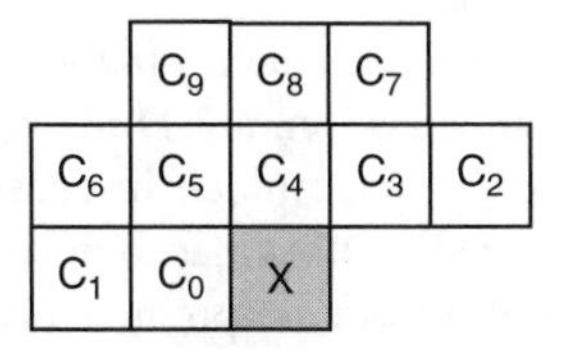

FIGURE 2.12A Template for defining the context of the pixel, X, to be coded in the intra mode.

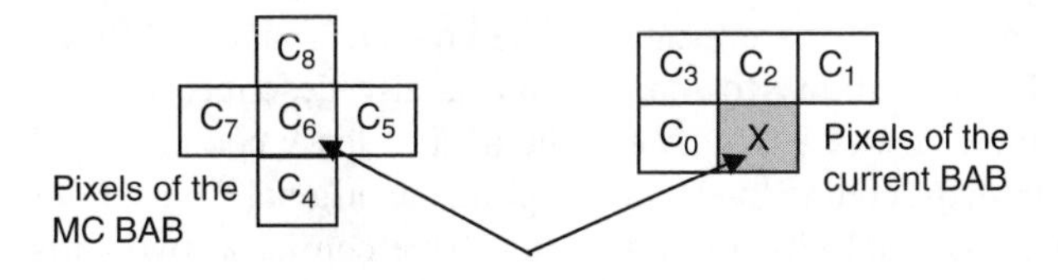

FIGURE 2.12B Template for defining the context of the pixel, X, to be coded in intra mode.

The process for coding a given pixel includes three steps: 1) compute a context number, 2) index a probability table using the context number, and 3) use the indexed probability to drive an arithmetic encoder. For the pixels in the top and left boundaries of the current macroblock, the template of causal context will contain the pixels of the already transmitted macroblocks on the top and on the left side of the current macroblock. For the two rightmost columns of the VOP, each undefined pixel, such as C_7, C_3, and C_2, of the context is set to the value of its closest neighbor inside the macroblock; that is, C_7 will take the value of C_8, and C_3 and C_2 will take the value of C_4. This causal context is used to predict the shape value of the current pixel.

For encoding the state transition, a context-based arithmetic encoder is used. The probability table of the arithmetic encoder for the 1,024 contexts was derived from sequences that are outside of the test set. Two bytes are allocated to describe the symbol probability for each context; the table size is 2,048 bytes. In order to increase coding efficiency and rate control, the algorithm allows lossy shape coding. In lossy shape coding, a macroblock can be down-sampled by a factor of 2 or 4, resulting in a subblock of size 8×8 pixels or 4×4 pixels, respectively. The subblock is then encoded using the same method as for the full-size block. The downsampling factor is included in the encoded bit stream and then transmitted to the decoder, which decodes the shape data and then up-samples the decoded subblock to full macroblock size according to the down-sampling factor. Obviously, it is more efficient to code shape using a high down-sampling factor, but coding errors may occur in the decoded shape after up-sampling.

However, in the case of low-bit-rate coding, lossy shape coding may be necessary since the bit budget may not be enough for lossless shape coding. Depending on the up-sampling filter, the decoded shape can look somewhat blocky. Several up-sampling filters were investigated. The best-performing filter in terms of subjective picture quality is an adaptive nonlinear up-sampling filter. Note that the coding efficiency of shape coding also depends on the orientation of the shape data. Therefore, the encoder can choose to code the block as described above or transpose the macroblock prior to arithmetic coding. Of course, the transpose information has to be signaled to the decoder. For shape coding in a P-VOP or B-VOP, the inter mode can be used to exploit the temporal redundancy in the shape information with motion compensation. For motion compensation, a 2-D integer pixel motion vector is estimated using full search for each macroblock in order to minimize the prediction error between the previous coded VOP (video object plane) shape and the current VOP shape. The shape motion vectors are predictively encoded with respect to the shape motion vectors of neighboring macroblocks. If no shape motion vector is available, texture motion vectors are used as predictors. The template for inter mode differs from the one used for intra mode. The inter mode template contains nine pixels, among which five are located in the previous frame and four are the current neighbors. The probability for one symbol is also described by 2 bytes, giving a probability table size of 1,024 bytes. The idea of lossy coding can also be applied to inter mode shape coding by down-sampling the original BABs. For the inter mode shape coding, the total bits for coding the shape consist of two parts, one part for coding motion vectors and the other for prediction residue. The encoder may decide that the shape representation achieved by just using motion vectors is sufficient;

thus, bits for coding the prediction error can be saved. Actually, there are seven modes to code the shape information of each macroblock: transparent, opaque, intra, inter with and without shape motion vectors, inter with and without shape motion vectors, and prediction error coding. These modes with optional down-sampling and transposition allow for encoder implementations of differing coding efficiency and implementation complexity. Again, this is a problem of encoder optimization, which does not belong to the standard.

2.4 ENHANCEMENT TOOLS FOR IMPROVING FUNCTIONALITY AND CODING EFFICIENCY

In this section, we present tools that are not included in the basic tools for video coding standards and that are used for increasing functionality, thus improving coding efficiency and reducing the complexity.

2.4.1 TOOLS FOR INCREASING FUNCTIONALITY

Several tools have been developed for increasing the functionality for digital video coding standards. Among them, the most important include object-based coding, scalability, and error resilience. In the following subsections, we briefly introduce these tools.

2.4.1.1 Object-Based Coding

The MPEG-4 standard became the first object-based video coding standard, and is capable of separately encoding and decoding video objects. To clearly explain the idea of object-based coding, we should review the definitions of video object (VO)–related definitions specified in the MPEG-4 video standard. In video sequences, an image scene of frames may contain several video objects, as shown in Figure 2.13.

In Figure 2.13, the scene contains three VOs: two moving VOs and one VO of background scene. The time instant of each VO is referred to as the *video object plane* (VOP). Each video object is described by the information of texture, shape, and motion vectors. Therefore, in the object-based coding, the encoder mainly consists of two parts: the shape coding and the texture coding of the input VOP. Texture coding

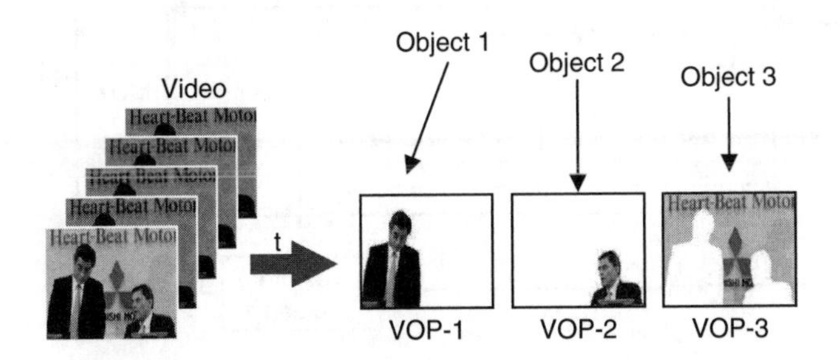

FIGURE 2.13 Video object and video object plane definition and format.

is based on DCT coding with traditional motion-compensated predictive coding. The VOP is represented by means of a bounding rectangular, as depicted in Figure 2.13.

For shape coding, the shape information is classified as binary shape or gray scale. Both binary and gray scale shapes are referred to as an *alpha plane*. The alpha plane defines the transparency of an object. Multilevel alpha maps are frequently used to blend different images. A binary alpha map defines whether or not a pixel belongs to an object. The binary alpha planes are encoded by modified content-based arithmetic encoding (CAE) as described in the previous section, whereas the gray scale alpha planes are encoded by motion-compensated DCT, which is similar to texture coding. The block diagram of the encoding structure is shown in Figure 2.14.

The concept of a video object provides a number of functionalities that are either impossible or very difficult in MPEG-1 or MPEG-2 video coding. For example, the video sequence can be encoded in a way that will allow the separate decoding and reconstruction of the objects and allow the editing and manipulation of the original scene by simple operations performed in the compressed bit stream domain. Object-based coding is also able to support such functionalities as the warping of synthetic or natural text, textures, image, and video overlays on reconstructed video objects.

2.4.1.2 Scalability

The scalability framework is referred to as *generalized scalability* and includes spatial, temporal, SNR, and fine granular scalability (FGS). The purpose of scalable coding is to satisfy the need of various applications, which include Internet video, wireless LAN video, mobile wireless video, video on demand (VOD), live broadcasting, multichannel content production and distribution, storage applications, layered protection of content, and multipoint surveillance systems. For such applications,

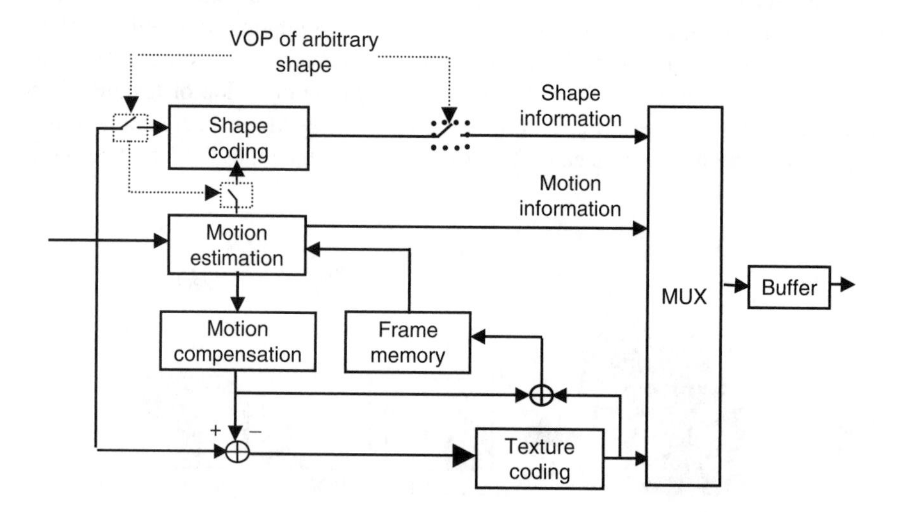

FIGURE 2.14 Block diagram of object-based encoding.

scalable video should possess several important features. First of all, it should be easily adaptive to bandwidth variations because available bandwidth may vary due to interference, overlapping wireless LANs, competing traffic, mobility, multipath fading, etc. The second is robustness against data losses because, depending on the channel condition, partial data losses may occur. These losses can occur also during hand-over between two access points. Specifically, for best-effort networks, all packets are treated equally and any packet may be lost. Prioritized video can alleviate this problem and provide prioritized transmission of video using different priority queues. For those networks that are QoS (Quality of Service)-enabled, packets can be treated in a prioritized or preferential manner. The third feature is the support for bandwidth and device scalability because various clients require differing data rates and transmission methods that are optimized for their particular connections and capabilities. Finally, it provides scalable power requirements because most portable wireless video devices are battery-powered and tradeoffs should be possible between longer battery power and lower quality video. Beyond these technical requirements, the copy protection or digital rights management (DRM) of content is also very important. Scalable video coding is able to meet this requirement since the copyright protection information can be included in the encoding process. We also should note that, in the applications of video transmission over WLAN, efficient spatial scalability is necessary, since the same content can be viewed on various wireless devices with different capabilities and display sizes.

Both MPEG-2 and MPEG-4 video coding standards have been developed to provide scalability functionality. The major difference between MPEG-2 and MPEG-4 is that MPEG-4 extends the concept of scalability to be content based. This unique functionality allows MPEG-4 be able to resolve objects into different VOPs. Using the multiple VOP structure, different resolution enhancements can be applied to different portions of a video scene. Therefore, the enhancement layer can be only applied to a particular object or region of the base layer instead of entire base layer. This is a feature that MPEG-2 does not have. In spatial scalability, the base layer and the enhancement layer can have different spatial resolutions. The base layer is encoded in the same way as the nonscalable encoding technique described previously. The enhancement layer is encoded as P-or B-pictures as shown in Figure 2.15.

The current picture in the enhancement layer can be predicted from either the up-sampled base layer picture or the previously decoded picture at the same layer as well as both of them. The down-sampling and up-sampling processing in spatial scalability is not a part of a standard and can be defined by the user. The spatial

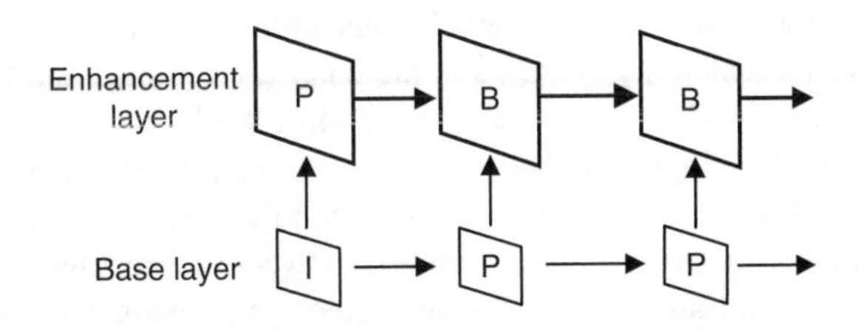

FIGURE 2.15 Illustration of spatial scalability.

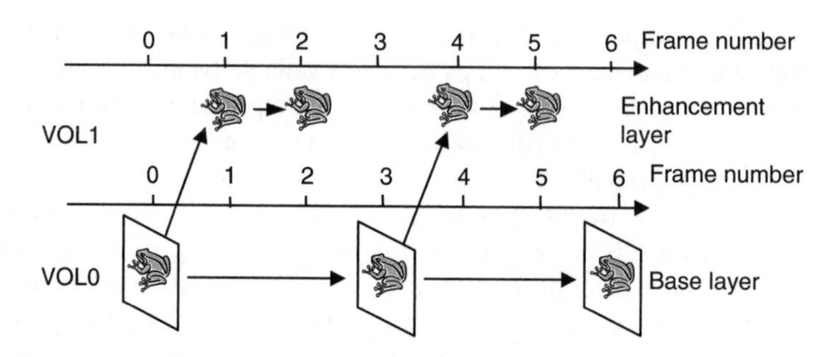

FIGURE 2.16 An example of temporal scalability.

scalability of MPEG-2 and MPEG-4 are able to support a variety of resolutions, including QCIF, CIF, SD, and HD, and also higher resolutions.

In temporal scalability, we use MPEG-4 as an example. A subsequence of subsampled VOP in the time domain is coded as a base layer. The remaining VOPs can be coded as enhancement layers. In this way, the frame rate of a selected object can be enhanced so that it has a smoother motion than other objects. An example of the temporal scalability is illustrated in Figure 2.16.

In Figure 2.16, the VOL0 is the entire frame with both an object and a background, whereas VOL1 is a particular object in VOL0. VOL0 is encoded with a low frame rate and VOL1 is the enhancement layer. The high frame rate can be reached for the particular object by combining the decoded data from both base layer and enhancement layer. Of course, the B-VOP is also used in temporal scalability for coding the enhancement layer, which is another type of temporal scalability. As in spatial scalability, the enhancement layer can be used to improve either the entire base layer frame resolution or only a portion of the base layer resolution. The temporal scalability supports decoding of moving pictures with a large range of frame rates. For instance, for multichannel content production and distribution, the same stream will be viewed on a variety of devices having different temporal resolutions. For example, 7.5, 15, and 30 Hz should be supported for certain applications.

The SNR scalability supports decoding the video having quality varying between acceptable and high quality with the same spatial resolution at different bit rates. The base layer is coded with a coarse quantization step at lower bit rates for low-capacity channels, whereas, in the enhancement layer, the difference between original and decoded base layer signal is then coded with a finer quantizer to generate an enhancement bit stream for high-capacity channel applications with higher quality video. An application for this scalability could be in the context of storage and transmission.

In the MPEG-4 video, a new scalable coding tool—fine granular scalability (FGS)—has been developed to allow the coverage of a wide range of bit rates for the distribution of video on Internet and other applications. The basic idea of FGS scalability coding is to encode the difference between the reconstructed base layer signal and the original signal at the enhancement layer using a bit plane representation of the DCT coefficients. The absolute values of the DCT coefficients are first represented by a binary format and then are grouped by bit planes starting from the most significant bit plane to the least significant bit plane. The run length coding is then applied to the each bit plane. In such a way, the enhancement bit stream can be truncated at any point according to the capability of transmission channels or decoders and the granular scalability can be achieved. The technical details of MPEG-4 FGS can be found in [2-13].

2.4.1.3 Tools for Error Resilience

Error robustness is very important for certain video transmission applications, such as wireless video transmission and terrestrial broadcasting television. In the MPEG-2 video coding standard, the method of error resilience encoding is to transmit the motion vector for the I-picture. In the case of losing intra-coded macroblocks packets, the lost macroblocks can be recovered by motion compensation with its motion vectors, which are transmitted with the macroblock data in the above-neighboring macroblocks. In the MPEG-4 video coding standard, the error resilience tool development effort is divided into three major areas, which include resynchronization, data recovery, and error concealment. As with other coding standards, MPEG-4 uses a large number of variable length coding to reach high coding performance. However, if one bit is lost or damaged, entire bit stream becomes undecodeable due to a loss of synchronization. The resynchronization tools attempt to enable resynchronization between the decoder and the bit stream after a transmission error or errors have been detected. Generally, the data between the synchronization point prior to the error and the first point, where synchronization is reestablished, is discarded. The purpose of resynchronization is to effectively localize the amount of data discarded by the decoder; the other methods such as error concealment can then be used to conceal the damaged areas of a decoded picture. Currently, the resynchronization approach adopted by MPEG-4 is referred to as a *packet* approach. This approach is similar to the group of blocks (GOB) structure used in H.261 and H.263. In the GOB structure, the GOB contains a start code, which provides the location information of the GOB. MPEG-4 adopted a similar approach in which a resynchronization marker is periodically inserted into the bit stream at the particular macroblock locations. The resynchronization marker is used to indicate the start of a new video packet. This marker is distinguished from all possible VLC codewords as well as the VOP start code. The packet header information is then provided at the start of a video packet. The header contains the information necessary to restart the decoding process. These include the macroblock number of the first macroblock contained in this packet and the quantization parameter necessary to decode the first macroblock. The macroblock number provides the necessary spatial resynchronization while the quantization parameter allows the differential decoding process to be resynchronized. It should be noted that some tools used for improving compression efficiency need to be modified when the error resilience tools are used within MPEG-4. For example, all predictive encoded information must be contained within a video packet to avoid error propagation. In conjunction with the video packet approach to resynchronization, MPEG-4 has also adopted fixed interval synchronization method, which requires that VOP start codes and resynchronization markers appear only at legal fixed interval locations in the bit stream. This will help to avoid the problems associated with start code emulation. When fixed interval synchronization is utilized, the decoder is only required to search for a VOP start code at the beginning of each fixed interval. The fixed interval synchronization method extends this approach to be any predetermined interval.

After resynchronization is reestablished, the major problem is recovering lost data. A new tool called *reversible variable length codes* (RVLC) is developed for the purpose of data recovery. In this approach, the variable length codes are designed such that the codes can be read both in the forward as well as the reverse direction. An example of such code includes codewords like 111, 101, and 010. All of these codewords can be read reversibly. However, it is obvious that this approach will reduce the coding efficiency that is achieved by the entropy coder. Therefore, this approach is used only in the case where error resilience is important.

Error concealment is an important component of any error-robust video coding. The error concealment strategy is highly dependent on the performance of the resynchronization technique. Basically, if the resynchronization method can efficiently localize the damaged data area, the error concealment strategy becomes much more tractable. Error concealment is actually a decoder issue if there is no additional information provided by the encoder. There are many approaches to error concealment, which can be found in [2-14].

2.4.2 Tools for Increasing Coding Efficiency

Several enhancement tools have been developed in the video coding standards for further improving coding performance.

2.4.2.1 Interlace video

Interlaced video format has been extensively used in the TV broadcasting industry for a long time. The main reason for using interlaced video format is to reach a good compromise between frame rate and spatial resolution. In interlaced video, each frame consists of two fields that are separated in capture time. This allows representing the video data with either the complete frames or the individual fields. Consequently, the video can be adaptively encoded by selecting frame coding or field coding on a picture-by-picture basis; also can be on a more localized basis within a coded frame. Frame encoding is typically preferred when the video scene contains significant detail with limited motion, while field encoding works better when there is fast picture-to-picture motion. In the field coding, the second field can be predicted from the first field. One of the main differences between MPEG-1 and MPEG-2 is that the MPEG-2 has two tools to code interlaced video: frame/field DCT and frame/field motion compensation. The frame/field DCT may be performed at the macroblock level for the interlaced video as shown in Figure 2.17.

At the MB level, the field DCT can be selected when the video scene contains less detail and large motion. Since the difference between adjacent fields can be large when there is large motion between fields, it may be more efficient to group the fields together, rather than the frames. In this way, the pixels within a block are highly correlated and more energy would be concentrated to fewer high-order

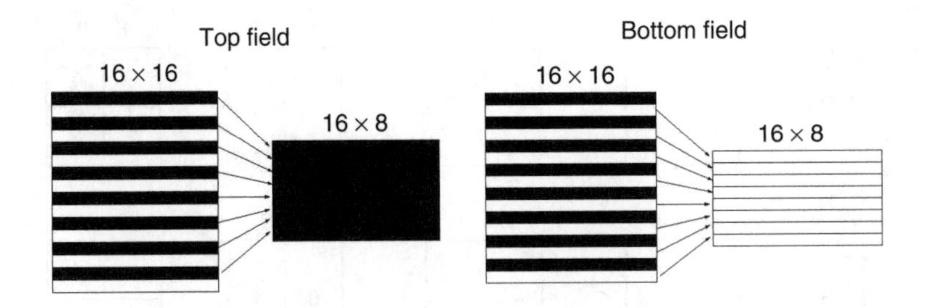

FIGURE 2.17 Frame and field DCT for interlaced video.

coefficients resulted by the transformation, especially if there is not much detail contained in the scene. This may greatly increase coding efficiency. In the frame/field motion compensation for the P-picture, the two reference fields for predictions are the most recently decoded top field and bottom field. Three cases of prediction are illustrated in Figure 2.18.

In Figure 2.18, the 16×16 luminance macroblock can be converted to two 16×8 blocks from top field and bottom field, respectively. The field DCT can be applied to these field blocks.

2.4.2.2 Adaptive Block Size Motion Compensation

In many video coding standards, a macroblock consisting of a 16×16 block of luma pixels and two corresponding blocks of chroma pixels is used as the basic processing unit of the video decoding process. A macroblock can be further partitioned for inter prediction. The selection of the block size using for inter prediction partitions is a compromised result between the bit savings provided by using motion compensation with smaller blocks and the increased number of bits needed for coding motion vectors. In MPEG-4 there is an advanced motion compensation mode. In this mode, the inter prediction process can be performed with adaptive selection of a 16×16 or 8×8 block. The purpose of the adaptive selection of the matching block size is to further enhance coding efficiency. The coding performance can be improved at low bit rates since the bits for coding prediction difference could be greatly reduced at the limited extra cost for increasing motion vectors. Of course,

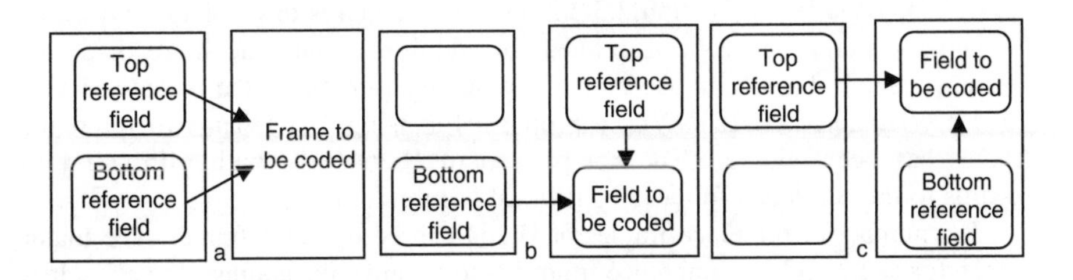

FIGURE 2.18 (a) Prediction of the first field or field prediction in a frame picture; (b) Prediction of the second field picture when it is the bottom field; (c) Prediction of the second field picture when it is the top field.

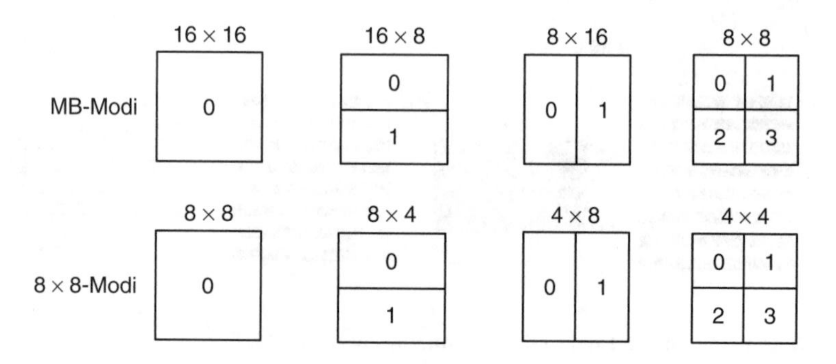

FIGURE 2.19 Macroblock partitioning in H.264.

if the cost for coding motion vectors becomes too high, the mode for using a small block size will not be selected. The decision made in the encoder should be very careful. If the 8×8 prediction is chosen, there are four motion vectors for the four 8×8 luminance blocks in a macroblock that will be transmitted. The motion vectors for coding two chrominance blocks are then obtained by taking an average of these four motion vectors and dividing the average value by a factor of 2. Since each motion vector for the 8×8 luminance block has half-pixel accuracy, the motion vector for the chrominance block may have sixteenth-pixel accuracy. The issues of motion estimation process in the encoder and the selection of whether to use inter prediction for each region of the video content are not specified in the standards. The encoding issues are usually described in the informative parts of the standards. In the recently developed MPEG and ITU joint standard, H.264/AVC, the 16×16 macroblock is further partitioned into even smaller blocks as shown in Figure 2.19.

In Figure 2.19, it can be seen that a total of eight kinds of blocks can be used for adaptive selection of motion estimation/compensation. With optimal selection of motion compensation mode in the encoder, coding efficiency can be greatly improved for some sequences. Of course, this is again an encoding issue, and an optimal mode selection algorithm is needed.

2.4.2.3 Motion Compensation with Multiple References

As we discussed in the Section 2.3.2.2, in most standards three picture types, I-, P-, and B-pictures, have been defined. In addition, usually no more than two reference frames have been used for motion compensation. In the recently developed new standard, H.264/AVC, a proposal for using more than two reference frames has been adopted. The comparison of H.264/AVC with MPEG-2/4 in regards to the reference frames is shown in Figure 2.20.

The number of reference frames of H.264 can be up to 15 frames. The major reason for using multiple reference frames is to improve the coding efficiency. It is obvious that better matching would be found by using multiple reference frames than by using two or fewer frames in the motion estimation. Such an example is shown in Figure 2.21.

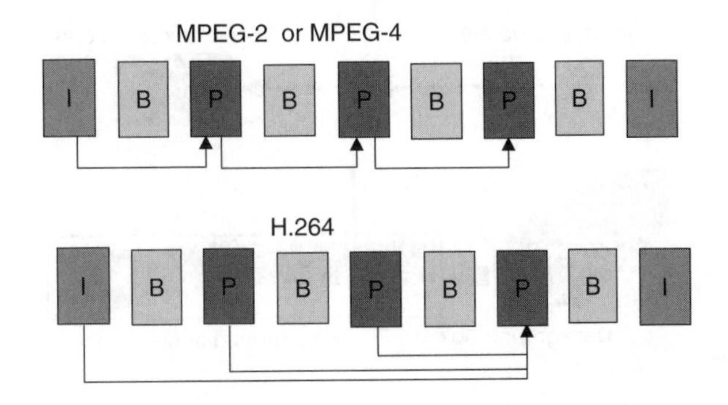

FIGURE 2.20 Comparison on reference frames between MPEG-2/4 with H.264.

2.4.2.4 Sprite Coding

Sprite coding is a tool in MPEG-4 video that attempts to improve the coding efficiency at low bit rates for a certain kind of video sequence. The "sprite" is an image, which is especially composed of pixels belonging to a video object, that is visible throughout an entire piece of video sequence [2-15]. From its definition, it is seen that the sprite is the video object that usually represents the background of a video segment . The background sprite contains all pixels that are at least visible once throughout the video sequence. However, certain portions of the background may not be visible in certain frames due to the occlusion of the foreground objects or the camera motion. An example of sprite coding is shown in Figure 2.22.

In the following, we are going to discuss two problems. The first is how to use sprite to improve the coding efficiency for certain segments of video sequences. The second is how to generate the sprite. The first problem has been addressed by the normal part of MPEG-4 visual standard. In the standard, the sprite encoding syntax allows transmitting a sprite as a still image progressively either spatially or hierarchically. In the spatially progressive transmission, the sprite is reconstructed at the

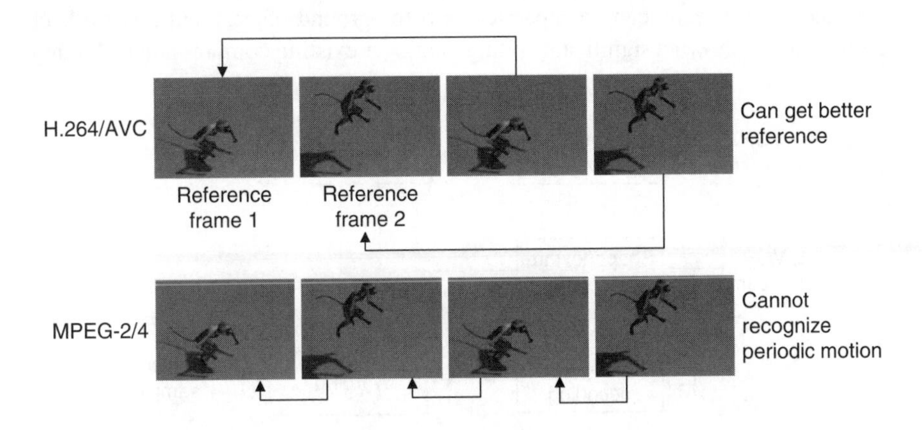

FIGURE 2.21 An example to explain the benefit for using multiple reference frames.

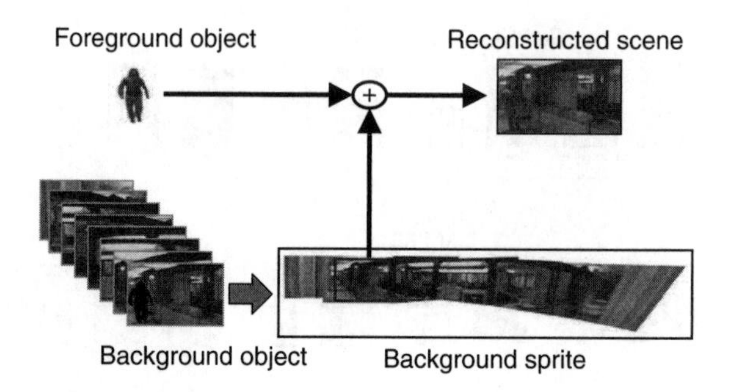

FIGURE 2.22 An example of sprite coding.

decoder a piece at a time, whereas in the hierarchical transmission, the sprite image quality is updated with residual images. How do we use sprite to increase the coding efficiency for some sequences? Since the sprite contains all visible background scenes of a video segment, where the changes within the background content are mainly caused by camera parameters, the sprite can be used for direct reconstruction of the background video object planes (VOPs) or as the prediction of the background VOPs within the video segment. In the sprite coding, the background sprite is first efficiently transmitted to the decoder and then is stored in a frame at the decoder. The camera parameters are then transmitted to the decoder for each frame so that the appropriate part of the background scene can be either used as the direct reconstruction or as the prediction of the background VOP. Both cases can significantly save the coding bits and increase the coding efficiency. This procedure is shown in Figure 2.22 and a block diagram of sprite coding is shown in Figure 2.23.

There are two types of sprites, static and dynamic. The static sprite is generated with copying the background from a video sequence. This copying includes the appropriate warping and cropping. Therefore, a static sprite is always built offline and it is coded and transmitted as a first I-VOP for coding a video sequence. The offline static sprites are particularly suitable for coding a video sequence in which the objects in a scene can be separated into foreground objects and a static background. It has shown a significant coding gain over existing compression technology

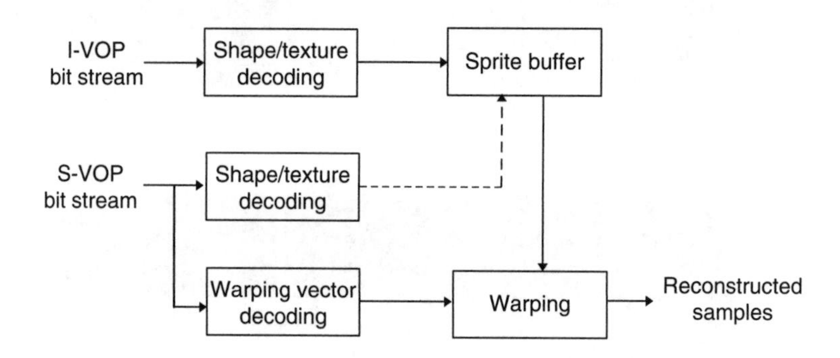

FIGURE 2.23 A block diagram of sprite decoding.

for this kind of video sequence. In contrast, the dynamic sprite is dynamically built during predictive coding. It can be built either online or offline. The dynamic sprite is more complicated in real-time applications due to the difficulty of updating the sprite during the coding. Therefore, only the static sprite is adopted by the MPEG-4 standard.

Sprite generation is not the normative part of the standard, but it is an important tool for sprite coding and it is described in the informative part of standard. The algorithm for sprite generation is described in MPEG-4 video verification model (VM) [2-16]. Fast and robust sprite generation is provided in the informative section of the standard, Part 7 [2-3].

When a sprite is generated offline, the entire segment of video is assumed to be available. In sprite generation, the global motion field is first estimated for each VOP in the sequence; the VOP then is registered with the sprite by warping the VOP to the sprite coordinate system based on global motion information. For a natural video object, sprite refers to a representative view collected from a video sequence, from which the video can be reconstructed. The effectiveness of this approach depends on whether the object motion can be effectively represented by a global motion model, e.g., translationa;, zooming, affine, and perspective. A block diagram of sprite generation is shown in Figure 2.24. In Figure 2.24, the new image is incrementally blended to the previous sprite and an equal weight is used for all of the images contributing to the sprite. This could suppress the noise in individual images. The detailed procedure of sprite generation can be found in Appendix D of the MPEG-4 Video VM [2-16].

The algorithm of fast and robust sprite generation described in the Part 7 was originally proposed by Lu et al. [2-17]. The purpose of fast and robust sprite generation is to accelerate the sprite generation process and reduce the complexity of the encoder. The algorithm is based on Appendix D of MPEG-4 VM; however, in order to achieve speed and robustness, several novel features have been introduced. One feature of the proposed algorithm is that it first warps the previous sprite and then calculates the global motion, referencing the warped sprite image instead of estimating the global motion of the current image directly from the previous sprite.

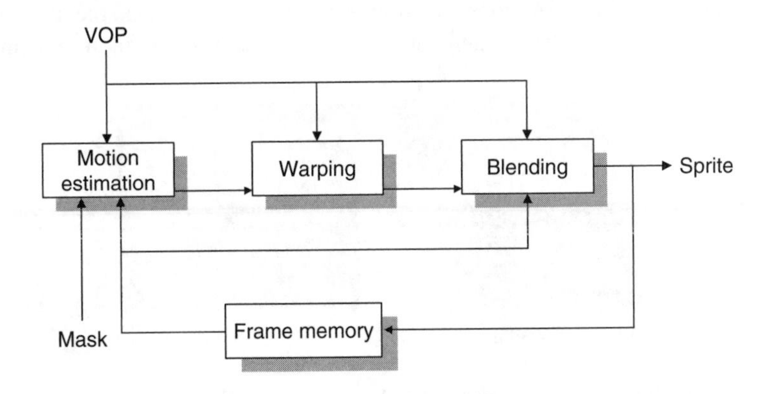

FIGURE 2.24 A block diagram of sprite generation.

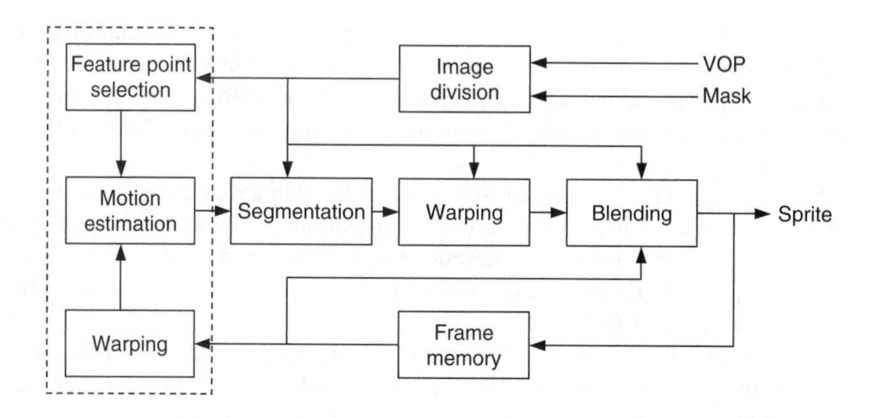

FIGURE 2.25 Fast and robust sprite generation.

This long-term motion estimation method can greatly decrease the error accumulations caused by the individual frame. The extra cost in memory is also reasonable because the size of warped sprite is the same as that of current frame. The other feature is to incorporate image segmentation in the sprite generation. Static sprite coding is normally used for object-based video coding; however, sometimes auxiliary segmentation information is either unavailable or not accurate enough to mask out all of the moving objects from the scene. The segmentation technique developed in [2-17] is incorporated into the proposed sprite generation, which is usually used in this algorithm when no auxiliary segmentation masks are available. A block diagram of fast and robust sprite generation is shown in Figure 2.25. The algorithm includes five parts: image region division, fast and robust motion estimation, image segmentation, image warping, and image blending. The function of image region division is to divide the frame into three regions, the reliable region, unreliable region, and undefined regions. An example of image division is shown in Figure 2.26.

The image division can contribute to sprite generation in two ways. First, only the reliable region participates in motion estimation, which can speed up motion estimation processing and eliminate the effect of foreground objects and frame borders. Second, all three regions are differently dealt with in image blending, which improves the visual quality of generated sprite. Fast and robust motion estimation

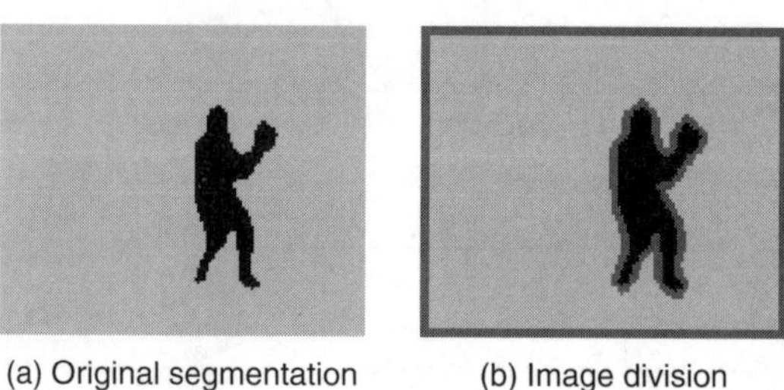

FIGURE 2.26 Illustration of image division. Light gray, reliable region; dark gray, unreliable region; black, undefined region.

partly aims at obtaining the motion parameters between the current image and the sprite image. In this part, preprocessing is used to accelerate motion estimation and eliminate the effect of foreground objects. Preprocessing mainly focuses on two aspects. First, feature point selection is performed in order to decrease the number of pixels involved in motion estimation. The selection is based on the fact that those pixels with large gradient values dominate the contributions on motion estimation in the background object, rather than those located in smooth areas. Second, the sprite is warped using the previously estimated motion parameters. The great displacement between the current image and the sprite may decrease the accuracy of motion estimation, or even lead to wrong estimation. Using the warped sprite as a reference of the current image can increase the robustness of global motion estimation, and speed up motion estimation.

Segmentation information is important for sprite generation. However, the auxiliary segmentation masks are sometimes unavailable. The segmentation method for sprite generation used here is based on the rough segmentation proposed in [2-17]. Although the segmentation is not accurate with some pixels in the background masked, it is enough only for the purpose of eliminating the effect of foreground objects in the proposed sprite generation. After image segmentation, the background areas are marked as reliable image regions, and the foreground areas are marked as unreliable regions. The results of image division will contribute to the following image blending process.

In image blending, reliable and unreliable regions take different roles in the process of sprite updating. The pixels in reliable regions are used to update the corresponding sprite pixels, whereas the pixels in the unreliable regions are used only for updating the sprite pixels that have never been updated by reliable region pixels. Undefined regions do not participate in this process.

2.4.2.5 Global Motion Compensation

Global motion compensation is useful for increasing coding efficiency at low bit rates for a certain kind of video sequences, containing global motion in scenes due to camera motion. In the case of low–bit rate coding, the cost of coding motion vectors cannot be ignored. If we could use one or a few motion vector sets to perform motion-compensated prediction for many macroblocks, it would greatly increase coding efficiency. The concept of global motion compensation is based on this very above observation. If video sequences contain global motion, a large number of macroblocks may share the same motion vectors and it may be reasonable to use one motion vector set to do motion-compensated prediction for those macroblocks in the VOP. The MPEG-4 syntax allows using up to four global motion vectors. The syntax also allows that each macroblock can be predicted either from the previous VOP by global motion compensation (GMC) using warping parameters described in the MPEG-4 standard [2-18] or from the previous VOP by local motion compensation (LMC), using local motion vectors as in the classical scheme. The mode selection is determined based on which predictor leads to the lower prediction error. Although macroblocks with GMC mode do not have their own block motion vectors, they have pel-wise motion vectors for sprite warping obtained from global motion

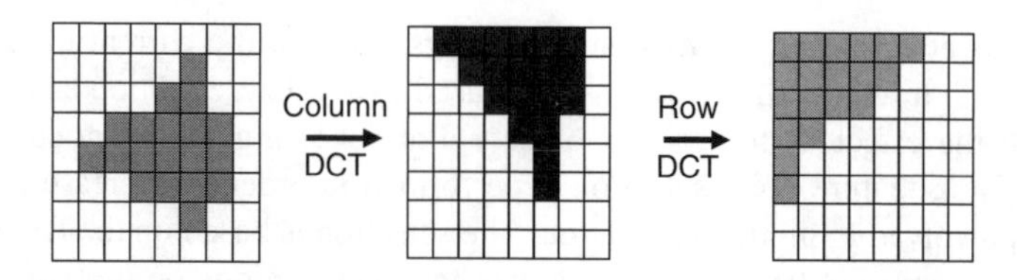

FIGURE 2.27 Illustration of SA-DCT.

parameters. The candidate motion vector predictor from the reference macroblock with GMC mode is obtained as the averaged value of the pelwise motion vectors in the macroblock. For the macroblocks with GMC mode, the local motion vectors are not transmitted.,

2.4.2.6 Shape-Adaptive DCT

Shape-adaptive DCT is applied in MPEG-4 only to those 8 × 8 blocks that are located on the object boundary of an arbitrarily shaped VOP. The idea of the SA-DCT is to apply one-dimensional DCT transformation vertically and horizontally according to the number of active pixels in the row and column of the block, respectively. The size of each vertical DCT is the same as the number of active pixels in each column. After vertical DCT is performed for all columns with at least one active pixel, the coefficients of the vertical DCTs with the same frequency index are lined up in a row. The DC coefficients of all vertical DCTs are lined up in the first row, the first-order vertical DCT coefficients are lined up in the second row, and so on. After that, horizontal DCT is applied to each row. As is the same for the vertical DCT, the size of each horizontal DCT is the same as the number of vertical DCT coefficients lined up in the particular row. The final coefficients of SA-DCT are concentrated into the upper-left corner of the block. This procedure is shown in the Figure 2.27.

The final number of the SA-DCT coefficients is identical to the number of active pixels of image. Since the shape information is transmitted to the decoder, the decoder can perform the inverse shape-adapted DCT to reconstruct the pixels. The regular zigzag scan is modified so that the inactive coefficient locations are neglected when counting the runs for the run length coding of the SA-DCT coefficients. It is obvious that for a block with all 8 × 8 active pixels, the SA-DCT becomes a regular 8 × 8 DCT and the scanning of the coefficients is identical to the zigzag scan. All SA-DCT coefficients are quantized and coded in the same way as the regular DCT coefficients employing the same quantizers and VLC code tables.

2.5 BRIEF SUMMARY OF VIDEO CODING STANDARDS

In the previous sections, we presented the tools for video coding standards. After we have learnt the basic principles of these tools, it is much easier to understand the standard itself. In the following we briefly introduce the standards developed by ISO/IEC and ITU independently or jointly. These standards include JPEG,

JPEG2000 for still image coding, MPEG-1, MPEG-2, MPEG-4, H.261, H.263, and H.264/AVC for video coding.

2.5.1 Summary of ISO/IEC Standards of Image and Video Coding

2.5.1.1 JPEG

The acronym *JPEG* stands for the Joint Photographic Experts Group standard, a standard for storing and compressing digital still images. JPEG was jointly developed by ISO/IEC and ITU-T in 1992 as international standard 10918-1 or ITU-T Recommendation T.81. The JPEG standard allows for both lossy and lossless encoding of still color images. From the algorithmic point of view, JPEG includes four coding modes: sequential DCT-based coding, progressive DCT-based coding, lossless coding, and e hierarchical coding [2-19].

The DCT-based coding is the baseline algorithm of JPEG, which includes sequential DCT-based and progressive DCT-based coding modes. In the sequential DCT-based coding mode, an image is first partitioned into blocks of 8×8 pixels. The blocks are then encoded according to the scanning order from left to right and top to bottom. In the progressive DCT-based coding mode, the DCT coefficients are first stored in a buffer before the encoding is performed. The stored DCT coefficients are then encoded with a multiple scanning process. In each scan, the quantized DCT coefficients are partially encoded by either spectral selection or successive approximation. In the method of spectral selection, the quantized DCT coefficients are first divided into multiple spectral bands according to a zigzag order. In each scan, a specified band is then encoded. In the method of successive approximation, a specified number of most-significant bits of the quantized coefficients are encoded first and the least-significant bits are encoded in subsequent scans. The difference between sequential coding and progressive coding is that in the former an image is encoded part-by-part according to the scanning order, whereas in the latter an image is encoded by a multiple scanning process and in each scan the full image is encoded to a certain quality level.

The baseline of JPEG is sufficient for many applications. However, to meet the needs of applications that cannot tolerate loss, such as in the compression of medical images, a lossless coding scheme is also provided and is based on a predictive coding scheme. In this scheme, the three nearest neighboring pixels are used to predict the pixel to be coded. The prediction difference is entropy coded using either Huffman coding or arithmetic coding. Note that the prediction difference in this scheme is not quantized; therefore, the coding is lossless.

In the hierarchical coding mode, an image is first spatially down-sampled into a multi-layered pyramid, resulting a sequence of frames. This sequence of frames is encoded by the predictive coding scheme. Except for the first frame, predictive coding is applied to the difference between the frame to be coded and the reference frame. The reference frame is equivalent to the previous frame that could be reconstructed in the decoder. The difference frames can be coded with DCT-based coding, lossless coding, or DCT-based coding with final lossless processing. Down-sampling and up-sampling filters are used in the hierarchical coding mode. The hierarchical coding mode provides a progressive presentation similar to that of the progressive DCT-based mode, but it can provide multiple resolution applications and the progressive DCT-based mode cannot.

The JPEG standard is mainly used for coding still images. Video is usually coded by the MPEG standards. But the extended version of JPEG — Motion-JPEG — can be also used for coding video sequences. In Motion-JPEG, each frame in the video sequence is encoded and stored as a still image with the JPEG format.

2.5.1.2 JPEG-2000

JPEG2000 is a new standard for still image coding that was jointly developed by ISO/IEC and ITU-T in 2001 as international standard ISO/IEC 15444-1 or ITU-T Recommendation T.800 [2-20]. A JPEG2000 overview is given in [2-21]. Before JPEG2000, the image standard generally used for still images was JPEG, a standard used for more than a decade and a tool of proven value. However, the JPEG standard cannot fulfill the advanced requirements of many of today's applications, such as in network and mobile environments. JPEG2000 was developed to respond to these advanced needs and to break through the limitations of the JPEG standard. The JPEG2000 system is optimized not only for coding efficiency, but also for scalability and interoperability. These new features of JPEG2000 are designed for many high-end and emerging applications, including applications for Internet, wireless devices, digital cameras, image scanning, and client/server imaging systems. The main difference between JPEG2000 and conventional JPEG is that the former uses the wavelet transform as a core technology whereas the latter uses traditional DCT-based coding technology. In the JPEG2000 system, the encoding process consists of several parts, including tiling, DC level shifting, component transformation, wavelet transform, partitioning into code blocks and precincts, entropy coding, and formation of packets and layers as shown in Figure 2.28.

In the encoding process of JPEG2000, the input image is first decomposed into rectangular nonoverlapped blocks called *tiles*. The tiles are the basic units of the

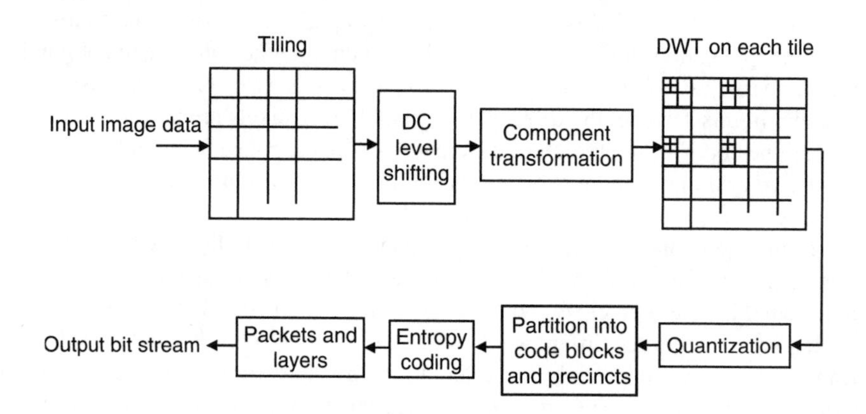

FIGURE 2.28 Encoding processing of JPEG2000.

original or reconstructed image, and they are compressed independently. The advantage of tiling is the reduction of the memory requirements because the tiles are reconstructed independently and can be used for decoding specific parts of an image instead of the whole image. It is obvious that the smaller tile uses less memory size but causes more image quality degradation, especially in the cases of low-bit rate coding. The second part of the encoding processing is the DC level shifting. The DC level shifting is performed on the pixels of image tile by subtracting a quantity $^{P-1}$, where P is the pixel's precision. The DC level is shifted before the transformation, and it does not change the variances of tiles. It actually converts an unsigned representation to a 2's complement representation, or vice versa.

JPEG2000 supports two component transformations, irreversible and reversible component transformations. Component transformations improve compression and allow for visually relevant quantization. After component transformation, the discrete wavelet transform (DWT) is applied to each tile of the image.

The irreversible DWT is implemented with a 9/7 filter [2-22], while the reversible DWT is implemented with a 5/3 filter [2-23]. After the wavelet transformation, a quantizer is applied to all coefficients. Each subband of coefficients is allowed to have its own quantization step size. The dynamic range depends on the number of bits used to represent the original image tile component and on the choice of the wavelet transform. After quantization, each subband is divided into nonoverlapped rectangular blocks. Since the tiles are decomposed into three layers by the transformation, three spatially consistent blocks are used to form a packet partition location or precinct as shown in Figure 2.29. Each precinct is further divided into nonoverlapping rectangulars, called *code blocks*, that are the input to the entropy encoder. The size of code block is typically 64×64 and no less than 32×32.

For each code block, a separate bit stream is independently generated. Each block is coded by the bit plane method with entropy coding. Entropy coding is achieved by an arithmetic coding system that compresses binary symbols relative to an adaptive probability model associated with each of 18 different coding contexts.

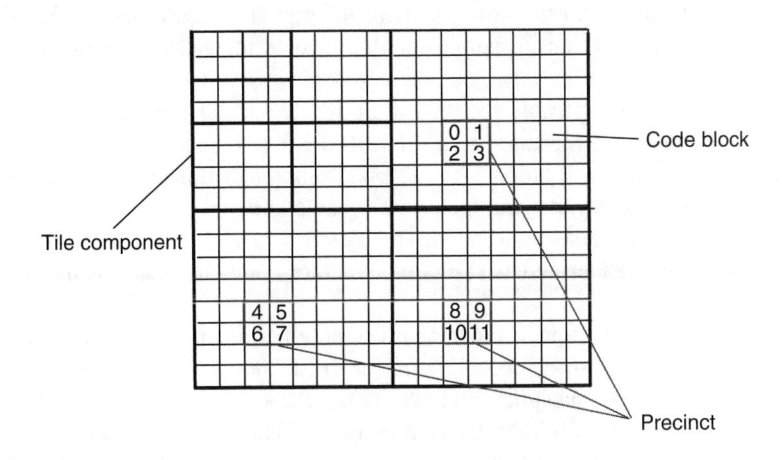

FIGURE 2.29 Partition of a tile into code blocks and precincts [2-21]

The idea of bit plane coding is to code the most significant bit first for all quantized coefficients in the code block and to send the resulting bits. Then the next most significant bit is coded and sent. This process is continued until all bit planes are coded and sent. During this procedure, rate distortion optimization is used to allocate truncation points to each code block. Therefore, the bit stream can be truncated at a variety of points, where the distortion of reconstructed image from the truncated bit stream is estimated and denoted by the mean squared error. The resulting bit streams from all code blocks in a precinct are used to form the body of a packet. Therefore, the packet can be interpreted as one quality increment for one resolution level at one spatial location because precincts are formed according roughly to the spatial locations. A collection of packets, one from each precinct of each resolution level is used to form a layer. Similar to the packets, a layer can be interpreted as one quality increment for the entire full-resolution image.

It should be noted that firstly, the encoding could be done in such a way that certain regions of interest can be coded at a higher quality than the background. Secondly, for the formation of bit stream, the markers are added to the bit stream to allow for error resilience. Finally, a header is added at the beginning of the bit stream for describing the original image, the various decomposition and coding styles that are used to locate, extract, decode, and reconstruct the image with the desired resolution, fidelity, region of interest, or other characteristics.

We summarize several advantages of JPEG2000 over JPEG:

- Better image quality at the same file size or 25-35% smaller file sizes at comparable image quality.
- Good image quality even at very high compression ratios (over 80:1).
- Having low complexity option for devices with limited resources.
- Scalable image files — no decompression needed for reformatting. With JPEG 2000, the image that best matches the target device can be extracted from a single compressed file on a server. Options include image sizes from thumbnail to full size; grayscale to full, three-channel color; low-quality image to lossless (identical to original image); and progressive rendering and transmission through a layered image file structure.

The following is an example for describing the feature of scalability. An original 512×512 image is compressed to a single 100-Kbyte compressed image file. As a starting point, a low-resolution 32×32 pixel thumbnail image can be transmitted by sending only 10 Kbytes, and then sending an additional 15 Kbytes increases the resolution to 64×64 pixels, and so on. The layered structure provides for progressive transmission and rendering based on quality, color component, and spatial location in the image.

The JPEG2000 compression standard is now being adopted by many applications. Hardware and software solutions for JPEG2000-enabled products such as printers, scanners, fax machines, digital cameras, PDAs, remote sensors, and wireless transmission devices will soon be commonplace. The public is ready to take full advantage of all that the JPEG2000 image compression standard has to offer. Finally, an examples of JPEG- and JPEG2000-encoded images are shown in Figure 2.30.

FIGURE 2.30 An example of JPEG2000 and JPEG images.

From the figure, it is clear that JPEG2000 has much better image quality than JPEG at compression ratio 125 to 1.

2.5.1.3 MPEG-1

MPEG-1 is a generic audio/video standard for digital storage media at bit rate up to about 1.5 Mbps. In MPEG-1 video, both intra-frame and inter-frame redundancies have been exploited. The MPEG coding algorithm is a full-motion-compensated discrete cosine transform (DCT) and DPCM hybrid coding algorithm. The DCT coding is used to remove the intra-frame redundancy, and the motion compensation is used to remove the inter-frame redundancy. A typical video encoding block diagram of MPEG-1 is shown in Figure 2.31.

With regard to input picture format, MPEG-1 allows only progressive pictures, but it offers great flexibility in the size, up to 4,095 × 4,095 pixels. However, the coder itself is optimized to the extensively used SIF video picture format. SIF is a simple derivative of the ITU-R 601 video format for digital television applications. According to ITU-R 601, a color video source has three components: a luminance component (Y) and two chrominance components (C_b and C_r), which are in the 4:2:0

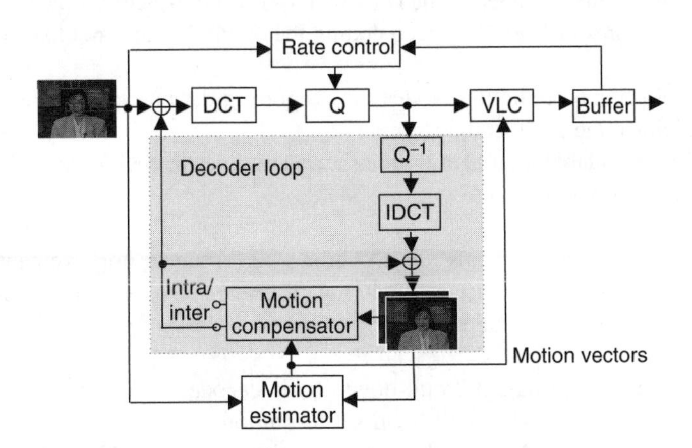

FIGURE 2.31 A typical video encoder.

subsampling format. In MPEG coding, the GOP structure is used as described previously. Each GOP can include three types of pictures or frames: intra-coded (I), predictive-coded (P), and bidirectionally predictive-coded (B) pictures or frames. The I-picture usually needs more bits to code it because no temporal redundancy is used. The P-picture uses fewer bits than I-pictures but more than coding B-picture. For the low-delay coding mode, there is no B-picture used. Regardless of the type of picture, each is optionally divided into slices; each slice consists of several macroblocks (MBs). If a slice contains all MBs in a picture, there is no slice layer. The distance between nearest I-pictures is as GOP size. The decision of selecting the GOP size and slice size is an encoding issue. Since I-pictures spend more bits than P- and B-pictures, small GOP size favors error prevention or drift propagation and an increase in random access points, but it will decrease coding performance. The issue is the same for selecting slice size: small slice size favors error resilience but causes coding performance degradation due to higher overhead. The MB containing four 8×8 luminance blocks and two spatially corresponding chrominance blocks is the basic coding unit for adaptive quantization and frame/field motion compensation. The 8×8 DCT is performed on each 8×8 block. The motion vector for each MB is estimated from luminance pictures using a block-matching algorithm. Usually, a decoder is included in the encoder loop because the motion estimation should be performed on the reconstructed pictures instead of on the original pictures. In such a way, the motion compensation in the decoder can obtain the same performance as in the encoder because in the decoder only the reconstructed pictures are available, but no original pictures are available. The criterion for the best match between the current MB and an MB in the anchor frame is not mean squared error, but is the minimum mean absolute difference, which is simple for computation and but with similar coding performance.

2.5.1.4 MPEG-2

The basic coding structure of MPEG-2 video is the same as that of MPEG-1 video, namely, intra-frame and inter-frame DCT with I-, P-, and B-picture structure. The most important features of MPEG-2 video coding that MPEG-1 does not have include:

- Field/frame prediction modes for supporting the interlaced video input
- Field/frame DCT coding syntax
- Downloadable quantization matrix and alternative scan order
- Scalability extension

The above enhancement items are all coding performance improvements that are related to the support of interlaced material. There are also several noncompression enhancements, which include:

- Syntax to facilitate 3:2 pull-down in the decoder
- Pan and scan codes with $^1/_{16}$ pixel resolution
- Display flags indicating chromaticity, subcarrier amplitude, and phase (for NTSC/PAL/SECAM source material)

The specifications of this coding can be found in the document proposed by the MPEG video committee [2-1].

2.5.1.5 MPEG-4

The goal of the MPEG-4 standard is to provide the core technology that allows efficient content-based storage, transmission, and manipulation of video, graphics, audio, and other data within a multimedia environment. The work of MPEG-4 started in July 1993. The first working draft (WD) was completed in November 1996 and the committee draft (CD) of version 1 was reached in November 1997. The draft international standard (DIS) of MPEG-4 was completed in November 1998. The international standard (IS) of MPEG-4 version 1 was completed in February 1999.

MPEG-4 has many interesting features; some of these features are focused on improving coding efficiency, and some are used to provide robustness of transmission and interactivity with end user. However, among these features, the most important one is the content-based coding. MPEG-4 is the first standard that supports the content-based coding of audio-visual objects. For content providers or authors, the MPEG-4 standard can provide the greater reusability, flexibility, and manageability of the content that is produced. For network providers, MPEG-4 offers transparent information, which can be interpreted and translated into the appropriate native signaling messages of each network. This can be accomplished with the help of relevant standards bodies that have the jurisdiction. For end users, MPEG-4 can provide many more capabilities for interaction with the content.

Contents such as audio, video or data are represented in the form of primitive audio-visual objects (AVOs). These AVOs can be natural scenes or sounds, which are recorded by video camera or synthetically generated by computers. The AVOs can be composed together to create compound AVOs or scenes. The data associated with AVOs can be multiplexed and synchronized so that they can be transported through network channels with certain quality requirements.

Since the MPEG-4 standard is targeted mainly for multimedia applications, there are many requirements to ensure that several important features and functionalities are offered. These features include the allowance of interactivity, high compression, universal accessibility, and portability of audio-video content. From the MPEG-4 video requirement document, the main functionalities can be summarized with the following three aspects: content-based interactivity, content-based efficient compression, and universal access.

To achieve content-based video coding, MPEG-4 uses the concept of the video object plane (VOP). It is assumed that in the object-based coding mode, each frame of an input video sequence is first segmented into a set of arbitrarily shaped regions, or VOPs. Each such region could be a particular image or video object in the scene. Therefore, the input to the MPEG-4 encoder can be a VOP, and the shape and the location of the VOP can vary from frame to frame. A sequence of VOPs is refereed to as a *video object* (VO). The concept of the VO provides a number of functionalities of MPEG-4 that are either impossible or very difficult in MPEG-1 or MPEG-2 video coding. The different VOs can be encoded into separate bit streams. MPEG-4 specifies demultiplexing and composition syntax that provide the tools for the receiver to decode the separate VO bit streams and composite them into a frame. In this way, the decoders have more flexibility to edit or rearrange the decoded video objects. Each video object is described by the information of texture, shape, and motion vectors. The video sequence can be encoded in a way that will allow separate decoding and reconstruction of the objects and allow editing and manipulation of the original scene by simple operation on the compressed bit stream domain. The feature of object-based coding is also able to support functionality such as warping of synthetic or natural text, textures, image, and video overlays on reconstructed video objects.

Since MPEG-4 aims at providing coding tools for multimedia environment, these tools not only allow one to efficiently compress natural video objects, but also compress synthetic objects, which are a subset of the larger class of computer graphics. The tools of MPEG-4 video include:

- Motion estimation and compensation, which includes block-based motion estimation and compensation, advanced motion estimation, and compensation with adaptive block sizes and global motion compensation
- Texture coding, which includes DCT-based coding for video and wavelet-based coding for still image
- Shape coding, which includes binary shape coding and gray-level shape coding
- Sprite coding
- Interlaced video coding
- Generalized temporal and spatial as well as hybrid scalability
- Error resilience, which includes synchronization markers, variable slice length, and reversible variable length coding

The specifications of MPEG-4 can be found in [2-3].

2.5.2 ITU-T Standards

2.5.2.1 H.261

The H.261 video coding standard has many features in common with MPEG-1. However, since they target different applications, there exist many differences between the two standards such as data rates, picture quality, end-to-end delay, and others. The major similarity between H.261 and MPEG-1/2 is as follows. First, both standards are used to code the similar video format. Standard H.261 is mainly used to code video with CIF or QCIF spatial resolution for teleconference application. MPEG-1 is used to code CIF, SIF, or greater spatial resolution videos for CD-ROM application. The original motivation of developing the H.261 standard was to provide a standard that can be used for both PAL and NTSC television signals. However, later H.261 became used mainly for video conferencing and the MPEG-1/2 used for digital television (DTV), VCD (Video CD), and DVD (Digital Video Disk). Second, the key coding algorithms of H.261 and MPEG-1 are very similar. Both H.261 and MPEG-1 use DCT-based coding to remove intra-frame redundancy and motion compensation to remove inter-frame redundancy.

The main differences between the MPEG-1/2 and H.261 with respect to coding algorithms are as follows:

- H.261 uses only I- and P-macroblocks but no B-macroblocks, whereas MPEG-1 uses three macroblock types—I, P, and B—and also three picture types—I, P, and B—as defined in MPEG-1/2 standard.
- A constraint of H.261 is that for every 132 inter-frame-coded macroblocks, which corresponds to 4 GOBs (group of blocks) or to one-third of CIF pictures, it requires at least one intra-frame-coded macroblock. In order to obtain better coding performance at low–bit rate applications, most encoding schemes of H.261 prefer not to use intra-frame coding on all of the macroblocks of a picture but only few macroblocks in every picture with a rotational scheme. MPEG-1 uses the GOP (group of pictures) structure, where the size of GOP (the distance between two I-pictures) is not specified.
- The end-to-end delay is not a critical issue for MPEG-1 but is critical for H.261. The video encoder and video decoder delays of H.261 need to be known to allow audio compensation delays to be fixed when H.261 is used in interactive applications. This will allow lip synchronization to be maintained.
- The accuracy of motion compensation in MPEG-1 is up to a half-pixel, but only a full pixel in H.261. However, H.261 uses a loop filter to smooth the previous frame. This filter attempts to minimize the prediction error.
- In H.261, a fixed aspect ratio of 4:3 is used. In MPEG-1, several aspect ratios can be used, and they are defined in the picture header. Finally, in H.261, the encoded picture rate is restricted to allow up to three skipped frames. This would allow the control mechanism in the encoder some flexibility to control the encoded picture quality and satisfy the buffer regulation. Although MPEG-1 has no restriction on skipped frames, the encoder usually does not perform frame skipping. Rather, the syntax for B-frames is exploited because B-frames require much fewer bits than P-pictures.

2.5.2.2 H.263

The basic configuration of the video source coding algorithm of H.263 is based on the H.261. Several enhancement features that are added in H.263 include the following new options: unrestricted motion vectors, syntax-based arithmetic coding, advanced prediction, and PB-frames. All these features can be used together or separately for improving the coding efficiency. The H.263 video standard can be used for both 625-line and 525-line television standards. The source coder operates on the noninterlaced pictures at picture rate about 30 pictures per second. The pictures are coded as luminance and two color difference components (Y, C_b, and C_r). The source coder is based on a CIF. Actually, there are five standardized formats, which include sub-QCIF, QCIF, CIF, 4CIF, and 16CIF. The H.263 encoder structure is similar to the H.261 encoder, with the exception that there is no loop filter in the H.263 encoder. The main components of the encoder include block transformation, motion-compensated prediction, block quantization, and variable length coding. Each picture is partitioned into groups of blocks (GOBs). A GOB contains a multiple number of 16 lines, k*16 lines, depending on the picture format ($k = 1$ for sub-QCIF, QCIF; $k = 2$ for 4CIF; $k = 4$ for 16CIF). Each GOB is divided into macroblocks that are the same as in H.261; each macroblock consists of four 8×8 luminance blocks and two 8×8 chrominance blocks. Compared with H.261, H.263 has several new technical features for the enhancement of coding efficiency for very low bit rate applications. These new features include picture-extrapolating motion vectors (or unrestricted motion vector mode), motion compensation with half-pixel accuracy, advanced prediction (which includes variable-block size motion compensation and overlapped block motion compensation), syntax-based arithmetic coding, and PB-frame mode.

In MPEG-1/2 and H.261, motion compensation is limited within the coded picture area of anchor frames. Standard H.263 allows the motion vectors to point outside the pictures. This is called the *unrestricted motion vector mode*. In this mode, when the values of motion vectors exceed the boundary of the anchor frame, the picture-extrapolating method is used. The values of reference pixels outside the picture boundary will take the values of boundary pixels. In this mode, the motion vector range is also extended.

In H.263 video coding, motion compensation of half-pixel accuracy is used. The half-pixel values are found using bilinear interpolation. Standard H.263 uses subpixel accuracy for motion compensation instead of using a loop filter to smooth the anchor frames as in H.261. This is similar to MPEG-1 and MPEG-2. In MPEG-4 video, quarter-pixel accuracy for motion compensation has been adopted as a tool for the Version 2.

Generally, the decoder will accept no more than one motion vector per macroblock for baseline algorithm of H.263 video coding standard. However, in the advanced prediction mode, the syntax allows up to four motion vectors to be used per macroblock. The decision of using one or four vectors is indicated by the macroblock type and coded block pattern for chrominance (MCBPC) codeword for each macroblock. How to make this decision is the task of the encoding process. In the advanced prediction mode, the overlapped block motion compensation can also be used for reducing the block artifacts.

As in other video coding standards, H.263 uses variable length coding and decoding (VLC/VLD) to remove the redundancy in the video data. The basic principle of VLC is to encode a symbol with a specific table based on the syntax of the coder. The symbol is mapped to a table entry in a table look-up operation, and then the binary codeword specified by the entry is sent to a bit stream buffer for transmitting to the decoder. In the decoder, an inverse operation, VLD, is performed to reconstruct the symbol by the table look-up operation based on the same syntax of the coder. The tables in the decoder must be the same as the one used in the encoder for encoding the current symbol. In order to obtain the better performance, the tables are generated in a statistically optimized way (such as with a Huffman coder), with a large number of training sequences. This VLC/VLD process implies that each symbol must be encoded into a fixed integral number of bits. An optional feature of H.263 is to use arithmetic coding to remove the restriction of fixed integral number bits for symbols. This syntax-based arithmetic coding mode may result in bit rate reductions.

The PB-frame is a new feature of H.263 video coding. A PB-frame consists of two pictures, one P- and one B-picture, being coded as one unit as shown in Figure 2.10.

Version 2 [2-24] of the H.263 standard, also known as *H.263+*, was approved in January of 1998 by the ITU-T. Version 2 of H.263includes new optional features based on the H.263 video coding standard. These features are added in order to broaden the application range of H.263 and to improve its coding efficiency. Among these options are five that include advanced intra coding mode, alternative inter VLC mode, modified quantization mode, deblocking filter mode, and improved PB-frame mode, which are all intended to improve coding efficiency. Slice structured mode, reference picture selection mode, and independent segment decoding mode are used to meet the need of mobile video application. The others provide such functionalities as spatial, temporal, and SNR scalability. The technical details of H.263 can be found in [2-7].

2.5.3 MPEG/ITU Jointly Developed H.264/AVC

Recently, the Joint Video Team (JVT) of ISO/IEC's MPEG and ITU-T's VCEG (Video Coding Expert Group) have developed new video coding standards, referred to formally as ITU-T Recommendation H.264 and ISO/IEC MPEG-4 (Part 10), Advanced Video Coding. The standard is referred to in short as H.264/AVC [2-8][2-25]. The real work of H.264/AVC actually started in early 1998 when the VCEG issued a call for proposals on a project called *H.26L*. The target of H.26L was to greatly improve the coding efficiency of existing standards. The first draft of H.26L was adopted in October 1999. The JVT was formed in December of 2001, with the mission of finalizing the new coding standard. The draft of new video coding standard was submitted for formal approval as H.264/AVC in March 2003.

The new video coding standard mainly features for high coding efficiency, which may be two times better than current existing video coding standards for a given level of fidelity. With high coding efficiency, the H.264/AVC can provide a technical solution for many applications, including broadcasting over different media, video storage on optical and magnetic devices, and DVD. To address the variety of applications and networks, the H.264/AVC design covers two layers: VCL (video coding layer) and NAL (network abstraction layer). The VCL is designed to efficiently compress the video content whereas the NAL is used to format the VCL representation of the video and to provide the header information for handling a variety of transport layers or storage media. The structure of H.264/AVC video encoder is shown in Figure 2.32.

Compared to other video coding standards, except for many common tools, H.264/AVC includes many highlighted features that greatly improve coding efficiency and increase error robustness and flexibility of operation over a variety of network environments.

Features for improving coding efficiency can be classified into two parts: the first is to improve the accuracy of prediction for the picture to be encoded and the

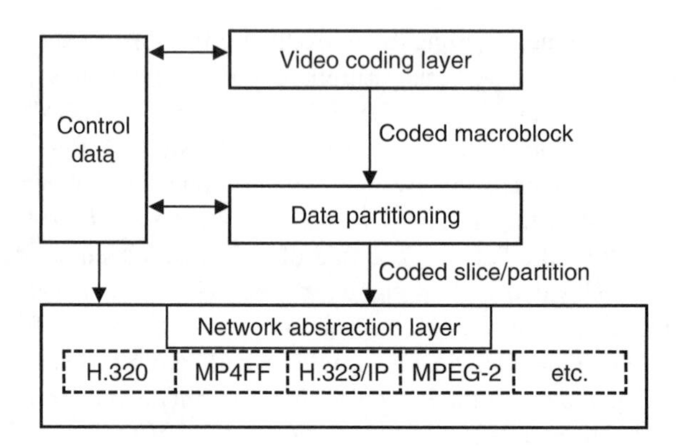

FIGURE 2.32 Structure of H.264/AVC video encoder.

second is the method of transform and entropy coding. Several tools have been adopted in H.264/AVC to improve the inter and intra prediction, and they include:

- Variable block size for motion compensation with small block sizes: as shown in Figure 2.19, a total of 7 selections of block size are used for motion compensation in H.264/AVC and among those the smallest block size for luma motion compensation can be as small as 4×4. The use of small block size enables increased prediction accuracy of motion compensation in the busy areas of the picture to be encoded and, consequently, saves bits for coding predictive differences for those areas.
- Quarter-pel accurate motion compensation: the quarter-pel accurate motion compensation has been used in the advanced profile of MPEG-4, Part 2, but H.264/AVC further reduces the complexity of the interpolation process.
- Multiple reference pictures for motion compensation and weighted prediction: in H.264/AVC, multiple reference pictures are used to predict the P-pictures and B-pictures. The number of reference pictures can be up to 15 for level 3.0 or lower and four reference pictures for levels higher than 3.0 [2-8]. It is obvious that better prediction can be obtained with more choices of reference pictures, as shown in the example given by Figure 2.21. When the multiple reference picture is used for motion-compensation prediction, the contribution of prediction from different references should be weighted and offset by amounts specified by the encoder. This can greatly improve coding efficiency for scenes containing fades.
- Directional spatial prediction for intra coding: a new intra prediction technique has been adopted for improving coding efficiency. In this technique, the intra-coded regions are predicted with the references of the previously coded areas, which can be selected from different spatial directions in the same frame, as shown in Figure 2.7. In such a way, the edges of the previously decoded areas of the current picture can be extrapolated into the current intra-coded regions.

- Skip mode in P-picture and direct mode for B-picture: to alleviate the problem for using too many bits for coding motion vectors in the inter-frame coding, H.264/AVC uses the skip mode for P-picture and direct mode for B-pictures. In these modes, the reconstructed signal is obtained directly from the reference frame, with the motion vectors derived from previously encoded information by exploiting either spatial (for skip mode) or temporal (for direct mode) correlation of the motion between adjacent macroblocks or pictures. In such a way, there is no need for encoding and transmitting any additional motion information.

Another set of tools for improving coding efficiency includes:

- In-the-loop deblocking filtering: the deblocking filters are used to reduce the block artifacts and, consequently, to improve both objective and subjective video quality. The difference compared with MPEG-1/2 is that in H.264/AVC the deblocking filter is brought within the motion compensation loop, so that it can be used for improving the inter-frame prediction and therefore improving the coding efficiency.
- Small block-size transform and hierarchical block transform: H.264/AVC uses a transform block size of 4×4 instead of 8×8, as in most video coding standards. The merit of using the small transform block size is to be able to encode the picture in a more local adaptive fashion, which would reduce such coding artifacts as ringing noise. However, using a small transform block size may cause coding performance degradation due to the correlations of large area may not be exploited pictures. H.264/AVC uses two ways to alleviate this problem, one is by using a hierarchical transform to extend the effective block size of inactive chroma information into an 8×8 block, and another is by allowing the encoder to select a special coding type of intra coding, which enables the extension of the length of the luma transform for an inactive area to a 16×16 block size. As mentioned previously, the basic function of integer transform used in H.264/AVC do not have equal norm. To solve this problem, quantization table size has been increased.
- Arithmetic entropy coding and content-adaptive entropy coding: two very powerful entropy coding methods — content-adaptive variable length coding (CAVLC) and content-adaptive binary arithmetic coding (CABAC) — are used in H.264/AVC for further improving coding performance.

The features of H.264/AVC for increasing the capability of error robustness include:

- Flexible slice size: the slice sizes in H.264/AVC are highly flexible. This allows the encoder to adaptively select the slice size for increasing the capability of error robustness.
- Flexible macroblock ordering (FMO): The FMO allows partitioning the macroblocks into slices in a flexible order. Since each slice is an independently decodable unit, the FMO can significantly enhance error robustness by managing the spatial relationship between the macroblocks in the slice. There are several features that are used to increase the flexibility for operation over a variety of network environments.

- Parameter set structure: the parameter structure is used to provide more flexible way to protect the key header information and increase error robustness.
- NAL unit syntax structure: the NAL unit syntax structure allows for carrying video content in a manner appropriate for each specific network in a customized way.
- Arbitrary slice ordering (ASO): the ASO is used to improve end-to-end delay in real-time application, particularly for the applications on the Internet protocol networks.
- SP/SI (switching P and I) slices: SP and SI slices are specially coded slices that allow exact, synchronization of the decoding process of some decoders with an ongoing video stream produced by the other decoders without penalty in efficiency, which may result from sending an I-picture in the conventional way. This feature can be used for efficiently switching a decoder to decode different bit streams with different bit rates.

Finally, we would like to give brief description about H.264/AVC profiles. It is the same as in MPEG-2; a profile defines a set of coding tools or algorithms used in generating a compliant bit stream with this profile. If a decoder is claimed to conform to a specific profile, it must support all tools and algorithms in that profile. There are three profiles defined in the H.264/AVC, which are the baseline, main, and extended profiles. The baseline profile supports all features in H.264/AVC except B slices, weighted prediction, CABAC, interlacing support, SP/SI slices, and slice data partitioning. The main profile supports all features except SP/SI slices, slice partitioning, FMO, ASO, and redundant pictures. The extended profile supports all features except CABAC. Therefore, three profiles are not subsets of each other and targets for different applications. The baseline profile is used for videophone, mobile communications, and low delay applications. The main profile targets interlaced video, broadcast, and packaged media applications. The extended profile mainly targets streaming video and wireless transmission applications. Compared with MPEG-2 video, H.264/AVC can provide 50 to 100% improvement in the coding efficiency but can increase decoder complexity about 2.5 to 4 times and encoder complexity 8 to 9 times.

An overview of H.264/AVC video coding standard can be found in [2-8] and the specifications can be found in [2-25].

2.6 VIDEO COMPRESSION ENCODING TECHNOLOGIES

Although the MPEG and ITU-T video coding standards recommended a general coding methodology and syntax for the creation of a legitimate compressed bit stream, there are many areas of research left open regarding how to generate high-quality compressed bit streams. This allows the designers of an encoder great flexibility in developing and implementing their own MPEG or H.26x-specific algorithms, leading to product differentiation on the marketplace. In order to design a performance-optimized encoder system, several major areas of research have to be considered. These include the topics of, but not limited to, preprocessing, motion estimation, optimal coding mode decisions, and rate control. Algorithms for all of these areas in an encoder should aim to minimize subjective distortion for a prescribed bit rate and operating delay constraint.

2.6.1 Pre-Processing

In the area of digital video coding, many researchers have studied the benefits of noise reduction prior to coding of video signal [2-26][2-27][2-28]. The affect of noise on video coding can be seen in both intra and inter frame coding. For the intra coding, after the block transformation, say DCT, the noise results in more nonzero coefficients, which the encoder cannot distinguish from the real picture data. We have to either increase the number of bits to code these coefficients or we have to increase the quantization steps to keep the budget of bits. In either case, the overall coding performance is reduced. For inter frame coding, the noise would reduce the similarities between successive frames, and this would result in increasing the predictive residues or causing improper prediction. It is obvious that this will reduce the coding efficiency. Therefore, it may be better to remove the noise or entropy of the input video signal before encoding it. This can be achieved by using prefiltering, which may be in the form of two-dimensional lowpass filters or other filters.

Actually, the strategies of prefiltering should be different according to the bit rates. At high bit rate for high-quality coding, the coding algorithm should be designed to produce very small coding errors, which are not visible to human eyes. In this case, prefilters should be designed to remove only the invisible noise and not introduce any observable distortion. In [2-28], the simulation results have shown that compared with error-free JPEG results, the proposed corner-preserving filter obtained about 43% improvement.

In the case of low–bit rate coding, the coding distortion is usually quite visible. Prefiltering may introduce distortion if this results in simplifying the image data so that the encoder is able to use fewer bits to code it. In other words, prefiltering is used to control the distribution of the distortions so that the distortion caused by prefiltering would greatly increase the coding performance. Therefore, in the low-bit rate case, we have to consider the trade-off between the factors: the distortion introduced by prefiltering, the distortion introduced by the encoder, and the available bit rate. In this scenario, prefiltering can be claimed to be as effective only if the combined distortion due to prefiltering and encoding is lower than the one due to encoding alone without prefiltering at the same bit rate. Anyway, there is an optimization problem between the extent of prefiltering and overall coding performance for a given bit rate.

2.6.2 Motion Estimation

In most video coding standards, motion compensation is used to remove the temporal redundancy between frames in a video sequence. The principle of motion compensation is based on the fact that if the motion trajectory of each pixel could be measured, then only the initial values of the pixels in the reference frame and the motion information need to be coded. To reproduce the pixel value, one can simply propagate each pixel along its motion trajectory. Since there is also a cost for transmitting motion vectors, in practice, we have to do motion compensation based on a group of pixels or a block of pixels, which share the cost for transmission of the motion information. However, in the block-based motion compensation, the pixels in the same block are assumed to have the same motion information. This is not always true since the pixels in the block may move in different directions, or some of them belong to the background. Therefore, both motion vectors and the prediction difference must be transmitted.

The procedure to obtain motion vector information is referred to as *motion estimation*, which is performed in the video encoders. Usually, block matching can be considered as the most practical method for motion estimation due to less hardware complexity. In the block matching method, the image frame is divided into fixed-size small rectangular blocks, such as 16×16 or as small as 4×4 in the video coding standards. Each block is assumed to undergo a linear translation, and the displacement vector of each block and the predictive errors are coded and transmitted. The related issues for motion estimation include motion vector searching algorithm, matching criteria, searching range, and coding method.

Finding the best-matching block from the reference frame requires optimizing the matching criterion over all possible candidate displacement vectors at each pixel in the block. The conventional search algorithm used to accomplish this target is the so-called full search algorithm, which conducts an exhaustive search, i.e., compares all possibilities within a search window and finds the best matching block. It is obvious that the full search can get the best matching block but it is very expensive on the computational complexity. In order to reduce the computational complexity for searching the best matching block, several fast search algorithms, such as logarithmic [2-29], hierarchical [2-30], and diamond search algorithms [2-31], have been proposed and developed to replace the full search algorithm at the price of light coding performance degradation. The simulation results have shown that using the diamond search algorithm can reduce the search time to 1% of the full search algorithm and the image quality degradation is only less than 0.5 dB [2-31]. The matching of the blocks can be determined according the various criteria, including maximum cross-correlation, minimum mean square error (MSE), minimum mean absolute difference (MAD), and maximum matching pixel count (MPC). For MSE and MAD, the best-matching block is reached if the MSE or MAD is minimized at that location. In practice, we use MAD instead of MSE as matching criterion due to its computational simplicity. The minimum MSE criterion is not commonly used in hardware implementations because it is difficult to realize the square operation. It is clear that the large search window provides more opportunities to get the best matching block, but the computational complexity is exponentially proportional to the search window size. Therefore, the encoders have to consider the trade-off between coding performance and complexity. To reduce the number of bits for coding motion vectors, predictive coding is used to code the motion vectors of the neighboring blocks and Huffman coding or variable length coding is used to code the predictive differences.

2.6.3 Mode Decision and Rate-Distortion Optimization

For a given bit rate, how to design a reasonable bit allocation strategy to obtain the optimal coding performance is an important topic for encoder design. The optimal bit allocation strategy can considered in two levels: frame or picture level and macroblock level. The problem of bit allocation in picture level includes how to select the GOP size, i.e., the distance between I-pictures, the number of B-pictures, and the proper locations of I-pictures. The decision on these items is not easy because not only must the related problem of coding performance be considered, but also the error robustness, random access, and computational complexity. As we know, the I-picture needs more bits to code, but a proper number of I-pictures is also needed for random access and preventing error propagation. Therefore, for most encoders the GOP size is selected from 10 to 15. The location selection for I-pictures really depends on the nature of the video sequences. Usually, it is better to allocate the I-pictures in the positions where the scene cuts occur. In such a way, the better prediction for P- or B-pictures can be obtained. However, the scene change detection may need the process of preanalysis that will increase the encoding complexity. Also, the optimal bit allocation on the frame level is complicated due to the temporal dependency across I-, P-, and B-frames and the requirement of an unwieldy number of preanalysis encoding passes over the window of dependent frames (for example, one MPEG GOP). To overcome these computational burdens, more pragmatic solutions that can realistically be implemented have been considered in [2-32]. In this work, the major emphasis is not on the problem of bit allocation among I-, P-, and B-frames; rather, the authors choose to utilize the frame level allocation method provided by the Test Model [2-9]. In this way, frame-level coding complexities are estimated from past frames without any knowledge of future frames. This type of analysis forms the most reasonable set of assumptions for a practical real-time encoding system. Anothe method extends the basic test model idea to alter frame budgets heuristically in the case of scene changes, use of dynamic GOP size, and temporal masking effects and can be found in [2-33]. These techniques also offer very effective and practical solutions for implementation. Given the chosen method for frame-level bit budget allocation, the focus of this section is to jointly optimize macroblock coding modes and quantizers within each frame.

At the macroblock level of coding optimization, there exists many choices for the macroblock coding mode, for example, under the MPEG-2 standard for P- and B-pictures, including intra mode, no motion compensation mode, frame/field/dual-prime motion compensation inter mode, forward/backward/average inter mode, and field/frame DCT mode. It is very important for the encoder to determine how to select the macroblock coding mode and quantizer scale to accomplish the optimal coding performance at a given bit rate. A constrained optimization problem can be formulated based on the rate-distortion characteristics, or $R(D)$ curves, for all of the macroblocks that compose the picture being coded. Distortion for the entire picture is assumed to be decomposable and expressible as a function of individual macroblock distortions, with this being the objective function to minimize. According to the above assumption, the cost function can be formatted as:

$$C = \sum_{i=1}^{N} D_i + \sum_{i=1}^{N} \lambda_i R_i \tag{2.7}$$

where C is the cost function, D_i is the distortion of i-th macroblock, R_i is the number of bits for coding i-th macroblock, λ_i is the constant, and N is the number of macroblocks in the slice. Traditionally, the optimization solution can be obtained by the Lagrange multiplier method. However, the determination of the optimal solution becomes very complicated due to the differential encoding of motion vectors and dc coefficients used in the standards, which introduce dependencies that carry over from macroblock to macroblock for a duration usually equal to the slice length. Finding the optimal set of coding modes for the macroblocks in each slice entails a search through a trellis of dimensions N stages × M states per stage, with N being the slice size and M being the number of coding modes being considered as shown in Figure 2.33.

This trellis structure arises because there are M^2 distinct rate distortions, $R_{\text{mode|previous-mode}}(D)$, characteristic curves corresponding to each of M coding modes, with each in turn having a different dependency for each of M coding modes of the previous macroblock. We now consider populating the trellis links with values by sampling the set of these M^2N rate-distortion curves at a specific distortion level. For a given fixed macroblock distortion level, each link on the trellis is assigned a cost equal to the number of bits to code a macroblock in a certain mode given the mode from which the preceding macroblock was coded. For any group of links entering a node, the cost of these links differs only because of the difference in bits caused by the motion vector and dc coefficient coding dependency on the prior macroblock.

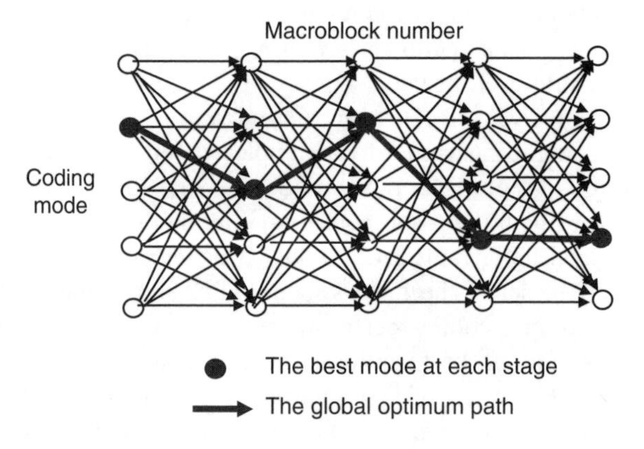

FIGURE 2.33 Full search trellis, M^N (M is number of modes at each stage and N is the length of slice) searches needed to obtain the best path.

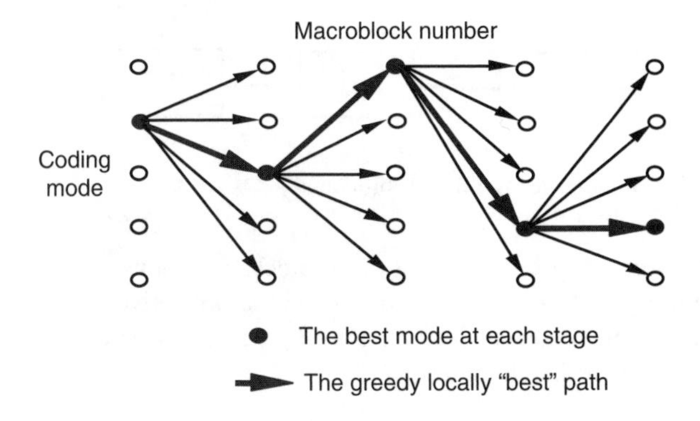

FIGURE 2.34 Greedy approach, M × N comparisons needed to obtain the locally "best" path.

The computational requirements per slice involve 1) determining link costs in the trellis, the number of "code the macroblock" operations (i.e., DCT + Quantization + RLC/VLC) is equal to M^2N; and 2) after determining all trellis link costs, the number of path searches is equal to M^N. From the results of the full search trellis, the optimal coding performance can be obtained. This result can be seen as the upper bound of the coding performance.

From this analysis, it can be seen that the solution from the full exponential-order search requires an unwieldy amount of computations. In order to avoid the heavy computational burden, we can use a greedy approach [2-34] to simplify and sidestep the dependency problems of the full search method. In the greedy algorithm, the best coding mode selection for the current macroblock depends only on the best mode of the previous coded macroblock. Therefore, the upper bound we obtain is a near-optimum solution instead of a global optimum. Figure 2.34 illustrates the greedy algorithm.

After coding a macroblock in each of the M modes, the mode resulting in the least number of bits is chosen to be "best." The very next macroblock is coded with dependencies to that chosen "best" mode. The computations per slice are reduced to $M \times N$ "code the macroblock" operations and $M \times N$ comparisons. As an approximation, a near-optimum greedy algorithm can be developed. The result of the near-optimum greedy algorithm can be seen as an approximation of the upper bound of coding performance. Once the upper bound in performance is calculated, it can be used to assess how well practical suboptimum methods perform.

It is obvious that the near-optimal solution discussed in the previous section is still not a practical method because of its complexity. In order to determine the best mode, we have to know how many bits are used to code each macroblock in every mode with the same distortion level without full encoding procedure. For this purpose, we have developed a model that can be used to obtain a close estimation of the above desired number of bits without full encoding, this estimation can then be used as the criterion for optimal mode selection. The following is the analysis for the model development.

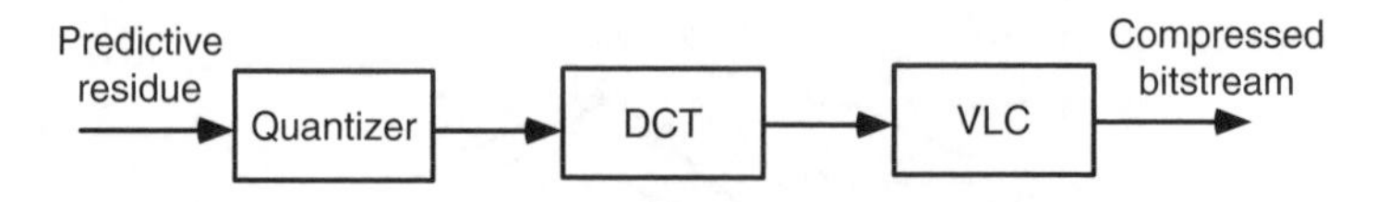

FIGURE 2.35 Coding stages to find out bit count.

The total number of bits for each macroblock, R_{MB}, consists of three parts: bits for coding motion vectors, R_{mv}; bits for coding the predictive residue, R_{res}; and bits for coding macroblock header information, R_{header}; such as macroblock type, quantizer scale, and coded-block pattern.

$$R_{MB} = R_{mv} + R_{\text{res}} + R_{\text{header}} \tag{2.8}$$

The number of bits for motion vectors, R_{mv}, can be easily obtained by VLC table look-up. But to obtain the number of bits for coding the predictive residue, one has to go through the three-step coding procedure: 1) DCT, 2) quantization, and 3) VLC as shown in Figure 2.35. At Step 3, R_{res} is obtained with a look-up table according to the run length of zeros and the level of quantized coefficients; i.e., R_{res} depends on the pair of values of run and level:

$$R_{\text{res}} = f(run,\ level) \tag{2.9}$$

As stated above, to obtain the upper-bound coding performance, all three steps are needed for each coding mode, and then the coding mode resulting in the least number of bits is selected as the best mode. To obtain a much less computationally intensive method, it is preferred to use a statistical model of DCT coefficient bit usage versus variance of the prediction residual and quantizer step size. This will provide an approximation of the number of residual bits, R_{res}. For this purpose, we assume that the run and level pair in (2.9) is strongly dependent on values of the quantizer scale, q_s, and the variance of the residue, V_{res}, for each macroblock. Intuitively, we would expect the number of bits to encode a macroblock is proportional to the variance of the residual and inversely proportional to the value of quantizer step size. Therefore, a statistical model can be constructed by plotting R_{res} versus the independent variables V_{res} and q_s over a large set of representative macroblock pixels from images typical of natural video material. This would result in a scatter plot showing tight correlation, and hence a surface can be fit through the data points. It was found that Equation (2.9) could be approximately expressed as:

$$R_{\text{res}} \approx f(q_s, V_{\text{res}}) = (K/(C\ q_s + q_s^2))V_{\text{res}} \tag{2.10}$$

where K and C are constants found through surface-fitting regression. If we assume R_{header} is a relatively fixed component that does not vary much with macroblock coding mode and can be ignored, then Formula (2.10) can be approximately replaced by:

$$R_{MB'} = R_{mv} + (K/(C\ q_s + q_s^2))V_{\text{res}} \tag{2.11}$$

The value of $R_{MB'}$ reflects the variable portion of bit usage that is dependent on coding mode and can be used as the measure for selecting the coding mode in our encoder. For a given quantizer step size, the mode resulting in the smallest value of $R_{MB'}$ is chosen as the "best" mode. It is obvious that in the use of this new measurement to select the coding mode, the computational complexity increases over the test model method is very slight (the same identical calculation for V_{res} is made in the test model).

2.6.4 Rate Control Algorithms

The purpose of rate control includes several aspects. First of all, it is to achieve a given target average bit rate by controlling the allocation of the bits. The second is to optimize the overall perceived picture quality at the given bit rate. The rate control has to ensure no buffer overflow and underflow at the encoder and decoder. In addition, the rate control scheme has to be applicable to a wide variety of sequences and bit rates. At a group of pictures level, the total number of available bits is allocated among the various picture types, taking into account the constraints of the decoder buffer, so that the perceived quality is balanced. Within each picture, the available bits are allocated among the macroblocks to maximize the visual quality and to achieve the desired target bits for encoding the whole picture. Various rate control schemes have been developed for video standards, including MPEG-2 Test Model [2-9], MPEG-4 Verification Model (VM) [2-35] and H.264/AVC Joint Model (JM) [2-36]. We will give a brief introduction of these rate control schemes.

2.6.4.1 MPEG-2 Rate Control

The typical MPEG-2 encoding rate control is presented in TM5 [2-9]. The TM5 rate control algorithm consists of three steps to adapting the macroblock quantization parameter for controlling the bit rate: target bit allocation, rate control, and adaptive quantization.

Target bit allocation is the first step of rate control, which is the picture-level bit allocation. Before coding a picture, we need to estimate the number of bits available for coding the picture. The estimate is based on a linear model between the bit budget and the reciprocal of quantization step. This model is very intuitive since the number of bits generated by encoder is inversely proportional to the quantization step. Based on this model, the estimation is determined by several factors: the picture type, buffer fullness, and picture complexity. The picture complexity is measured by the product of the number of bits and the quantization parameter used for coding the same type of previous picture in the GOP, which are:

$$Xi = Si\,Qi, \quad Xp = Sp\,Qp \quad Xb = Sb\,Qb \tag{2.12}$$

where the subscript i, p, and b stand for picture types I-, P-, and B-picture; X_i, X_p, and X_b are the complexity measure of I-, P-, and B-picture; S_i, S_p, and S_b are the number of bits generated by encoding this picture; and Q_i, Q_p, and Q_b are the average quantization steps computed the actual quantization values used during the encoding of all the macroblocks including the skipped macroblocks, respectively. We denote the N_i, N_p, and N_b to be the numbers of I-, P-, and B-pictures, respectively, and then the total number of frames, N, equal to $N = N_i + N_p + N_b$. The goal for the rate control algorithm to achieve is

$$N_i \cdot S_i + N_p \cdot S_p + N_b \cdot S_b = R = R_t \cdot N/f \tag{2.13}$$

with the quality balance between picture types as:

$$Q_i = \frac{Q_p}{K_p} = \frac{Q_b}{K_b} \tag{2.14}$$

where f is the frame rate, R_t is the target bit rate, and R is the number of bits assigned for coding the GOP, and K_p and K_b are "universal" constants dependent on the quantization matrices. For the matrices of TM5, $K_p = 1.0$ and $K_b = 1.4$. Based on the above conditions, the target number of bits for the next picture in the GOP (T_i, T_p, and T_b) is computed as follows:

$$T_i = \max\left\{ \frac{R}{1 + \frac{N_p X_p}{X_i K_p} + \frac{N_b X_b}{X_i K_b}},\ \ R_t/(8*f) \right\}$$

$$T_p = \max\left\{ \frac{R}{N_p + \frac{N_b K_p X_b}{X_b K_p}},\ \ R_t/(8*f) \right\} \tag{2.15}$$

$$T_b = \max\left\{ \frac{R}{N_b + \frac{N_p K_b X_p}{X_p K_b}},\ \ R_t/(8*f) \right\}$$

In picture-level bit allocation, the initial complexity values are given according to the type of picture as follows in the TM5:

$$\begin{aligned} X_i &= 160 * R_t/115 \\ X_p &= 60 * R_t/115 \\ X_b &= 42 * R_t/115 \end{aligned} \tag{2.16}$$

The second step of rate control is the macroblock-level bit allocation. It is first to determine the number of bits available for each macroblock within a picture according to the rate control algorithm. The rate control should ensure that no buffer overflow and underflow occur. A quantizer step is then derived from the number of bits available for the macroblock to be coded. The following is an example of rate control for the P-picture. Before encoding the j-th macroblock, the virtual buffer fullness is adjusted during the encoding according to the following equation for P-picture:.

$$d_j^p = d_0^p + B_{j-1} - \frac{T_p(j-1)}{MB-cnt} \tag{2.17}$$

The quantization step is then computed with the equation:

$$Q_j^p = \frac{d_j^p}{r} \tag{2.18}$$

where d_0^p is initial virtual buffer fullness, T_p is the target bits for the P-picture, B_j is the number of bits generated by encoding all macroblocks in the picture up to and including j-th macroblock, $MB - cnt$ is the number of macroblocks in the picture, the "reaction parameter" r is given by $r = 2$ * bit rate/picture rate, and d_j^p is the fullness of the appropriate virtual buffer. The fullness of the virtual buffer for the last macroblock is used for encoding the next picture of the same type as the initial fullness. The above example can be extended to the general case for all I-, P-, and B-pictures.

Adaptive quantization is the last step of the TM5 rate control. It is noted that for active areas or busy areas, the human eyes are not so sensitive to the quantization noise, while the smooth areas are more sensitive to the quantization noise. Based on this observation we modulate the quantization step obtained from the previous step in such a way to increase the quantization step for active areas and reduce the quantization step for the smooth areas. In other words, we use more bits in the smooth areas and fewer bits for the active areas. Experimental results have shown that the subjective quality is higher with adaptive quantization step than without this step. The adaptive quantization in TM5 is performed with modulating the quantization step with a normalized spatial activity of the macroblock. The normalized activity factor N_act_j is:

$$N\_act_j = \frac{2 \cdot act_j + avg\_act}{act_j + 2 \cdot avg\_act} \tag{2.19}$$

where act_j is the spatial activity of j-th macroblock, which is the variance of luminance pixels in the macroblock, and avg_act is the average activity value of the last picture to be encoded. Finally, we can obtain the modulated quantization step for j-th macroblocks

$$\text{mquant}_j = Q_j \cdot N\_act_j \tag{2.20}$$

where Q_j is the reference quantization step value obtained in the last step. The final value of mquant$_j$ is clipped to the range of [1, 31] and is used and coded as described in the MPEG standard.

It should be noted that, the TM5 rate control provides only a reference model. It is not optimized in many aspects such as not working well for scene changes. Therefore, there is still a lot of room for improving the rate control algorithm, such as the optimized mode decision presented in Section 2.6.3.

2.6.4.2 MPEG-4 Rate Control

The MPEG-4 rate control was first proposed in [2-35]. Instead, using a linear model as in MPEG-2 TM5 rate control, the MPEG-4 VM uses quadratic model for rate control. Let R represent the texture bits spent for a frame, Q denote the quantization parameter, (X_1,X_2) the first- and second-order model parameters and S, the encoding complexity. The relation between R and Q is given by:

$$R = S \cdot \left(\frac{X_1}{Q} + \frac{X_2}{Q^2} \right), \tag{2.21}$$

In an encoder, the complexity measure, S, is typically given by the mean absolute difference of residual blocks. This value can be obtained from motion estimation part during the encoding. However, for compressed-domain transcoding architectures; this measure must be computed from the DCT coefficients. We adopt a DCT-based complexity measure, $\tilde{S}$, which was presented in [2-37] is given by

$$\tilde{S} = \frac{1}{M_c} \sum_{1}^{63} \rho(i) \cdot \left| B_m(i) \right|^2, \tag{2.22}$$

where $B_m(i)$ are the AC coefficients of a block, m is a macroblock index in the set M of inter-coded blocks, M_c is the number of blocks in that set, and $\rho(\mathrm{i})$ is a frequency dependent weighting, e.g., a quantizer matrix. Given the above model and complexity measures, the rate control works in two main stages: pre-encoding and post-encoding. In the pre-encoding stage, the target estimate for the frame is obtained based on available bit-rate and buffer fullness, and the complexity is also computed. Then, the value of Q is determined based on these values and the current model parameters. After encoding, the post-encoding stage is responsible for calibrating the model parameters, (X_1, X_2), based in the actual bits spent and corresponding Q. This can be done by linear regression using the results of the past n frames.

2.6.4.3 H.264/AVC Rate Control

The rate control algorithm used in H.264/AVC is proposed in [2-38]. The basic algorithm of H.264/AVC rate control is similar to MPEG-4 VM rate control, which uses the quadratic rate distortion model as in (2.21). Since the H.264/AVC is a rather new video coding standard, we would like to give more detail about its rate control scheme. This algorithm consists of three parts: GOP level rate control, picture level rate control, and the optional basic unit level rate control.

The GOP level rate control has two functions. The first is to calculate the total number of bits for the rest pictures in the GOP. The second is to compute the initial quantization parameter of instantaneous decoding refresh (IDR) picture. In H.264/AVC, the IDR picture is the first coded picture of the video sequence to be coded. It contains only I or SI slices. Therefore, after decoding an IDR picture, all following coded pictures in decoding order can be decoded without inter prediction from any picture decoded prior to the IDR picture.

The number of bits for coding the first picture in the i-th GOP is calculated as:

$$B_i(1) = \frac{R_i(1)}{f} N_i - V_i(1) \tag{2.23}$$

where f is the predefined coding frame rate, N_i is the total number of pictures in i-th GOP, and $R_i(1)$ and $V_i(1)$ are the instant available bit rate and occupancy of the virtual buffer, respectively, when the first picture in i-th GOP is coded. For the other picture in the i-th GOP, the number of bits is calculated as:

$$B_i(j) = B_i(j-1) + \frac{R_i(j) - R_i(j-1)}{f}(N_i - j + 1) - b_i(j-1) \tag{2.24}$$

where $b_i(j-1)$ is the actual generated bits in the $(j-1)$-th picture. For the dynamic channels, the value of $R_i(j)$ may vary at different frames and GOPs. In the case of constant bit rate, $R_i(j)$ is always equal to $R_i(j-1)$, then

$$B_i(j) = B_i(j-1) - b_i(j-1) \tag{2.25}$$

The virtual buffer level is updated as

$$V_i(1) = \begin{cases} 0, & i = 1 \\ V_{i-1}(N_{i-1}), & \text{other} \end{cases}$$

$$V_i(j) = V_i(j-1) + b_i(j-1) - \frac{R_i(j-1)}{f}, \qquad j = 2,3,\ldots,N_i \tag{2.26}$$

In the JM of H.264/AVC, the initial quantization parameter, $QP_i(1)$ for the IDR picture and the first stored picture of the first GOP are set as:

$$QP_1(1) = \begin{cases} 40 & bpp \le l1 \\ 30 & l1 < bpp \le l2 \\ 20 & l2 < bpp \le l3 \\ 10 & bpp > l3 \end{cases} \tag{2.27}$$

where $bpp = \frac{R_1(1)}{f \cdot N_{\text{pixel}}}$; N_{pixel} is the number of pixels in a picture; and $l1 = 0.15$, $l2 = 0.45$, $l3 = 0.9$ are recommended for QCIF/CIF. For the i-th GOP, the quantization parameter is:

$$QP_i(1) = \max\left\{QP_{i-1}(1) - 2, \min\left\{QP_{i-1}(1) + 2, \frac{SumPQP(i-1)}{N_p(i-1)} - \min\left\{2, \frac{N_{i-1}}{15}\right\}\right\}\right\} \tag{2.28}$$

where $N_p(I - 1)$ is the total number of stored picture in the $(I - 1)$-th GOP and $SumPQP(I - 1)$ is the sum of average picture quantization parameters for all stored pictures in the $(I - 1)$-th GOP. This value can be further adjusted by:

$$QP_i(1) = QP_i(1) - 1, \quad \text{if} \quad QP_i(1) > QP_{i-1}(N_{i-1} - L) - 2 \tag{2.29}$$

where $QP_{i-1}(N_{i-1} - \mathrm{L})$ is the quantization parameter of the last stored picture in the previous GOP, and L is the number of successive unstored pictures between two stored pictures.

In the picture level rate control, two stages: pre-encoding and post-encoding are involved. In the pre-encoding, the quantization parameter for each frame has been computed, but the different methods are used for the unstored pictures and stored pictures. For unstored pictures, the quantization parameters are obtained by linear interpolation of the quantization parameters between two adjacent stored pictures. Assume that two stored pictures are j-th and $(j + L + 1)$-th pictures in i-th GOP. If there is only one unstored picture between these two stored pictures, i.e., $L = 1$, the quantization parameter for the $(j + 1)$-th unstored picture, $QP_i(j + 1)$, is:

$$QP_i(j+1) = \begin{cases} \dfrac{QP_i(j) + QP_i(j+2) + 2}{2}, & \text{if} \quad QP_i(j) \neq QP_i(j+2) \\ QP_i(j) + 2, & \text{otherwise} \end{cases} \tag{2.30}$$

If there is more than one unstored picture between two stored pictures, i.e., $L > 1$, then the quantization parameters for $(j + k)$-th $(1 \le k \le L)$ unstored picture is calculated as:

$$QP_i(j+k) = QP_i(j) + \alpha + \max\left\{\min\left\{\frac{(QP_i(j+L+1) - QP_i(j)}{L-1}, 2(k-1)\right\}, -2(k-1)\right\} \tag{2.31}$$

where α is given by:

$$\alpha = \begin{cases} -3, & QP_i(j+L+1) - QP_i(j) \le -2L-3 \\ -2, & QP_i(j+L+1) - QP_i(j) \le -2L-2 \\ -1, & QP_i(j+L+1) - QP_i(j) \le -2L-1 \\ 0, & QP_i(j+L+1) - QP_i(j) \le -2L \\ 1, & QP_i(j+L+1) - QP_i(j) \le -2L+1 \\ 2, & \text{otherwise} \end{cases} \tag{2.32}$$

The value of the quantization parameter, $QP_i(j+k)$, is further bounded by 0 and 51. For the stored picture, before calculating the quantization parameters, we have to calculate the target bits allocation. The bits allocated to the j-th stored picture depend on several factors including target buffer level, the frame rate, the available channel bandwidth, and the actual buffer occupancy. This is similar to the MPEG-4 rate control. In order to determine the target bits for the stored picture, we have to first calculate the target buffer level. After coding the first stored picture in i-th GOP, the initial value of target buffer level is set to $S_i(2) = V_i(2)$, the occupancy of the virtual buffer. The target buffer value for the subsequent stored picture is calculated as:

$$S_i(j+1) = S_i(j) - \frac{S_i(2)}{N_p(i)-1} + \frac{\overline{W}_{p,i}(j)\cdot(L+1)\cdot R_i(j)}{f\cdot(\overline{W}_{p,i}(j)+\overline{W}_{b,i}(j)\cdot L)} - \frac{R_i(j)}{f} \tag{2.33}$$

where $\overline{W}_{p,j}(j)$ is the average complexity weight of stored pictures and $\overline{W}_{b,j}(j)$ is the average complexity weight of unstored pictures; their values are calculated as:

$$\begin{aligned}
\overline{W}_{p,i}(j) &= \frac{W_{p,i}(j)}{8} + \frac{7\cdot\overline{W}_{p,i}(j-1)}{8} \\
\overline{W}_{b,i}(j) &= \frac{W_{b,i}(j)}{8} + \frac{7\cdot\overline{W}_{b,i}(j-1)}{8} \\
W_{p,i}(j) &= b_i(j)\cdot QP_{p,i}(j) \\
W_{b,i}(j) &= \frac{b_i(j)\cdot QP_{b,i}(j)}{1.3636}
\end{aligned} \tag{2.34}$$

where the $b_i(j)$ is defined as in (2.24). Based on the value of the target buffer, the target bits for the j-th stored picture in the i-th GOP is determined as:

$$\tilde{T}_i(j) = \frac{R_i(j)}{f} + \gamma\cdot(S_i(j) - V_i(j)) \tag{2.35}$$

where γ is a constant and its typical value is 0.5 when there is no unstored picture and 0.25 otherwise. The remaining bits are then calculated as:

$$\hat{T}_i(j) = \frac{W_{p,i}(j-1)\cdot B_i(j)}{W_{p,i}(j-1)\cdot N_{p,r} + W_{b,i}(j-1)\cdot N_{b,r}} \tag{2.36}$$

where $N_{p,r}$ and $N_{b,r}$ are the numbers of the remaining stored pictures and unstored pictures, respectively. Finally, the target bits are a weighted combination as:

$$T_i(j) = \beta\cdot\hat{T}_i(j) + (1-\beta)\cdot\tilde{T}_i(j) \tag{2.37}$$

where β is a constant and its typical value is 0.5 when there is no unstored picture and 0.9 otherwise. It should be noted that a tight buffer regulation can be achieved by choosing a small value of b. To conform with the hypothetical reference decoder (HRD), the target bits are further to be bounded by

$$T_i(j) = \max\{Z_i(j),\ T_i(j)\}$$

$$T_i(j) = \min\{U_i(j),\ T_i(j)\} \tag{2.38}$$

where $Z_i(j)$ and $U_i(j)$ are calculated by

$$Z_i(j) = \begin{cases} B_{i-1}(N_{i-1}) + \dfrac{R_i(j)}{f} & j = 1 \\ Z_i(j-1) + \dfrac{R_i(j)}{f} - b_i(j) & \text{other} \end{cases}$$

$$U_i(j) = \begin{cases} (B_{i-1}(N_{i-1}) + t_{r,1}(1)) \times \varpi & j = 1 \\ U_i(j-1) + \left(\dfrac{R_i(j)}{f} - b_i(j)\right) \times \varpi & \text{other} \end{cases} \tag{2.39}$$

$t_{r,1}(1)$ is the removal time of the first picture from the coded picture buffer and ϖ is a constant with typical value of 0.9. These boundary values were proposed in (2.38).

When the target bits for the stored picture are determined, the quantization parameter can be calculated according to the quadratic model:

$$T_i(j) = c_1 \times \frac{\tilde{\sigma}_i(j)}{Q_{\text{step},i}(j)} + c_2 \times \frac{\tilde{\sigma}_i(j)}{Q^2_{\text{step},i}(j)} - m_{h,i}(j) \tag{2.40}$$

where $m_{h,j}$ is the total number of header bits and motion vector bits, c_1 and c_2 are two coefficients, and $\tilde{\sigma}_i(j)$ is the complexity; here we use the mean absolute differences (MADs) as the distortion measure, and it should be the actual MAD of the current stored picture. In order to simplify the encoding complexity, the value of $\tilde{\sigma}_i(j)$ is predicted with a linear model using the actual MAD of the previous stored picture, $\sigma_i(j - 1 - L)$.

$$\tilde{\sigma}_i(j) = a_1 \times \sigma_i(j-1-L) + a_2 \tag{2.41}$$

where a_1 and a_2 are two coefficients. The initial value of a_1 and a_2 are set to 1 and 0, respectively. They are updated by a linear regression method after coding each picture or each basic unit, which is similar to that of MPEG-4 VM. The principle for updating the model coefficients c_1 and c_2 is the least square estimation using a set of actual data on the target bits, complexity, and quantization steps as stated in [2-37]. The corresponding quantization parameter $QP_i(j)$ is computed by using the relationship between the quantization step and the quantization parameter of H.264/AVC, which is given in (2.9):

$$Q_{\text{step},i}(j) = 2^{QP_i(j)/6} \cdot d(QP_i(j) \cdot \%6) \tag{2.42}$$

where $d(0) = 0.675$, $d(1) = 0.6875$, $d(2) = 0.8125$, $d(3) = 0.785$, $d(4) = 1$ and $d(5) = 1.125$. To maintain the smoothness of visual quality among successive frames, the quantization parameter $QP_i(j)$ is further adjusted by

$$QP_i(j) = \min\{QP_i(j-L-1)+2, \quad \max\{QP_i(j-L-1)-2, \quad QP_i(j)\}\} \tag{2.43}$$

The final quantization parameter is further bounded with 51 and 0. The quantization parameter is then used to perform the rate-distortion optimization for each MB in the current frame.

In the post-encoding stage of the picture level rate control, two parameter sets, (a_1, a_2) and (c_1, c_2), are updated after encoding the picture by linear regression method. The actual bits generated for encoding the picture are added to the buffer to ensure that the updated buffer occupancy is not too high.

The basic unit is defined to be a group of continuous macroblocks. If a picture contains several basic units, not only one basic unit, the rate control for the basic unit level is needed. The method for basic unit rate control is similar to the picture-level rate control. The purpose of the basic unit level rate control is to determine the quantization parameters for all basic units so that the sum of generated bits is equal or close to the picture target bits.

At the first step of basic unit-level rate control, the complexity or MAD of the current basic unit is predicted with a linear model using the actual MAD of the co-located basic unit in the previous stored picture as in (2.41).

The second step is to calculate the target bits for encoding the current basic unit according to the complexity ratio of the current basic units and the rest of basic units in the picture. The target bits for encoding the texture is then obtained by subtracting the bits for coding the header information from the total target bits.

The third step is to calculate the quantization step by using the quadratic rate-distortion model as in (2.40).

The rest steps of the basic unit rate control include performing the rate-distortion optimization on each macroblock in the basic unit, updating two parameter sets in (2.40) and (2.41) and number of remaining bits.

The JM of H.264/AVC rate control scheme recommends the size of basic unit to be the number of MBs in a row for field coding, adaptive field/frame coding, or MB-AFF coding and to be 9 for other cases in order to achieve a good trade-off between average PSNR and bit fluctuation.

2.7 SUMMARY

In this chapter, the basic concepts of video coding standards have been first introduced. Then the algorithms and tools used in most video coding standards were introduced. A brief summary for each video coding standard was introduced. Finally, the video encoding issues have been discussed.

2.8 EXERCISES

2-1. What are the basic tools for video coding standards for removing the spatial redundancy and temporal redundancy? Explain why these tools are useful.

2-2. What is the difference between sequential coding and progressive coding in JPEG? Conduct a project to encode an image with sequence coding and progressive coding. Use JPEG lossless mode to code several images and explain why different bit rates are obtained.
2-3. Generate a Huffman code table using a set of images with 8-bit precision (approx. 2 or 3) using the method presented in Annex C of the JPEG specification. This set of images is called the *training set*. Use this table to code an image within the training set and an image that is not in the training set, and explain the results.
2-4. Describe the principles of Huffman coding and arithmetic coding, and describe the differences between Huffman coding and arithmetic coding.
2-5. In most video coding standards, block-based transformation is used. The correlation between blocks is always not full exploited. What tools are used to remove the redundancy between blocks? Briefly describe these tools.
2-6. MPEG-2 video coding is used for DVD application. The method used for trick mode, i.e., fast forward and reverse, is to display only the I-frame in some DVD players. Propose an encoding scheme that can perform the trick modes more smoothly, such as in the analog VCR and without effects of frame skip. Use a software decoder to demonstrate your scheme.
2-7. Give the detailed description about the linear regression method used in the MPEG-4 VM rate control by using the results of the past n frames.
2-8. In block-based motion compensation, what kind of motion model is assumed? If the real motion does not follow this model, what would happen? What motion model can the global motion estimation handle?
2-9. Conduct a project to make a comparison of coding performance between MPEG-2, MPEG-4 (choose a profile, such as the advance simple profile), and H.264/AVC (main profile) at a range of bit rate from 1 to 12 Mbps for SDTV sequences. Draw rate distortion curves (bit rate vs. PSNR) and explain the results. (Suggestion: use $N = 15$ and $M = 1$).
2-10. In H.264/AVC, a 4×4 integer transform is used; its basic functions do not have equal norm. How to solve this problem in H.264? Please explain.
2-11. Describe the advantages and disadvantages by using small transform block size compared with using large transform block size at bit rate about 3 to 6 Mbps for SD sequences. Explain the difference between the cases at low bit rates and high bit rates. (For example, for SD sequences, assume that a low bit rate is lower than 1.5 Mbps, and high bit rate is higher than 10 Mbps).

REFERENCES

[2-1] ISO/IEC JTC1 IS 11172 (MPEG-1), Coding of moving picture and coding of continuous audio for digital storage media up to 1.5 Mbps, 1992.
[2-2] ISO/IEC JTC1 IS 13818 (MPEG-2), Generic coding of moving pictures and associated audio, 1994.

[2-3] ISO/IEC JTC1 IS 14386 (MPEG-4), Generic coding of moving pictures and associated audio, 2000.

[2-4] ISO/IEC JTC1 15938 (MPEG-7), Information technology — multimedia content description interface — multimedia description schemes, 2001.

[2-5] Requirements Group, MPEG-21 Overview v4, J. Bormans, Ed., ISO/IEC JTC1/SC29/WG11 N4801, Fairfax, VA, May 2002.

[2-6] ITU-T Recommendation H.261, Video codec for audiovisual services at p × 64 Kbit/s, March 1993.

[2-7] ITU-T Recommendation H.263, Video coding for low bit rate communication, Draft H.263, May 2, 1996.

[2-8] T. G. Wiegand, J. Sullivan, G. Bjontegaard, and A. Luthra, Overview of the H.264/AVC Video Coding Standard, *IEEE Transactions on Circuits and Systems for Video Technology*, 13, 560, 2003.

[2-9] MPEG-2 Test model 5, ISO-IEC/JTC1/SC29/WG11, April, 1993.

[2-10] F.W. Mounts, A video encoding system with conditional picture-element replenishment, *Bell System Technical Journal,* 48, 2545, 1969.

[2-11] B.G. Haskell, F.W. Mounts, and J.C. Candy, Interframe coding of video telephone pictures, Proceedings of the IEEE, 60, 792, 1972.

[2-12] M. Nelson, Arithmetic coding + statistical modeling = data compression, Part 1 — arithmetic coding, *Dr. Dobb's Journal,* February, 1991.

[2-13] W. Li, Overview of fine granularity scalability in MPEG-4 video standard, *IEEE Transactions on Circuits and Systems for Video Technology*, 11, 301, 2001.

[2-14] Y. Wang and Q.-F. Zhu, Error control and concealment for video communication: A review, *Proceedings of the IEEE*, 86, 974, 1998.

[2-15] N. Grammalidis, D. Beletsiotis, and M. G. Strintzis, Sprite generation and coding in multiview image sequences, *IEEE Transactions on Circuits and Systems for Video Technology*, 10, 302, 2000.

[2-16] ISO/IEC WG11 MPEG Video Group, MPEG-4 Video verification model version 16.0, ISO/IEC JTC1/SC29/WG11 MPEG00/N3312, Noordwijkerhout, the Netherlands, March, 2000.

[2-17] Y. Lu, W. Gao, F. Wu, H. Lu, and X. Chen, A robust offline sprite generation approach, ISO/IEC JTC1/SC29/WG11 MPEG01/M6778, Pisa, Italy, January 2001.

[2-18] Y. He, W. Qi, S. Yang, and Y. Zhong, Feature-based fast and robust global motion estimation technique for sprite coding, ISO/IEC JTC1/SC29/WG11, MPEG00/M6226, Beijing, July 2000.

[2-19] Digital compression and coding of continuous-tone still images — requirements and guidelines, ISO-/IEC International Standard 10918-1, CCITT T.81, September, 1992.

[2-20] ISO/IEC 15444-1 or ITU-T Rec. T.800, March 2001.

[2-21] Skodras, C. Christopoulos, and T. Ebrabimi, The JPEG2000 still image compression standard, *IEEE Signal Processing Magazine*, 18, 36, 2001.

[2-22] R. Calderbank, I. Daubechies, W. Sweldens, and B.-L Yeo, Lossless image compression using integer to integer wavelet trnsform, *Proceedings of the IEEE International Conference on Image Processing,* 1, 596, 1997.

[2-23] L. Gall and A. Tabatabai, Subband coding of digital images using symmetric short kernel filters and arithmetic coding techniques, Proceedings of IEEE International Conference on ASSP, New York, 1988.

[2-24] ITU-T Recommendation H.263, Video coding for low bit rate communication, Draft H.263, January 27, 1998.

[2-25] ISO/IEC 14496-10 AVC or ITU-T Rec. H.264, September 2003.

[2-26] K. Katsaggelos, R.P. Kleihorst, S.N. Efstratiadis, and R.L. Lagendijk, Adaptive image sequence noise filtering methods, Proceedings of SPIE Visual Communication and Image Processing, Boston, Nov. 1991.

[2-27] M.K. Ozkan, M.I. Sezan, and A.M. Tekalp, Adaptive motion-compensated filtering of noisy image sequences, *IEEE Transactions on Circuits and Systems for Video Technology,* 3, 277, 1993.

[2-28] V.A. Algazi, T. Reed, G.E. Ford, and R.R. Ester, Image analysis for adaptive noise reduction in super high definition image coding, SPIE VCIP Proceedings, Vol. 1818, 1992.

[2-29] V.-N. Dang, A.-R. Mansouri, and J. Konrad, Motion estimation for region-based video coding, IEEE International Conference on Image Processing, 1995, Washington DC.

[2-30] M. Bierling, Displacement estimation by hierarchical block matching, *SPIE Proceedings on Visual Communication and Signal Processing,* 1001, 942, 1988.

[2-31] P. I. Hosur and K.-K. Ma, Motion vector field adaptive fast motion estimation, second international conference on information, Communications and Signal Processing (ICICS'99), Singapore, Dec. 1999.

[2-32] H. Sun, W.Kwok, M. Chien, and C.H. J. Ju, MPEG coding performance improvement by jointly optimization coding mode decision and rate control, *IEEE Transactions on Circuits and Systems for Video Technology,* 7, 449, 1997.

[2-33] L. Wang, Rate control for MPEG-2 video coding, SPIE Conference on Visual Communications and Image Processing, Taipei, May 1995.

[2-34] J. Lee and B.W. Dickerson, Temporally adaptive motion interpolation exploiting temporal masking in visual perception, *IEEE Transactions on Image Processing,* 3, 513, 1994.

[2-35] ISO/IEC WG11 MPEG Video Group, MPEG-4 video verification model version 16.0, ISO/IEC JTC1/SC29/WG11 MPEG00/N3312, Noordwijkerhout, the Netherlands, March, 2000.

[2-36] Joint Video Team (JVT) of ISO/IEC MPEG and ITU-T VC e.g., Document JVT-H036, Joint model reference encoding methods and decoding concealment methods, August 30, 2003.

[2-37] T. Chiang and Y.-Q. Zhang, A new rate control scheme using quadratic rate-distortion modeling, *IEEE Transactions on Circuits and Systems for Video Technology,* Feb 1997. Vol. 7, no. 1, pp. 246–250.

[2-38] S. Ma, Z. Li, and F. Wu, Joint Video Team (JVT) of ISO/IEC MPEG and ITU-T VC e.g., Document JVT-H017, Proposed draft of adaptive rate control, May 20, 2003.

[2-38] A.Puri, X. Chen, and A. Luthra, Video coding using the H.264/MPEG-4 AVC video compression standard, to be published at *Signal Processing: Image Communication Journal*, 2004

3 Video Transcoding Algorithms and Systems Architecture

In this chapter, we will review state-of-the-art transcoding algorithms and systems architecture. In the first section, we will discuss the general concept of transcoding and its applications. The applications scenarios of transcoding will be studied in detail without descriptions on the specific techniques or algorithms. The second section will be focused on the various functionalities of a transcoder, and the third section will describe techniques to achieve those functionalities. The optimization of each technique will be studied in the next chapter.

3.1 GENERAL CONCEPTS FOR THE TRANSCODER

The transcoder is a device or system that converts one bit stream into another bit stream that possesses a more desirable set of parameters [3-29]. The parameters can be syntax, format, spatial resolution, temporal resolution, bit rate, functionality, or hardware requirements. Such a device or system provides a bridge to fill the gap between server/client and encoder/decoder. We will now discuss several applications for the transcoder.

3.1.1 Transcoder for SDTV to HDTV Migration

As the terrestrial broadcasting of HDTV is becoming popular, we are faced with the matter of how to provide smooth migration from the legacy analog television to the new digital television era. Obviously, there is a need to enable existing analog televisions to receive the new high-quality digital television service. The task is not trivial because the spatial resolution and frame rate may be different and the embedded information, such as closed captioning, need to be preserved in the new bitstream. One of the ideas is to provide a low-cost TV set–top box that receives the digital signal and a transcoder can convert it to an appropriate format. Such a device must be very cost effective because the alternative is to replace the old television set with a new one that has digital tuner. In the case of British digital television promotion, the TV set–top boxes are distributed freely and the cost is recovered through advertisement revenue.

3.1.2 Multi-Format and Compatible Receiver and Recorder

Even when the receiver is already digitally enabled, there is another issue with multiple format decoding. For example, the existing DVD receivers or digital cameras have to support multiple formats, including VCD, Photo CD, JPEG, and Motion

JPEG. In order to stay competitive in the market, it is imperative that the receiver supports multiple formats at minimum cost in both software and hardware requirements. To that end, common hardware/software modules should be reused for the supported format or syntax. For example, the DCT module should be reused whenever possible.

For devices with only playback functionality, such as a DVD player, it is easier since the transcoder involves converting multiple formats into a single target format. The transcoder needs to adapt the bitstreams in the digital domain to appropriate spatiotemporal resolutions suitable for the display. The transcoder becomes more complex when the number of display devices increases. For example, both the standard/high definitions and interlaced/progressive displays will coexist in the near future. The displays nowadays are mostly equipped with format or standards conversion. However, better visual quality can be achieved if such conversion is done at the source coding level. For example, motion parameters of the objects are available in the compressed bitstream, which can be readily used for de-interlacing or spatial interpolation.

For devices with recording functionality, such as DVD recordable and digital cameras, the transcoder needs to provide arbitrary conversion between one format and another format upon the user's selection. This poses a challenge because the combination matrix grows dramatically. Currently, the trancoding is performed offline on a powerful PC but it is foreseeable that such functionality will eventually supported by embedded systems.

3.1.3 Transcoder for Broadcasting and Statistical Multiplexing

For the direct broadcasting services (DBS) using satellite or terrestrial broadcasting, in the early days the approach was to install an array of real-time hardware encoders with tape as the primary storage device for programs. However, this poses logistic problems as well as difficulty in simultaneously supporting commercial insertion, editing, and live programs. Thus, a cost-effective alternative is to precompress the program at high-quality compressed format and to transcode in real-time for broadcasting. The advantages are that the transcoder is cheaper, more reliable, and easier to manage than a real-time encoder. The transcoder can facilitate real-time slicing and multiplexing easily for commercial insertion and bit stream splicing.

For the broadcasting applications, the satellite transponder may transmit wide bandwidth that contains multiple programs. These programs are multiplexed together statistically on a single channel so there are potential statistical multiplexing gains available. The bandwidth can be allocated dynamically among different programs so that the overall visual quality is optimal. For example, the news and sports programs may require different channel bandwidth to achieve acceptable visual quality. Chances are that materials require high bandwidth may not occur at the same time, so the multiplexer may allocate the bandwidth adaptively. The goal is to provide best overall quality for a given aggregate bandwidth of the transponder. The transcoder can provide bit rate adaptation when the aggregate bandwidth is exceeded. For terrestrial applications, there is a similar situation where multiple standard resolution signals are transmitted over one high-definition channel to generate more revenue for the broadcaster [3-1].

3.1.4 Multimedia Server for Communications Using MPEG-7

During the past 20 years, various video standards such MPEG-1/2/4 and H.261/2/3 have been proposed and implemented for video communications over the networks. The multimedia server needs to address devices with various decoding format capabilities. While serving various formats of multimedia information, the server must be built with not only transcoding capability but also tools for performing such transcoding effectively. In [3-2], Mohan et al. focused on an effective distribution of multimedia-rich web pages using proprietary content analysis. However, the construction of multimedia server with transcoding requires not only a proprietary solution but also a standards-based approach such that functionality can be enhanced with a commonly supported protocol at both the server and client sides. The common protocol is referred to as the transcoding hints in the context of the MPEG-7 standards. The MPEG-7 standard is commonly known as the *multimedia content description interface,* which aims at providing a set of low-level descriptors and a high-level description scheme for content description and management.

Besides using the MPEG-7 metadata for search and indexing, Vetro et al. proposed a novel idea to use metadata for guiding the transcoding process [3-3] at the server side. The server receives client requests and accesses the external network conditions and client device capability, which are combined with the transcoding hints for making the adaptations. In particular, the hints indicate the importance of the picture area and video object with its associated bit rate in the context of MPEG-4 video bit stream. The hints were used to guide the transcoder so that it can 1) allocate the bit rate wisely, 2) identify important objects in the scene, and 3) choose the best temporal resolution of the video [3-5] for given exterior conditions. In an aggressive transcoding scenario, certain less-important video objects or information will be dropped to make room for more important information.

3.1.5 Universal Multimedia Access

Universal multimedia access is defined by the capability that content can be adapted and delivered over the Internet to diverse client devices with heterogeneous networks. The client devices may have different bandwidths, screen sizes, and video/audio/image display or playback capability. Sometimes, we refer to the overall applications as "ubiquitous," or "pervasive," computing, for which users connect to the multimedia service through various devices and for different applications. The critical issue is to present the materials and allocate adequate resources to support the client device for a successful experience. A set of rules need to be executed based on a certain "value" optimization, where "value" is defined by the service provider as a measure for the client'sperceived value given a certain bit rate

In [3-2], Mohan et al. described such a system with the architecture shown in Figure 3.1. The goal of this system is to serve multimedia-rich web pages over

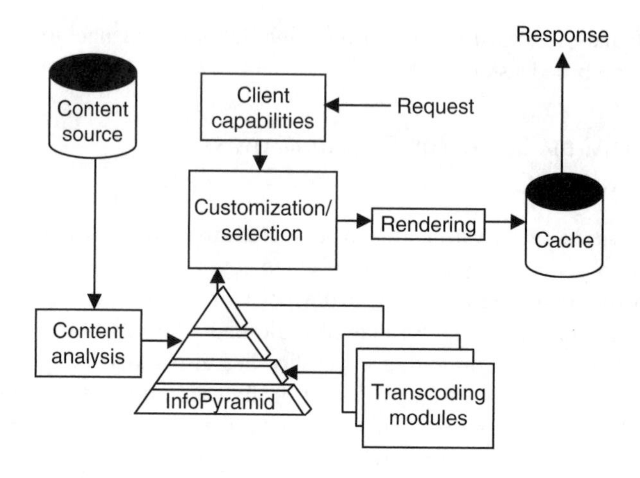

FIGURE 3.1 Internet content adaptation system architecture [3-2].

a wide range of networks and devices, including hand-held computers, personal digital assistants, TV set–top boxes, screen telephones, smart cellular phones, network computers, and personal computers. The system consists of two major modules: the infoPyramid and the Customizer. The infoPyramid offers a multimodal, multiresolution hierarchy for multimedia abstraction as shown in Figure 3.2. The key concept in this architecture is that video information can be transcoded into information of different modalities of perception, display resolutions, bandwidths, compression

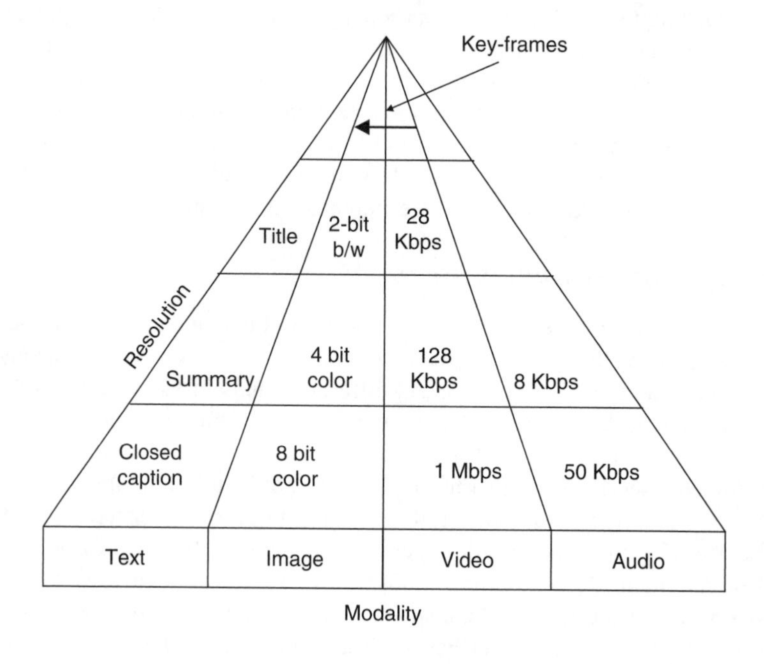

FIGURE 3.2 Information pyramid [3-2].

formats, and color depth. The transcoding is offline performed with automatic content analysis with occasional human intervention. The content is authored in XML and is converted into HTML seamlessly before transmission. The cache stores the web pages when they were delivered to the client. The design goal is to maximize the hit ratio through grouping of clients with similar capabilities.

The transcoding of composite multimedia web information over the Internet for a wide range of devices poses a challenging task. The adaptation can be achieved at the server, the client, or an intermediate proxy. The client side transcoding requires computational power and inefficient networks usage since partial information is discarded at the client, which can be prevented by a smart server. For example, the color information may not be useful if the display can only display gray level pictures.

The intermediate proxy allows transparent content adaptation for both the content server and end clients. However, the content providers may have concerns about how their content is presented for the clients. The implementation may not be straightforward because HTML tags contain mostly formatting information and lack semantics and transcoding hints. The real-time adaptation or transcoding may be difficult to implement for sophisticated multimedia content such as video and audio. The adaptation is somewhat rigid because of its complexity. For example, the format transcoding may limit the ratio of size reduction for certain bit streams for practical implementation. Another innovative approach for transcoding is to adapt the material into a different modality of representation. For example, a web page that contains mostly text may be difficult to display on a cellular phone but is ideal for speech presentation.

The Customizer selects adapted material based on content analysis and transcoding and sends it to the client devices [3-2]. The optimization of the Customizer involves the mix and match between content and client capability including various modalities. The Customizer selects from the infoPyramid to satisfy the constraints including content size, display size, network bandwidth, color depth, compression format, support for audio and image, and lastly ease of use. For example, speech is the favorable modality for cellular phone, which is not the case for desktop PC as demonstrated in Figure 3.3 described in [3-2].

The concept of universal multimedia access and its relationship with transcoding, which is used for content adaptation, will be revisited in Chapter 10.

3.1.6 Watermarking, Logo Insertion, and Picture-in-Picture

There is a commonality in technology for insertions of watermark [3-4][3-6], logo [3-6] or picture-in-picture, where the transcoder needs to compose two bit streams into a single compliant bit stream. One bit stream is embedded either visibly (logo and picture-in-picture) or invisibility (watermarking) into the final bit stream. These three applications all involve the addition of information to the video bit stream in either the compressed or uncompressed domain. From the perspective that a transcoder modifies one bit stream to another bit stream, one can conclude that such a functionality is achievable with a transcoder. Furthermore, these functionality

In the case of watermarking, the watermark is inserted invisibly so that authentication is possible later and is preferably achievable without knowledge of the

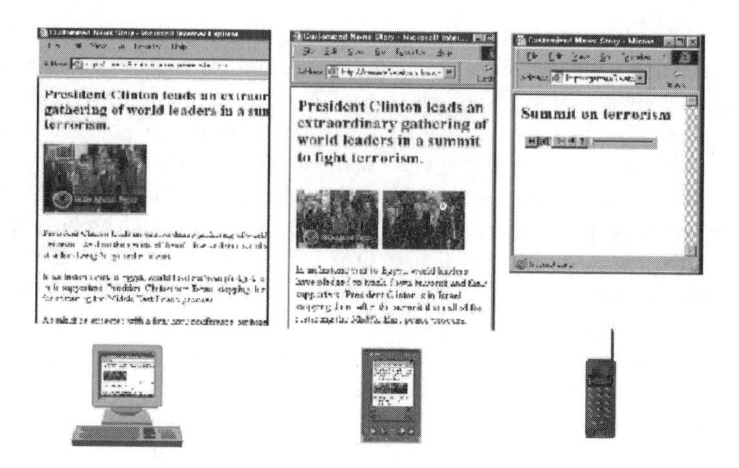

FIGURE 3.3 Multimodal content transcoding and adaptation [3-2].

original sequence [3-4][3-6]. In the case of logo insertion, the logo should be visible and robust against re-encoding and editing. In the case of picture-in-picture as shown in Figure 3.4, there is a need for a continuous presence of a reduced size video in a given time interval. All of these applications require modifications and manipulations of the original content or bit stream so the transcoder can be used to achieve such a purpose. If the operations require real time, then the complexity needs to be considered while the quality is maintained.

3.1.7 Studio Applications

In the television studio, there is a need to mix or "splice" multiple bit streams into a single bit stream in the postproduction process. Since the studio typically handles high-quality contribution-quality video, real-time splicing demands challenging high-speed switching and the transcoder has less tolerance in any visual quality loss. Another challenge is to maintain compliance with the virtual buffer verifier requirements in the standards specifications.

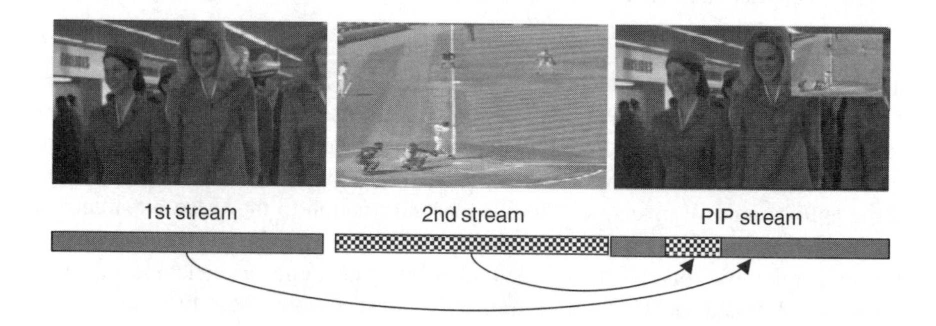

FIGURE 3.4 PIP video service with a transcoder at the server.

3.2 TRANSCODER FOR BIT RATE AND QUALITY ADAPTATION

Now we will focus on the most fundamental functionality of a transcoder, namely, bit rate adaptation or transcoding. The bit rate adaptation transcoder converts a high–bit rate stream (e.g., 15 Mbps) into a low–bit rate stream (e.g., 3 Mbps). Such a conversion can be used to transmit bit streams over heterogeneous networks where the bandwidth of the transmission path is not identical for each link. The buffering mechanism can smooth the traffic irregularity but the quality of service for a timely delivery of content is sacrificed. When the bit rate is scaled down, the adapted visual quality is also reduced. The transcoder needs to maintain visual quality to suit the target applications. For example, the multimedia-on-demand server may store the source material in precompressed format and perform quality adaptation on the fly to simulate a real-time encoder at much reduced complexity.

The output from the transcoder needs to meet several prescribed requirements, such as bit rates, minimum visual quality, constrained delay, limited buffer size, and periodic random access points. Here we will discuss approaches to meet such requirements. Since the source bit stream has been precompressed, it will experience initial quality loss from the encoding process. Furthermore, it is expected that more quality degradation results from transcoding depending on the architecture.

3.2.1 CASCADED TRANSCODER

A straightforward approach to implement transcoding is to decode fully so that the reconstructed sequence is obtained and then re-encode the sequence with a new set of coding parameters. Such an implementation is typically referred to as the *cascaded* approach and employs maximum complexity since it includes the motion estimation module as shown in Figure 3.5. There are various ways to reduce the complexity of the cascaded transcoder. Mostly, the methods for improvement are to partially decode the source bit stream up to dequantized DCT coefficients and requantize the coefficients with a coarser quantization step.

For the intra-frame of a video bit stream, the complexity is reduced with no significant penalty in coding efficiency by itself. Since the intra-frame is used as anchor frame for future prediction, the reference frame has been changed due to the requantization process. Therefore, for the predictive frames, the displaced frame difference needs to be recomputed in order to have a closed prediction loop. Such a process may involve recomputation of the motion vectors, coding mode decisions, and residuals that require similar complexities of an encoder. Here we will introduce four architectures to improve the cascaded transcoder by removing the motion estimation module.

The four architectures are listed from low to high complexities as follows [3-8]:

Architecture 1: Transcoder with truncation of the high-frequency coefficients.
Architecture 2: Transcoder with requantization of the DCT frequency coefficients.
Architecture 3: Transcoder with re-encoding of the reconstructed pictures using the old motion vectors and mode decision as the original bit stream.
Architecture 4: Transcoder with re-encoding of the reconstructed pictures using the old motion vectors but new mode decision.

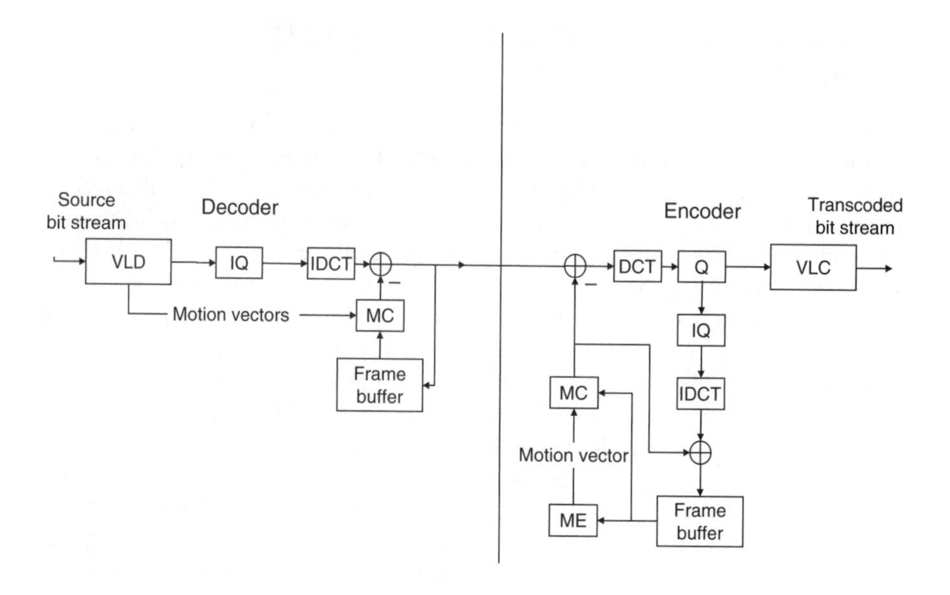

FIGURE 3.5 Spatial domain cascaded transcoder [3-26].

The first two architectures are often referred to as *open loop* transcoders because the encoder loop is not closed in the re-encoding process. The first architecture has the simplest form, with only two modules: bit stream parser and variable length decoder, as shown in Figure 3.6. As shown in Figure 3.7, the bit rate is scaled in a macroblock-by-macroblock fashion. The second architecture added two more modules: inverse quantization, quantization modules as shown in Figure 3.8. The inverse quantization module reconstructs the block back to the DCT coefficients. The quantization module requantize it into suitable resolution according to the bit allocation analysis. These two architectures significantly reduce the complexity with loss of picture quality. The first two architectures are suitable for the trick modes and extended play recording of the digital video tape recorder.

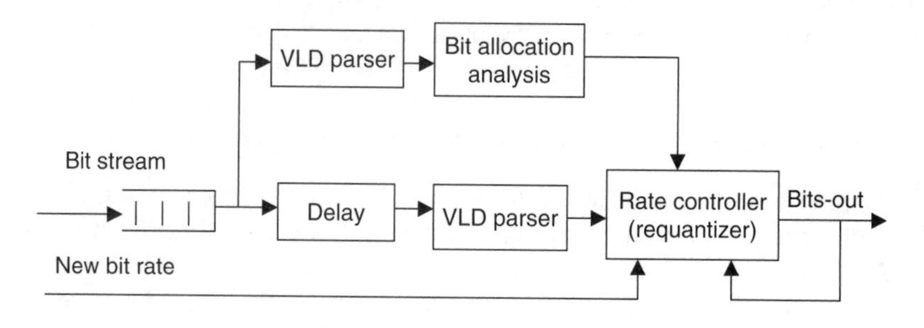

FIGURE 3.6 Architecture 1: truncation of the high-frequency coefficients [3-8].

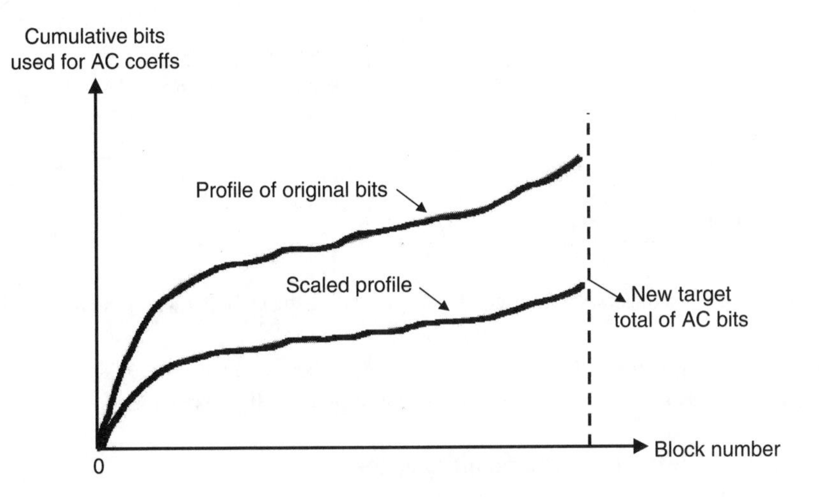

FIGURE 3.7 Scaling of the cumulative bit u sage profile [3-8].

The third and fourth architectures as shown in Figure 3.9 close the encoding loop and have no "error drift," which significantly improves the picture quality. However, this is at the expense of increased complexity. These two architectures are useful for video on demand and statistical multiplexing for multiple channels.

3.2.1.1 Architecture 1: Truncation of the High-Frequency Coefficients

As illustrated in Figure 3.6, bit rate adaptation is achieved by removing the high-frequency coefficients. The VLD parser does not require any decoding capability and simply determines the codeword lengths. With the bit stream parser, the analyzer produces a cumulative bit usage profile as shown in Figure 3.7. The transcoder needs to define a scaled cumulative bit usage profile that meets the target bit rate. With the newly scaled profile, the rate controller simply discards the DCT coefficients that exceed the scaled profile. All of the codewords other than the ac coefficients should be kept to achieve a compliant bit stream. The advantage of this approach is its simplicity since it does not need to perform variable length decoding and inverse quantization.

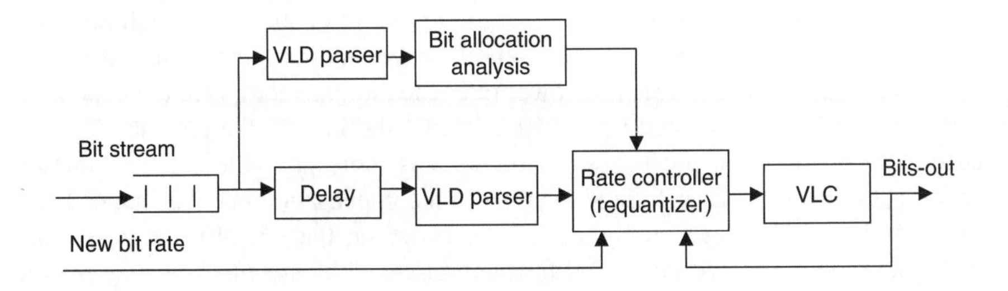

FIGURE 3.8 Architecture 2: transcoder by requantizing the DCT frequency coefficients [3-8].

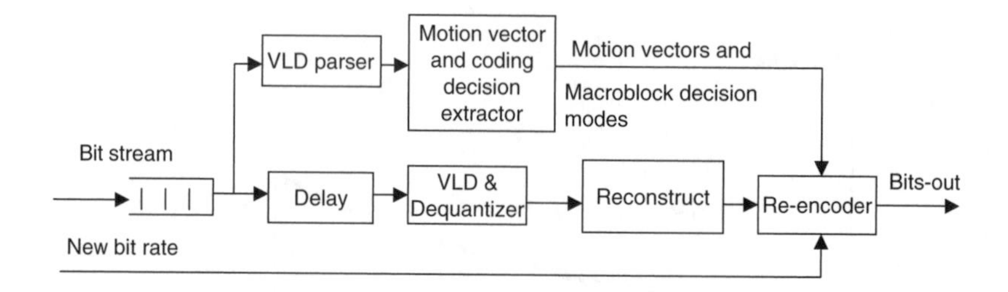

FIGURE 3.9 Architecture 3: transcoder by re-encoding the DCT frequency coefficients [3-8].

Thus, the complexity is lower than that of a decoder. However, there is a serious drift since the decoding is not closed, as will be shown by the experiments.

3.2.1.2 Architecture 2: Requantizing the DCT Frequency Coefficients

The second architecture slightly increases the complexity by adding the variable length decoder, inverse quantization and quantization modules, and variable length encoder. There are various techniques to select the quantization parameters so that the requantization process can achieve good rate-distortion performance. One example is to quantize the DCT coefficients according to the scaled cumulative bit usage profile. The profile can be optimized for each transcoder.

3.2.1.3 Architecture 3: Re-Encoding with Old Motion Vectors and Mode Decisions

The third architecture is to extract the motion vectors and mode decisions from the original stream and use them to re-encode the reconstructed pictures. The motion-compensated residual is recomputed at the encoder so there is no error drift. The benefit from this architecture is that no motion estimation and mode decisions will be computed. These two modules will constitute about 50 and 20% of the encoder complexity.

3.2.1.4 Architecture 4: Re-Encoding with Old Motion Vectors

The fourth architecture is a variation of the third architecture with additional complexity. In compressing the video source, the encoder selects the coding mode based on the rate-distortion behavior of the operating point of encoding. At high bit rates, the encoder can afford bandwidth for sending more motion vectors to reduce the residual energy. On the contrary, at lower rates the motion vectors are too expensive to be encoded and it is more important to spend the bits on the residual. The re-encoder is operating at much lower rates so it is more probable that less motion vector is favorable. That is, bidirectional motion compensation may be changed into forward or backward motion compensation. However, there is also the possibility that more motion vectors are needed for the transcoded bit stream. This may require ancillary data containing the other motion vectors precomputed at the server. Thus, this ancillary data will occupy significant overhead storage, which can only be justified

TABLE 3.1
Hardware Complexity Savings over the Cascaded Transcoder, Which is Denoted as Architecture "C"

Architecture	1	2	3	4	C
VLD bit stream parser and rate control	✓	✓	✓	✓	✓
Quantizer/dequantizer/VLC/DCT		✓	✓	✓	✓
Decoding/encoding loops/frame memory/motion comp.			✓	✓	✓
Mode decision				✓	✓
Motion estimation					✓

when performance is critical. In video-on-demand applications, it is critical to achieve high video quality while the storage on the server is of less concern.

At this point, the complexity is approaching the cascaded transcoder, except for the motion estimation. To further enhance the quality, the transcoder can reestimate the motion vector. The motion reestimation can use the extracted motion vectors as the starting point and refine the motion vector based on the reconstructed picture. Such variation slightly increases the complexity but with improved quality.

3.2.1.5 Summary and Experimental Results

In Table 3.1, we provide a summary of the hardware complexity for the four architectures and the cascaded approach. The cascaded transcoder is denoted as architecture "C." It is obvious that there are quality and complexity tradeoffs for various architectures. As one can observe from the table, the first architecture has minimum complexity but suffers from a serious drift problem. The requantization needs four additional modules, but it still maintains the operation in the DCT domain with moderate drift problem. To completely remove the drift problem and close the prediction loop, the transcoder needs to perform the decoding and re-encoding loops. This is achieved at the expense of frame memory and motion compensation module.

To understand the quality improvement from additional complexity, we perform seven experiments as shown in Table 3.2. The first three experiments are used as

TABLE 3.2
Transcoding Experiments

	Experiments
A	Compressed at 15 Mbps using the original sequence
B	Compressed at 4 Mbps using the original sequence
C	Cascaded transcoder from 15 Mbps to 4 Mbps
D	Architecture 1
E	Architecture 2
F	Architecture 3
G	Architecture 4

TABLE 3.3
Transcoding from 15 Mbps to 4 Mbps [3-8]

	PSNR (dB)		
	Flower Garden	**Bicycle**	**Bus**
A	37.02	35.12	37.74
B	29.60	27.26	30.44
C	29.34	27.14	30.22
D	20.89	20.18	21.87
E	27.41	25.04	28.44
F	29.19	27.02	30.14
G	29.32	27.11	30.20

benchmarks for the performance and the resultant performance is shown in Table 3.3. The experiments are performed with an MPEG-2 encoder for video sequence of 150 frames at CCIR-601 720 × 480 4:2:0 resolution.

As one can observe in Table 3.3, the first significant quality improvements come from requantization in the second architecture as opposed to simply truncating the DCT frequency coefficients in the first architecture. The next level of improvement comes from closing the loop in the third architecture. The advantage of using the fourth architecture is not very significant because the difference lies in the mode decision process. As explained earlier, the performance gap comes from adjustment of the overhead bits for the motion vectors at various bit rate operation points. At the same time, we can observe that the fourth architecture is already very close to the theoretical optimal case, which is to compress at 4 Mbps from the original sequence. Thus, there is little room for improvement from the fourth architecture.

3.2.1.6 Optimized Spatial Domain Transcoder

In the experimental results from the Table 3.3, the first architecture presents a significant loss in quality, which seems to dismiss its usefulness. However, Eleftheriadis and Anastassiou described a dynamic rate-shaping approach to select the set of DCT coefficients to be kept in the transcoding process [3-23]. There are two methods to select the set. The first approach is referred to as *constrained dynamic rate shaping* (CDRS), wherein a set of coefficients at the end of each block is removed from the bit stream. The second approach is referred to as *general*, or *unconstrained dynamic rate shaping* (GDRS).

In the first approach, the transcoder needs to decide the breakpoint of each block similar to the data partitioning tool in the MPEG-2 video specification. In the second approach, there is a 64-element binary vector indicating which coefficients will be kept. The optimization process consists of iterative minimization of the Lagrangian cost computed as $L = D + \lambda R$. The iteration starts with an all-zero breakpoint configuration so the Lagrangian cost is maximal. The incremental cost reduction is considered by including new run length code of DCT coefficients. In the full search approach, the set contains all the preceding DCT run lengths. A fast approach prunes the combinations using the monotonic property of Huffman run length codes in MPEG and JPEG. That is, we have longer codes for longer zero run lengths. Detail algorithms can be found in [3-23]. The dynamic rate-shaping technique has a less than 1 dB loss in quality for a transcoding from 4 Mbps to 3.2 Mbps. The CDRS and GDRS have only a 0.5-dB difference in performance. Thus, the DCT coefficient truncation can be further optimized to make the performance acceptable.

3.2.2 Frequency Domain Transcoder

As discussed in the four architectures that simplify the cascaded transcoder, the methodology is to eliminate modules progressively at the expense of drift or coding efficiency. Another approach for simplification is to identify and combine the decoder and encoder and to merge common modules based on assumed linearity of the transform and motion compensation. Such approximation does not completely eliminate the drift but does significantly reduce it.

As shown in Figure 3.10, the first simplified architecture is achieved with two steps. First, the DCT module at the encoder loop is moved before the summation process assuming such a process is linear [3-11]. The relocated DCT and the IDCT module at the decoder cancel out each other. Therefore, the IDCT module at the decoder loop is now removed. The second simplification is to merge the two frame memories and loops (i.e., decoding and encoding loops) so that motion compensation is done only once.

With these two simplifications, we have the simplified spatial domain transcoder (SSDT) as shown in Figure 3.10. The SSDT performs the motion compensation in the spatial domain, which is why it is referred to as the *spatial domain transcoder.*

The SSDT has multiple benefits in complexity reduction as compared to the cascaded transcoder. In particular, there is only one loop, one DCT/IDCT module, and one frame memory with much less drift.

To further simplify the SSDT, we reexamine the block diagram in Figure 3.10 and realize that the constraint to operate in the spatial domain comes from the fact that motion compensation needs to be performed in the spatial domain. If the motion compensation were to perform in the DCT domain, the DCT/IDCT modules in the

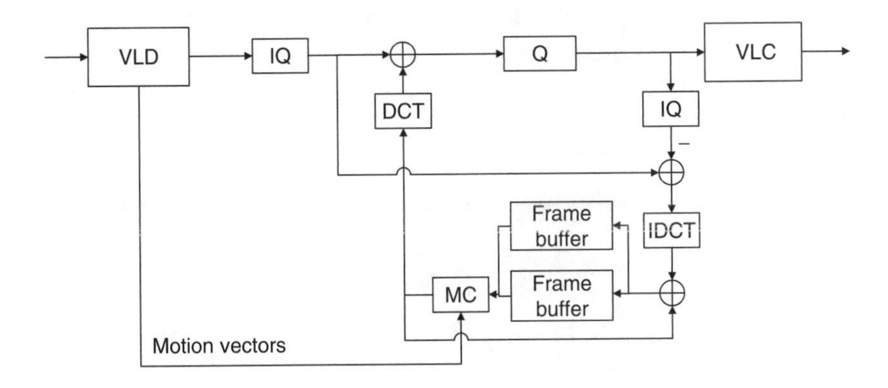

FIGURE 3.10 Simplified spatial domain transcoder [3-11].

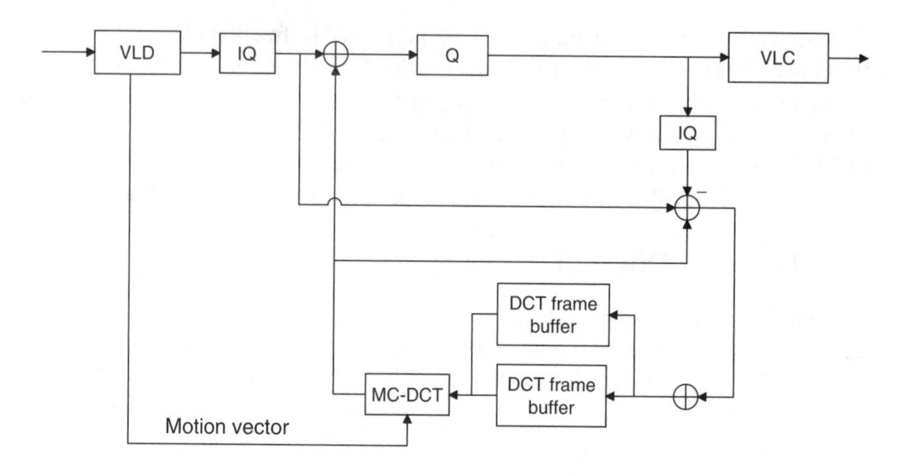

FIGURE 3.11 Simplified frequency domain transcoder [3-11].

encoder loop can be cancelled out when the IDCT is moved after the motion compensation module as shown in Figure 3.11. The motion vector can be derived from the precompressed bit stream. The tricky part is to perform the motion compensation in the DCT domain as shown in Figure 3.12.

The new motion-compensated residuals need to be derived from the four DCT blocks without returning to the spatial domain. It is further complicated by the frame/field adaptation used in MPEG-2 and later standards. Such a problem can be solved as a whole with approximate matrices to compute the MC-DCT residuals. The approximate matrices have been reported to achieve about 81% reductions as compared to the processing in the spatial domain [3-11].

Further optimization can be done with the selection of the quantization parameters for the encoder loop. The problem can be formulated as

$$\min \ D = \sum_i d_i(q_i) \ \text{subject to} \quad R = \sum_i r_i(q_i) \leq R_t$$

The distortion and rate are computed as the summation of distortion and rate for each block dependent on the quantization parameter q_i. The target bit rate is set as R_t.

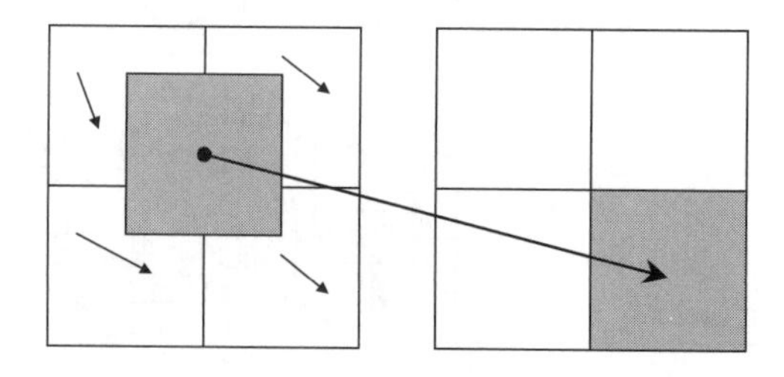

FIGURE 3.12 Frequency domain motion compensation.

Such a constrained problem is readily solved by converting it into an unconstrained minimization problem with a Lagrange multiplier. The new problem can be formulated as the minimization of the Lagrangian cost L_i as follows:

$$L_i = \min_{q_i} \{d_i(q_i) + \lambda r_i(q_i)\}$$

Such optimization can significantly improve the picture for the simplified frequency domain transcoder. In [3-11], the transcoder with Lagrangian optimization can outperform the cascaded approach given the same complexity. The transcoder can now move the resources to perform the Lagrangian optimization for selection the quantization parameters of the transcoder.

3.3 FINE GRANULARITY SCALABILITY

Another technique to implement transcoding is to use scalability, where the adaptations are done with an embedded bit stream and the need of transcoding is significantly reduced. However, such a desirable feature comes with a penalty in coding efficiency. We will describe techniques to achieve both coding efficiency and ease of transcoding in the next section but first we will describe fine granularity scalability as defined in the MPEG-4 standard.

3.3.1 MPEG-4 FGS

To provide the bandwidth adaptation functionality, there are many alternatives and one of them is scalability, wherein the bit stream is pre-encoded in an embedded format. The bit rate adaptation can be performed on the fly by the network or the decoder. In view of this desirable functionality, the MPEG-4 committee has developed the fine granularity scalability (FGS) profile, which provides a scalable approach for streaming video applications. [3-24]

The MPEG-4 FGS representation starts by separating the video frames into two layers with identical spatial resolutions, which are referred to as the *base layer* and the *enhancement layer*. The base layer is compressed with a nonscalable MPEG-4 advanced simple profile (ASP) format, while the enhancement layer is obtained by coding the difference between the original DCT coefficients and the coarsely quantized base layer coefficients in a bit-plane fashion. The enhancement layer can be truncated at any location and can provide fine granularity of reconstructed video quality proportional to the number of bits actually decoded. There is no temporal prediction for the FGS enhancement layer, which provides an inherent robustness for the decoder to recover from any errors. However, the lack of temporal redundancy removal at the FGS enhancement layer decreases the coding efficiency as compared to that of the single-layer nonscalable scheme at high rate.

As shown in Figure 3.13, the base layer is encoded with a nonscalable encoder. In fact, there is no limitation on the specific standard compliance for the base layer. The differences between the reconstructed and original blocks are transformed into DCT transform confidents. These coefficients are encoded with a bit-plane approach similar to the wavelet technique.

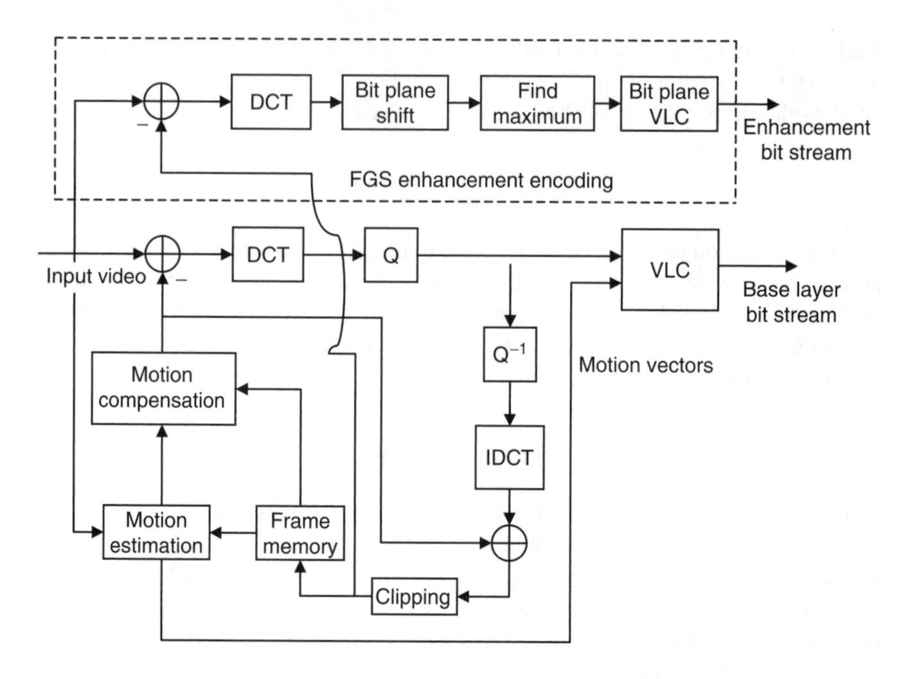

FIGURE 3.13 Fine granularity scalable encoder [3-14].

As shown in Table 3.4, we show an example of the bit-plane code method. The DCT coefficients are shown in the first row, and the sign information is shown in the second row. The absolute value of the coefficient is converted into the bit-plane format in the third through sixth columns. Now we can examine the third row from left to right and convert the sequence of 1's and 0's into 2-D variable length code symbols. The 2-D symbol has the first component as the number of consecutive zeros or RUN before the next nonzero bit. The second component represents the binary decision on whether the current bit plane has reached its last nonzero bit or end of plane (EOP). Thus, the 2-D symbol is now represented as the (RUN, EOP) vector. The sign bit needs to be sent only once with the (RUN, EOP) symbol when the most significant bit (MSB) of a particular coefficient is transmitted. That is, the

TABLE 3.4
Bit-Plane Coding of FGS

DCT coefficient	11	0	0	−5	0	1	Symbols: (RUN, EOP),S
Sign bit (S)	0	x	x	1	x	0	
Bit-plane 4 (MSB)	1	0	0	0	0	0	(0,1),0
Bit-plane 3 (MSB-1)	0	0	0	1	0	0	(3,1),1
Bit-plane 2 (MSB-2)	1	0	0	0	0	0	(0,1)
Bit-plane 1 (MSB-3)	1	0	0	1	0	1	(0,0),(2,0),(1,1),0

sign bit is transmitted only when the coefficient becomes significant and will not be sent again for the less significant bit plane of the same coefficient. For the same DCT coefficient, the coding efficiency of the bit-plane coding method outperforms up to 18% as compared to the nonscalable (RUN, LEVEL) coding method commonly used in JPEG and MPEG. The results are not surprising since wavelet has been shown to outperform JPEG with the scalability functionality.

3.3.2 Advanced FGS

Although FGS provides the desirable property of continuous SNR scalability, it suffers from a serious loss of coding efficiency because there is no temporal redundancy removal at the enhancement layer. To improve the MPEG-4 FGS, a motion compensation-based FGS technique (MC-FGS) with a high-quality reference frame was proposed for removing the temporal redundancy for both the base and enhancement layers [3-25]. In MC-FGS, the encoder combines the base and enhancement layer to form a very high quality reference frame, which is used to perform the motion compensation. Thus, the motion-compensated residual will be significantly reduced. The advantage of MC-FGS is that it can achieve high compression efficiency, close to that of the nonscalable approach in an error-free transport environment. However, the MC-FGS suffers from the disadvantage of error propagation or drift when part of the enhancement layer is corrupted or lost.

Similarly, progressive fine granularity scalability (PFGS) improves the coding efficiency of FGS and provides a means to alleviate the error drift problems simultaneously [3-15]. To remove the temporal redundancy, the PFGS adopts a separate prediction loop that contains a high-quality reference frame where only partial temporal dependency is used to encode the enhancement layer video. As shown in Figure 3.14, at low bit rate the decoder uses only the base layer as reference frames to compute the motion-compensated residual. However, the efficiency is close to that of the baseline FGS. When the bit rate is higher and reaches the third bitplane,

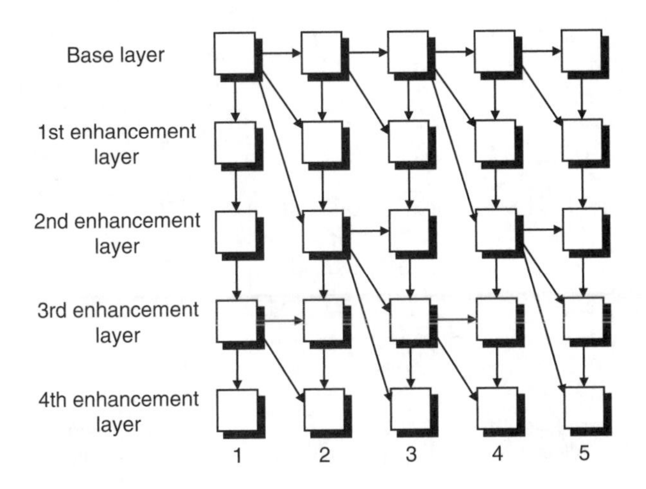

FIGURE 3.14 Simplified progressive fine granularity scalability [3-15].

the decoder uses the base layer and the first three bit planes as reference frames. In this case, the coding efficiency is significantly improved. Thus, the PFGS trades coding efficiency for a certain level of error robustness. In order to address the drift problem, the PFGS keeps a prediction path from the base layer to the highest bit planes at the enhancement layer across several frames to make sure that the coding schemes can gracefully recover from errors over a few frames. The PFGS suffers from loss of coding efficiency whenever a lower quality reference frame is used. Such a disadvantageous situation occurs only when a limited number of bit planes are used or a reset of the reference frame is invoked.

To prevent the error propagation due to packet loss in variable bit rate channels, the leaky prediction technique was used for the prediction loop in DPCM and subband coding systems. Based on a fraction of the reference frame, the prediction is attenuated by a leak factor of value between zero and unity. The leaky prediction strengthens the error resilience at the cost of coding efficiency since only part of the known information is used to remove the temporal redundancy. For a given picture activity and bit error rate (BER), there exist an optimal leak factor to achieve balance between coding efficiency and error robustness. In [3-16], a flexible robust fine granularity scalability (RFGS) framework was proposed to allow the encoder to select a tradeoff point that simultaneously improves the coding efficiency and maintains adequate video quality for varying bandwidth or error-prone environments.

As shown in Figure 3.15, the leaky prediction technique scales the reference frame by a factor α, where $0 \leq \alpha \leq 1$, as the prediction for the next frame. The leak factor is used to speed up the decay of error energy in the temporal directions. In RFGS, we use the leak factor to scale a picture that is constructed based on the concept of partial prediction, as detailed in the next paragraph.

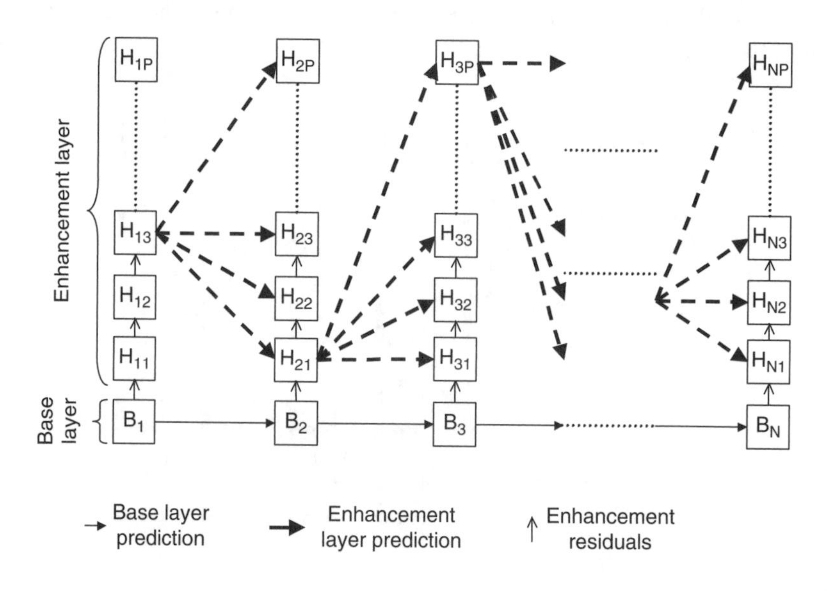

FIGURE 3.15 Simplified progressive fine granularity scalability [3-16].

As described in Figure 3.15, the RFGS is constructed with two prediction loops for the base and enhancement layers. The base layer loop is coded with a nonscalable approach for all frames F_i. The enhancement layer loop uses an improved quality reference frame that combines the base layer reconstructed image and partial enhancement layer. Thus, the enhancement layer loop can be built with an adaptive selection of number of bit planes for the reference picture. The combinations of selections for each frame constitute multiple prediction paths.

Let's assume that each frame has P maximal number of bit planes for the enhancement layer. As the number of bit planes (denoted as β) used is increased, the residuals will be decreased, which translates into improved coding efficiency. On the other hand, the reconstruction errors will accumulate and propagate if the bit planes used for the reference frame are not available at the decoder. Thus, the parameter β can be used to control the tradeoff between coding efficiency and error robustness.

Combining the concepts of partial and leaky predictions, the first β bit planes will be scaled by a leak factor. Consequently, if any information at the first β bit planes is lost, the error will be attenuated by α times for each frame at the enhancement layer. Since the value of α is smaller than unity, the drift will be eliminated in a few frames. Thus, the RFGS is implemented by defining a set of the parameters for each frame t:

$$\{M_t(\alpha,\beta)\}, \quad t = 0,\cdots,(N-1)$$

where the parameter α denotes the leak factor and the parameter β denotes the number of the bit planes used to construct the reference frame. The symbol N is the total number of frames in the video sequence. As compared to the PFGS, the periodic reset of the reference frames can be simulated with a periodic selection of the parameter α as zeros. The MPEG-4 FGS is equivalent to the case of setting α to zero through the whole sequence. As compared to the MC-FGS [3-25], the use of high-quality reference frames can be simulated with α equal to unity for all reference frames. Thus, RFGS provides a flexible MC prediction scheme that can be adapted to achieve various tradeoffs as proposed by PFGS and MC-FGS.

As shown in Figure 3.16, the error resilience property of RFGS is demonstrated by dropping the first P-picture when the GOP size is 60. This will be the worst-case scenario since the first P-picture affects the rest of frames in a GOP. It is obvious that the error drift causes at least a 2-dB loss for all of the frames when α is set to 1. The problem is significantly alleviated when α is set to 0.5. Thus, the RFGS offers a flexible architecture to balance the coding efficiency and error robustness.

3.4 FGS TO MPEG-4 SIMPLE PROFILE TRANSCODING

In this section, we will study a sample transcoding system design for Internet multimedia streaming applications. We will consider such issues as processing complexity, the visual quality of the transcoded bit stream, and the archived bit stream format.

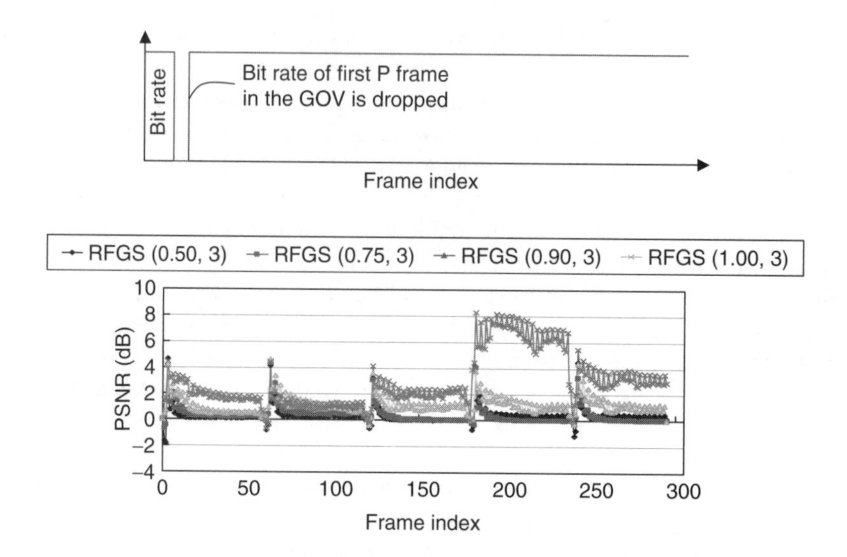

FIGURE 3.16 Error resilience of robust fine granularity scalability.

3.4.1 Application Scenario for an FGS-to-SP Transcoding

At the server, the archived or precompressed bit stream must be compactly stored and yet flexible enough to deliver the required visual quality on the fly at minimal complexity. To meet these requirements, the video sequences are archived in the MPEG-4 FGS profile bit stream as described in the previous section. The lack of best coding efficiency is not a major concern since the storage is not a big concern at the server whereas the coding efficiency of the transcoded bit stream is critical. The FGS bit streams can support different visual qualities and frame rates for the streaming video. The fully decoded FGS bit stream has almost lossless quality whereas the partially FGS bit stream can adapt to the required network bandwidths and provide proper video quality with continuous SNR scalability.

As for the client side, there are a wide variety of receivers with different formats, e.g., MPEG, MPEG-2, MPEG-4, QuickTime, and Windows Media. In this study, MPEG-4 Simple Profile is selected as an example because it has been recommended by the organizations such as the Internet Engineering Task Force (IETF) and the Third Generation Partnership Project 2 (3GPP2).

As shown in Figure 3.17, system architecture of an FGS-to-SP transcoding for video streaming over heterogeneous networks [3-26]. The source bit streams are stored in MPEG-4 FGS format in the server database and can be retrieved via the network by the users with devices capable of decoding either MPEG-4 FGS or MPEG-4 Simple Profile bit streams. For MPEG-4 FGS-enabled devices, the server truncates the enhancement layer bit stream to fit the transmission rate. For MPEG-4 SP-enabled devices, the server performs the FGS-to-Simple-Profile transcoding to deliver the Simple Profile bit stream. Such a video delivery system can minimize the archival space and achieve the highest interoperability with most mobile devices. In the proposed video content delivery system, the FGS-to-SP transcoder plays a

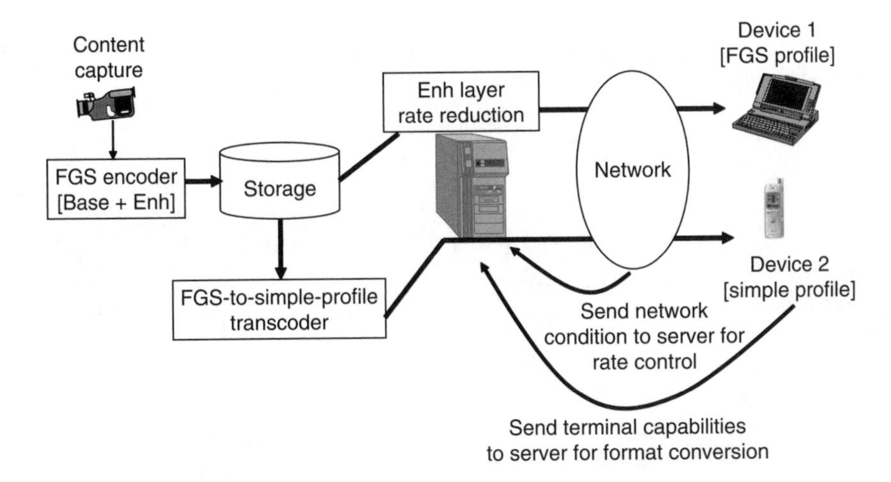

FIGURE 3.17 System architecture of an FGS-to-SP transcoding application scenario [3-26].

key role in adapting the source to match different device capabilities and channel bandwidths.

To demonstrate the usefulness of such a system, both static and dynamic experiments are performed to test the transcoder. The static experiment is to verify the coding efficiency of the transcoder and its effectiveness in adapting to the channel bandwidth. The dynamic experiment is demonstrated on a test bed with real-time streaming capability, namely, the MPEG-21 21000-12 Multimedia Test Bed for Resource Delivery. In Chapter 11, the test bed will be described in detail, and it can provide diverse channel conditions that can be duplicated for fair comparisons. The transcoder can also shape the bit rate of the transmitted bit stream to fit the time-varying channel rates in real time. The rate shape is processed with the aforementioned rate control model using the ρ-domain source model [3-18].

3.4.2 Architectures for an FGS-to-SP Transcoding

We will now investigate the design process of a transcoder starting from a cascaded transcoder, which is a cascade of FGS decoder and MPEG-4 SP encoder referred to as the first architecture as shown in Figure 3.18. The first step of simplification is to remove the motion estimation from the SP encoder by re-using the motion vectors extracted from the archived bit streams.

Adopting the techniques discussed in Section 3.2, the cascaded transcoder architecture will be further simplified by exploiting the linearity of DCT and motion compensation. As shown in Figure 3.18, the reconstructed DCT coefficients of the base layer, the enhancement layer, and the final output bit stream are denoted as B^*, E^*, and R^*, respectively. The prediction residue ΔX_n of the P-VOP X_n is defined as

$$\Delta X_n = X_n - MC(X_{n-1}) \tag{3.1}$$

where $MC(\cdot)$ denotes the motion compensation process.

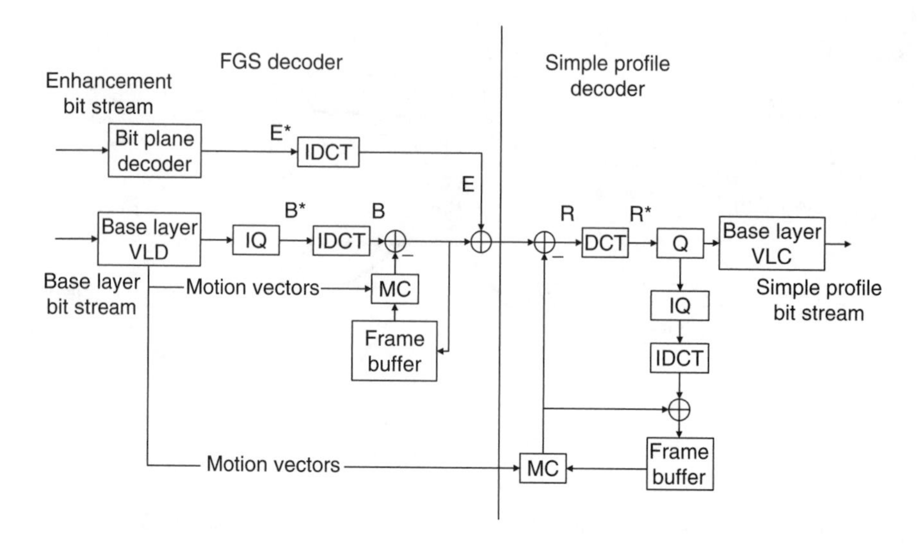

FIGURE 3.18 Architecture 1 of an FGS-to-SP transcoding by removing the motion estimation module [3-26]

The bit stream with the P-VOP that carries the DCT coefficients of the prediction residue is

$$B^* + E^* = DCT\,(\Delta X_n) \tag{3.2}$$

The frame buffer stores the reconstructed VOP (Y_{n-1}) to synchronize with the decoder. After motion compensation, the output bit stream carries the DCT coefficients as

$$R^* = DCT(X_n - MC(Y_{n-1})) + \Delta \tag{3.3}$$

where Δ is the quantization error for the quantizer Q2. After substituting X_n, the output bit stream can be expressed as

$$R^* = B^* + E^* + DCT(MC(X_{n-1}) - MC(Y_{n-1})) + \Delta \tag{3.4}$$

The intermediate architecture corresponding to (3.4) is illustrated in Figure 3.19

Becuase the motion compensation can be done after taking the frame difference, which can be performed in the DCT domain, Equation (3.4) is reduced to

$$R^* = B^* + E^* + DCT(MC(D_{n-1})) \tag{3.5}$$

$$\text{where } D_n = MC(D_{n-1}) + B_n^* - R_n^* + \Delta_n \tag{3.6}$$

Based on (3.5) and (3.6), the intermediate architecture as shown in Figure 3.19 is transformed into the second architecture in Figure 3.20. As shown in Table 3.5, the two architectures are compared based on the used modules. The simplifications of the two architectures have resulted in savings in complexities for modules including

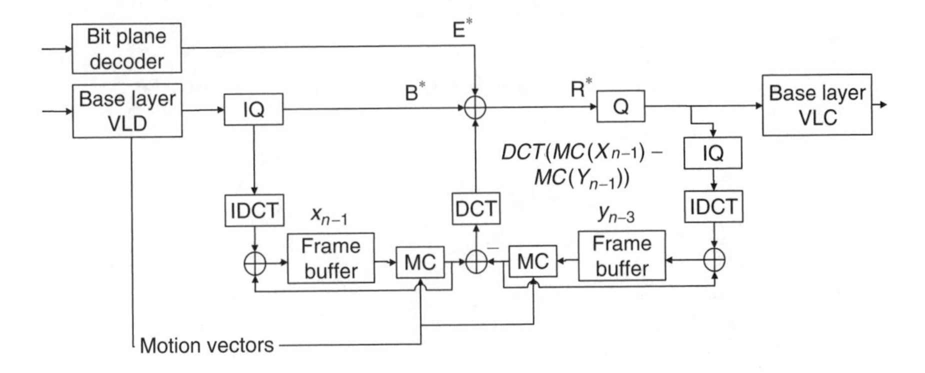

FIGURE 3.19 Intermediate block diagram of the first architecture [3-26].

motion estimation, motion compensation, DCT/IDCT, and frame memory. The second architecture can be further simplified by performing the motion compensation in the DCT domain as shown in Figure 3.21. The third architecture completely removes the need for DCT/IDCT, which makes it an efficient implementation. However, this is achieved at the expense of a more complex motion compensation module.

3.4.2.1 Rate Control for Transcoding

In order to achieve high transcoding performance, rate control is necessary to allocate the bits for different picture types. As shown in Equation (3.7), the target bit rate for each picture type is computed as a weighted sum of the available bandwidth

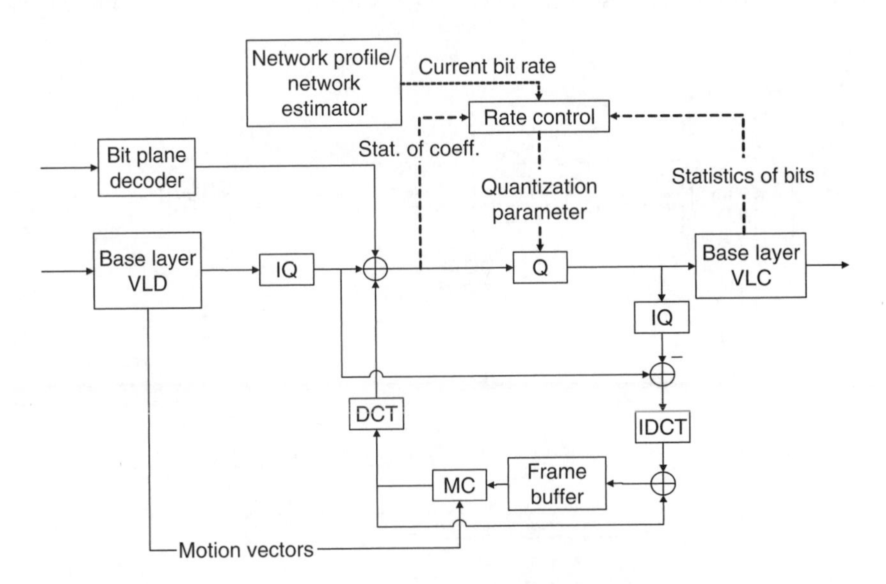

FIGURE 3.20 Architecture 2 of an FGS-to-SP transcoding [3-26].

TABLE 3.5
Modules Used in the Three Architectures [3-26]

Architecture	1	2	3
Bit Plane VLD	1	1	1
Base Layer VLD	1	1	1
Quantizer/Inverse Q.	3	3	3
Motion Compensation	2	1	1
Frame Buffer	2	1	1
DCT/IDCT	4	2	0
Base Layer VLC	1	1	1

$R(t)$. The weightings for different types of VOPs are found empirically. The values are set as $W_I = 1$, $W_p = 1$, and $W_B = 0.6$ for I-pictures, P-pictures, and B-pictures, respectively.

$$Rx = \frac{R(t)}{W_I N_I + W_P N_P + W_B N_B} \times W_x, \text{ where } N_x \text{ is the number of } x\text{-picture}, \tag{3.7}$$

The rate control as illustrated in Figure 3.18 adjusts the quantization parameter in the quantizer (Q) and monitors the output bits. In our proposed transcoder module, two rate control methods, including the TM5 and ρ-domain source model rate control, are realized.

The rate control module has four input parameters: the current bit rate, the statistics of coefficients, the quantization parameter, and the statistics of output bits. In the transcoding process, the rate control module fetches the current bit rate and

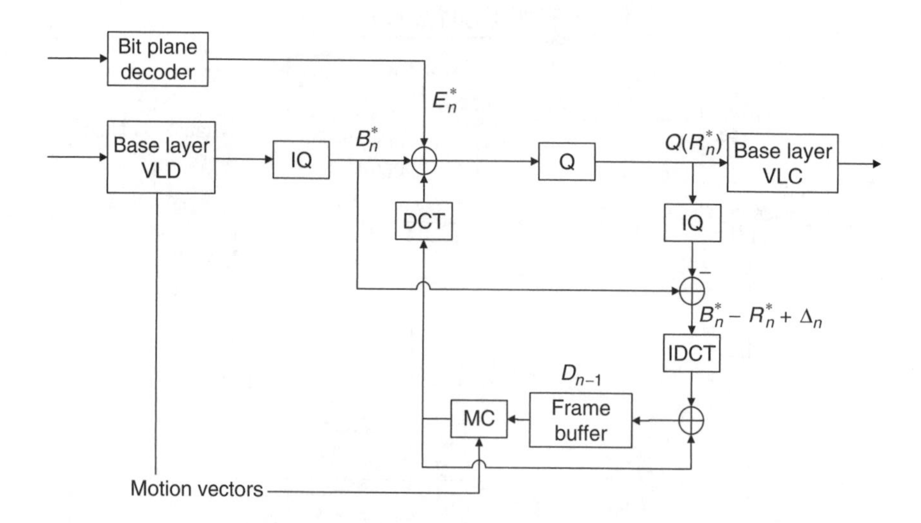

FIGURE 3.21 Architecture 3 of an FGS-to-SP transcoding [3-26].

performs bit allocation according to the current picture type. Subsequently, the rate control module starts to assign the quantization parameters macroblock by macroblock with the feedback information from the buffer fullness. The buffer fullness is computed based on the statistics of output bits and the current channel rate. For ρ-domain source model rate control, the statistics of coefficients are required to assign the quantization parameters for rate-distortion optimization.

3.4.3 Experimental Results

We will perform both the static and dynamic tests. The static test evaluates only the coding efficiency. The dynamic test of the transcoder is test based on the modified multimedia test bed. In the dynamic test, the transcoder performs the rate shaping in real time to fit the channel conditions. The network attributes and the received video quality are monitored and logged. The network attributes include the packet loss ratio, the bandwidth usage, the buffer fullness, etc.

3.4.3.1 Static Test Without Rate Control

Three experiments are performed to test the coding efficiency of the three architectures as listed in Table 3.6. The experiments A should have the best performance because they use the original sequence as source and perform the motion estimation for the specific bit rate. The experiments B should have the next best performance since they are a cascade of a decoder and encoder. The only cause of degradation may come from less-accurate motion vectors. The B experiments are for the transcoder with the second architecture that has lowest complexity. As one can observe in Figure 3.22, the loss in Peak Signal-to-Noise Ratio is quite small at low rates and slightly higher at high rates. This may be caused by reusing the mode selection at the higher rates, as discussed in Section 3.2. For subjective verification, the snapshots of the three experiments are shown in Figure 3.23 for the high and low bit rates. The transcoders of the first two architectures result in similar performance in visual quality.

3.4.3.2 Static Test with Rate Control

The previous test focuses on the coding efficiency so the rate control is not activated for comparison. However, the comparison may not be practical because a real-time encoder uses rate control for adapting to the channel bandwidth. Thus, we will

TABLE 3.6
Transcoding Experiments

	Experiments
A	Compressed at the same bit rate using the original sequence
B	Architecture 1
C	Architecture 2

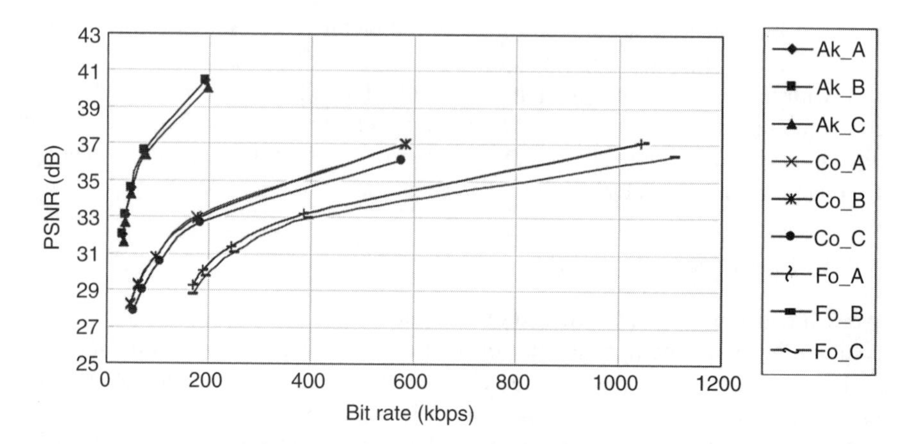

FIGURE 3.22 PSNR results for the experiments described in Table 3.6 [3-26] (A) Snapshots of encoded and transcoded sequences at high bit rate. (B) Snapshots of encoded and transcoded sequences at low bit rate.

compare the transcoder with rate control. Here we use two different rate control schemes as discussed earlier.

A network profile that describes the time-varying channel conditions is loaded to the rate control modules within the transcoder. In the network profile, the bit rates are changed every 2 s. After transcoding, the plots of the physical bit rate and the received quality are given for advanced analysis. As shown in Figure 3.24 and Figure 3.25, the target bit rates and PSNRs on a frame basis for the transcoder are

(A) Snap shots of encoded and transcoded sequences at high bit rate.

(B) Snap shots of encoded and transcoded sequences at low bit rate.

FIGURE 3.23 Snap shots of encoded and transcoded video sequences for experiments A, B, and C in the order from left to right [3-26].

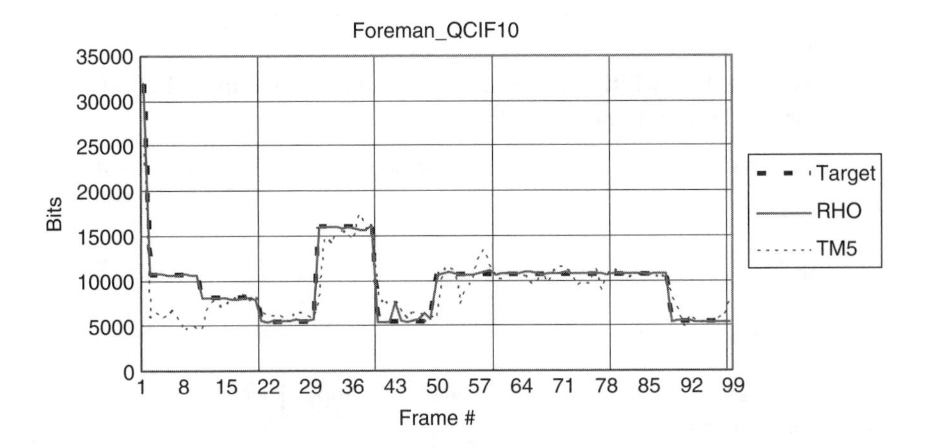

FIGURE 3.24 Target bit rates for transcoder using two rate control schemes [3-26].

tested for the specified network profile. The results show that ρ-domain source model rate control can provide better rate adaptation compared to TM5 rate control.

3.4.3.3 Dynamic Test Using the MPEG-21 Multimedia Test Bed

Now we will test the transcoder over the MPEG-21 multimedia test bed to simulate the real-time behavior of the transcoder with second architecture [3-27]. In the multimedia test bed, the server sends archived bit streams in real time as media packets to the client. While streaming FGS bit stream, the server truncates the FGS enhancement layer bit stream based on the rate control. For the MPEG-4 SP client, the server transcodes in real time using the rate control module to match the network conditions. The network emulator intercepts the packets between the client and the

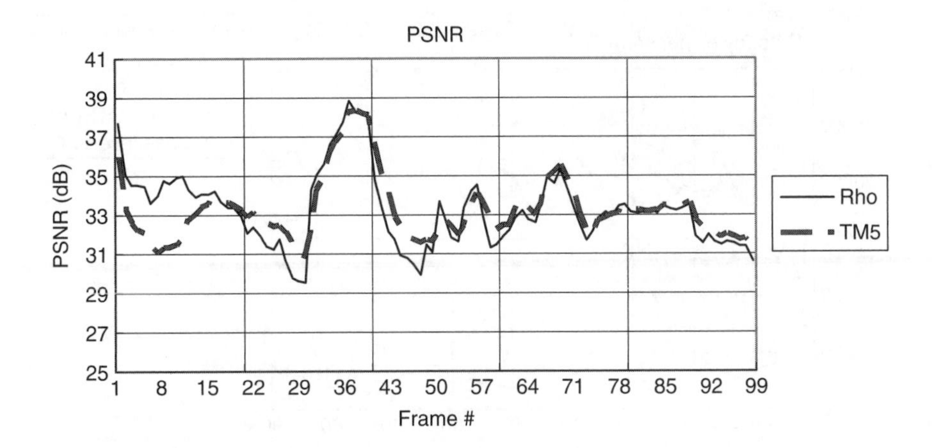

FIGURE 3.25 PSNR plots for transcoder using two rate control schemes [3-26].

TABLE 3.7
Average Bandwidth Utilization for Different Streaming Methods for 3G-1x EVDO Network Profiles

Experiment	Bandwidth Utilization for 3G-1x EVDO
Pseudo R-T Enc.	97.05%
Transcoding	91.65%
FGS	86.47%
FGS Base Layer	12.23%

server. The intercepted packets are rescheduled or dropped according to the network profile and transmission status.

As shown in Table 3.7, we compare four types of streaming techniques including transcoders of the second architecture.

The first experiment is an MPEG-4 SP encoding from the original sequence for the given network profile using the ρ-domain source rate control. This is assuming a *priori* knowledge of the network profile so that the performance is considered as the upper bound for the transcoder. The second experiment is the transcoding output based on the second architecture. The third and fourth experiments are the FGS bit stream and FGS base layer bit stream. These two results are used as the lower bounds for performance.

The dynamic network profile changes the channel rate every 5 s to simulate 3G 1x-EVDO traffic [3-28]. As shown in Table 3.7, the average bandwidth utilization percentages of the four methods for the 3G-1x EVDO network profile and the dynamic bandwidth utilization are shown in Figure 3.26. The delivered video qualities of the four strategies are shown n Figure 3.27. The overall average PSNR is tabulated in Table 3.8. Based on the comparisons, the transcoding approach has the

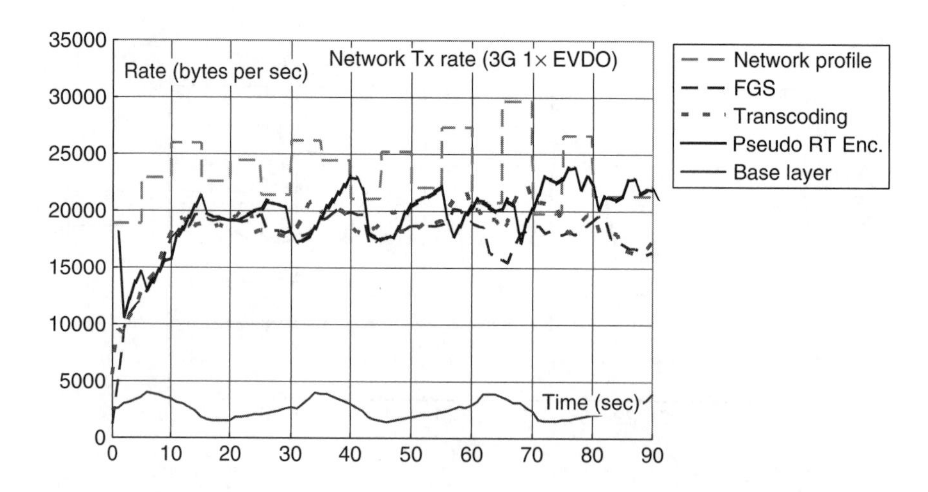

FIGURE 3.26 Bandwidth utilization comparison of various coding schemes. [3-26].

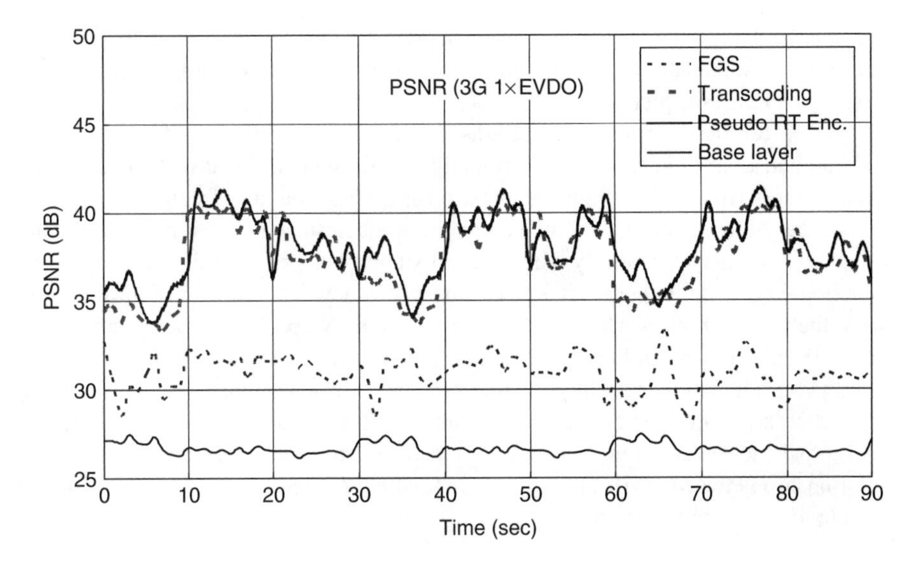

FIGURE 3.27 The PSNR (averaged over 1 s) of the four experiments [3-26].

performance close to that by the pseudo-real-time encoding, with about 0.3 to 0.5 dB of PSNR loss.

From the results of dynamic testing, the transcoder of the second architecture can provide good channel utilization and better video quality simultaneously. The FGS scalable approach provides good channel utilization and improves video quality as compared to that of the nonscalable approach. Since the FGS, transcoding, and pseudo-real-time methods have bandwidth adaptation, they can provide high bandwidth utilization. Due to its lack of adaptation, the nonscalable approach has the lowest quality and lowest bandwidth utilization.

3.5 SUMMARY

In this chapter, we discussed various aspects of transcoding and its system architecture. The first section discussed how the transcoder is used for various applications and scenarios, including compatible DTV decoding; multiformat compatible

TABLE 3.8
Averaged PSNR Values of the Four Methods for the 3G-1x EVDO Network Profile

Experiment	Y	U	V
Pseudo RT Enc.	37.99	41.95	42.16
Transcoding	37.41	40.92	41.08
FGS	30.92	36.94	36.97
FGS Base Layer	26.64	34.07	34.17

receiver and recorder; broadcast transcoder; statistical multiplexing; multimedia server using metadata, universal multimedia access; and watermarking, logo insertion, and picture-in-picture.

Based on the described applications, in the second section we focused on the bit rate and quality adaptation functionality of transcoder. We investigated several architectures from low to high complexity for both spatial and frequency domains.

In the third section, we studied another approach to achieve bit rate and quality adaptation, namely, scalability. The scalability approach provides simple transcoding functionality at the expense of low coding efficiency. Therefore, it is desirable to have the best of both worlds. In the fourth section, we provide a case study to use scalable solution for archiving the source material and a transcoder to achieve the bandwidth adaptation with high coding efficiency. The resultant system is demonstrated with a static test for coding efficiency and a dynamic test for capability to make real-time adaptation. The basic concept of universal multimedia access and relationship between UMA and transcoding, which is used as strong tool for content adaptation, will be revisited in Chapter 10.

3.6 EXERCISES

3-1. Explain why there is a drift for the frequency domain transcoder.

3-2. Explain why there is no drift for the spatial domain cascaded transcoder.

3-3. Explain why there is a need for motion reestimation in a transcoder when spatial resolution is reduced.

3-4. Explain why there is a need to make new mode decisions for transcoding. In particular, explain the mode selection process for the low bit rate and high bit rate.

3-5 Explain the major differences between spatial and frequency domain transcoders.

REFERENCES

[3-1] J. Xin, M.T. Sun, and K. S. Kan, Bit allocation for joint transcoding of multiple MPEG coded video streams, in *Proceedings of the IEEE International Conference on Multimedia and Expo,* pp. 8–11, 22–25, Aug. 2001.

[3-2] R. Mohan, J. R. Smith, and C.-S. Li, Adapting multimedia internet content for universal access, *IEEE Transactions on Multimedia,* Vol. 1, Issue 1, pp. 104–114, March 1999.

[3-3] A. Vetro, H. Sun, and Y. Wang, Object-based transcoding for adaptable video content delivery, *IEEE Transactions on Circuits and Systems for Video Technology,* Vol. 11, Issue 3, pp. 387–401, March 2001.

[3-4] F. Hartung, and B. Girod, Watermarking of uncompressed and compressed video, *IEEE Transactions on Signal Processing,* Vol. 66, No. 3, pp. 283–301, 1998.

[3-5] A. Vetro, H. Sun, and A. Divakaran, Adaptive object-based transcoding using shape and motion-based hints, ISO/IEC M6088, Geneva, 2000.

[3-6] J. Meng and S. F. Chang, Embedding visible video watermarks in the compressed domain, *Proceedings of the IEEE International Conference on Image Processing,* Vol. 1, pp. 474–477, Oct. 1998.

[3-7] J. Youn, J. Xin, and M.-T. Sun, Fast video transcoding architectures for networked multimedia applications, *Proceedings of the IEEE International Symposium on Circuits and Systems,* Vol. 4, pp. 25–28, 28–31 May 2000.

[3-8] H. Sun, W. Kwok, and J. Zdepski, Architectures for MPEG compressed bit stream scaling, *IEEE Transactions on Circuits and Systems for Video Technology,* Vol. 6, pp. 191–199, April 1996.

[3-9] J. Youn and M.-T. Sun, A fast motion vector composition method for temporal transcoding, *IEEE International Symposium on Circuits and Systems,* Vol. 4, pp. 243–246, June 1999.

[3-10] B. Shen, I. K. Sethi, and B. Vasudev, Adaptive motion-vector resampling for compressed video downscaling, *IEEE Transactions on Circuits and Systems for Video Technology,* Vol. 9, Issue 6, pp. 929–936, Sept. 1999.

[3-11] P. A. A. Assuncao and M. Ghanbari, A frequency-domain video transcoder for dynamic bit-rate reduction of MPEG-2 bit streams, *IEEE Transactions on Circuits and Systems for Video Technology,* Vol. 8, Issue 8, pp. 953–967, Dec. 1998.

[3-12] N. Bjork and C. Christopoulos, Transcoder architecture for video coding, *IEEE Transactions on Consumer Electronics,* Vol. 44, Issue 1, pp. 88–98, Feb. 1998.

[3-13] P. A. A. Assuncao and M. Ghanbari, Transcoding of single-layer MPEG video into lower rates, *IEEE Proceedings—Vision, Image and Signal Processin,* Vol. 144, Issue 6, pp. 377–383, Dec. 1997.

[3-14] W. Li, Overview of Fine Granularity Scalability in MPEG-4 Video Standard, *IEEE Transactions on Circuits and Systems for Video Technology,* Vol. 11, Issue 3, pp. 301–317, March 2001.

[3-15] F. Wu, S. Li, and Y.-Q. Zhang, A Framework for Efficient Progressive Fine Granularity Scalable Video Coding, *IEEE Transactions on Circuits and Systems for Video Technology,* Vol. 11, Issue 3, pp. 332–344, March 2001.

[3-16] H.-C. Huang, C.-N. Wang, and T. Chiang, A Robust Fine Granularity Scalability Using Trellis-Based Predictive Leak, *IEEE Transactions on Circuits and Systems for Video Technology,* Vol. 12, Issue 6, pp. 372–385, June 2002.

[3-17] S.-F. Chang and D. G. Messerschmitt, Manipulation and Compositing of MC-DCT Compressed Video, *IEEE Journal on Selected Areas in Communications,* Vol. 13, Issue 1, pp. 1–11, Jan 1995.

[3-18] Z. He and S. K. Mitra, A linear source model and a unified rate control algorithm for DCT video coding, *IEEE Transactions on Circuits and Systems for Video Technology,* Vol. 12, Issue 11, pp. 970–982, Nov. 2002.

[3-19] W. Tan and A. Zakhor, Real-Time Internet video using error resilient scalable compression and TCP-friendly transport protocol, *IEEE Transaction on Multimedia,* Vol. 1, Issue 2, pp. 172–186, June 1999.

[3-20] H. Kasai, T. Hanamura, W. Kamayama, and H. Tominaga, Rate control scheme for low-delay MPEG-2 video transcoder, in *Proceedings of the IEEE International Conference on Image Processing,* Vol. 1, pp. 10–13, Sept. 2000.

[3-21] Y. Nakajima, H. Hori, and T. Kanoh, Rate conversion of MPEG coded video by re-quantization process, *Proceedings of the IEEE International Conference on Image Processing,* Vol. 3, pp. 408–411, 23–26 Oct. 1995.

[3-22] O. Werner, Requantization for transcoding of MPEG-2 Intraframes, *IEEE Transactions on Image Processing,* Vol. 8, Issue 2, pp. 179–191, Feb. 1999.

[3-23] A. Eleftheriadis and D. Anastassiou, Constrained and general dynamic rate shaping of compressed digital video, *Proceedings of the IEEE International Conference on Image Processing,* Vol. 3, pp. 396–399, 23–26 Oct. 1995.

[3-24] MPEG Video Group, Information technology — coding of audio-visual objects — Part 2: Visual ISO/IEC 14496-2: 2001, International Standard, ISO/IEC JTC1/SC 29/WG 11 N4350, July 2001.

[3-25] M. Schaar and H. Radha, Motion-compensation based fine-granular scalability (MC-FGS), ISO/IEC JTC1/SC29/WG11, MPEG00/M6475, Oct. 2000.

[3-26] Y.-C. Lin, C.-N. Wang, T. Chiang, A. Vetro, and H. Sun, Efficient FGS-to-single layer transcoding, in *Proceedings of the IEEE International Conference on Consumer Electronics,* pp. 134–135, 18–20 June 2002.

[3-27] C.-J. Tsai, M. Shaar, Y.-K. Lim, Working draft 2.0 of ISO/IEC TR21000-12 multimedia test bed for resource delivery, ISO/IEC JTC1/SC29/WG11, N5640

[3-28] TIA/EIA/IS-856 CDMA2000, High Rate Packet Data Air Interface Specification.

[3-29] J. Xin, C.-W. Lin, and M.-T. Sun, Digital Video Transcoding, *IEEE Proceedings, preprint manuscript,* to appear in 2005.

4 Topics on the Optimization of Transcoding Performance

In this chapter, we will focus on the optimization of video transcoder performance. The performance of a video transcoder can be evaluated from several aspects, such as implementation complexity, dynamic range of bit rate, subjective visual quality, types of formats it can support, and the quality of service. An ideal video transcoder should have very low complexity, wide dynamic bit rate range, and superior visual quality. Such a transcoder should support multiple formats seamlessly. To optimize the performance of transcoder, there is a vast literature to address these issues. We will review recent important results categorically based on the functions of the transcoder.

The transcoder contains functionalities such as multistandards and multiresolution support. In an end-to-end streaming system, there are other needs to optimize the cost-effectiveness and performance for a given feature, such as the implementation complexity and dynamic range of the bit rates. Such optimization will be covered in this chapter.

4.1 INTRODUCTION

In Section 4.2, we will study transcoders that can support various spatial resolutions. There is a need to support different spatial resolutions for various devices, where one spatial resolution bit stream may be converted to a different spatial resolution bit stream. This is particular useful when the target receiver has limited capability for display and decoding, for example, as with SDTV-compatible decoding from a HDTV bit stream.

We will study techniques to reduce the implementation complexity. Typical modules that will be optimized include down-conversion, motion vector remapping, motion vector refinement, and requantization of the transform coefficients. Integration of redundant modules is used to reduce complexity while maintaining similar visual quality. We will introduce eight architectures that solve the problems in both spatial and temporal domains. The issue of visual quality will be addressed. Various techniques to preserve visual quality for given constraints are studied. For example, drift corrections are widely studied for motion-compensated predictive coding, such as MPEG-1/2/4 and ITU H.261/263/H.264.

In Section 4.3, we will discuss various approaches for temporal adaptation, where the temporal frame rate is reduced or increased to fit the target display. For example, cellular phone handsets have limited temporal resolution. To save bandwidth, the transcoder can extract only partial bit streams or transcode them into lower temporal bit streams. Temporal adaptation involves techniques that provide motion vector remapping and refinement. The requantization of coefficients are also discussed.

In Section 4.4, we will review techniques to provide a wide variety of supports of standards and syntax, wherein compressed bit streams are converted from one coding standard (e.g., MPEG-1) to another (e.g., JPEG). This is typical for desktop-based software and communications systems where one user may have different standards support for similar content. We will review several examples of syntax conversion, including JPEG/MPEG-2 to MPEG-1/DV transcoding, MPEG-2 to MPEG-4 transcoding, and MPEG-4 FGS transcoding.

In Section 4.5, we will focus on various types of networks designed to optimize the quality of service, such as delay, scalability, and user experience. In particular, error-resilient transcoding for GPRS wireless networks and the Internet are studied. The application of logo insertion and watermarking are reviewed in Section 4.6 Logo Insertion and Watermarking. We discuss both spatial and frequency domain architectures for embedding information into a bit stream.

In Section 4.7, we discussed the concept of considering transcoding to be a second-pass encoding, and thus there is a possibility of achieving higher quality through transcoding from a high-quality source bit stream. In Section 4.8 Switched Picture, we will review the results using a switch picture for the H.264/AVC standard. This is an alternative for the server to achieve the same functionality as the transcoder. Lastly, we introduce an example that achieves H.264/AVC picture-in-picture functionality using the techniques reviewed in the previous sections. Lastly, we give a summary of this chapter in Section 4.11 Summary, comparing the advantages and disadvantages of various techniques.

4.2 REDUCED SPATIAL RESOLUTION TRANSCODER

There are several application scenarios that would require spatial format conversion, for example, multimedia servers that provide video streaming over Internet or wireless links for clients with different display capability. In the case of video browsing or streaming with picture-in-picture or logo insertion features, the server needs to down-convert the spatial resolution on the fly and to transcode multiple bit streams to single stream for transmission. A straightforward technique is the conventional cascaded transcoder requiring significant complexity. Thus, there are various techniques to address this functionality in the compressed domain with reduced complexity.

In a simplified case, the transcoder needs to down-scale from video of spatial resolution of size $N \times N$ to size of $(N/2) \times (N/2)$. The cascaded transcoder approach has several steps: 1) decompression of the source bit stream, 2) down-sampling in the spatial domain, 3) reestimation of motion vectors, and 4) compression into the target bit stream.

To achieve spatial down-scaling in the compressed domain for motion-compensated DCT video bit streams, such as MPEG-2 or H.261, there are two major technical challenges [4-2], as shown in Figure 4.1. The first issue is to compose the DCT block in the frequency domain that has four blocks of the same size but may contain different quantization parameters [4-3] [4-4]. The second issue is to provide down-scaled

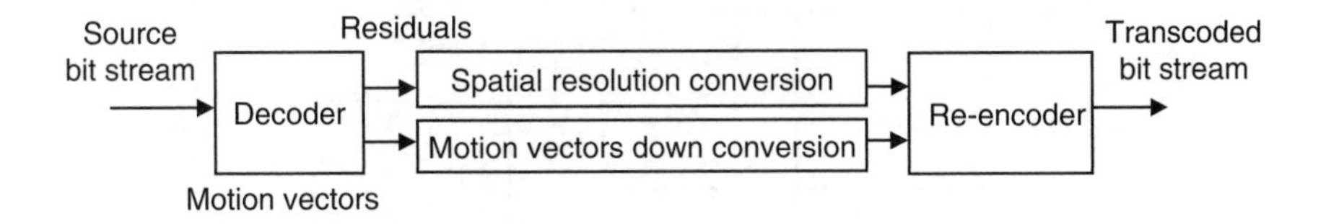

FIGURE 4.1 Spatial and temporal transcoding.

motion vector that can offer the best residual for the down-scaled resolution. Furthermore, the new motion vectors may change the residual information since the prediction loop has been modified with a new vector. The "drift" errors cause the video quality degrades gradually until the next intra-frame. We will review solutions for both issues in the following two subsections.

4.2.1 Spatial Down-Scaling in the Compressed Domain

Now we will address the first issue: spatial down-conversion in the frequency domain. The challenge to perform down-conversion or resizing in the compressed domain is to find efficient ways to compute the four IDCT and one DCT. In [4-23] Chang and Messerschmitt describe such a technique that performs the down-conversion in the DCT domain using only the low-frequency coefficients of each block and recompose the new block all in the DCT domain. For a down-conversion by a factor of 2, the four 4 × 4 DCT coefficients of each 8 × 8 block can be used to reconstruct the new block. The four 4 × 4 DCT coefficients are filtered by a set of frequency domain filters achieve the equivalent operations in the spatial domain.

Let $\mathbf{x}$ and $\mathbf{X}$ be 8 × 8 matrices representing blocks in the spatial and DCT domains. Define the DCT transformation matrix as

$$\mathbf{T}(m,n) = c(m)/2 \cos((2n + 1)\, m\pi/16)$$

where $c(0) = 1/\sqrt{2}$ and $c(m) = 1$ for $m > 0$. Thus, the DCT transformation can be represented as $\mathbf{X} = \mathrm{T}\mathbf{x}\mathrm{T}^t$ and the superscript t denotes the matrix transposition. We also have $\mathbf{x} = \mathrm{T}^t\mathbf{X}\mathrm{T}$ using the unitary transformation property.

In the case of spatial down conversion without motion compensation or zero motion vector, we are given four 8 × 8 blocks $\mathbf{x}_1, \ldots, \mathbf{x}_4$. The down-conversion can be done by averaging the four pixels in the spatial domain and we need to identify the equivalent operations in the DCT domain. As shown in Figure 4.2, the goal is to compute $\mathbf{X}_c$, the DCT of $\mathbf{x}_c$, directly from the $\mathbf{X}_1, \ldots, \mathbf{X}_4$, the DCT of $\mathbf{x}_1, \ldots, \mathbf{x}_4$.

In the case of spatial down-conversion with motion compensation vector, we are given the motion vectors and the four 8 × 8 blocks, $\mathbf{X}_1, \ldots, \mathbf{X}_4$, the DCT of $\mathbf{x}_1, \ldots, \mathbf{x}_4$. As shown Figure 4.3, we are also given $\mathbf{D}$, the DCT of the residual block $\mathbf{d}$, where $\mathbf{d} = \mathbf{x}_c - \mathbf{x}_p$ and the block $\mathbf{x}_p$ is the predicted block for the current block $\mathbf{x}_c$. Using the linearity of the transform, we have $\mathbf{X}_c = \mathbf{X}_p + \mathbf{D}$, where $\mathbf{X}_c$ and $\mathbf{X}_p$ are the DCT blocks of the current block $\mathbf{x}_c$ and the predicted block $\mathbf{x}_p$, respectively. Chang

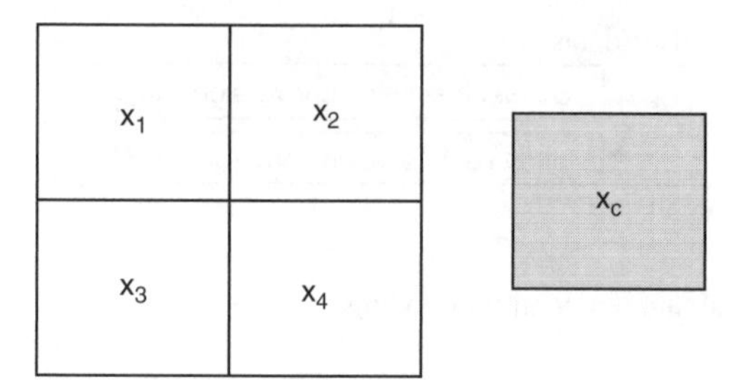

FIGURE 4.2 DCT domain down-conversion [4-23].

and Messerschmitt [4-23] show that $\mathbf{x}_p$ can be expressed as a linear combination of $\mathbf{x}_1, \ldots, \mathbf{x}_4$, with appropriate windowing as shown in the following equation

$$x_p = \sum_{i=1}^{4} c_{i1} x_i c_{i2}$$

where $\mathbf{c}_{ij}$ are sparse 8×8 matrices of zeros and ones and $i = 1, \ldots, 4$ and $j = 1, 2$. Use the fact that $T^t T = I$, we have

$$x_p = \sum_{i=1}^{4} c_{i1} T^t T x_i T^t T c_{i2}$$

By premultiplying and postmultiplying both sides by T and T^t, we have

$$X_p = \sum_{i=1}^{4} C_{i1} X_i C_{i2}$$

where $\mathbf{C}_{ij}$ are the DCT of the blocks $\mathbf{c}_{ij}$. Chang et al. [4-23] propose to precompute the $\mathbf{C}_{ij}$ for possible combination of h and w, as shown in Figure 4.3. Since the DCT blocks of the source content are mostly sparse matrix, Chang et al. can develop a fast algorithm based on the sparseness of the block. In [4-3], a more efficient algorithm that is not based on the sparseness constraint is developed. The algorithm

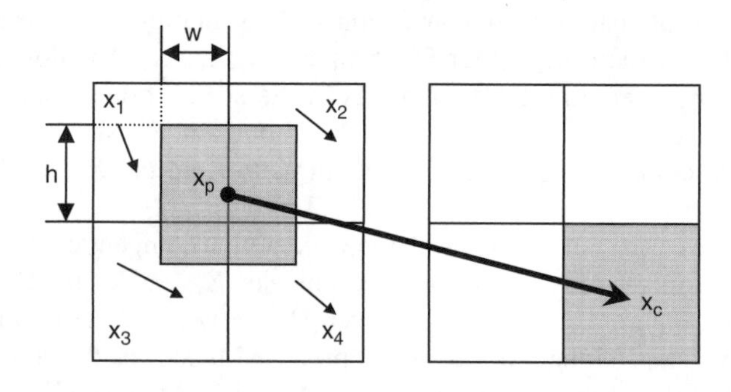

FIGURE 4.3 DCT-domain motion compensation [4-23].

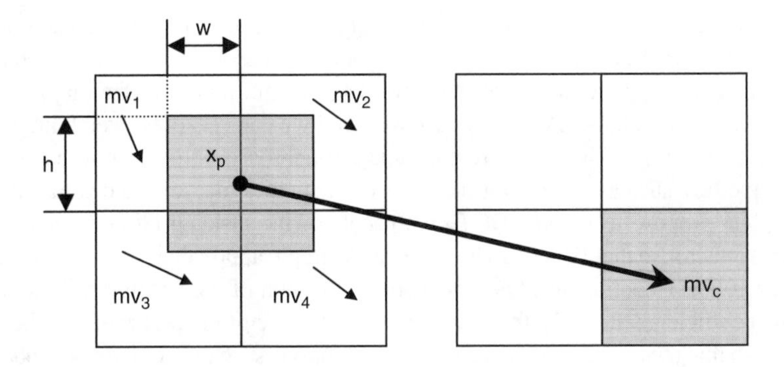

FIGURE 4.4 Motion vector downscaling.

is developed based on the matrix factorization and a fast eight-point Winograd DCT/IDCT. Merhav and Bhaskaran can achieve computation saving from 37 to 50%.

4.2.2 Motion Vector Aadaptation

After the spatial down-conversion in the frequency domain is resolved, we will now address the second issue of down-scaling of the motion vectors.

As shown in Figure 4.4, the transcoder needs to down-scale the motion vector by half based on the four adjacent motion vectors. There are several ways to compute the new motion vector.

The first method is to compute the average of the four motion vectors and halve the result, which is referred to as align-to-average (AAW) in [4-2]. This technique works well when the motion vector field is homogeneous but can result in bad results when the macroblock contains two objects moving in different directions.

The second method is to select the one motion vector out of the four candidates as the new vector based on the prediction errors. One could select the motion vector with minimal or maximal prediction errors, which are referred to as either the align-to-best (ABW) approach or the align-to-worst (AWW) approach [4-2]. It turns out that we should prefer the one with maximal prediction errors (AWW) because it dominates the overall prediction errors and should be close to the optimal case at the reduced resolution. However, one can find the situation that the object boundary is perfectly predicted and the ABW should be used. All these scenarios are caused by the imperfection of the block-based motion estimation.

The second scheme is a weighted average of the four corresponding vectors that can be used as the new vector [4-2], and is referred to as the *adaptive motion vector resampling* (AMVR) approach with the following formula.

$$mv_{\text{new}} = \frac{1}{2}\left(\frac{\sum_{i=1}^{4} w_i mv_i}{\sum_{i=1}^{4} w_i}\right)$$

There are several ways to compute the weightings. For example, one can select the one with maximal or minimal prediction errors. The rationale for using the maximal prediction errors is to assume that the new vector should minimize the majority of the errors. This will make AMVR close to AWW and we can make the weighting close to ABW for uniform motion. A more sophisticated approach is to adapt w_i according to certain metrics, such as the reconstruction errors, activities measure, and variances. Shen et al. [4-2] propose to use the DCT coefficients of the residual block to compute the weighting factor so that IDCT can be avoided. A simple approach is to count the number of nonzero DCT coefficients. One can also use the sum of the absolute values of all of the AC coefficients. In [4-2], the downscaling of the picture is performed in the spatial domain so that the transcoder has savings in the motion estimation of the re-encoder only.

In [4-32], Wong and Au developed a predictive motion estimation (PME) to perform the motion vector adaptation. The algorithm uses the four motion vectors as candidates and examines the homogeneity of the motion vector field. The results will be used to guide the motion vector refinement process.

4.2.3 Spatial Domain Reduced Spatial Resolution Architectures

Now that we have understood the issues for each module, we can focus on the overall architectures that can achieve reduced spatial resolution transcoding. In [4-11] and [4-26], Yin et al. proposed several architectures that can achieve such transcoding in the both the spatial and frequency domains. We will refer to these architectures as *reduced spatial resolution architectures* (RSRA). There are also other reduced spatial resolution transcoders, as proposed in [4-25].

4.2.3.1 Reduced Spatial Resolution Architecture 1

As shown in Figure 4.5, this is a cascaded transcoder with motion vector reuse and down-converter in the spatial domain between the decoder and re-encoder. It can use all of the techniques we mentioned in Subsection 4.2.2. The only complexity savings with RSRA-1 as compared to the cascaded transcoder is the motion estimation in the re-encoder, and the benefit is that there is no drift and the resolution is arbitrary. However, the performance should be slightly below the cascaded transcoder unless motion is reestimated.

4.2.3.2 Reduced Spatial Resolution Architecture 2

As shown in Figure 4.6, this is a cascaded transcoder with only motion vector reuse and down-conversion in the frequency domain. This architecture, which is referred to as RSRA-2, has minimal complexity since all the processing is done in the DCT domain, including the down-conversion, and there is no need for frame memory.

With careful analysis of the drift RSRA-2, it was found that drift for reduced spatial resolution transcoding comes from two sources. The first source of drift errors comes from requantization and arithmetic errors caused by integer truncation. This first source is common from the transcoder for the same spatial resolution. The second source of drift errors comes from the noncommutative property of the motion

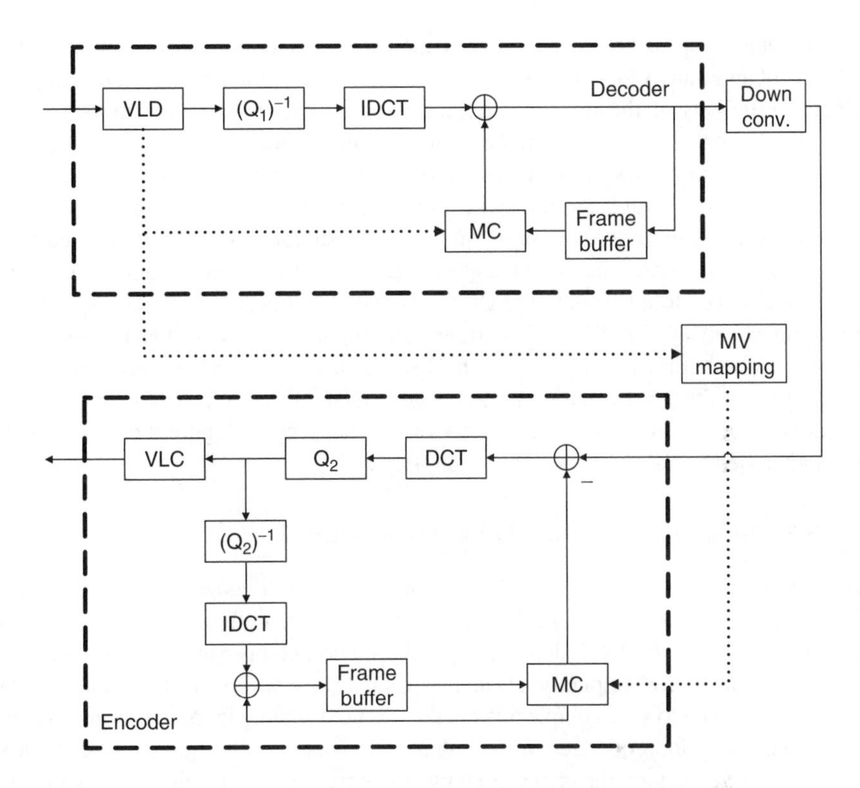

FIGURE 4.5 Reduced spatial resolution architecture 1 (RSRA-1) [6-24].

compensation and down-conversion. This is uniquely present for reduced resolution transcoders. The two modules that contribute to this source are motion vector adaptation and down-conversion, in which both modules operate on a block basis, so it poses additional constraints since processing cannot be overlapped between blocks.

As shown in Figure 4.6, there is a need for what Yin et al. referred to as *mixed block processing* as one can observe from Figure 4.2. In addition to down-conversion of the pixel information, the coding mode for a macroblock needs to be determined from four macroblocks at the original resolution. Thus, the transcoder needs to choose the appropriate mode based on the four given modes because so far video standards cannot support mixed mode within the same macroblock. In particular, the transcoder needs to decide whether intra-mode or inter-mode should be selected.

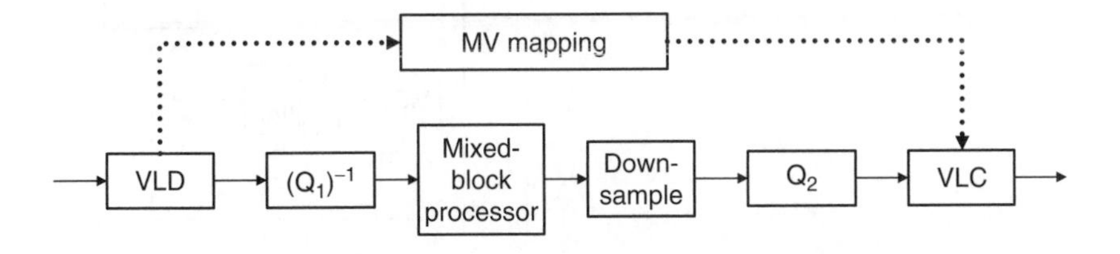

FIGURE 4.6 Reduced spatial resolution architecture 2 (RSRA-2) [6-24].

Yin et al. proposed three techniques which are ZeroOut, IntraInter, and InterIntra. The ZeroOut strategy selects the inter-mode and it resets the motion vectors and the DCT coefficients of the intra-macroblock to zero. This means that the transcoder will simply copy the pixels from the co-located macroblock of the reference frame. The IntraInter strategy selects the inter-mode and derives the motion vectors according a weighted average from the neighboring blocks. There are many ways to compute the weightings, e.g., one can use the residual to decide which vector is more reliable and deserves more weighting. Based on the derived motion vector, the transcoder needs to recompute the DCT residual. The InterIntra strategy selects the intra-mode and the DCT coefficients are recomputed based on the reconstructed picture. Note that for the second and third strategies, the decoding loop is necessary to reconstruct the full-resolution picture so that the conversion can be made between intra-mode and inter-mode. The motion vector mapping in Figure 4.6 can be done with approaches discussed in Subsection 4.2.2.

4.2.3.3 Reduced Spatial Resolution Architecture 3

As shown in Figure 4.7, this is a DCT-domain cascaded transcoder without the decoder loop. This architecture, which is referred to as RSRA-3, allows the re-encoder to perform the differencing in the DCT domain while the motion compensation is performed in the spatial domain for simplicity. Thus, the spatial down-conversion is performed in the frequency domain to save complexity, but there is no decoding loop, so drift may occur. The re-encoding loop operates in the reduced resolution so the frame requirement is less. In this architecture, the drift caused by requantization will be eliminated while the other drift caused by down-conversion is not completely removed.

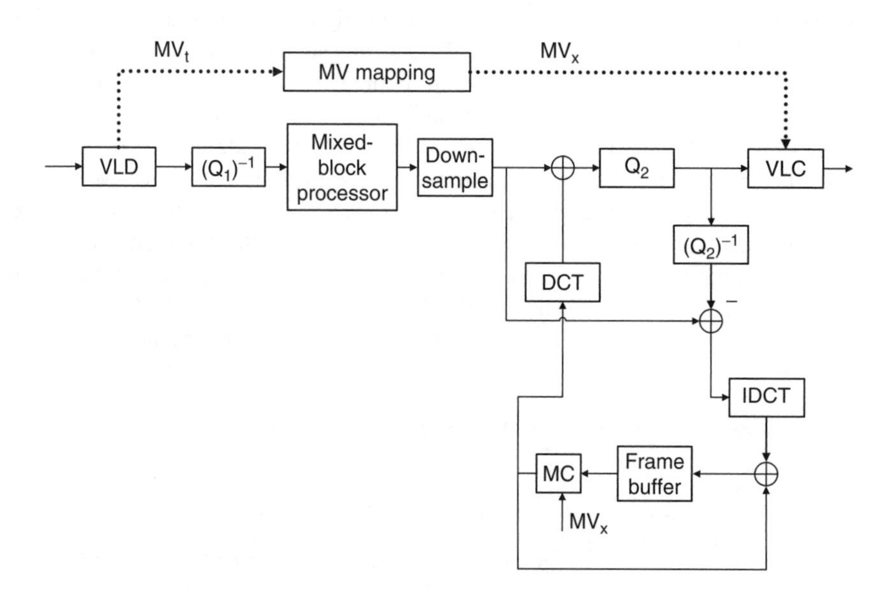

FIGURE 4.7 Reduced spatial resolution architecture 3 (RSRA-3) [6-24].

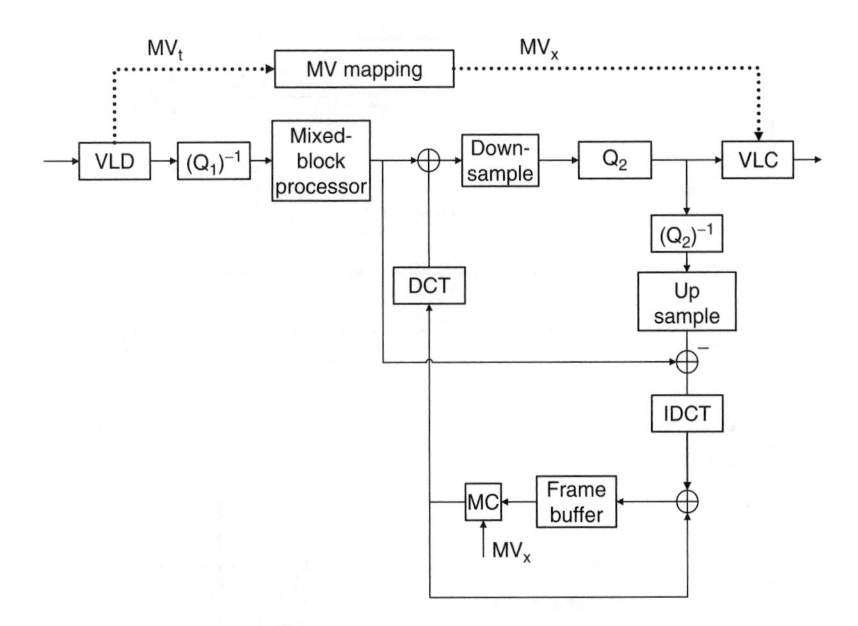

FIGURE 4.8 Reduced spatial resolution architecture 4 (RSRA-4) [6-24].

4.2.3.4 Reduced Spatial Resolution Architecture 4

As shown in Figure 4.8, this is a DCT-domain cascaded transcoder. This architecture, which is referred to as RSRA-4, allows the re-encoder to perform the down-conversion within the encoding loop. Thus, there is an additional need for up-conversion in the encoding loop. The differencing of the re-encoder is performed at the frequency domain so that down-conversion is simpler. It maintains spatial representation at the original resolution for the reference frame storage.

The RSRA-4 uses motion vectors and frame memory at the original resolution in order to reduce the drift with additional complexity. It aims to reduce both drifts caused by requantization and down-conversion. The up-conversion in the re-encoding loop is also performed in the DCT domain. Although it has been optimized in complexity, it is still a new module for the transcoder.

4.2.3.5 Reduced Spatial Resolution Architecture 5

As shown in Figure 4.9, this is a spatial domain cascaded transcoder. This architecture, which is referred to as RSRA-5, allows the frame memory of the re-encoder to be updated from the frame memory of the front decoder instead of the reconstructed picture using the re-encoding loop. This is referred to as a *partial encoder* since the motion estimation and reconstruction loop are both eliminated. There is an obvious benefit in complexity and drift reduction.

In addition to reduced complexity at the re-encoding loop, it does not have the problem of mixed block processing. For the two aforementioned architectures (RSRA-3 and RSRA-4), if the mixed mode processing uses either the InterIntra or

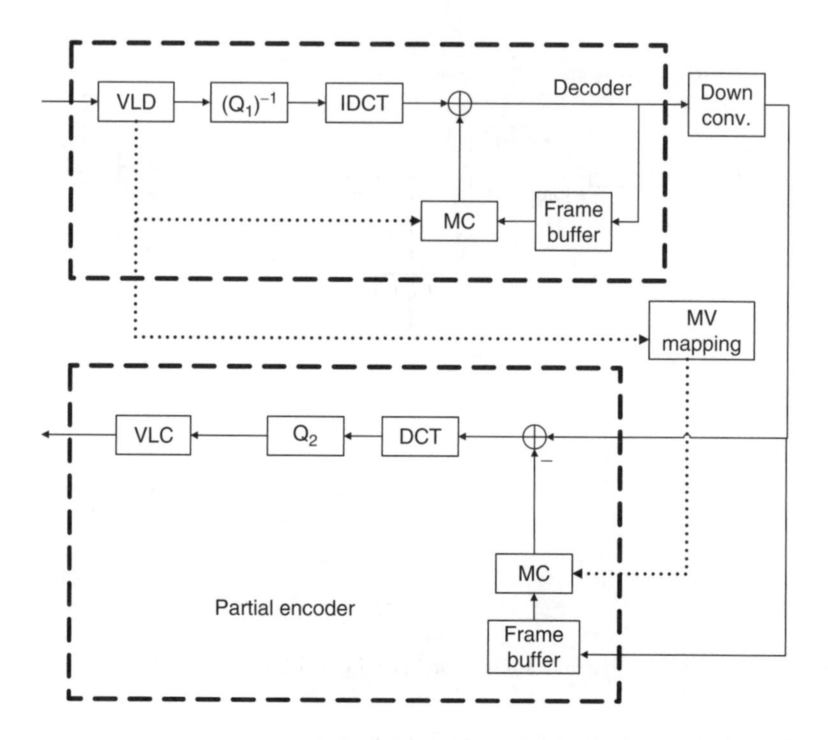

FIGURE 4.9 Reduced spatial resolution architecture 5 (RSRA-5) [6-24].

IntraInter strategy, the transcoder needs to perform the reconstruction of the picture at the original resolution in the front decoder. Thus, there is additional complexity savings when mixed block is not an issue. However, the drift still occurs since the reference frame is not synchronized.

4.2.3.6 Reduced Spatial Resolution Architecture 6

As shown in Figure 4.9, this is a frequency domain cascaded transcoder. This architecture, which is referred to as RSRA-6, eliminates the re-encoding loop completely but retains the front decoder. As shown in Figure 4.10, the front decoder is preserved to perform an intra-refresh by switching the macroblock to intra-mode systematically to control the error drift. The rate control module needs to estimate the drift based on motion vector truncation errors, residual energy, motion activity, and requantization error. After the estimate, it derives an intra-refresh rate for the transcoder to perform the inter-to-intra conversion. This technique comes with a penalty in coding efficiency.

4.2.3.7 Complexity and Performance Analysis

For RSRA-1 through -6, the complexity is compared in Table 4.1. Obviously, there is significant complexity from the third to sixth architecture. However, there is a need to verify the penalty in coding efficiency as shown in Table 4.2.

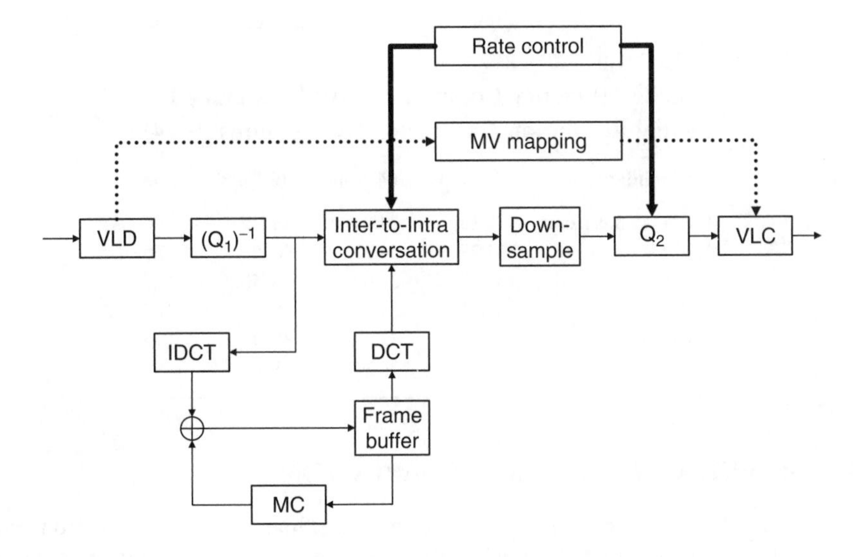

FIGURE 4.10 Reduced spatial resolution architecture 6 (RSRA-6) [6-24].

4.2.3.8 Frequency Domain Reduced Spatial Resolution Architectures

As shown in Figure 4.11 and Figure 4.12, Shanableh and Ghanbari proposed in [4-31] two other architectures that are referred to as RSRA-7 and RSRA-8. The down-conversion is performed in the DCT domain but the results of the down-converter can be in either the spatial or the frequency domain. The down-converter uses matrix-based filters in the frequency domain for optimization in complexity reduction.

In summary, it is a cascaded transcoder with optimized down-converter. Another distinction from the previous six architectures is that the motion compensation is performed in the frequency domain for both architectures. Thus, we should classify these two architectures as spatial domain transcoders.

TABLE 4.1
Complexity Comparisons of the Reduced Spatial Resolution Transcoder Architectures [6-24]

Architecture (RSRA-x)	1	2	3	4	5	6
DCT/IDCT	4	0	4	4	2	2
MC	2	0	2	2	2	1
Frame memory	2	0	2	2	2	1
Up-sample		✓				
Requantization drift		✓			✓	✓
Down-conversion drift		✓	✓	✓		✓
Drift amount	Low	High	Med.	Med.	Low	Low

TABLE 4.2
Coding Efficiency Comparisons of the Reduced Spatial Resolution Transcoder Architectures [6-24]

Architecture	32 Kbps	64 Kbps	96 Kbps	Time
Cascaded transcoder	32.13	34.83	36.12	—
RSRA-1	31.77	34.78	36.26	17.6
RSRA-3	31.12	34.02	35.08	12.9
RSRA-4	31.52	34.73	36.09	21.7
RSRA-5	31.83	34.40	35.66	10.5
RSRA-6	31.30	34.31	35.34	5.2

4.3 TEMPORAL RESOLUTION ADAPTATION

The temporal adaptation transcoder performs the frame rate conversion from the original bit stream to a new bit stream with lower frame rate. For example, the source is encoded as a 60-Hz interlaced format while the target needs to be a progressive 30-Hz video. Another example could be that the source is 30 Hz but the target is 10 Hz. A straightforward approach is to use a cascade of a decoder and an encoder. However, various techniques can be used to reduce the complexity.

In the second example, the transcoder needs to reduce the frame rate by an integer factor of 3. For the cascaded transcoder, it need only retain one frame out of three, so it is natural to consider transcoding in the compressed domain. One simple approach is to compress the source format with a GOP structure that has periodic P-frames with a temporal distance of three. To extract lower frame rate, the transcoder only needs to examine the header information and extracts only the P-frame bit streams. The transcoder is simply implemented with a bit stream parser. Care

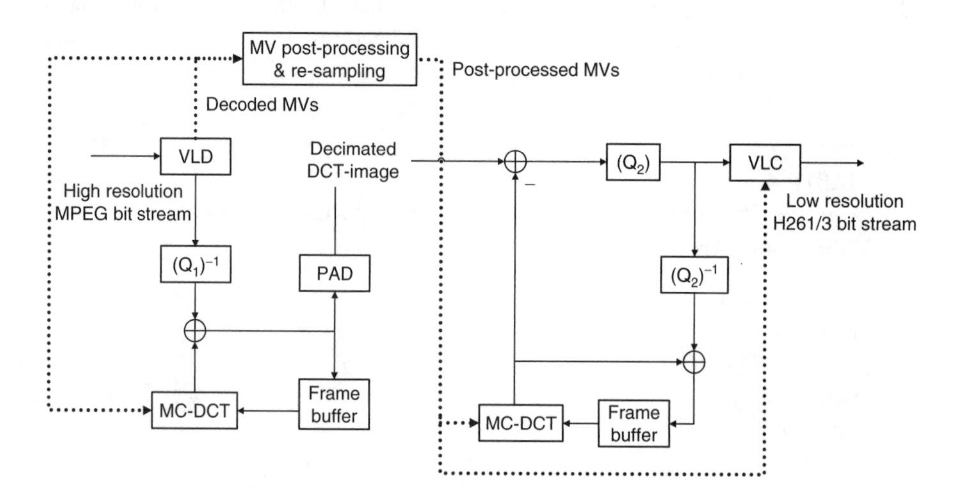

FIGURE 4.11 Reduced spatial resolution architecture 7 (RSRA-7) [6-31].

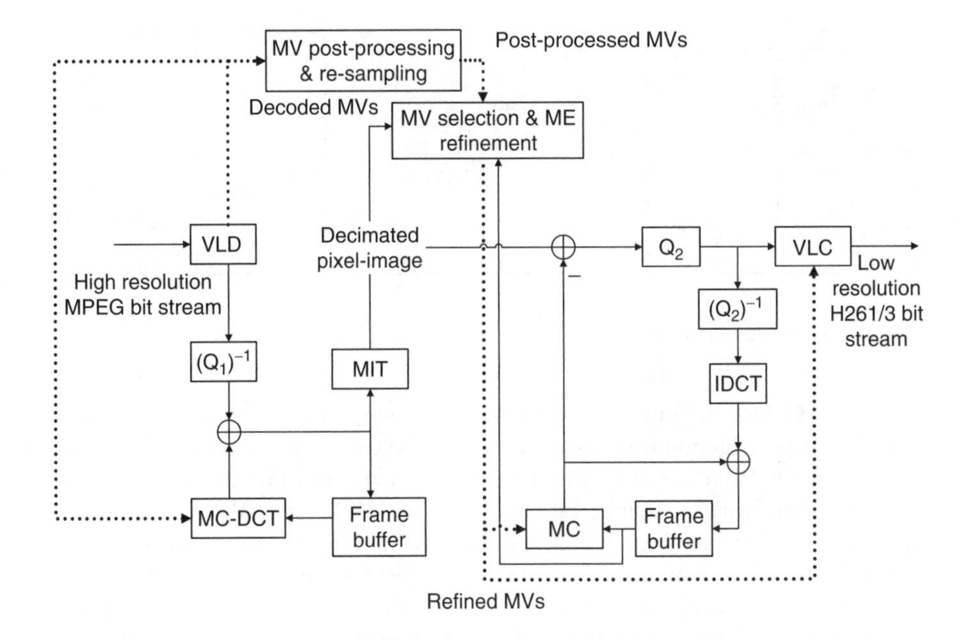

FIGURE 4.12 Reduced spatial resolution architecture 8 (RSRA-8) [6-31].

must be taken with the decoder buffer management because the virtual buffer may overflow or underflow since the encoder has no control or prior knowledge of the buffer level at the decoder. For broadcasting over terrestrial or satellite link, the buffer consideration is critical due to constrained memory since consumers are cost sensitive. In this case, the transcoder can adjust the buffer level with selective requantization of certain frames at the price of error drift.

4.3.1 Motion Vector Refinement

In, [4-14], Wee et al. proposed such an approach to transcode from MPEG-2 to H.263 bit stream. It provides a field-to-frame transcoding with both spatial and temporal down-sampling. The bit stream is first parsed so that all the B-frames are removed and the motion vectors and coding modes are extracted for the transcoder. The spatial down-conversion is performed based on the decoded pictures. The extracted motion vectors are reused for motion refinement search and the extracted coding modes are adapted for reencoding into H.263 bit stream.

For frame rate conversion transcoding, new motion vectors need to be derived from the motion vector field of anchored and dropped frames as shown in Figure 4.13.

To solve the frame rate conversion transcoding, it is desirable from the aspect of low complexity to use such techniques as motion vector reuse, DCT-domain down-conversion, and open-loop transcoding. However, the quality degradation due to dropped frames may not justify the reduced complexity. Thus, the next alternative is to reduce the motion estimation complexity with a combination of motion vector re-use and refinement. The additional search can significantly improve the motion accuracy and that may justify the additional complexity.

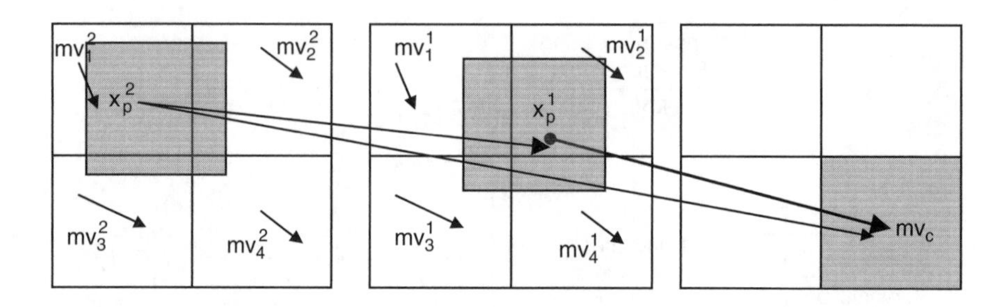

FIGURE 4.13 Motion vector refinement.

In [4-16] [4-27], Youn et al. proposed a motion vector refinement algorithm for frame rate reduction transcoders based on the cascaded transcoder. The issue is how we can derive the new motion vector predictor based on the motion vector field extracted from the source bit stream. In [4-27], Youn et al. considered two approaches: bilinear interpolation and forward dominant vector selection. The bilinear interpolation approach uses the motion vectors available from the previous frames and uses bilinear interpolation to compute the new vector [4-26]. The major disadvantage of this approach is that the transcoder is required to store all the relevant motion vectors for derivation. It was also shown that the motion vectors become divergent after a few frames, so the resultant motion vectors are unusable.

Youn et al. proposed an alternative, which is referred to as forward dominant vector selection (FDVS), that selects the motion vector based on the motion vector of the dominant macroblock. The dominant macroblock is defined as the macroblock that has largest overlapping area pointed by the motion vector. In the case of Figure 4.13, mv_1^2 and mv_2^1 will be selected. One of the advantages of this approach is that only one table is needed for all the dropped frames, which saves significant storage requirements. Experimental results have shown about 1.7 dB and 0.8 dB [4-27] for the "foreman" and "carphone" sequence, respectively. This coincides with the observation since the "foreman" sequence contains significant random motion that is hard to interpolate. In [4-29], Chen et al. proposed an activity dominant vector selection (ADVS) algorithm, which uses the activity of a macroblock as a new selection criterion to replace the overlapped area in FDVS.

The transcoding performance can be further enhanced with motion estimation of small search range. The search range can be limited to ±2. In order to reduce the motion estimation complexity, adaptive motion vector refinement is proposed. The idea comes from an observation that the benefit of using motion vector refinement increases when the bit rate difference between the source and transcoded bit streams becomes significant. This is because reference frames have lower reconstruction quality when the target transcoding bit rates are lower. Therefore, there is more need to search for more appropriate reference frames. Thus, Youn et al. use criteria based on the difference in QP and prediction errors to determine if motion vector refinement is necessary. This would reduce the motion estimation complexity significantly.

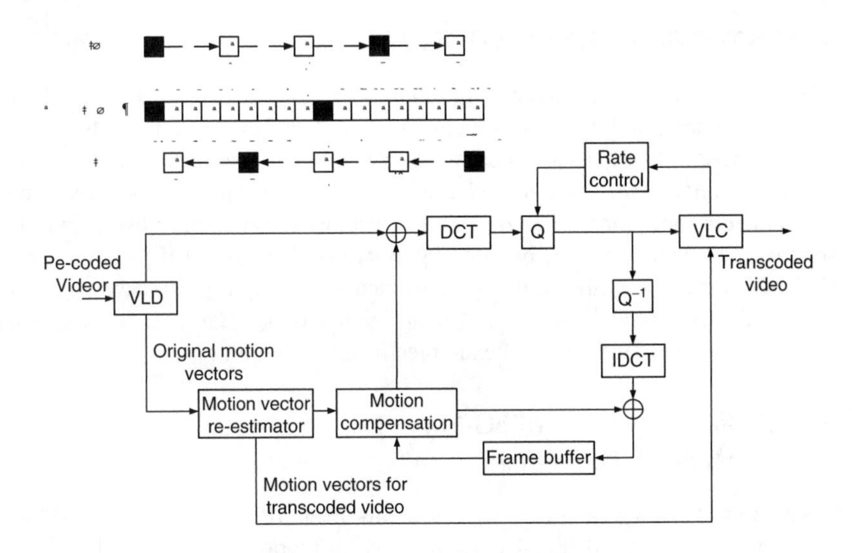

FIGURE 4.13A Architecture for fast forward/reverse playback [4-41].

In [4-9] and [4-28], Fung et al. proposed a dynamic frame-skipping transcoder having a novel architecture that operates in the frequency domain. There is direct addition of DCT coefficients and an error compensation feedback loop, so the re-encoder errors are significantly reduced. The frame buffer stores the transformed and quantized residual information. Furthermore, the transcoder dynamically adjusts the frame rate according to the motion vector fields and re-encoding errors so that the transcoded bit stream enjoys smooth motion. There is an obvious complexity savings and speedup although the technique is more complicated.

4.3.2 Requantization

As to the transcoder that can fit the served quality with the dynamic bandwidth [4-1], the output visual quality cannot be better than that derived from the input bit stream. Consequently, these transcoders can only adapt to variable channel rates slower than those used to transmit the input bit stream. The algorithms for quality-degrading transcoding can be classified into the schemes covering the complicated cascaded approaches as described in Chapter 3, the efficient DCT-domain transcoding, and the simplest requantization approaches. The requantization scheme is the simplest scheme of quality adjustment, but it has a serious error drift problem. The cascaded decoder-encoder transcoder scheme has better quality, but this scheme has the highest complexity.

4.33 Transcoding for Fast Forward/Reverse Playback

As shown in Figure 4-13A [4-39], Tan et al. proposed a transcoding architecture with motion reestimation. The interesting feature is that the GOP structure is modified and sometimes resorts back to intra-mode coding.

4.4 SYNTACTICAL ADAPTATION

So far we have studied approaches of transcoding and adaptation for various bit rates and spatiotemporal resolutions within the same standards. It is more challenging to support transcoding across standards since the coding parameters may not exist in the source bit stream. The transcoder needs to use new coding tools to reevaluate the parameters, and sometimes the motion vectors have to be reestimated. The cascaded transcoder seems to be a straightforward solution but the complexity is high. Thus, we can use all the tools we studied so far to design a new transcoder based on the source and target bit stream syntax or standard. In the following subsections, we will study several examples from the literature.

4.4.1 JPEG/MPEG-2 TO MPEG-1 AND DV TO MPEG-2 TRANSCODING

We will now discuss two transcoding systems from JPEG/MPEG-2 to MPEG-1 syntax. In the JPEG to MPEG-1 case, there is no motion vector available for the transcoder. As for the MPEG-2 to MPEG-1 transcoder, there is a need to address the three issues: spatial down-sampling, motion reestimation, and mode decisions. We will now see the examples studied in the literature.

In [4-15], a JPEG to MPEG-1 transcoder was studied for a target application that allows editing capability with the JPEG syntax and allows high compression efficiency for distribution purposes through transcoding to the MPEG-1 syntax. Wu et al. have proposed several transcoding algorithms and GOP structures including 1) I-frame only; 2) I- and P-frames; and 3) I-, P-, and B-frames. All of the algorithms are implemented with a frequency domain transcoder. That is, the JPEG decoder only decodes up to the inverse quantization module and then performs the MPEG-1 transcoding in the DCT domain. There are several variations that perform adaptive selection of a subset of the DCT coefficients and requantization of the DCT coefficients depending on the picture types. However, motion estimation was simulated with zero motion vectors, which can result in a certain level of inefficiency in compression; the tradeoff was implementation complexity and coding efficiency. For editing purposes, the best coding efficiency is not a major concern but ease of use and speed may be critical. The approach presented in [4-15] shows an interesting system that meets the requirements for studio applications.

MPEG-2 to MPEG-1 transcoding requires both bit rate conversion and resolution reduction with the same temporal resolution. In [4-6], an approach is described for performing a transcoding from 4 Mbps to either 1.2 or 2.4 Mbps. The proposed approach uses a simple mode decision and an efficient motion vector resampling. It does not use any frequency domain operation so the source bit stream is first decoded into the pixel domain and spatially down-sampled into the target resolution.

For the mode decision module, the macroblock mode needs to be determined again since MPEG-1 has less encoding modes. In [4-6], the authors use a majority rule (two out of the four macroblocks) for common codes available for both standards. Otherwise, the modes will be reselected based on the prediction residual. As for the motion vector computation, motion vector rescaling was used to reduce the complexity. The quantization parameter (QP) was used as "spatial activity measures" to provide weightings for each macroblock. The justifications for using QP are simplicity and similar usage in the rate control algorithm. Thus, the new motion vector is computed as

$$MV_{MPEG-1} = \frac{1}{2} \frac{\sum_{i=1}^{4} QP_i \times MV_i}{\sum_{i=1}^{4} QP_i}$$

where QP_i and MV_i are quantization parameters and motion vectors of the four adjacent macroblocks for the i-th macroblock. This means the motion vectors from high activity are emphasized since they may contain an edge of an object. Such a transcoder removes most of the complexity for mode selection and motion estimation and provides an excellent tradeoff between picture quality and complexity

In [4-17], an application to transcode between digital video (DV) and MPEG-2 intra coding is described. The key issues are 1) conversion from a 2-4-8 DCT transform (DV) to 8 × 8 DCT transform (MPEG-2) and 2) the selection of the quantization parameter (QP) for MPEG-2 bit stream. Since there is no motion vector necessary for the target syntax, the transcoding is performed completely in the DCT domain. The transform adaptation is performed by matrix decomposition. As for the QP selection, an equivalent formula to compute the variance in the frequency domain is derived for the TM5 rate control. Such a transcoding provides a high-quality yet low-complexity solution.

4.4.2 MPEG-2 TO MPEG-4 TRANSCODING

The MPEG-2 syntax has been widely used for the entertainment and broadcasting industries, in which available bandwidths range from 3 Mbps to 20 Mbps for interlaced CCIR-601 resolution at 30 frames per second. However, MPEG-4 is typically coded at 128 kbps of QCIF resolution at 15 frames per second. There is an obvious need to transcode from MPEG-2 syntax bit stream to MPEG-4 simple profile syntax bit stream for mobile and low-bandwidth applications.

The transcoding has high complexity since it involves different syntax and spatiotemporal resolution. It is challenging to make a conversion from an interlaced source to a progressive format even in the pixel domain. The transcoding is further complicated by the fact that B-pictures are supported in MPEG-2 whereas MPEG-4 Simple Profile supports only P-pictures. Thus, the GOP structure is quite different, which makes the derivation of motion vectors nontrivial. Motion reestimation can resolve the issue at the expense of high complexity.

In [4-19], Xin et al. describe a novel technique to solve the problem in two steps to derive the motion vectors for the target spatiotemporal resolution and GOP structure. The first step is to convert the motion vectors from 30 frames per second to 15 frames per second using the extracted motion vectors from the current and adjacent frames of the source bit stream while the spatial resolution remains unchanged. In the second step, the motion vectors are derived for reduced spatial resolution progressive frames. The interlaced-to-progressive conversion and spatial down-sampling are handled in the second step. All the processing occurs in the motion vector field so the complexity is minimal as compared to the alternative of motion reestimation.

MPEG-4 is making progress into the advanced simple profile (ASP), where three additional tools are introduced: 1) interlaced B-pictures, 2) quarter-pixel motion compensation, and 3) global motion compensation. The introduction of interlaced B-pictures actually reduces the transcoding complexity. The use of quarter-pixel motion compensation is simple because the half-pixel movement corresponds to quarter-pixel movement. The only tricky issue is the derivation of the global motion parameters.

In [4-20], Su et al. described an algorithm that determines the global motions from a coarsely sampled motion vector field extracted from the source bit stream. The global motion parameters are computed by fitting the transform coefficients using least square estimator. The matching errors between the source motion vectors and the estimated motion vectors are minimized. The model parameters are searched based on the Newton-Raphson method with outlier rejection. The computation is obviously much lower in two orders of magnitude since it operates only in the motion vector fields while the picture quality is well preserved. In addition to the transcoding application, this innovative technique is also applicable for error concealment when bit streams are corrupted such that complete motion vector fields are unavailable to the decoder. It also can be used for a fast global motion estimator for an MPEG-4 ASP encoder. The technical details of a MPEG-2 to MPEG-4 transcoder will be further discussed in Chapter 9.

4.4.3 MPEG-4 FGS Transcoding

In Section 3.4, we discussed a transcoder from an MPEG-4 FGS bit stream to an MPEG-4 SP bit stream. In [4-21], Liang and Tan describe a more flexible transcoder that allows both spatial and bit rate reductions. The overall architecture is a cascaded transcoder without motion estimation, so spatial down-sampling is performed in the spatial domain. Because motion estimation is removed, there is a need to derive the motion vector fields for reduced spatial resolution as described in the following formula:

$$MV_{\text{transcoder}} = \begin{bmatrix} sfx & 0 \\ 0 & sfy \end{bmatrix} \times \frac{\sum_{i=1}^{4} A_i B_i \times MV_i}{\sum_{i=1}^{4} A_i B_i}$$

where A_i, B_i, and MV_i are weight factors for relevance and motion vectors of the relevant macroblocks from the full spatial resolution bit stream for the i-th macroblock. The factors sfx and sfy represent the scaling factors in the spatial resolution reduction. The motion derivation formula makes it a flexible transcoder for arbitrary downsizing.

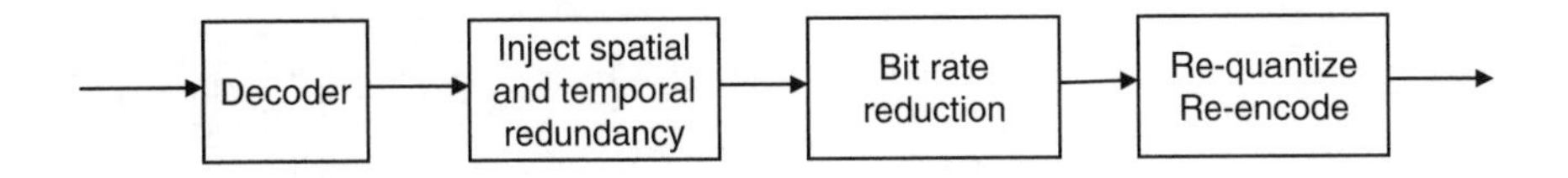

FIGURE 4.14 Architecture 1 for error resilience.

In [4-22], Barrau describes a transcoder from a nonscalable syntax to the MPEG-4 FGS syntax for "elastic" storage of video content. The concept is to allow a few compressed bit streams to be stored with high quality in the beginning and to overwrite the bit stream when the storage device is full. The author provides both closed-loop and open-loop architectures. The open-loop approach suffers from the error-drift problem for the benefit of low complexity.

4.5 ERROR-RESILIENT TRANSCODING

Transcoding can be very useful for the error-resilient transmission of wireless networks. The transcoder now aims to match the robustness of the bit stream with the channel. The process starts with the front decoder and injects the redundancy into the bit stream (Figure 4.14). It uses the typical error-resiliency tools available in the video coding standards, such as repeated headers and frequent slice headers. Such operations achieve spatial error localization but are easy to implement with bit stream parsers. To control temporal error propagation, the transcoder can limit the range of motion vectors and limit the reference within the same slice. The transcoder can use more complicated tools, such as adaptive intra refresh, but it requires more complexity to implement. However, this still does not require any motion estimation. Finally, the bit rate will be higher than the target bit rate since additional redundancy was added. Thus, the transcoder needs to reduce the bit rate accordingly. In [4-33], the transcoder needs to reduce the bit rate accordingly. In [4-33], the transcoder does the bit rate reduction by throwing away the coefficients because of its simplicity. G. De Los Reyes et al. [4-33] proposed a methodology to derive the rate distortion function for a given bit error rate. Such a rate distortion curve is used to guide the transcoder on how many intra-macroblocks should be used and how many coefficients should be discarded. Detailed technical discussion will be given in Section 9.2.

As shown in Figure 4.15, Dogan et al. [4-34] has proposed another transcoding system using adaptive intra refresh and feedback control signal (FCS) for video streaming over general packet radio service (GPRS) mobile access networks.

4.6 LOGO INSERTION AND WATERMARKING

The process of adding watermarks and logos can be considered as the same process since it tries to embed information into the source either invisibly or visibly. The process can be represented with the following formula:

$$f_{\text{transcoded}}(x, y) = \alpha \times L + \beta \times f_{\text{orig}}(x, y)$$

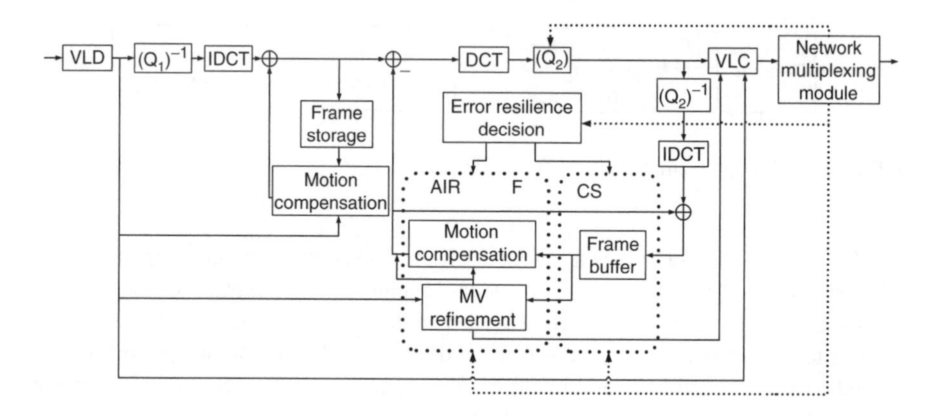

FIGURE 4.15 Architecture 1 for error-resilient transcoding over GPRS [6-34].

where the $f_{\text{transcoded}}(x, y)$ and $f_{\text{orig}}(x, y)$ are the original and transcoded sequences, respectively. The L represents the logo or watermark to be inserted. The scaling factors α and β are used to control the strength of the watermark.

As shown in Figure 4.16 and from [4-18], Panusopone et al. described a technique to insert a logo on the fly. The proposed architecture can be seen as a spatial domain cascaded transcoder. The basic requirements are for minimal quality degradation and translucent insertion. As compared to the conventional transcoder, the logo insertion transcoder is easier to implement since most of the area is unchanged. The transcoder only needs to find the area that is affected and recompute the residual of those macroblocks. Panusopone et al. experimented with two methods that give separate tradeoff points.

The first method reuses the motion vectors and recomputes the residual for the inserted logo area. The complexity is extremely low since no motion estimation is necessary and only a small affected area needs to be recomputed. However, the reuse of motion vectors does not make sense since the logo typically remains in the same location. An improvement is proposed to adaptively set the motion vectors to zero

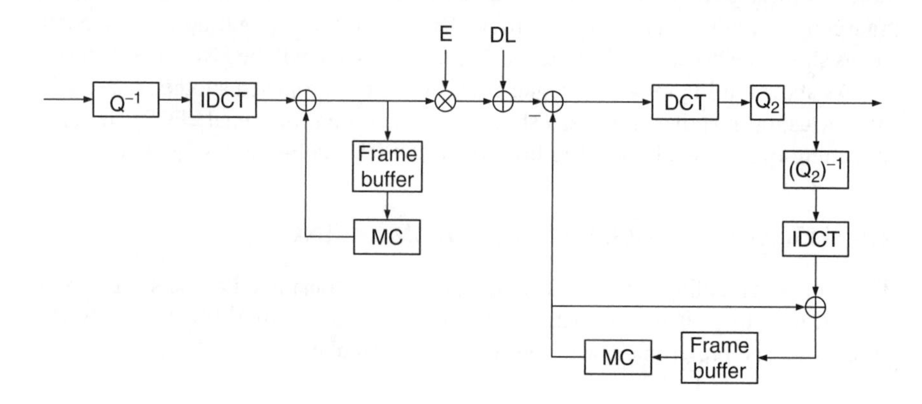

FIGURE 4.16 Architecture 1 for logo insertion [4-18].

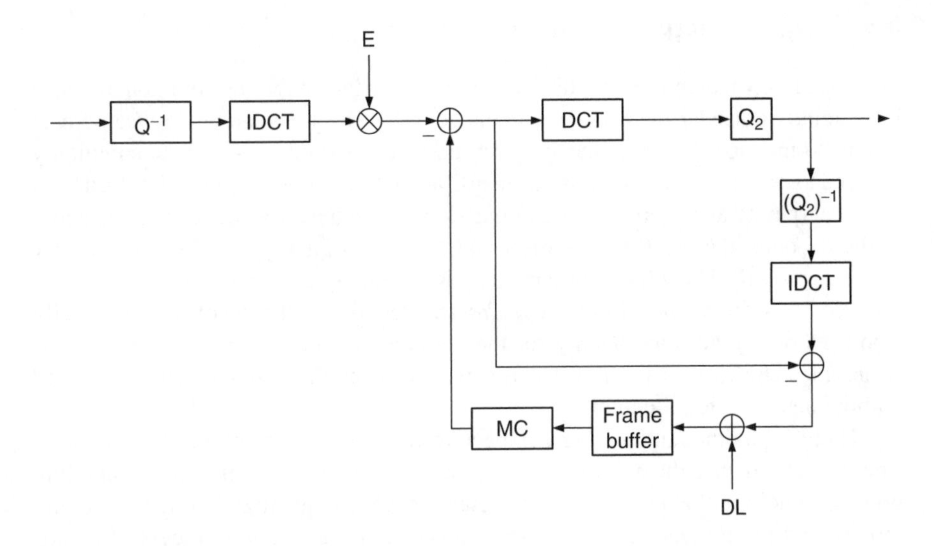

FIGURE 4.17 Intermediate simplified architecture for logo insertion [4-16].

depending on a predetermined threshold. If the macroblock is intra-coded, a change of coding mode to inter-mode with zero motion is useful. In this way, the coding efficiency of the logo insertion transcoder is greatly improved.

As shown in Figure 4.17, the first logo insertion architecture can be simplified into an intermediate architecture using the techniques derived in [4-16]. After we reach the intermediate architecture, we can use a similar simplification methodology to convert the transcoder that performs motion compensation in the DCT domain. As shown in Figure 4.18, we have the resultant DCT-domain watermarking or logo insertion architecture.

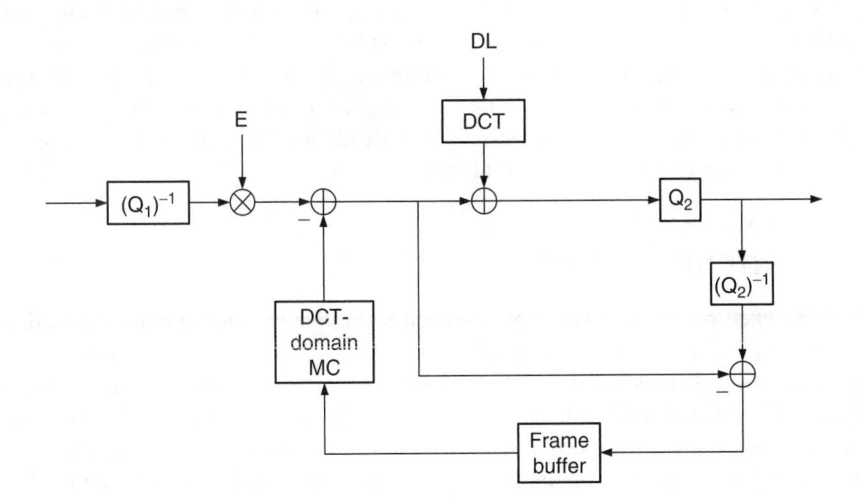

FIGURE 4-18 Final simplified architecture for logo insertion [4-16].

4.7 QUALITY IMPROVEMENT

There are many techniques to reduce the complexity by combining common modules if we jointly consider the decoder and encoder. However, such a process can create desynchronization between the encoder and decoder loop, which is commonly referred to as "drift." The obvious visual artifact of the drift is the gradual degradation of visual quality as the errors are accumulated. When the intra-frame is encountered at the decoder, the error drift is reset and the visual quality returned to a quality without any drift. The viewers observe a periodic quality degradation that is typically referred to as the *temporal pulsing artifacts*. Sometimes, the decoder intentionally smoothes or degrades the quality for the pictures that are closer to the intra-frame in order to achieve a consistent visual quality. However, the degradation is introduced without any bit rate reduction.

There is another interesting perspective to view such a transcoding process. The concept is that the original bit streams can be seen as a first-pass encoding process wherein the encoding parameters, such as quantization level, coding modes, and motion vectors, are known. In typical bit allocation process of a real-time encoder, the estimation of picture complexity requires additional frame buffer dependent on the number of looks ahead, or preanalysis. In the extreme case of a variable bit rate encoding, it is possible borrow bit rate from the future frames to improve the quality of the current frames over the time interval of complete video sequence. Such encoding is typically done with a two-pass approach where the first-pass encoding collects all the relevant complexity information and the encoder can take advantage of the knowledge of future frames. If we consider the precompressed original bit stream of high bit rates as the first-pass encoder, we realize that we have access to the relevant coding parameters as if we had done a first-pass encoding. Thus, we can now view the transcoding process as a two-pass encoding. In this case, we can achieve variable bit rate encoding with no buffering requirement. Therefore, it is possible to achieve better encoding performance than a real-time encoder that originates from uncompressed source material. Generalization of this concept for statistical multiplexing for satellite broadcasting service has proven very effective where it is impossible to perform two-pass encoding. The transcoder can efficiently multiplex programs since all of the complexity information for the complete program is already known. The transcoder can borrow bits from other channels or future pictures.

4.8 SWITCHED PICTURE

The switched picture is a recently proposed technique to achieve bandwidth adaptation, random access, and coding efficiency with a prestored secondary bit stream for adaptation. This is a hybrid, efficient solution of scalability and transcoding to address the same issue. The switch picture in H.264 allows identical SP-frames that can be reconstructed from different reference frames [4-30]. The main applications are bit stream switching, splicing, random access, error resilience, and video redundancy coding.

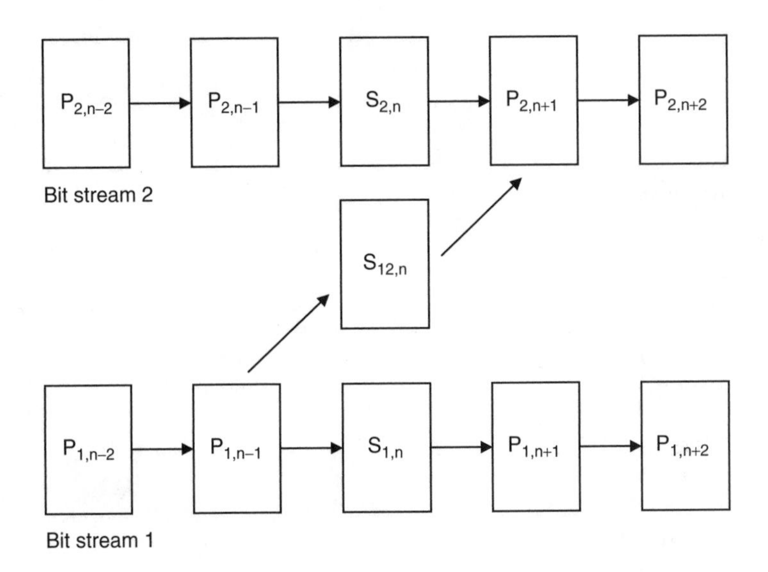

FIGURE 4.19 Bit stream switching [4-30].

4.8.1 Bit Stream Switching

As shown in Figure 4.19, there are two bit streams of different bit rates for the same source video. For a video-streaming application, the server may choose to switch between different bit streams based on the feedback information from the networks. In the traditional compression format, the server has to wait for an I-picture to switch to a different bit rate so that there is no drift. However, there is a tradeoff between ease of random access and coding efficiency because an I-picture takes many more bits to encode.

The SP-frame provides a technique that can achieve the best of both worlds. Through a special encoding loop, the encoder provides identical pictures $S_{1,n}$ and $S_{2,n}$ from different reference frames $P_{1,n-1}$ and $P_{2,n-1}$ as shown in Figure. 4.19. The frames $S_{1,n}$ and $S_{2,n}$ are referred to as the *primary SP-frames* and the frame $S_{12,n}$ is referred to as the *secondary SP-frame*. The secondary SP-frame will be transmitted only when the server switches from bit stream 1 to bit stream 2. In this case, the sequence of bit streams are $P_{1,n-1}$, $S_{12,n}$, and $P_{2,n+1}$ as shown in Figure 4.6. Therefore, the server only needs to use the secondary bit stream for switching when necessary. The expensive I-picture bit stream can be skipped most of the time for better coding efficiency.

4.8.2 Splicing and Random Access

As shown in Figure 4.20, there are two bit streams of different content that need to be spliced together. The contents can be different scenes or programs that need to be sliced together. In this scenario, the prediction from $P_{1,n-1}$ is not useful and may

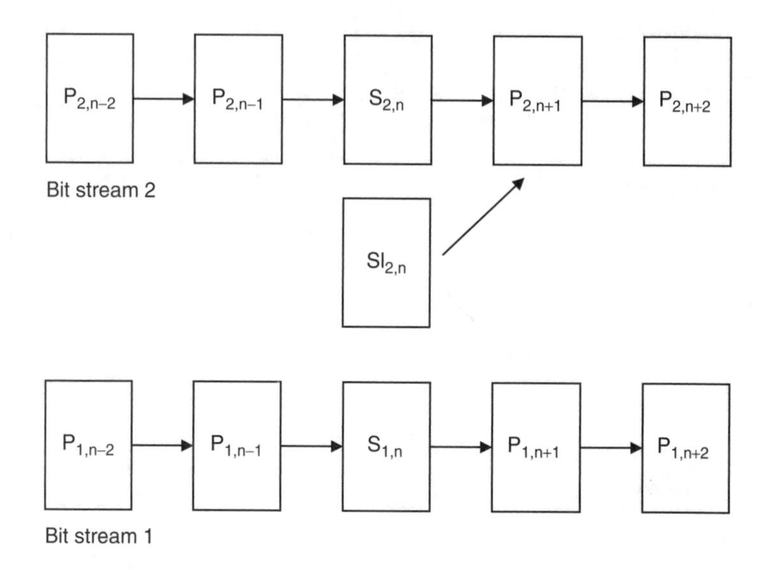

FIGURE 4.20 Slicing and random access [4-30].

cause the residual to increase. Therefore, the SI-frame is introduced to allow seamless transition from bit stream 1 to bit stream 2 using only $SI_{2,n}$ bit stream. Through special encoding, the $SI_{2,n}$ and $S_{2,n}$ are identical so that they can generate identical reference pictures for reconstructing $P_{2,n+1}$. Similar to the previous case, the $SI_{2,n}$ need not be sent unless a splicing point or random access point is determined to be at that instant. Thus, on-the-fly splicing and random access are possible with SI-frame.

4.8.3 Error Resilience

As shown in Figure 4.21, in the process of video streaming, packets of certain frames are lost and the client has a mechanism to notify the server which frames are received correctly ($P_{1,n-3}$) and which frames ($P_{1,n-2}$ $P_{1,n-1}$) are lost. Thus, the server can send the secondary SP-frame ($S_{21,n}$) to the client such that reconstruction is possible although multiple frames ($P_{1,n-2}$ $P_{1,n-1}$) are lost in between. Thus, the server is able to provide efficient error recovery. One can use the same trick to realize the intra refresh on the fly without costing the server unnecessary transmission of I-slices, which is inefficient.

4.8.4 SP-Frame Encoding

The key idea for SP-frame is to perform processing in the DCT domain with proper quantization. As shown in Figure 4.22, the SP-frame achieves an identical reconstructed picture by transforming the motion-compensated prediction to the frequency domain and performs the differencing in the DCT domain. When the frame memory needs to be updated, the reconstructed DCT coefficients are quantized, inverse quantized, and inverse transformed so that the encoder frame memory is synchronized with the decoder. To achieve identical reconstruction, the quantization parameters

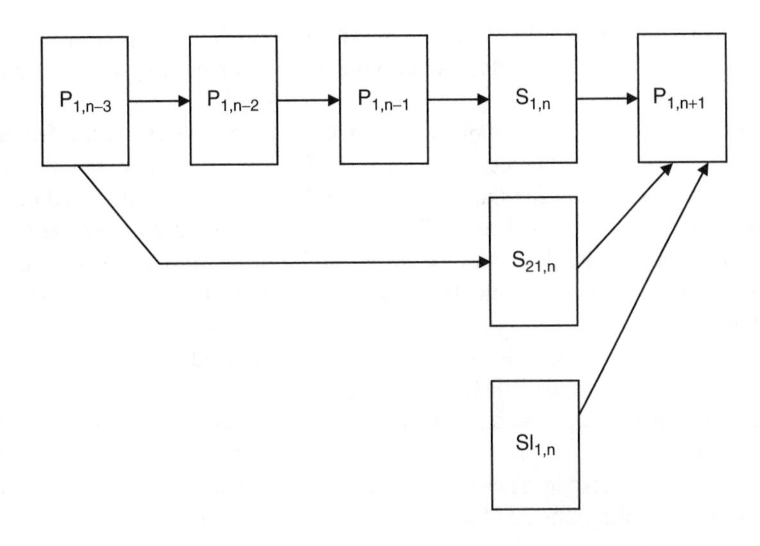

FIGURE 4.21 Error recovery and resilience [4-30].

SPQP of the primary SP-frame and PQP of the secondary SP-frame should be made identical. There are obvious additional complexities for more quantization and inverse quantization modules.

4.9 H.264/AVC PICTURE-IN-PICTURE TRANSCODING

Now we will use an example to demonstrate how all the transcoding techniques we have discussed so far can be used for a transcoding from multiple H.264 bit streams to form a single H.264 bit stream. As we know, the state-of-the-art video coding

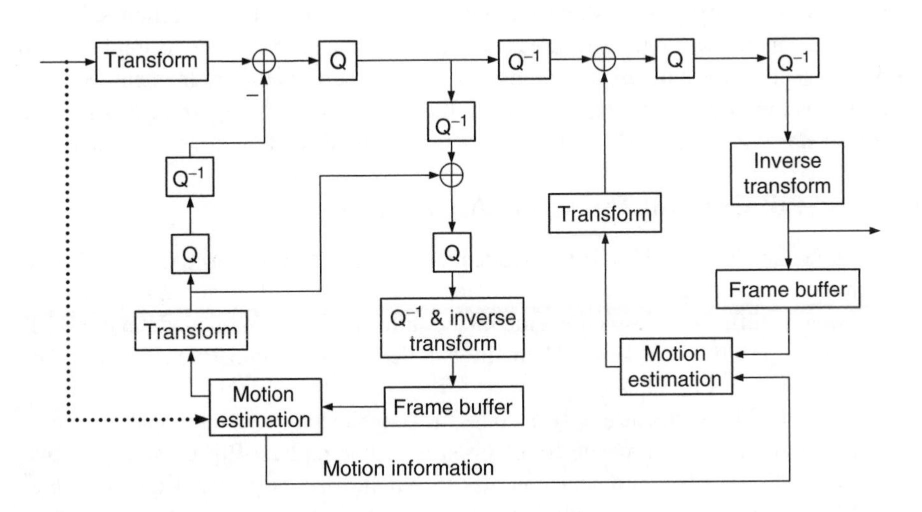

FIGURE 4.22 Block diagram of a SP-frame [4-30].

standard is H.264/AVC [4-35]. With its significant improvement in coding efficiency, H.264/AVC is expected to replace and dominate the storage format for future multimedia applications.

For the video streaming applications, one of the frequently found features in analog TV systems is the picture-in-picture (PIP) functionality. For the analog system, it typically requires the receiver to add an additional tuner and compose the embedded picture in real time. This poses an additional cost for the receiver. Other applications for PIP include building and home surveillance systems, digital television, preview guides, and television walls, which are the inverse of PIP functionality.

The PIP for compressed bit stream can be implemented at either the client side or at the server side. If the PIP functionality is implemented at the client modules, in addition to being less cost effective for the receiver, the server needs to transmit multiple bit streams via transmission channels, which requires inefficient bandwidth usage. The solution also lacks complexity scalability because each client would need N decoders if N programs need to be multiplexed simultaneously.

Thus, it is more cost effective to implement the digital PIP functionality at the server side and it should be transparent and requires no additional hardware or software cost. We will now discuss a practical implementation of PIP based on the transcoding technology that produces compliant bit streams that offer transparent PIP functionality to the user.

In [4-36] and [4-37], there are several PIP transcoding approaches based on the coding standards, such as H.263, and MPEG-4 that has lower complexity than H.264. The new technical challenges are that many coding tools take advantage of spatial and temporal redundancies, which makes the design of transcoder more complicated. For example, the use of multiple direction intra-frame spatial prediction and multiple reference frame prediction makes it difficult for the design of a DCT-domain motion compensation for the transcoder. Similar to the logo insertion problem discussed earlier, most of the area will not be changed, so the transcoder can be implemented with low complexity. It is slightly more complicated than logo insertion since the inserted content has larger pixel area and motion. Thus, the transcoder is implemented with a spatial domain cascaded transcoder with motion vector refinement. Since it performs the re-encoder, we will refer to the architecture as the *partial re-encoding transcoder architecture* (PRETA) using both inter-mode and intra-mode refinements.

4.9.1 PIP Cascaded Transcoder Architecture

As shown in Figure 4.23, a PIP cascaded transcoder (PIPCT) based on H.264/AVC is implemented with two source bit streams of CIF and QCIF formats. As discussed earlier, it is difficult to implement the mixed intra-mode and inter-mode in the DCT-domain as proposed in [4-36]. There are two decoders to reconstruct frames before transcoding.

The PIPCT architecture fully decodes the two bit streams into the spatial domain and compresses the composite frame into a single compliant PIP bit stream. Obviously, PIPCT needs significant complexity but has the best visual quality. Thus, PIPCT is used as an upper bound for quality evaluation.

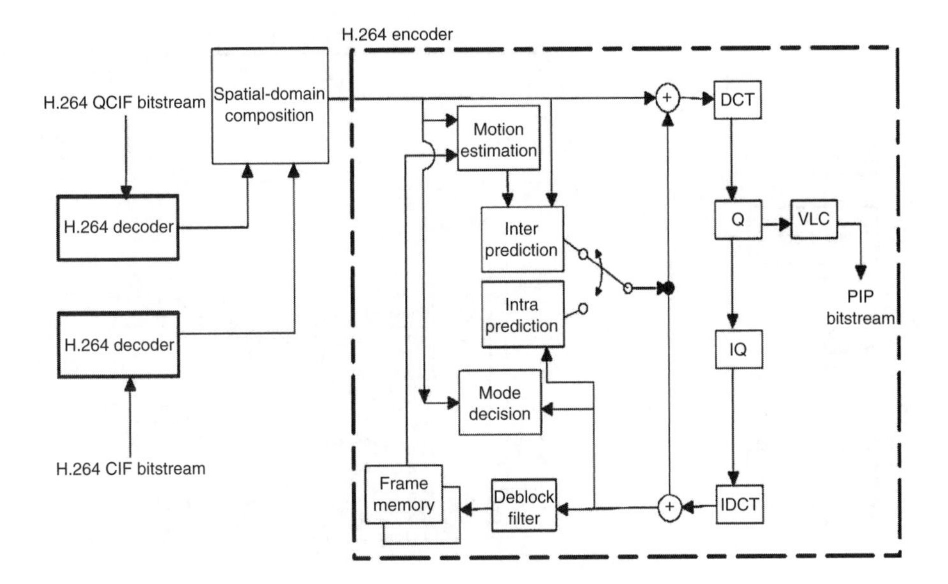

FIGURE 4.23 Picture-in-picture cascaded transcoder (PIPCT) architecture.

4.9.2 Partial Re-encoding Transcoder Architecture (PRETA)

As shown in Figure 4.24, the partial re-encoding transcoder architecture (PRETA) reduces the complexity in the following modules: motion estimation, intra-prediction, inter-prediction and mode decision using R-D optimization. It uses the idea of intra-mode and inter-mode refinement for each macroblock. The basic idea is to select the proper prediction for each MB in the boundary regions for both the successive frames and the current. For the inter-mode refinement, the new motion vectors are derived from the archived bit stream.

In PRETA, the coding parameters from the incoming bit streams are used to classify each MB into three categories as intra-mode, inter-mode and unchanged. The unchanged area will be completely skipped by the transcoder.

4.9.2.1 Intra-mode Refinement

In H.264, DCT coefficients can be predicted from neighboring intra-coded blocks, and the neighboring decoded pixels can be used as spatial prediction. When an area is replaced with the inserted picture, the macroblocks at the border need to be refined since the assumed decoded pixels are different now, a situation that is similar to the drift problem in the temporal direction.

If this is not handled properly, the reconstruction error will propagate over the whole frame spatially. Thus, macroblocks at the boundary regions need to recompute the residual for the intra-mode refinement and then re-encode it into the PIP bit stream. Note that we need not recalculate all kinds of prediction residue for PIP since not all reference blocks are modified.

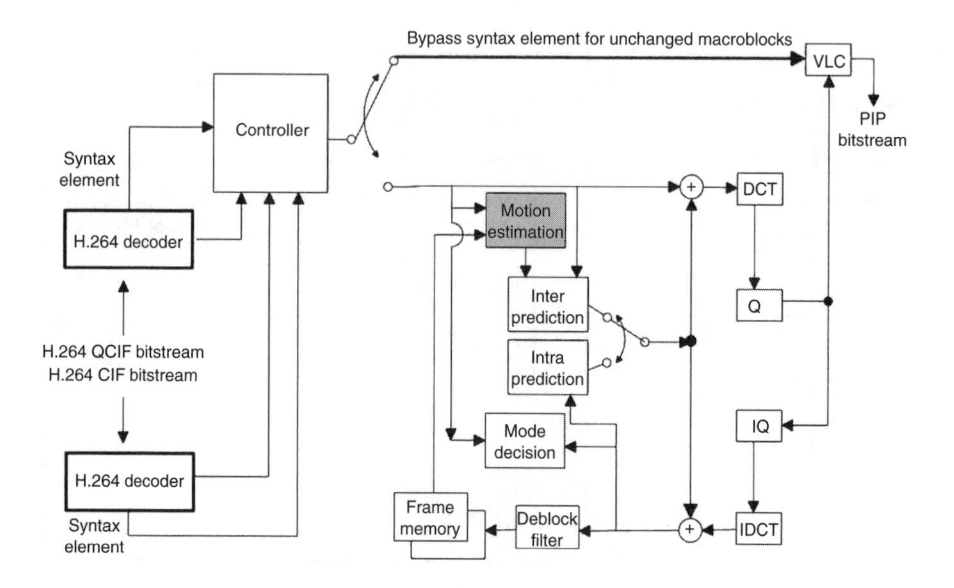

FIGURE 4.24 Partial re-encoding transcoder architecture (PRETA).

4.9.2.2 Inter-mode Refinement

Since H.264/AVC allows unrestricted motion compensation and the PIP reference frame is a composite of different video sequences, the prediction of the original bit stream may point to two scenes as illustrated in Figure 4.25. Thus, the macroblocks of the affected area need to recompute both the motion vectors and the associated residual. However, the complexity is much lower since the affected area is only at the boundary regions.

As discussed in Section 4.2.2, motion vector remapping (MVR) can be used to obtain new motion vectors. In this MVR process, we first compute four distances

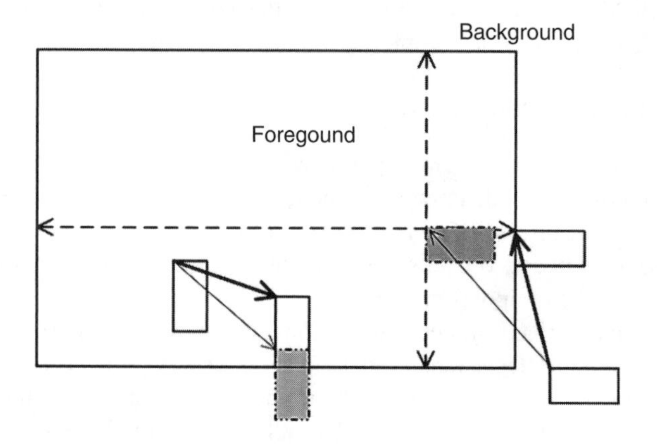

FIGURE 4.25 Motion vector remapping for PRETA.

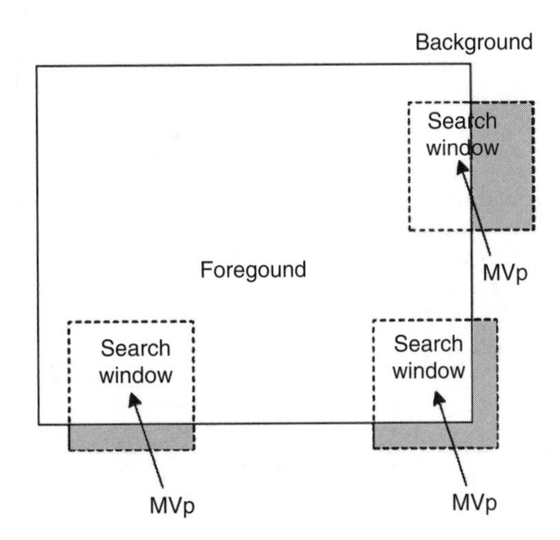

FIGURE 4.26 Motion vector refinement.

from four boundaries and map the original motion vectors to a new location that is closest to the boundary as illustrated in Figure 4.26. The residual needs to be recomputed to reduce drift. However, the remapped motion vectors may not yield minimum residual due to the simplicity of its derivation.

Thus, motion vector refinement technique can be used to improve the coding efficiency. For each MB, the search points that point to the area from the other source are removed as depicted in Figure 4.26 where only the shaded areas are searched. To further reduce the complexity for the motion refinement, an adaptive scheme based on the difference of the remapped and the original motion vectors ($MV_{\text{re-mapping}} - MV_{\text{original}} > Threshold$) is used to decide whether to perform the motion refinement search or just use the remapped motion vectors. The thresholds found empirically are 4 and 2 for the background and for foreground scenes as depicted in Figure 4.26, respectively. Based on the adaptive method, only less than 1% of the macroblocks need motion vector refinement. To prevent the fractional pixel search from falling into local minimum, the original motion vector is rounded to the nearest full pixel resolution before applying the adaptive scheme.

4.9.2.3 Simulation Results

The fast PIP transcoder has been implemented based on H.264 reference software of version JM7.3.

As shown in Figure 4.27, the proposed adaptive refinement approach can improve the R-D performance of blocks within the border areas of multiple frames. Based on proper motion vector refinement, the PRETA method can improve the coding efficiency of the motion vector remapping approach, and it achieves performance almost identical to that of the PIPCT scheme. PRETA has an average speedup of around 10 times over the PIPCT architecture.

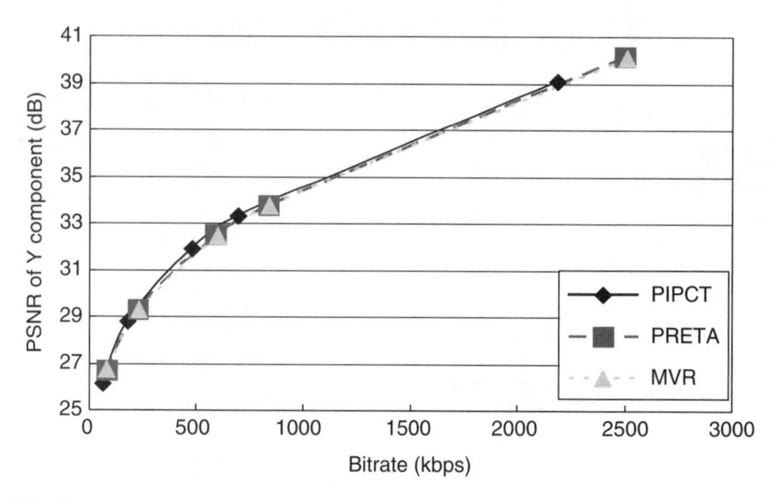

FIGURE 4.27 Motion vector refinement.

4.10 TRANSCODING FOR STATISTICAL MULTIPLEXING

In [4-20] and [4-40], Serial et al. describe the joint transcoding of multiple bit streams as shown in Figure 4.28. The proposed transcoder is a cascaded transcoder so there is no significant change in terms of transcoder architecture. However, the joint optimization of the requantization and rate control process provide a better utilization of overall channel bandwidth. Some examples of statistical multiplexers using transcoders will also be given in Chapter 5.

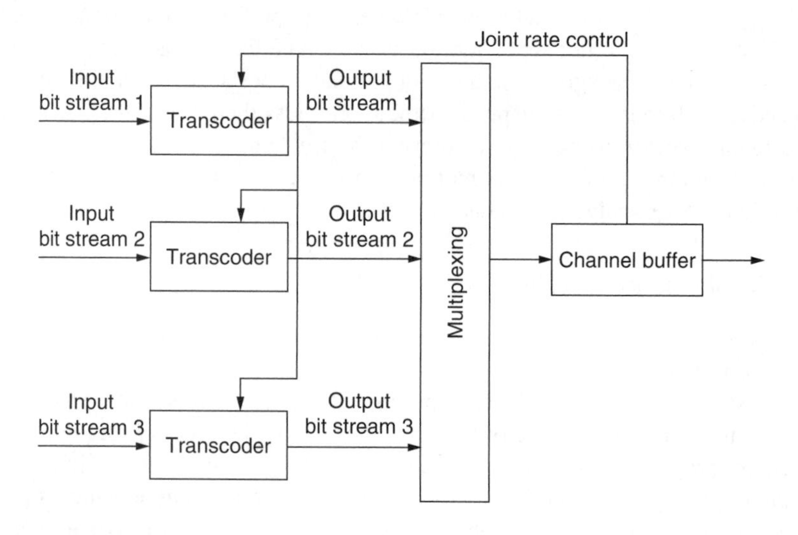

FIGURE 4.28 Joint transcoder for statistical multiplexing [4.40].

4.11 SUMMARY

In this chapter, we reviewed various techniques to optimize the transcoding performance and functionality. In particular, we reviewed recent works in reduced spatial resolution transcoding, reduced temporal resolution transcoding, syntactical transcoding, error-resilient transcoding, logo and watermarking transcoding, quality improvement, switch picture, and, lastly, picture-in-picture transcoding. Some transcoding techniques described in this chapter have been used for application of MPEG-2 to MPEG-4 transcoders, which will be discussed in detail at Chapter 9.

4.12 EXERCISES

4-1. In reduced spatial resolution transcoding, explain the concept of the mix mode processing.
4-2. In reduced spatial resolution transcoding, explain why there are the two sources of drift that are caused by quantization and down-conversion.
4-3. In reduced spatial resolution transcoding, explain the advantages and disadvantages of spatial and frequency domain transcoding.
4-4. Explain why transcoders can provide superior visual quality when the source bit stream is at very high bit rates.
4-5. Explain why a spatial domain transcoder was used for H.264/AVC picture-in-picture transcoding. Design a DCT domain transcoder if you can.

REFERENCES

[4-1] P.A.A. Assuncao and M. Ghanbari, A frequency-domain video transcoder for dynamic bit-rate reduction of MPEG-2 bit streams, *IEEE Transactions on Circuits and Systems for Video Technology,* Vol. 8, Issue 8, pp. 953–967, Dec.1998.
[4-2] B. Shen, I.K Sethi, and B. Vasudev, Adaptive motion-vector resampling for compressed video downscaling, *IEEE Transactions on Circuits and Systems for Video Technology,* Vol. 9, Issue 6, pp. 929–936, Sept. 1999.
[4-3] N. Merhav and V. Bhaskaran, A fast algorithm for DCT-domain inverse motion compensation, in Proceedings of ICASSP, Atlanta, GA, Vol. 4, pp. 2307–2310, 7–10 May 1996.
[4-4] B. Natarajan and V. Bhaskaran, A fast approximate algorithm for scaling down digital images in the DCT domain, in Proceedings of IEEE International Conference on Image Processing (ICIP), Washington, D.C., Vol. 2, pp. 241–243, Oct. 1995.
[4-5] M.-J. Chen, M.-C. Chu, S.-and Y. Lo, Motion vector composition algorithm for spatial scalability in compressed video, *IEEE Transactions on Consumer Electronics,* Vol. 47, Issue 3 Aug. 2001.
[4-6] M. Sugano, Y. Nakajima, H. Yanagihara, and Yoneyama, An efficient transcoding from MPEG-2 to MPEG-1, Proceedings of the 2001 International Conference on Image Processing, Vol. 1, pp. 417–420, 7–10 Oct. 2001.
[4-7] J.I Khan,.and Q. Gu, Network aware video transcoding for symbiotic rate adaptation on interactive transport, IEEE International Symposium on Network Computing and Applications, pp. 201–212, 8–10 Oct. 2001.
[4-8] K.-T. Fung, Y.-L. Chan, and W.-C. Siu, Dynamic frame skipping for high-performance transcoding, Proceedings of the 2001 International Conference on Image Processing, Vol. 1, pp. 425–428, 7–10 Oct. 2001.

[4-9] K.-T. Fung, Y.-L. Chan, and W.-C. Siu, Low-complexity and high quality frame-skipping transcoder, The 2001 IEEE International Symposium on Circuits and Systems, Vol. 5, pp. 29–32, 6–9 May 2001.

[4-10] Guobin Shen, B. Zeng, Y.-Q. Zhang, and M.L. Liou, Transcoder with arbitrarily resizing capability, The 2001 IEEE International Symposium on Circuits and Systems, Vol. 5, pp. 25–28, 6–9 May 2001.

[4-11] P. Yin, M. Wu, and B. Liu, Video transcoding by reducing spatial resolution, Proceedings of the 2000 International Conference on Image Processing, Vol. 1, pp. 972–975, 10–13 Sept. 2000.

[4-12] T. Shanableh, and M. Ghanbari, Transcoding of video into different encoding formats, Proceedings of the 2000 IEEE International Conference on Acoustics, Speech, and Signal Processing, Vol. 6, pp. 1927–1930, 5–9 June 2000.

[4-13] A. Vetro, H. Sun, and Y. Wang, Object-based transcoding for scalable quality of service, Proceedings of the 2000 IEEE International Symposium on Circuits and Systems, Geneva, Vol. 4, pp. 17–20, 28–31 May 2000.

[4-14] S.J. Wee, J.G. Apostolopoulos, and N. Feamster, Field-to-frame transcoding with spatial and temporal downsampling, Proceedings. 1999 International Conference on Image Processing, Vol. 4, pp. 271–275, 24–28 Oct. 1999.

[4-15] J.-L. Wu, S.-J. Huang, Y.-M. Huang, C.-T. Hsu, and J. Shiu, An efficient JPEG to MPEG-1 transcoding algorithm, IEEE Transactions on Consumer Electronics, Vol. 42, Issue 3, pp. 447–457 Aug. 1996.

[4-16] J. Youn, J. Xin, and M.-T. Sun, Fast video transcoding architectures for networked multimedia applications, Proceedings of the IEEE International Symposium on, Circuits and Systems, Vol. 4, pp. 25–28, 28–31 May 2000.

[4-17] D. Kim, B. Yoon, and Y. Choe, Conversion between DV and MPEG-2 intra coding, International Conference on Consumer Electronics, pp. 34–35, 19–21 June 2001.

[4-18] K. Panusopone, X, Chen, and F. Ling, Logo insertion in MPEG transcoder, Proceedings of the 2001 IEEE International Conference on Acoustics, Speech, and Signal Processing, Vol. 2, pp. 981–984, 7–11 May 2001.

[4-19] J. Xin, M.-T. Sun, and K. Chun, Motion Re-estimation for MPEG-2 to MPEG-4 Simple Profile Transcoding, in Proceedings of the International Packet Video Workshop, 2002.

[4-20] Y. Su, M.-T. Sun, and V. Hsu, Global motion estimation from coarsely sampled motion vector field and the applications, Proceedings of the IEEE International Symposium on Circuits and Systems, Vol. 2, pp. 628–631, 25–28 May 2003.

[4-21] Y.-P. Tan and Y.-Q. Liang, Methods and need for transcoding MPEG-4 fine granularity scalability video, Proceedings of the IEEE International Symposium on Circuits and Systems, Vol. 4, pp. 719–722, 26–29 May 2002.

[4-22] E. Barrau, MPEG video transcoding to a fine-granular scalable format, *Proceedings of the IEEE International Conference On Image Processing*, Vol. 1, pp. 717–720, 22–25 Sept. 2002.

[4-23] S.F. Chang and D.G. Messerschmidt, Manipulation and compositing of MC-DCT compressed video, *IEEE Journal of Selected Areas in Communications*, Vol. 13, Issue 1, pp. 1–11, Jan 1995.

[4-24] P. Yin, A. Vetro, B. Liu, and H. Sun, Drift compensation for reduced spatial resolution transcoding, *IEEE Transactions on Circuits and Systems for Video Technology*, Vol. 12, Issue 11, pp. 1009–1020, Nov. 2002.

[4-25] J. Xin, M.-T. Sun, B. S. Choi, and K.W. Chun, An HDTV to SDTV spatial transcoder, *IEEE Transactions on Circuits and Systems for Video Technology*, Vol. 12, pp. 998–1008, Nov. 2002.

[4-26] J.-N. Hwang, T.-D. Wu, and C.-W. Lin, Dynamic frame-skipping in video transcoding, Proceedings of the IEEE Workshop Multimedia Signal Processing, pp. 616–621, 7–9 Dec. 1998.

[4-27] J. Youn, M.-T. Sun, and C.-W. Lin, Motion vector refinement for high-performance transcoding, *IEEE Transactions on Multimedia*, Vol. 1, Issue 1, pp. 30–40, March 1999.

[4-28] K.-T. Fung, Y.-L. Chan, and W.-C. Siu, New architecture for dynamic frame-skipping transcoder, *IEEE Transactions on Image Processing*, Vol. 11, Issue 8, pp. 886–900, Aug. 2002.

[4-29] M.-J. Chen, M.-C. Chu, and C.-W. Pan, Efficient motion estimation algorithm for reduced frame-rate video transcoder, *IEEE Transactions on Circuits and Systems for Video Technology*, Vol. 12, Issue 4, pp. 269–275, April 2002.

[4-30] M. Karczewicz and R. Kurceren, The SP- and SI-Frames design for H.264/AVC, *IEEE Transactions on Circuits and Systems for Video Technology*, Vol. 13, Issue 7, pp. 637–644, July 2003.

[4-31] T. Shanableh, and M. Ghanbari, Hybrid DCT/pixel domain architecture for heterogeneous video transcoding, *Signal Processing: Image Communications.*, 18, 601, 2003.

[4-32] J. Wong and O. Au, Modified predictive motion estimation for reduced-resolution video from high-resolution compressed video, *ISCAS'99*, Vol 4, pp. 524–527, 30 May-2 Hune 1999.

[4-33] G. de los Reyes, A. R. Reibman, S.-F. Chang, and J. C.-I. Chuang, Error-resilient transcoding for video over wireless channels, *IEEE Journal of Selected Areas in Communications*, Vol. 18, Issue 6, pp.1063–1074, June 2000.

[4-34] S. Dogan, A. Cellatoglu, M. Uyguroglu, A. H. Sadka, and A. M. Kondoz, Error-resilient video transcoding for robust internetwork communications using GPRS, *IEEE Transactions on CSVT.* Vol. 12, Iussue 6, pp. 453–646, June 2002.

[4-35] ITU-T Rec. H.264/ISO/IEC 11496-10, Draft ITU-T recommendation and final draft international standard of joint video specification, Mar. 2003.

[4-36] D. G. Messerschmitt, Y. Noguchi, and S.-F. Chang, MPEG video compositing in the compressed domain, in Proceedings of ISCAS'96, Vol. 2, pp. 596–599, 12–15 May 1996.

[4-37] Y.-P. Tan, H. Sun, and Y. Liang, On the methods and applications of arbitrary downsizing video transcoding, in Proceedings of ICME'02, Vol. 1, pp. 609–612, 26–29 Aug. 2002.

[4-38] Z. He and S. K. Mitra, A linear source model and a unified rate control algorithm for DCT video coding, *IEEE Transactions on Circuits and Systems for Video Technology,* Vol. 12, Issue 11, pp. 970–982, Nov. 2002.

[4-39] Y.-P. Tan, Y.-Q. Liang, and J. Yu, Video transcoding for fast forward/reverse video playback, *Proceedings of the IEEE International Conference On Image Processing,* Vol. 1, pp. 713–716, 22–25 Sept. 2002.

[4-40] H. Sorial, W.E. Lynch, and A. Vincent, Joint transcoding of multiple MPEG video bitstreams, *in Proceedings of the IEEE International Symposium on Circuits and Systems,* Vol. 4, pp. 251–254, 30 May-2 Jung 1999.

[4-41] I. Koo, P. Nasiopoulos, and R. Ward, Joint MPEG-2 coding for multi-program broadcasting of pre-recorded video, *Proceedings of the IEEE International Conference on Acoustics, Speech and Signal Processing,* Vol. 4, pp. 2227–2230, 15–19 March 1999.

5 Video Transport Transcoding

In this chapter, we discuss video transcoding at the transport level. For this purpose, we first introduce the basic concept of MPEG-2 [5-1], a system that has specified two stream forms: the transport stream (TS) and the program stream (PS), which are designed for different sets of applications. The functions of the transport and program streams include methods for multiplexing the elementary audio and video streams to a signal stream with time synchronization and other important information. We are therefore going to present the transcoding techniques that are used to perform the conversions between the twostreams. The transcoding between TS and PS has an important application, which can be applied to convert digital television (DTV) broadcasting's audio and video bit streams into audio and video bit streams for digital video disc (DVD). An important technical issue for transcoding between PS and TS is the rate contro algorithm. Usually, for DVD applications, the video is encoded by variable bit rate (VBR), whereas for DTV applications, due to buffer requirements, the video should be encoded by constant bit rate (CBR). Finally, the technical details about CBR and VBR and transcoding algorithms for CBR to VBR will be introduced.

5.1 OVERVIEW OF MPEG-2 SYSTEM

In this section, we introduce the basic concept of system part of the MPEG-2 standard. The system specification is very important for real applications of video and audio compression since it provides the mechanism of multiplexing and demultiplexing video, audio, and other data with information for time synchronization. The output of the system encoder is the packetized bit stream.

5.1.1 INTRODUCTION

The problem addressed by the MPEG-2 System standard is to combine one or more elementary streams of video and audio, as well as other data, into single or multiple streams, that are suitable for storage or transmission. The specification of system standard is defined according to syntactical and semantic rules. The system encoding provides the mechanism, which is able to insert the time synchronization information into the system bit stream for decoding in decoder buffers over a wide range of retrieval or receipt conditions. The MPEG-2 system standard specifies two kinds of streams: transport and program. Each stream is optimized for a different set of applications, which will be discussed in the following section. The major functions of the streams include two aspects. The first is to provide the coding syntax necessary and sufficient to synchronize the decoding and presentation of the video, audio, and other data. The second is to provide the buffer regulation in the decoders, thus preventing buffer overflow and underflow. The time information is coded in the syntax by using time stamps for decoding and presentation of coded video and audio information and also for delivery of the system bit stream itself.

5.1.2 Transport Stream and Program Stream

In the system encoder, the video, audio, and data signals are first encoded with audio and video encoder, respectively. The coded data are referred to as *compressed bit streams*, or *elementary streams*. The elementary streams are generated according to the syntax rules specified by the video coding standard and audio coding standard of MPEG-2. The MPEG-2 system is to multiplex and synchronize the coded audio, video, and data into a single bit stream or multiple bit streams for storage or transmission. In other words, the digital compressed video, audio, and data all are first compressed and represented as binary formats or elementary bit streams, and then the system part is to mix elementary bit streams from video, audio, and data together to form the system bit streams with timing information for synchronization in the decoders. The two kinds of system bit streams are the transport and program streams. The block diagram of MPEG-2 system is described in Figure 5.1.

Packetization is an important feature of the MPEG-2 system. In MPEG-2, the compressed audio and video bit streams are first packetized to the packetized elementary stream (PES). The video PESs and audio PESs are then encoded with system encoder to either transport stream or program stream according to the requirements of the application. Both the transport and program streams are packet-oriented multiplexing. The function of system encoder is multiplexing the video and audio PESs into system packets by adding packet headers and optional adaptation fields, which contain the timing information according to the system syntax. This time information, as mentioned previously, is used for video, audio synchronization, and presentation in the decoders; at the same time, it also has to ensure that the decoder buffers do not overflow and underflow. Of course, the buffer regulation has to be considered by the buffer control or rate control mechanism in the encoder.

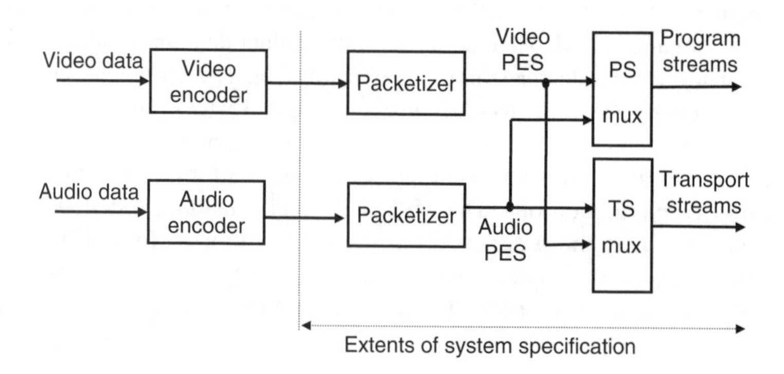

FIGURE 5.1 Block diagram of the MPEG-2 system.

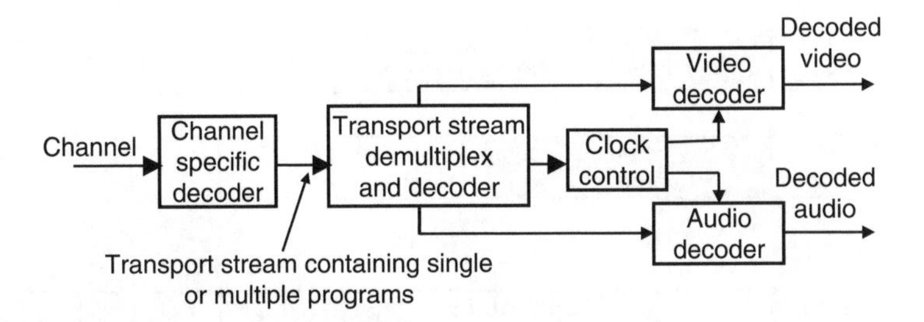

FIGURE 5.2 Example of transport demultiplexing and decoding.

The transport stream is designed for transmitting or storing the compressed video, audio, and other data in lossy or noisy environments where significant errors may occur, such as in the broadcasting environment. A transport stream combines one or more programs with one or more time bases into a single stream. It is not only possible to generate transport streams from elementary coded data steams, but also from program streams, or from other transport streams, which may themselves contain one or more programs. The basic function of the transport stream is to provide a data presentation format that can be easily demultiplexed and decoded by the system decoder, as shown in Figure 5.2.

The additional important feature of the transport stream is that it is designed in such a way that makes several transcoding operations possible with minimum effort. These transcoding operations will be discussed in the next section. The basic function of the program stream is the same as the transport stream. It is defined for the multiplexing of the coded audio, video, and other data into a single stream for communication or storage application. But the essential difference between the two streams is that the transport stream is designed for applications with noisy media, such as in terrestrial broadcasting or satellite applications, and the program stream is designed for applications in a relatively error-free environment, such as in the digital video disk (DVD) and digital storage applications. Therefore, the overhead used for packet headers in the program streams is much less than in the transport stream.

5.1.3 Transport Stream Coding Structure and Parameters

In this subsection, we present the coding structure and parameters in the Transport stream. The structure of the transport stream is shown in Figure 5.3. As described previously, in the MPEG-2 system the transport stream coding layer allows one or more programs to be combined into a single stream. In this process, the audio and video data are first encoded to the elementary streams. By choosing the time interval of anchors for prediction, the elementary streams consist of access units. The elementary streams are then packetized to PES packets. Each PES packet consists of a PES packet header followed by packet data, which contains a variable number of contiguous bytes from one elementary stream. The packet header starts with a 4-byte start code and is followed by a 2-byte PES packet length as well as an optional PES header, if necessary. The PES header is used to identify the stream or stream type to

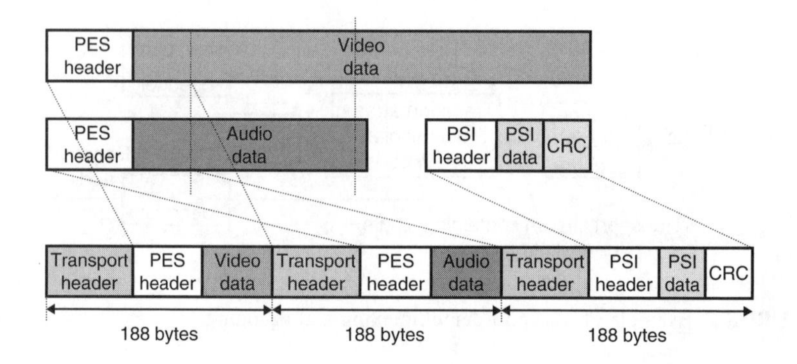

FIGURE 5.3 Structure of transport stream containing both PES packets and PSI packets.

which the packet data belongs. The optional PES header may contain decoding and presentation time stamps (DTS and PTS). The first byte of each PES packet header is located at the first available payload location of a transport stream packet.

The different kinds of PESs are multiplexed together and inserted into the transport stream packets with information that allows synchronized presentations of elementary streams within a program. Each packet of transport stream is a 188-byte-long packet starting with a packet header followed by data from elementary stream or program-specific information (PSI). The packet header begins with a 4-byte prefix followed by the adaptation field if necessary. The 4-byte prefix contains a 13-bit packet ID (PID) that is used to identify the contents of data contained in the transport stream via the following four PSI tables:

- Program association table
- Program map table
- Conditional access table
- Network information table

These tables contain the necessary and sufficient information used to demultiplex and present programs. The program map table is used to specify which PIDs, and therefore, which elementary streams are associated to form each program. It is also used to indicate which PID of the transport stream packets carries the PCR for each program. The conditional access table exists only if scrambling is employed. The network information table is optional and its contents are not specified by the MPEG-2 system standard.

Several things have to be noted for the transport stream. The first is that one PID of transport stream packets carries one and only one elementary stream. The second note is about null packets of the transport stream packets. The null packets are used for padding of transport stream to avoid the system buffer underflow. They may be inserted or deleted by remultiplexing processes and, therefore, the delivery of the payload of null packets to the decoder cannot be assumed. The third note is about conditional access. The standard provides only the mechanisms for conditional access but does not specify which coding method for conditional access. The mechanisms for conditional access provided by the standard are used for program service providers to transport and identify the conditional access data for decoder processing, and to reference correct data, which are specified by the standard. This type of support is provided both through transport stream packet structures and the conditional access table of the PSI.

5.1.4 Program Stream Coding Structure and Parameters

The program stream is similar to the transport stream; its function is to multiplex one or more elementary streams into a single stream. The structure of program stream is shown in Figure 5.4. Data from each elementary stream is multiplexed and encoded together with information that allows elementary streams to be replayed in synchrony. Data from elementary streams is also packetized to PES packets, which are same as in the transport stream.

In a program stream, the PES packets are organized in packs. The pack size may be quite large. In such a way, the overhead of Program stream can be much less than one in the transport stream. Since the program stream is used in the error-free or nearly error-free application environment, it does not spend extra bits in overhead for protection of the bit error or packet loss. A pack begins with a pack header, or called system header, followed by zero or more PES packets. The pack header, which begins with a 32-bit start-code, is used to store timing and bit rate information. The program stream begins with a system header that optionally may be repeated. The system header carries a summary of the system parameters defined in the stream. The conditional access part of the program stream is the same as in the transport stream, the standard does not specify the coded data, but provides mechanisms for content service providers to transport and identify this data for decoder processing, and to reference correct data which are specified by the standard.

5.2 MPEG-2 SYSTEM LAYER TRANSCODING

As we know, the most successful applications of the MPEG-2 video coding standards are digital television (DTV) and digital video disk (DVD). In the application of digital television, the video, audio, and other data are multiplexed to the transport stream. In the DVD application, the video, audio, and other data are multiplexed to the program stream. Except the syntax differences, there are also other different features for DTV and DVD applications; for example, DTV uses constant bit rate coding and DVD uses variable bit rate coding. When we design the transcoding between transport stream and program streams, we have to consider all these differences. We are going to discuss the related techniques, which will be used in this kind of transcoding.

5.2.1 Transcoding Features of Transport Stream

Except the basic function of multiplexing video, audio, and time information into a single bit stream, the transport stream has an additional feature. It is designed in such a way that makes several transcoding operations become possible with

Au: Figure not final

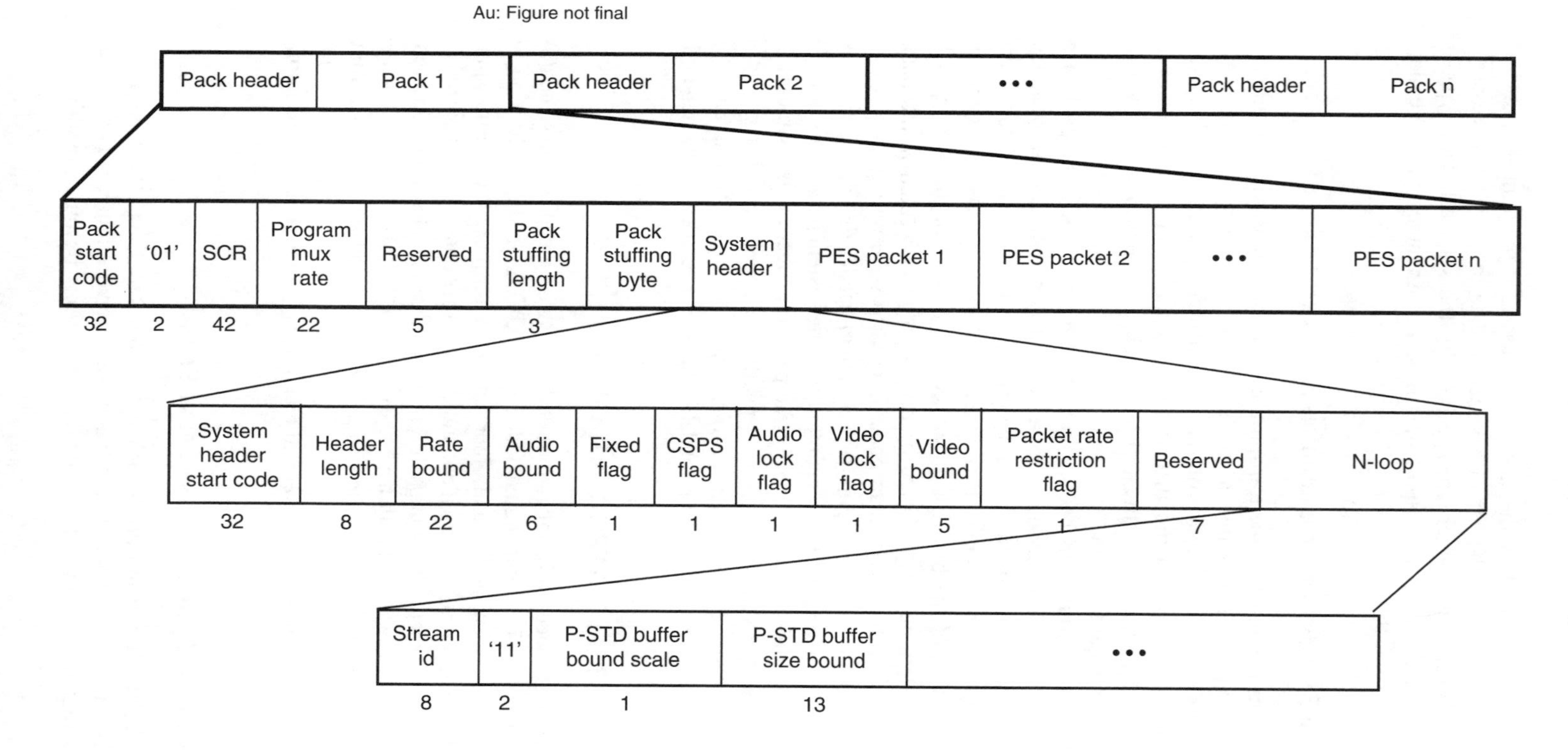

FIGURE 5.4 Structure of program stream.

minimum effort. The structure of transport stream supports the following transcoding operations.

In the first transcoding operation, the transport stream contains multiple programs, and we want to convert this kind of stream to one that contains only one program. This operation extracts the transport stream packets from one special program among the multiple programs in the transport stream and produces an output as a new transport stream that contains only that special program. This operation can be considered as the system layer transcoding that converts a transport stream containing multiple programs into a transport stream containing only a single program. In this operation, the value of program clock reference (PCR) may need to be corrected to take account for changes of the PCR locations in the bit stream. In the program stream, the timing information for a decoding system is the system clock reference (SCR), while in the transport stream, the timing information is given by the PCR. The SCR and PCR are time stamps that are used to encode the timing information of the bit stream itself. The 27-MHz SCR is the kernel time base for the entire system. The PCR is 90 kHz, which is $^1/_{300}$ of the SCR. In the transport stream, the PCR is encoded with 33 bits and is contained in the adaptation field of the transport stream. The PCR can be extended to the SCR with an additional 9 bits in the adaptation field. For the program stream, the SCR is directly encoded with 42 bits and it is located in the pack header of the program stream. Detailed information can be found in the MPEG-2 *System Document* [5-1].

In the second system-level transcoding operation, the input is multiple transport streams, and each may contain multiple programs. We want to extract one or more programs from one or more transport streams and produce a new transport stream that contains the extracted programs. This is a different kind of system-level transcoding compared with previous one. This transcoding converts the current transport streams into a different one that contains the selected programs extracted from the input transport streams.

The third transcoding operation is to convert a transport stream into a program stream for the applications of converting the DTV data format into the DVD data format. In this case, the contents of one program are extracted from the transport stream and used to produce output of the program stream.

Finally, the operation is to convert a program stream to a transport stream that can be used in a lossy communication environment.

To answer the question of how to define the transport stream and then make the above transcoding become simpler and more efficient, we have to look at the technical detail of the systems specification [5-1]. In the transport stream packet headers, there is a 13-bit field, which is referred to as the PID that provides information for multiplexing and demultiplexing by uniquely identifying each packet belonging to a particular bit stream. Therefore, the parser in the system decoder can easily extract the selected program, which would be converted to another transport stream. Since both the transport stream and program stream are built from PESs, it may not need to make any changes at the PES packet level for the conversion between the transport stream and program stream with some constraints.

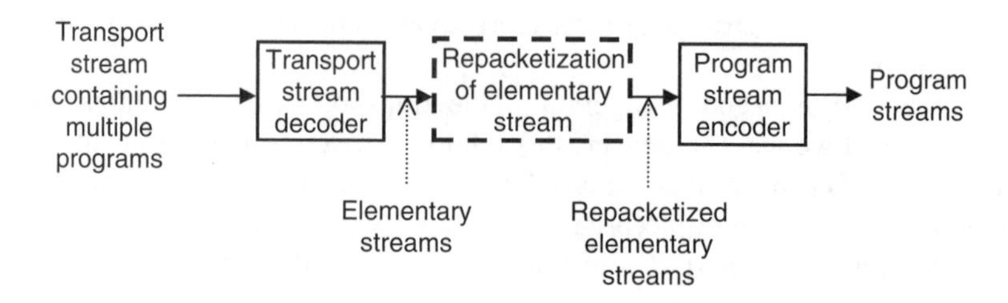

FIGURE 5.5 Conversions from transport stream to program stream.

5.2.2 Transcoding between Transport Stream and Program Stream

As we mentioned previously, we may need to convert the transport stream into the program stream or vice versa. In the first case, we want to convert the broadcasting digital television signal, which is the transport stream, into the digital video disk (DVD), which uses the program stream. This conversion is shown in Figure 5.5.

From Figure 5.5 it is noted that the transport stream is first demultiplexed to the elementary streams by the transport decoder and then these elementary streams such as video streams and audio streams are remultiplexed to the program stream by the program stream encoder. In such a way, the transcoding of the DTV signals to DVD signals at system level has been completed. Actually, in the procedure of this kind of transcoding, only the header part of the system stream (either transport stream or program stream) has been changed, the packetized elementary streams (PES) may be directly mapped. From the specification of transport stream and Program Stream, the PES packets may, with some constraints, be mapped directly from the payload of one multiplexed bit stream into the payload of another multiplexed bit stream. Be careful that only with some constraints, the PES packets of transport stream can directly be mapped to the packets of program stream. In general, after demultiplexing the video and audio elementary streams may need to be repacketized due to two reasons. First, the original packets may not be suitable to fit the packs of the program stream; we need to repacketize the elementary streams into PESs for satisfying the requirements of packs in the transport stream. Second, we may need to change the bit rate of video stream, mostly, to scale down the bit rate for DVD applications. After bit rate reduction or spatial resolution reduction, the video elementary stream needs to be repacketized to new PESs. Consequently, a video transcoder may be needed for the bit rate or resolution conversion between transport stream Decoder and program stream Encoder in Figure 5.5. The technical details of the video transcoder can be found in other chapters of this book.

During the packet mapping, the correct order of PES packets in a program can be identified by the assistant with the program_packet_sequence_counter. In the MPEG-2 syntax, the program_packet_sequence_counter is present in all PES packets, which is a 7-bit field. It is an optional counter that increments with each successive PES packet from a program stream or the PES packets associated with a single program definition in a transport stream, providing functionality similar to a continuity counter. This allows an application to retrieve the original PES packet sequence of a program stream. The counter will wrap around to 0 after its maximum value. Repetition of PES packets shall not occur. Consequently, no two consecutive PES packets in the program multiplex shall have identical program_packet_sequence_counter values. It is the same as in the case of conversion from program stream to transport stream; the program_packet_sequence_counter field is included within the header of each PES packet carrying data from a program stream. This allows the order of PES packets in the original program stream to be reproduced at the decoder.

Also there is certain other information necessary for the conversions, e.g., the relationship between elementary streams, which is available in the tables and the headers in both streams. Such data, if available, shall be corrected in any stream before and after conversion. For the details on those tables, please check the MPEG-2 system document [5-1].

Sometimes, we may need to use the contents from DVD for broadcasting application. In this case, it needs to convert the program stream to the transport stream. During this conversion, besides the header changes, other changes may be needed. For example, for DVD application, the video is encoded by variable bit rate (VBR) in order to provide constant quality over a whole sequence, whereas for DTV application, due to buffer requirements in the decoders, the video should be encoded with constant bit rate (CBR). Therefore, for this kind of transcoding we need to convert the VBR to CBR.

In the next section, we will discuss the technical details of transcoding between VBR coding and CBR coding.

5.3 TRANSCODING BETWEEN CBR AND VBR

We have described the applications of video coding with CBR and VBR in the previous section. In this section we first introduce the principle of CBR and VBR coding, and then indicate the fundamental difference between VBR and CBR coding. Afterwards, we present the algorithms for achieving a desired bit rate for VBR video coding. Both theoretical analysis and experimental results have shown that the VBR coding has significant bit savings potential over CBR coding for achieving the same coding quality. We will discuss how this fact would affect the transcoding design for the conversion between CBR to VBR coding. Finally, we will introduce the application of transcoding, which converts multiple VBR streams into a CBR stream and the fundamental requirements for CBR and VBR transcoding.

5.3.1 CBR VIDEO CODING ALGORITHM

Currently, most digital video broadcasting systems use the MPEG-2 video coding standard to encode raw video signals into the CBR bit streams. In CBR video coding, the encoder has to generate bit streams with a desired average bit rate over a certain time period, but the variation of the bit rate cannot exceed a limit due to the buffer constraints. The major reason for using CBR coding is that because its bit rate is constant, it is easy to be transported over conventional communications channels, which usually have constant bandwidth. Another reason for using CBR coding is required by those video decoders that have a limited buffer size. Since CBR coding rate control aims at balancing the desired average bit budget over a specific time period, this implies that the number of bits used for encoding the different segments of a video program is the same. This results in video coding quality variation with respect to video content. For many video broadcasting or transmission applications, higher bit rate is usually used to guarantee acceptable video quality for the most complex video content segments. At the same time, more bits are unnecessarily wasted for simple video content segments. To improve the efficiency of digital broadcasting networks, the migration of digital video coding technology from CBR to VBR coding is the next trend that could achieve substantial bit savings without sacrificing encoded video quality under some application scenarios such as for the multiple programs transmitted over a constant bandwidth channel with a statistical multiplexer.

For CBR coding, the rate control algorithm is designed with the following bandwidth-limiting requirements. First, the decoder's bit receiving buffer is fed with bits from a communication channel or storage media at a constant rate. Second, the encoder's bit transmitting buffer shall always have enough bits to sustain the requirement of constant bit rate transmission. Finally, the encoder should never spend more bits for a picture that could not be transmitted to the decoder's bit receiving buffer before its decoding time at the specified constant output bit rate. For example, for a video with 30 frames/s, the time interval for decoding a frame $^1/_{30}$ second; then the maximum number of bits for a frame is determined by the specified bit rate.

Based on the above three requirements, the simplest encoding rate control algorithm is to encode each input video picture with a fixed number of bits so that the total output bit rate is constant and easily controllable. However, this kind of algorithm is not suitable for video coding, it could only be used for video sequence coding based on a still picture coding algorithm, such as motion JPEG, JPEG2000, or MPEG-2 with all intra-coded frames. It is known that in the MPEG video coding standards, in order to remove temporal redundancy and to increase coding efficiency, the intra-coded frames (I-picture), the forward predictive (P-picture), and bipredictive coded frames (B-picture) have been employed. Motion estimation and motion compensation techniques have been used for P- and B-pictures and have improved the coding efficiency substantially. One of the impacts of such an inter-picture coding technique is the variation of resulted number of bits for coding a picture, which is no longer a constant. It is also well known that the I-picture consumes the most bits, the P-picture consumes fewer bits than I-picture, and the B-picture consumes the least bits on average. However the bit allocation differences among different coding picture types are constrained also by the requirement of a constant output bit rate for CBR coding. For most MPEG video encoder implementations, the constant bit rate requirement is usually enforced or balanced at the group of pictures (GOP) layer, which is typically set to 15 for 30-Hz video formats. This implies that the rate control algorithm has to spend the budgeted amount of bits within the GOP. This mechanism results in variation of coded video quality due to insufficient bits for the video segments with complicated content and too many bits for ones with simple and still video content. Under this sense, we could say that CBR encoding can generate the bit stream with constant bit rate but variable visual quality. The typical CBR rate control algorithm is presented in the Test Model 5 (TM5) of MPEG-2 video [5-2], which has been presented in Chapter 2.

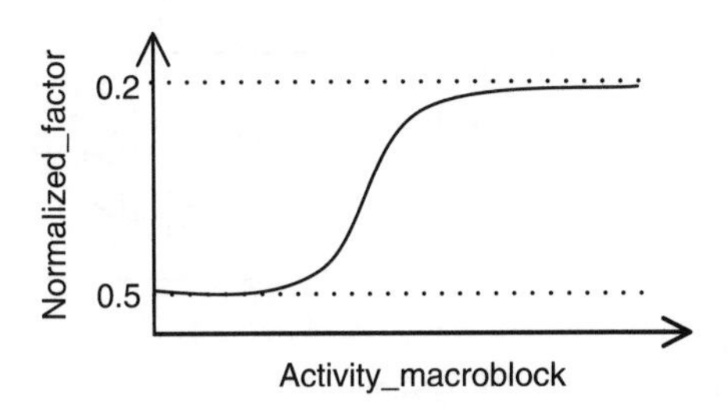

FIGURE 5.6 Normalization factor in the adaptive quantization vs. macroblock activity.

In the MPEG-2 TM5, the rate control is achieved by selecting a specific quantization step or scale for each macroblock in a picture. The rate control algorithm of TM5 includes three steps: target bit allocation, rate control, and adaptive quantization. In the step of target bit allocation, the number of bits available for coding the next picture is estimated before coding the picture. The rate control step sets the reference value of the quantization parameter for each macroblock by means of a virtual buffer. The step of adaptive quantization aims at improving the subjective quality. In this step, the reference value is modulated to get the real quantization scale according to the spatial activity in the macroblock. This relies on the fact that the smoothing areas visually are more sensitive to the coding errors than are active areas. The adaptive quantization step generates a normalized factor, which is used to modulate the reference value of the quantization scale according to the activity of the macroblock. The normalized factor is calculated by the following formula.

$$Normalized\_factor = \frac{2 \bullet activity\_macroblock + average\_activity}{activity\_macroblock + 2 \bullet average\_activity}$$

In Figure 5.6, it is noted that for the macroblock with lower activity its quantization scale is reduced up to half after normalization and for the one with higher activity the quantization scale is increased up to 2 times the reference value. In such a way, the subjective quality can be improved but the numerical SNR may be lower. The technical details of TM5 rate control can be found in Chapter 2.

Due to the use of different coding types of pictures, the number of bits spent for each frame may be quite different, but its variation should follow the buffer rule. The modeling of buffer fullness is shown in Figure 5.7.

In Figure 5.7, R is the bit rate for CBR coding, $B_{\max}$ is the buffer size, B_{init} is the initial buffer fullness, b_i is the number used to code i-th frame, and B_i is the buffer fullness after coding the i-th frame. The buffer rule can be described as follows.

$$\begin{aligned} B_0 &= B_{\text{init}} \\ B_{i+1} &= B_i - b_i + RT_i \\ b_i &\le B_i\ B_{\max} \end{aligned} \tag{5.1}$$

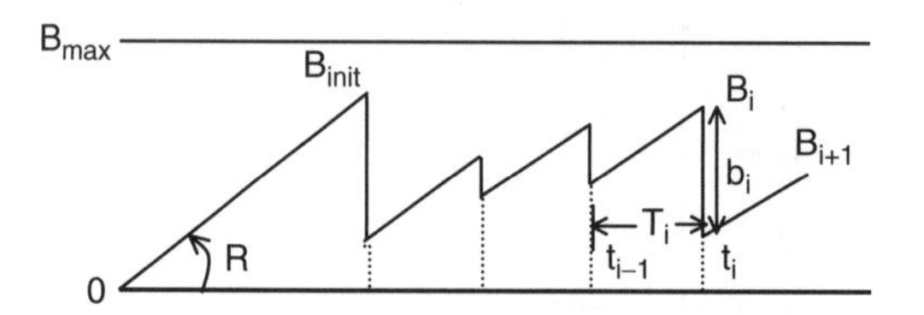

FIGURE 5.7 Buffer modeling of CBR coding.

If thees buffer rules are satisfied, the buffer overflow and underflow in the CBR video coding can be avoided.

5.3.2 VBR Video Coding Algorithm

The rate control strategies for CBR video coding have been investigated by many researchers and the strategies are quite mature. For example, in the MPEG-2 video coding algorithms the CBR coding has been studied in detail and used as the default mode of operation for most known applications [5-2]. The rate control algorithm for VBR coding has now attracted more attention [5-3], [5-4], [5-5], [5-6] due to its high compression performance over CBR coding. The VBR coding has been supported syntactically but not studied in detail. In this section, we first introduce the general principle of VBR video coding and then introduce two-pass VBR coding algorithms.

5.3.2.1 General Principles of VBR Coding

The definition of VBR video coding is often a cause of confusion. Generally, a VBR encoder would generate bit streams with nonconstant bit rate. In other words, the output data volume during a unit of time is not a constant. The basic principle of VBR coding is to encode the video sequences as much as possible to a constant picture quality. To reach this target, the number of bits used for coding each frame should be a function of spatial and temporal frame content complexity. A simple example of VBR coding is to use the same quantization scale to code a whole video sequence to achieve consistent video quality based on the fact that the same quantization scale usually leads similar distortion when the quantization scale is small enough. As we know, if the quantization scale is small enough, the quantization error can be considered as white Gaussian noise, the variance of quantization scale can be calculated as 12/(square of quantization scale), which can be easily converted to PSNR, which, as a numerical quality measure, is widely used in the image or video coding. This kind of VBR video coding is referred to as *free VBR* [5-6], or *unconstrained VBR* [5-3]. Free VBR coding has several problems in the practical application. First, the decoder buffer is assumed to be infinite because the peak number of bits generated by free VBR has no limit. In the MPEG encoding, there are usually three types of video pictures, I-, P-, and B-pictures. If the same quantization scale is used throughout the entire sequence by the VBR encoder, it may result in the similar distortion. However, due to different scene

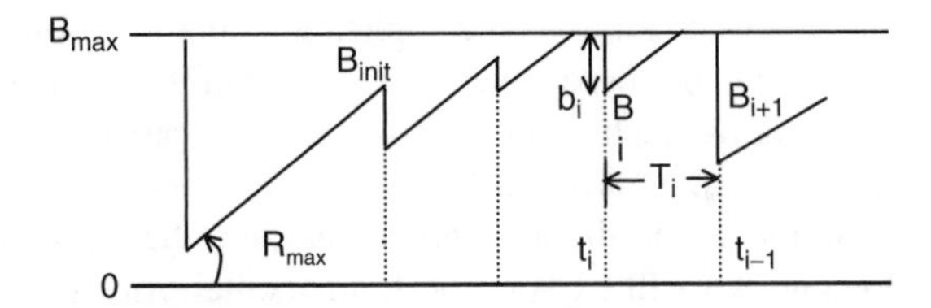

FIGURE 5.8 Buffer modeling of VBR coding.

activities and motion features in the different segments of video sequence, the bit rate of such encoding systems will vary from time to time. More bits will be generated for the more active video segments than for not so active video segments. Therefore, the decoder has to handle this problem with very large buffer size; theoretically, it should be infinite. Second, in the practical application, usually a defined output bit rate is expected due to the fixed communication channel capability or fixed size of storage devices, but it is hard for free VBR to satisfy this requirement since there is no rate control mechanism in the free VBR encoder. It may be achieved with a multiple-pass encoding process, which is very expensive. Therefore, in the practical application for designing a VBR encoder, we assume that the total budget for coding a video sequence is fixed such as for the fixed-size storage video application. Under this constraint, the bits generated by the VBR encoder may enter the decoder buffer at any bit rate, but with a given maximum rate $R_{\max}$. The bits enter the decoder buffer at a constant peak rate $R_{\max}$ until the buffer is full, at which point the bit rate is temporarily set to zero until the next picture is removed from the buffer. The buffer model of VBR coding is described in Figure 5.8.

In Figure 5.8, $B_{\max}$ is the buffer size, B_{init} is the initial buffer fullness, b_i is the number used to code i-th frame, and B_i is the buffer fullness after coding the i-th frame. The buffer rule for VBR video coding can be described in Equation (5.2).

$$B_0 = B_{\max}$$

$$B_{i+1} = \min(B_{\max,}\ B_i - b_i + R_{\max}\ T_i) \tag{5.2}$$

$$B_i \geq b_i$$

Here B_0 is the initial buffer fullness; in Figure 5.8, we set it to the buffer maximum buffer size. The term T_i is the time that takes to display i-th picture. In the MPEG-2 video standard, because of a possible repeat-first-field in the case of 3:2 pull-down, this time may vary from picture to picture. All bits for i-th picture must arrive in the buffer before it is removed to prevent the buffer underflow, but there is no buffer overflow constraint in the VBR video coding. Therefore, there is no need for padding in the VBR coding, which is needed in the CBR coding to prevent the buffer underflow.

Even though there are several difficulties in the VBR coding, the feature of high compression ratio of VBR coding has attracted many applications, such as DVD,

VCD, and video servers. Depending on the application configuration of encoder and multiplexer design, significant improvements on coding efficiency could be reached. In the practical applications, such as in the fixed storage cases, VBR coding algorithms should be designed quite differently than the free VBR algorithm; they need to satisfy both requirements, on the average bit rate and the finite size of decoder buffer. In the following, we will explore the fundamental principles and introduce several algorithms of two-pass VBR coding.

5.3.2.2 Two-Pass VBR Coding Algorithms

As previously described for the practical applications of VBR video coding, the problems addressed can be formatted as follows. In fixed-size video storage applications, the total budget for coding an entire video sequence is given, e.g., the average bit rate of the compressed video bit stream will not exceed a specific value; for the given bit budget or the average bit rate, we want to develop encoding algorithms that can generate the bit stream with optimized visual quality for the whole video sequence. This kind of VBR coding can be seen as a long-term CBR coding, for which the rate control algorithm is needed. In order to optimally allocate the given bit budget into different frames, we need to know the distribution of the amount bits for each frame in the entire video sequence generated under all possible quantization scales. From the distribution of bit allocation we can select the best combination of quantization scales, which use the least number of bits to obtain the best coding performance for coding the entire video sequence. This task is impossible in the real-time encoding, but it is possible with non–real time multiple-pass encoding. In the following, we present several two-pass coding schemes for VBR codin [5-5], [5-6]; the two-pass scheme is a kind of trade-off between coding performance and coding complexity.

The Two-Pass VBR Coding Algorithm Using Training Sequences [5-5]

One such two-pass VBR coding scheme proposed in [5-5] is shown in Figure 5.9. In this scheme, the video sequence is first encoded by the conventional CBR coding algorithm in the first pass. The purpose of the first pass coding is to obtain a set of statistics about the whole input video sequence. These statistical parameters include the average quantization scale for each picture, the number of bits for each picture, the picture coding type, the number of intra-coded macroblocks, the measure of spatial activity, and the measure of temporal activity.

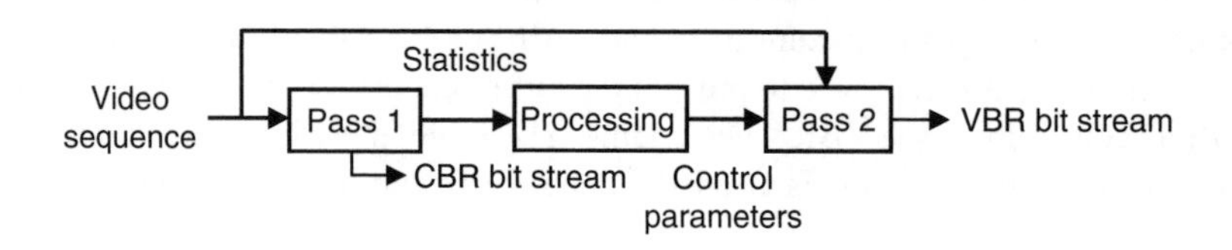

FIGURE 5.9 Two-pass VBR coding scheme proposed in [5-5].

The average quantization scale is an important parameter. It, together with the number of bits for each frame, can be used to determine the constant in the bit-production model. In MPEG-2 video coding, a linear model is suggested in the Test Model 5:

$$b_i(Q_i) = \frac{c_i}{Q_i} \tag{5.3}$$

Here, $b_i(Q_i)$ is the number of bits used to code the i-th picture, Q_i is the average quantization scale for coding i-th picture, and c_i is the constant for this bit-production model. The spatial activity of a picture is the averaged spatial activities of all macroblocks. The spatial macroblock activity is the absolute norm variance for each block, which is calculated as follows. For each macroblock, the average pixel luminance value is first calculated. The absolute difference between each pixel value and this average is then accumulated and averaged. This yields an absolute norm variance measure for each macroblock.

The temporal activity is based on the motion vectors. The motion vectors are first normalized with the temporal distance between the anchor frame and the predicted frame. Then the absolute norm variance for both the vertical and the horizontal directions of motion vector for each macroblock is calculated similarly as for the spatial macroblock activity. The picture temporal activity is obtained with the sum of the two variances in the macroblock activity.

The statistical parameters gathered in the first pass are then processed in the next step of processing in an offline way, since the processing cannot start until the first pass through whole video sequence has finished. In this offline processing, the control parameters obtained are used for the second-pass coding. The target bits for each picture in the video sequence are determined such that the objective of VBR coding can be achieved, which is to obtain the constant coding performance with a variable bit rate coding; at the same time, the constraints on the average bit rate and prevention of decoder buffer underflow are satisfied.

The task in the second pass of video coding is to redistribute the bits among the various scenes in the video sequence in the way that results in a constant quality throughout entire video sequence with a variable bit rate coding. To achieve this, we take the bits from the easier scenes of the video sequence and put those bits into the more complicated scenes. Now the problem becomes how to reallocate the bits among the different scenes so that the various scenes can be coded to the same quality. To solve this problem, perceptual experiments have been conducted in the scheme proposed in [5-5]. Experiments were carried out with a set of nine training sequences to determine constant visual quality across different scenes with different quantization scales. Based on the acquired training set data, a mathematical model has been developed. With this model, the quantization scales and bit rates for different kinds of scenes in the second-pass coding are determined on the basis of first-pass statistics.

The technical details of the scheme proposed in [5-5] are summarized as follows. First, a set of experiments with N different scenes has been conducted. In the experiments, each scene is coded with various fixed quantization scales and a fixed bit rate CBR coding. The bit rate for coding scene n in the second pass is calculated from the equation

$$R_{2,n} = k(Q_{1,n})^p \tag{5.4}$$

where parameter $p = 0.5$ and k is calculated from the constraint.

$$\frac{1}{N}\sum_{n=1}^{N} R_{2,n} = R_{2,\text{ave}} \tag{5.5}$$

The target bits for coding i-th picture are

$$b_{2,n,i} = \left(\frac{R_{2,n}}{R_{1,n}}\right) \bullet b_{1,n,i} \tag{5.6}$$

where $R_{1,n}$ and $b_{1,n,i}$ are the bit rate and number of bits for coding scene n in the first-pass coding, respectively.

Note that due to the limited capability of the first-pass CBR coding, some segments of statistics were found not suitable for immediate use by preprocessing. A filtering operation is then required, which includes scene change and fade detection and recalculation of the first-pass quantization scales. In addition, the buffer underflow is considered in the proposed scheme. The proposed two-pass VBR coding scheme has shown that the output of VBR coding has a higher overall quality than the ones coded with CBR coding at the same bit rate for the video sequence containing a mix of "easy" and "difficult" scenes.

The Two-Pass VBR Coding Algorithm with R-Q Function [5-6]

The problem with previously proposed method is that it is based on the bit-producing model, which is the empirical formula obtained by encoding many training sequences. In the following, we introduce the two-pass VBR algorithm without the need of training sequences [5-6]. The block diagram of the system proposed in [5-6] is shown in Figure 5.10.

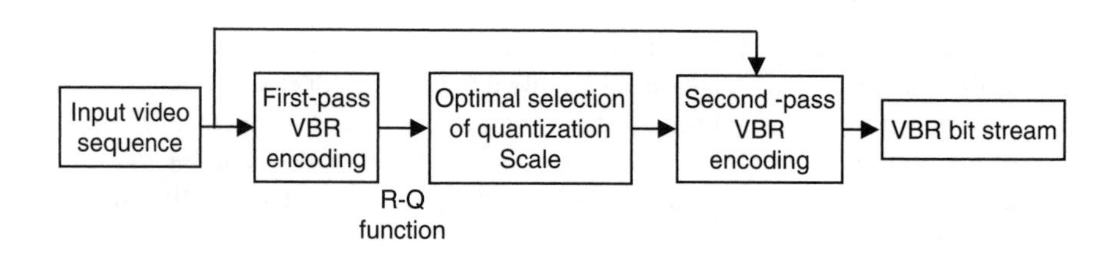

FIGURE 5.10 Two-pass VBR coding with R-Q function.

In the proposed algorithm, the function of the first-pass coding is to generate an R (rate) and Q (quantization scale) function. Intuitively, the averaged quantization scale over a frame determines the number of bits generated for each frame. If we can obtain the relationship between all possible quantization scales and the number of bits generated with these quantization scales for each frame, i.e., a R-Q function, we can determine the appropriate quantization scales for each frame of the entire sequence to satisfy the constraints on both the average bit rate and the finite decoder buffer size. However, it is very expensive to generate an R-Q function using all possible quantization scales through whole encoding processing. As we know, motion estimation consumes the largest part of encoding computation; especially, the decoding loop is usually involved in the process of the motion estimation for obtaining high coding performance. To alleviate this problem, in the proposed scheme the original video frames, instead of reconstructed frames from the decoder, are used as predictive reference frames for motion estimation in the process of generating the R-Q function. The computation of motion estimation and DCT is to consume about 80% of total computation at the encoder. In such a way, the computation of the R-Q function is approximately equivalent to one pass of a free VBR encoding.

After the R-Q function is generated in the first-pass encoding, the next step is to optimally choose the quantization scales, which can be used in the second-pass encoding to obtain the optimal coding performance and at the same time to satisfy the constraints on the fixed storage size and finite decoder buffer. The problem can be summarized as:

$$\{Q_I^*, Q_P^*, Q_B^*\} = \arg\min_{\{Q_I^*, Q_P^*, Q_B^*\}} \left| \sum_{i=1}^{N_I} R_i^I + \sum_{i=1}^{N_P} R_i^P + \sum_{i=1}^{N_B} R_i^B - R \right| \tag{5.7}$$

subject to Q_I, Q_P, $Q_B \in \{\text{all possible quantization scales}\}$ and

$$\sum_{i=1}^{N_I} R_i^I + \sum_{i=1}^{N_P} R_i^P + \sum_{i=1}^{N_B} R_i^B \leq R \tag{5.8}$$

where R is the total number of bits for encoding the whole video sequence; Q_I^*, Q_P^*, Q_B^* are the optimal quantization scales for I-, P-, and B-frames, respectively; N_I, N_P, N_B are the total numbers of I-, P-, and B-frames, respectively; and R_i^I, R_i^P, R_i^B are the number of bits used for encoding i-th I-, P-, or B-frames with quantization scales of Q_I, Q_P, Q_B, respectively.

According to MPEG-2 TM5, suitable ratios between the numbers of bits used to encode the I-, P-, and B-frames are approximately 4:2:1. Therefore, we have

$$\sum_{i=1}^{N_I} R_i^I : \sum_{i=1}^{N_P} R_i^P : \sum_{i=1}^{N_B} R_i^B = 4N_I : 2N_P : N_B \tag{5.9}$$

With the constraint

$$\sum_{i=1}^{N_I} R_i^I + \sum_{i=1}^{N_P} R_i^P + \sum_{i=1}^{N_B} R_i^B = R$$

we can obtain the values of

$$\sum_{i=1}^{N_I} R_i^I, \ \sum_{i=1}^{N_P} R_i^P \text{ and } \sum_{i=1}^{N_B} R_i^B$$

Then with R-Q function, the appropriate quantization scales can be selected for the second pass of VBR encoding. Of course, before the second pass encoding, the quantization scales selected need some modification to meet the decoder buffer regulation. To resolve the problem of decoder buffer overflow, the scheme proposes to pause the data-reading process at some specific time for the software decoder and to stop inputting data into the decoder buffer for the hardware decoder. The decoder buffer underflow may occur at the specific time; it is usually an accumulated effect due to several continuous big frames removed from the buffer. To avoid the buffer underflow, the scheme proposes to use coarser quantization scales for several frames. Experimental results have shown that the proposed two-pass VBR coding scheme is able to achieve high coding performance and more consistent visual quality than the conventional CBR video coding schemes.

5.3.3 Comparison of CBR and VBR

After we introduced the basic concept and algorithms associated with CBR and VBR video coding, we can summarize the differences between these two strategies. Compared with CBR encoding, VBR encoding is able to provide coding results with relative constant visual quality but variable bit rate, and higher coding efficiency for many video sequences.

Conside ring the principles of rate control algorithms of CBR and VBR coding, we can find the following differences.

1. CBR coding has a minimal bit spending constraint due to constant bit rate transmission needs. Even for still sequences, the specified bit rate is still maintained so that some bits are unnecessarily wasted.
2. VBR coding has no minimal bit spending constraint such that minimal bits can be used to code still sequences or other simple sequences.
3. VBR coding bit rate control is mainly modulated by the coding quality parameter such that minimal bits are used to achieve the target coding quality.
4. The bit savings on the simple video content effectively produces a coded bit stream with a lower average bit rate than the peak rate.

Finally, comparing VBR coded average bit rate to the constant bit rate of CBR coding, VBR coding could achieve higher video coding quality since it can use higher peak rate to encode the most complicated content. From the bit savings perspective, CBR coding will suffer video quality impairment when the bit rate is not enough for high complexity content. On the other hand, there will be significant bit overspending for lower complexity video content that will not improve the coded video quality significantly. The variation of coded video quality associated with video content variation will produce so-called pumping, or breathing effect.

In summary, the major difference between CBR and VBR is rate control scheme. The major parameter for rate control is the quantization scale. The finer quantization scale will generate more bits and better visual quality, whereas the coarser quantization scale will result in fewer bits but worse visual quality in general. In the MPEG-2 CBR encoder, the quantization scale is updated from macroblock to macroblock according to the buffer fullness and bit budget for the GOP. For the VBR encoder, usually the fixed quantization scales are used for I-, P-, and B-frames, respectively. In such a way, the problem of how to satisfy the bit rate constraint has to be addressed. Also, the decoder buffer size is assumed to be infinite in the VBR decoder. This is not practical, but how to add the constraint of buffer size is another topic for VBR encoding, which we have discussed in the two-pass VBR coding algorithms.

5.3.4 An Example of Transcoding between Transport Stream and Program Stream

In the following subsection, a system-level MPEG transcoding scheme for DVD recording applications is introduced. The purpose of this transcoding is to convert the MPEG-2 transport stream (TS) to program stream (PS). The major application of this transcoding is to convert digital television (DTV) broadcasting audio and video bit streams to the audio and video bit streams for digital video disc (DVD) applications. An important technical issue for transcoding between PS and TS is the rate control algorithm as discussed previously. Here, a novel VBR rate control algorithm is proposed for the online video transcoding process. In this algorithm, the bit allocation strategy at picture level, which is based on a bit producing model and linear *R-Q* function, is first discussed. Then, the rate control algorithm at macroblock level with *R-D* optimization is proposed. Experimental results have shown that the proposed VBR transcoding algorithm achieved the better image quality compared with CBR transcoding at the same bit rate.

The system architecture of the transcoding is shown in Figure 5.11. The transcoding process consists of three steps. First, the transport stream is demultiplexed into video and audio elementary streams. For digital television broadcasting, the video contents are encoded to MP@ML (standard-definition television) or MP@HL (high-definition television), and usually are encoded by constant bit rate (CBR). However, the video stream in recorded DVD contents must be MP@ML, and usually is variable bit rate stream. Therefore, it is necessary to develop an MPEG-2 video transcoder that is able to convert CBR video streams into VBR video streams in order to recorder the DTV program to DVD with better image

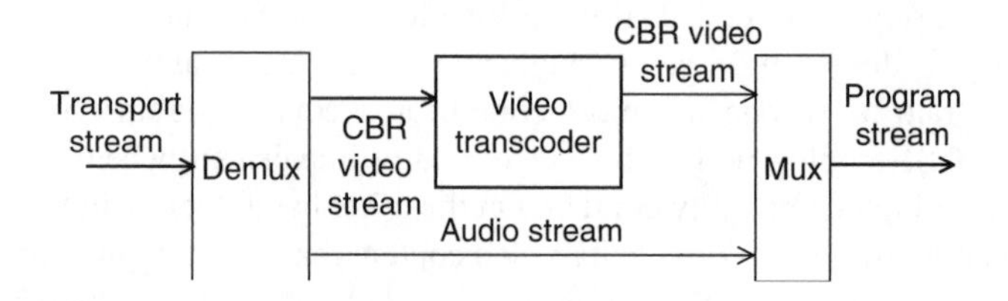

FIGURE 5.11 System architecture of MPEG transcoding for DVD recording.

quality. Finally, the transcoded video stream and the audio stream are multiplexed into MPEG-2 program stream for DVD recording.

In this example, the input of the transcoder is the video bit stream of *MP@ML* with the constant bit rate at 6 Mbps, and the output is the bit stream of *MP@ML* with variable bit rate at the average bit rate of 3 Mbps. Therefore, the function of the proposed transcoder aims at reducing the bit rate and at the same time converting a CBR video stream into a VBR video stream. The output VBR video stream can then be able to be encode to the DVD medium in an efficient way. During the past decade, there have been many papers published that address video transcoding problems [5-8], [5-9], [5-10], [5-11]. For bit-reduction video transcoding, usually there are three types of MPEG-2 video transcoder architectures: cascaded architecture, open-loop architecture, and closed-loop architecture [5-8]. In this example, the proposed video transcoder is based on the close-loop architecture. A VBR rate controller is proposed and implemented with this architecture to solve the problem of converting CBR video stream into VBR video stream and at the same time converting the high bit rate to the lower bit rate. During this process, we try to use the information contained in the CBR stream as much as possible.

The algorithm works in two-steps: target bit allocation and macroblock rate control. In the step of target bit allocation, the target number of bits for k-th picture in the transcoded VBR video stream (T_k) is set according to Equation (5.10) [5-5],

$$T_k = \alpha \times R_{c,k} \times \left(Q_{c,k}\right)^{\beta} \tag{5.10}$$

where $R_{c,k}$ and $Q_{c,k}$ are the produced number of bits and the average quantization scale, respectively, for the k-th picture in the CBR video stream; α and β are function parameters. In this work, the power parameter β is set to 0.5, and the parameter α is calculated as in Equation (5.11).

$$\alpha = \left[\frac{B - \sum_{i=1}^{k-1} R_{v,i}}{n-(k-1)} \times W\right] \Bigg/ \left[\sum_{i=k}^{k+W-1} R_{c,i} \times \left(Q_{c,i}\right)^{\beta}\right] \tag{5.11}$$

where $R_{v,i}$ is the produced number of bits for i-th picture in the transcoded VBR stream and W is the slide window size, which is set to the half of picture rate; B is the total bit budget that can be calculated with Equation (5.12):

$$B = \left(R_{v,\mathrm{avg}} \times N\right)/F \tag{5.12}$$

where $R_{v,\mathrm{avg}}$ is the VBR video stream average bit rate; N is the total picture number; and F is the picture rate, i.e., the number of pictures per second. According to the linear R-Q function, the reference value of the quantization scale $Q_{\mathrm{ref},k}$ for k-th picture in the VBR video stream can be calculated with Equation (5.13).

$$Q_{\mathrm{ref},k} = \left(R_{c,k} \times Q_{c,k}\right)/T_k \tag{5.13}$$

After we obtain the reference value of quantization scale, the next step is the macroblock-level rate control. The function of macroblock rate control step is to meet the target number of bits for each picture, and at the same time to achieve the best video quality possible. For each macroblock, the distortion d and the number of bits r can be represented as follows.

$$d_k(q_k) = q_k^2/12, \quad r_k(q_k) = x_k/q_k = \left(r_k' \times q_k'\right)/q_k \tag{5.14}$$

Here r_k' and q_k' are the number of bits and the value of quantization parameter for the same macroblock in the CBR bit stream, and x_k is the complexity for this macroblock. In order to achieve the minimum distortion, the problem can be formulated as

$$\min \sum_{k=1}^{M} d_k(q_k), \quad \text{subject to} \sum_{k=1}^{M} r_k(q_k) \leq T \tag{5.15}$$

here T is the target number of bits for the picture.

We can use the Lagrange optimization method to get the new quantization scale.

$$\begin{aligned} J(\lambda) &= \sum_{k=1}^{M} d_k(q_k) + \lambda\left(\sum_{k=1}^{M} r_k(q_k) - T\right) \\ &= \sum_{k=1}^{M} q_k^2/12 + \lambda\left(\sum_{k=1}^{M} x_k/q_k - T\right) \end{aligned} \tag{5.16}$$

Then, the reference value of quantization scale for each macroblock can be represented as

$$q_k = \left[\sqrt[3]{x_k} \times \sum_{i=1}^{M} \sqrt[3]{x_i^2}\right] \Big/ T, k = 1, 2, \ldots, M \tag{5.17}$$

TABLE 5.1
Experiment Results for Four Test Sequences

	CBR_3M	VBR_3M
Sequence 1	31.56	31.86
Sequence 2	33.44	33.77
Sequence 3	28.20	28.64
Sequence 4	39.68	40.35

Therefore, the reference value of quantization parameter for next macroblock q_j can be calculated as

$$q_j = \left[\sqrt[3]{x_j} \times \sum_{i=j}^{M} \sqrt[3]{x_i^2} \right] \Big/ \left[T - \sum_{i=1}^{j-1} r_i \right]$$

$$\Delta q. = q_j - Q_{\text{ref}} \tag{5.18}$$

$$\text{If } \Delta q < -2, \quad \Delta q = -2, \quad \text{if } \Delta q > 2, \quad \Delta q = 2$$

The value of quantization scale for next macroblock is then $Q_{\text{ref}} + \Delta q$.

To verify the results, we have conducted several experiments. In the experiments, four video sequences are used. All these sequences are firstly encoded to the stream of 6 Mbps with CBR rate control. Then, these 6-Mbps CBR streams are transcoded to 3-Mbps streams with VBR rate control. For comparison, transcoding CBR video stream into CBR video stream is are also implemented. The experiments results are listed in the Table 5.1 and Figure 5.12. From these results, we can conclude that our VBR rate control algorithm can achieve better video quality for DVD recording.

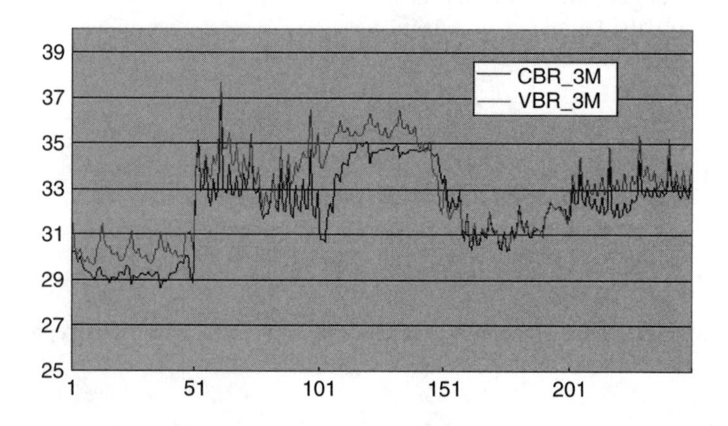

FIGURE 5.12 Simulation results of sequence 2 by using transcoding with CBR and VBR.

In this example, an MPEG-2 system-level transcoding scheme has been proposed. A novel algorithm of CBR-to-VBR conversion, which is the key technical issue of this transcoder, has been proposed. The simulation results have shown that the proposed VBR encoding algorithm can reach better image quality at the same bit rate comparing with the CBR encoding.

5.4 TRANSPORT OF VBR STREAMS OVER CONSTANT BANDWIDTH CHANNEL

As we have described, VBR video coding can always obtain coding performance that is at least equal to or better than CBR video coding. However, for some applications the constant bit rate is required at the transport level. For transport with a constant bandwidth channel, a problem that should be addressed is how to take advantage of better coding performance using VBR coding and at the same time satisfy the requirement on the constant bit rate or fully utilize the bandwidth. Three application scenarios have been proposed in [5-7]: VBR streams with available bit rate (ABR), open-loop intelligent multiplexers using multiple VBR streams, and closed-loop intelligent multiplexers with multiple VBR streams.

5.4.1 SIMPLE MULTIPLEXER WITH VBR STREAMS

It is noted that due to the bit-saving nature of VBR coding, if multiple VBR streams are multiplexed together under condition that the peak bit rate of the sum does not exceed the bandwidth of the transport channel, there will be more time when the sum of the input bit rate is much lower than the bandwidth of the transport channel. These unused bits are therefore used as an ABR data channel for various applications. The proposed simple multiplexer system is shown in Figure 5.13.

Figure 5.14 is an illustration of bit usage distribution over time of this simple multiplexer and ABR data service configuration. The question about ABR is how many bits can be used for ABR with the simple multiplexer. The answer to this question is provided by the experimental results. Some experiments have shown that a large percentage of bits could be used for ABR data channel. The reason for this is quite intuitive: among multiple VBR streams, the original video sequences may

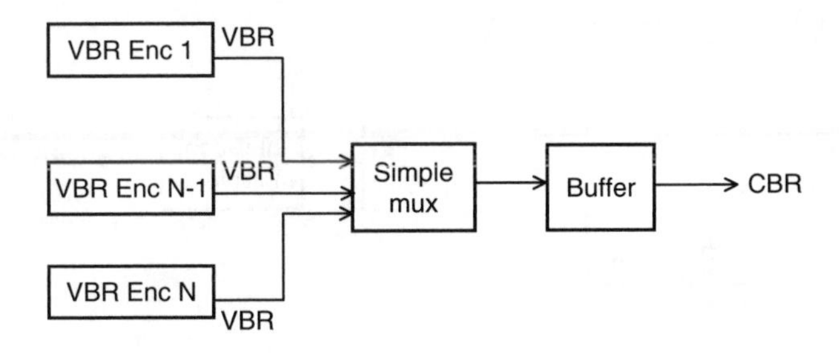

FIGURE 5.13 Simple multiplexer and ABR data configuration.

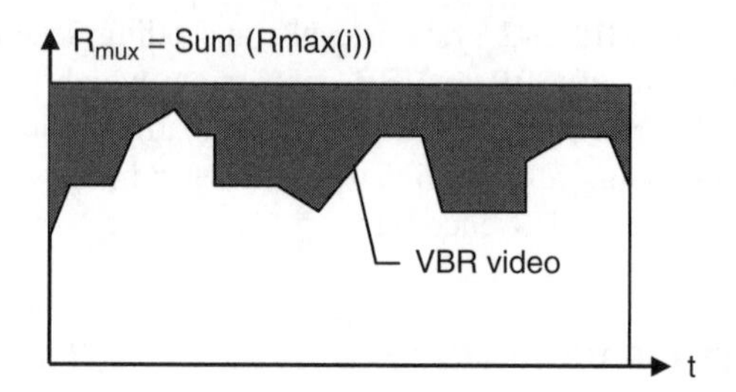

FIGURE 5.14 Transport bandwidth utilization for both VBR video and ABR data services [5-7].

have very different natures. Some are very active and may need more bits to code them, but some may be simple and easy to code. If an equal budget is assigned to each program, there should be extra bits available for ABR with the proposed scheme. The drawback of this multiplexing method is that the bit-saving capability of VBR coding is not used to either improve the video quality or increase the number of video programs with the same output bit rate.

5.4.2 Multiple VBR Streams for Open-Loop Intelligent Multiplexer

The simple multiplexer with multiple VBR streams provides an opportunity for using the saved bandwidth for available bit rate data channel. However, in many applications the major target is to improve the video quality or increase the number of programs with the same bandwidth. For this purpose, an open-loop intelligent multiplexer has been proposed, as shown in Figure 5.15.

Compared with the simple multiplexer, this system has a relatively large buffer, preanalysis processor, and intelligent multiplexer. The large buffer is needed so that enough time is available for the multiplexer to analyze the input bit stream. Based the results of preanalysis, the intelligent multiplexer understands the input VBR bit streams, it can conduct efficient statistical multiplexing of input VBR bit streams,

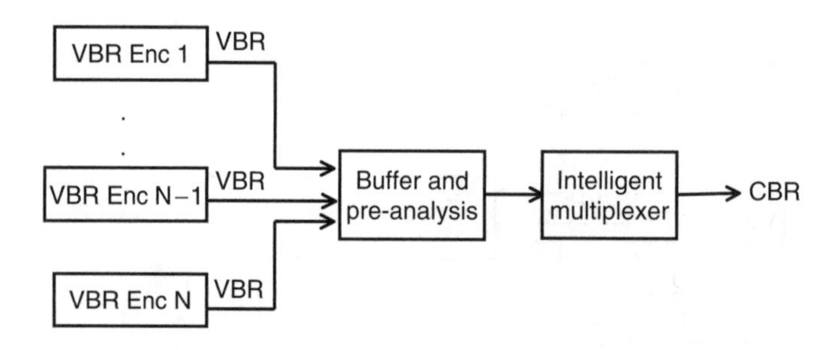

FIGURE 5.15 Open-loop intelligent multiplexer.

and has many other capabilities, such as bit stream splicing, routing, and remultiplexing. Because this intelligent multiplexer takes open-loop VBR bit stream feeds, there is no requirement to control the video encoders. This feature is especially important for some applications where more precompressed videos are directly used to feed the digital networks. The bit rates of these video are generally not controllable at the transport point of the networks to the video encoder layer.

5.4.3 Multiple VBR Streams for Closed-Loop Intelligent Multiplexer

As we mentioned, the open-loop intelligent multiplexer with VBR streams is very important for applications that have to use the precompressed video bit streams. It is obvious that the open loop can greatly improve the efficiency of channel bandwidth utilization, but it is still far from optimal due to the nature of uncontrollable variations on peak rates of VBR streams even with the use of intelligent multiplexer. To further utilize VBR bit-saving capabilities for video services, a closed-loop intelligent multiplexer can be used where the peak rate of each VBR encoder is coordinated by a statistical multiplexing rate control agent as shown in Figure 5.16. The function of the rate control agent includes changing the peak rate allocation for each VBR encoder frequently such that the sum of the peak rate is always equal to a higher percentage of the multiplexer output bit rate. With this closed-loop peak rate control mechanism, it is possible then to set the intelligent multiplexer output bit rate to be very close to 40% of the sum of the peak rate, which is the maximum bit-saving potential of VBR coding.

Of course, the closed-loop multiplexer is used for those applications where the video encoders can be controlled. From the viewpoint of applications, the bits saved by using statistical multiplexing VBR bit streams can be either used by the ABR data channel or used to improve the video coding quality through the dynamic peak rate allocation mechanism.

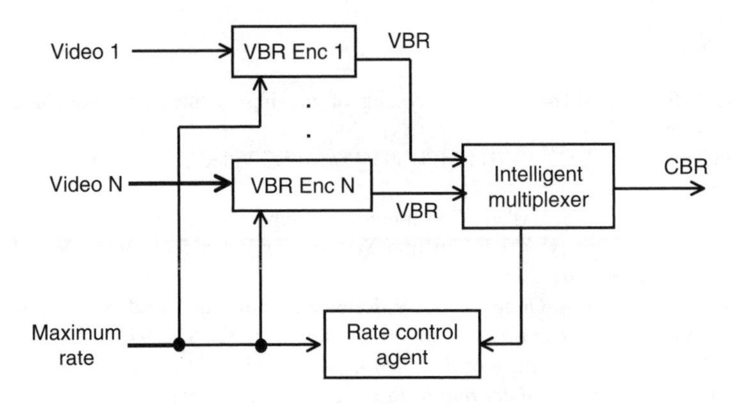

FIGURE 5.16 Closed-loop intelligent multiplexer for VBR stream statistical multiplexing.

5.5 SUMMARY

In this chapter, the basic concept of MPEG-2 system standards was first introduced. Then the transcoding strategy between the transport stream and program stream of MPEG-2 was discussed. In the subsequent sections, the CBR and VBR video coding algorithms and the issues of transcoding between CBR to VBR were described. Finally, applications of multiplexing the multiple VBR streams to single CBR stream were presented.

5.6 EXERCISES

5-1. A system layer transcoder converts a transport stream into a program stream: list all related items that need to be changed and discuss what technical issues must be considered.

5-2. In the program stream to transport stream transcoding, what technical issues have to be considered? Conduct an experiment to verify whether, to maintain the same video quality, any additional bits are needed.

5-3. Explain why the VBR coding scheme would provide better coding quality than CBR coding scheme at the same bit rate. Conduct an experiment to verify it.

5-4. If a VBR coded stream is converted to CBR stream with the same or very similar coding quality, adding extra bits is expected. Conduct an experiment to show how many additional bits are needed. Use two different video sequences, one active, such as football, and the other simple, such as the news.

5-5. Propose an algorithm for VBR coding using the multiple-pass method. Compare the results with CBR coding and two-pass VBR coding.

5-6. Conduct a project to convert a CBR bit stream from 6 Mbps to a CBR bit stream and a VBR bit stream at 3 Mbps; compare the results. Assume that the original video sequence is a SDTV format.

REFERENCE

[5-1] ISO/IEC IS 13818-1: Generic coding of moving pictures and associated audio, November 1994.

[5-2] ISO/IEC-JTC1/SC29/WG11 MPEG93/457 MPEG Video Test Model 5 (TM-5), April 1993.

[5-3] A. Reibman and B. Haskell, Constraints on variable bit-rate video for ATM networks, *IEEE Transactions on Circuits and Systems for Video Technology*, Vol. 2, no. 4, Dec. 1992, pp. 361–372.

[5-4] T. V. Laskhman, A. Ortega, and A.R. Reibman, VBR video: tradeoffs and potentials, *Proceedings of the IEEE*, Vol. 86, no. 5, May 1998, pp. 952–973.

[5-5] P. H. Westerink, R. Rajagopalan, and C.A. Gonzales, Two-pass MPEG-2 variable-bit-rate encoding, *IBM Journal of Research and Development*, 43, 471, 1999.

[5-6] Y. Yu, J. Zhou, Y. Wang, and C.W. Chen, A novel two-pass VBR coding algorithm for fixed-size storage application, *IEEE Transactions on Circuits and Systems for Video Technology*, Vol. 11, March 2001, pp. 345–356.

[5-7] S. J. Huang, Principles, benefits and applications of variable bit rate coding for digital video broadcasting, with statistical multiplexing extension, *Technical papers, NAB'99.* Las Vegas, NE.

[5-8] H. Sun, W. Kwok, and J. W. Zdepski, Architecture for MPEG compressed bit stream scaling, *IEEE Transactions on Circuits and Systems for Video Technology*, Vol. 6, Apr. 1996, pp. 191–199.

[5-9] P. Assuncao and M. Ghanbari, A frequency-domain video transcoder for dynamic bit-rate reduction of MPEG-2 bit streams, *IEEE Transactions on Circuits and Systems for Video Technology*, Vol. 8, Dec. 1998, pp. 953–967.

[5-10] Y. Nakajima, H. Hori, and T. Kanoh, Rate conversion of MPEG coded video by re-quantization process, *ICIP'*1995. Washington DC.

[5-11] P. Yin, A. Vetro, H. Sun, and B. Liu, Drift compensation for reduced spatial resolution transcoding, *IEEE Transactions on Circuits and Systems for Video Technology*, 12, 1009, 2002.

6 System Clock Recovery and Time Stamping

In many digital video applications, it is important for the signals to be stored, transmitted, or processed in a synchronized manner. This requires that the signals be time-aligned. For example, it is important that the video decoder clock locks to the video encoder (or transcoder) clock, so that the video signals can be decoded and displayed in the almost exact time instants. The action of controlling timing in this way is called *video (clock) synchronization.* In this chapter we will discuss video synchronization techniques, such as system clock recovery and time stamping for decoding and presentation.

6.1 BASICS ON VIDEO SYNCHRONIZATION

Video synchronization is often required even if the video signals are transmitted through synchronous digital networks because video terminals generally work independently of the network clock. In the case of packet-based transmission, packet jitter caused by packet multiplexing should also be considered. This implies that synchronization in packet transmission may become more different than with synchronous digital transmission. Hence, video synchronization functions that consider these conditions should be introduced into video codecs and streams. There are two typical techniques for video synchronization between transmitter and receiving terminals.

One video synchronization technique measures the buffer fullness at the receiving terminal to control the decoder clock. Figure 6.1 shows an example of such a technique that uses the digital phase-locked loop (PLL), activated by the buffer fullness. In this technique, a digital PLL controls the decoder clock so that the buffer fullness maintains a certain value. There is no need to insert additional information in the stream to achieve video synchronization.

The other technique requires the insertion of a time reference, e.g., a time stamp, into the stream at the encoder (or the encoding process of the transcoder). At the receiving terminal, the digital PLL controls the decoder clock to keep the time difference between the reference and actual arrival time at a constant value. The block diagram of this technique is shown in Figure 6.2.

The clock accuracy required for video synchronization will depend on video terminal specifications. For example, a television display generally may demand an accuracy of less than 10% of a pixel. Thus, for 720 pixels per horizontal video line, the required clock stability is about 7.2×10^{-5}. This accuracy can be achieved by using digital PLL techniques.

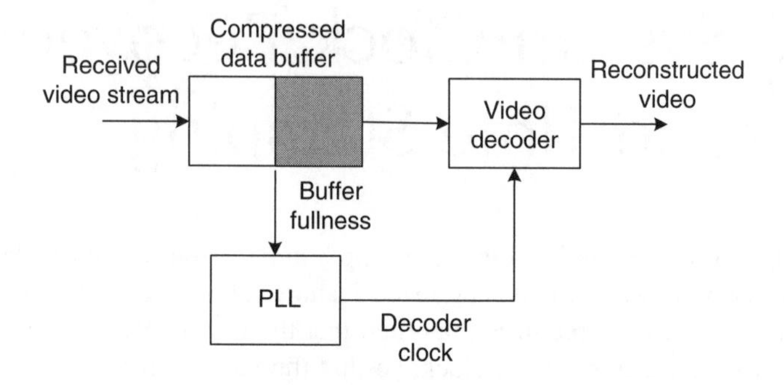

FIGURE 6.1 Video synchronization at decoder by using buffer fullness.

When a clock is synchronized from an external sync source, e.g., time stamps, jitter can be coupled from the sampling jitter of the sync source clock. It can also be introduced in the sync interface. Fortunately, it is possible, sometimes, to filter out sync jitter while maintaining the underlying synchronization. The resulting system imposes the characteristics of a low-pass filter on the jitter, resulting in jitter attenuation above the filter corner frequency.

When sample timing is derived from an external synchronization source, the jitter attenuation properties of the sync systems become important for the quality of the video signal.

In this chapter, we will discuss the technique of video synchronization at decoder through time stamps. As an example, we will focus on MPEG-2 transport systems to illustrate the most important function blocks of the video synchronization technique.

The MPEG-2 systems [6-1][6-2] specify two mechanisms to multiplex audio, video, or other data elementary streams for storage or transmission purposes. The first, which is called the *program stream*, is suitable for error-free while the second,

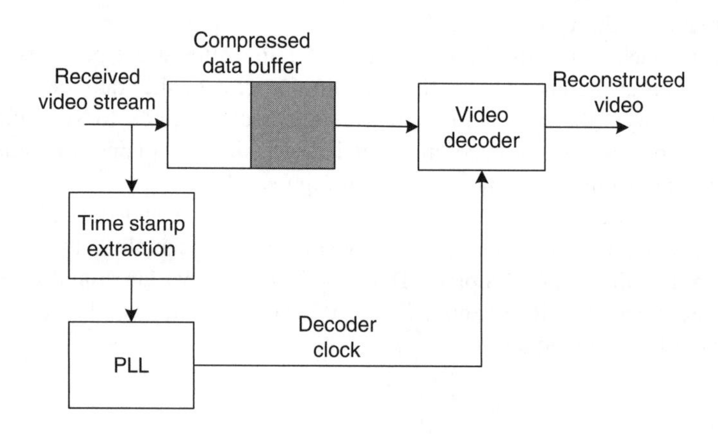

FIGURE 6.2 Video synchronization at decoder through time stamps.

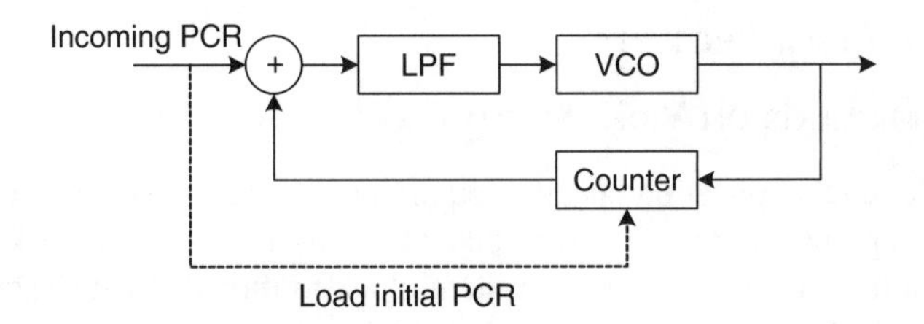

FIGURE 6.3 A digital PLL structure.

the *transport stream*, is the appropriate choice for data transmission in error-prone environments. An MPEG-2 transport stream is a single packetized-format multiplex of several programs each composed of video, audio, or private-data elementary streams that can be modulated and transmitted over communications networks. Several tables defined by the MPEG-2 standard carry program-specific information (PSI) on the transport stream to enable many functions at the receiver end. Decoding time stamps (DTSs) and presentation time stamps (PTSs) are carried in some transport (stream) packets to provide the time instants at which the video or audio pictures/frames should be decoded and displayed, respectively. Decoding time stamps and PTS's are all referenced to the encoder's (or transcoder's) system time clock (STC) which runs at 27 MHz. Samples of this clock, program clock references (PCRs), are also transmitted by the transport stream to the decoder, and the reference time is reconstructed by means of these clock samples through a digital PLL, which is shown in Figure 6.3.

In this block diagram, the counter contains the reconstructed time. The time error that is the difference between the counter value and the incoming PCR value is fed to the low pass filter (LPF) and then output to drive an instantaneous frequency of the voltage-controlled oscillator (VCO), which runs nominally at 27 MHz. Three effects contribute to the degradation of the reconstructed time. The first is the difference between the encoder clock frequency and the free-running frequency of the decoder's STC. The second is the network jitters due to the variations in queuing delays in network switches, and the third is the packetization process at the source, which may displace the time stamps within the stream. The first effect is simply vanished at the PLL by inserting a constant term to the input of the VCO, whereas the frequency variations of the reconstructed clock due to the network jitter and the packetization process should be restricted according to the required system clock specifications but at the expense of increasing the locking period. For example, frequency requirements of the NTSC specify a tolerance of ±10 Hz around the central subcarrier frequency of 3,579,545 Hz (which corresponds to ±3 parts per million (ppm)), whereas the corresponding tolerance for PAL signals is ±5 Hz around the central subcarrier frequency of 4,433,618 Hz. These specifications should be used to design the LPF that accordingly affects the locking period of the PLL. Alternatively, the VCO should be precise enough to meet the required clock specifications.

In the following sections, we will discuss an approach to reconstruct the clock time using the PCRs.

6.2 System Clock Recovery

6.2.1 Requirements on Video System Clock

At the decoder end, application-specific requirements such as accuracy and stability determine the approaches that should be taken to recover the system clock [6-3][6-4]. Some applications use the recovered system clock to directly synthesize a chroma subcarrier for the composite video signal. The system clock, in this case, is used to derive the chroma subcarrier, the pixel clock, and the picture rate. The composite video subcarrier must have at least sufficient accuracy and stability that any normal television receiver's chroma subcarrier PLL can lock to it, and the chroma signals that are demodulated by using the recovered subcarrier do not show any visible chrominance phase artifacts. There are often cases in which the application has to meet NTSC, PAL, or SECAM specifications for analog televisions [6-5], which are even more stringent. For example, NTSC requires a subcarrier accuracy of 3 ppm with a maximum long-term drift of 0.1 Hz/s.

The use of applications with stringent clock specifications requires careful design of the decoder since the decoder is responsible of feeding the TV set with a composite signal that satisfies the requirements. The demodulator of the TV set, as shown in Figure 6.4 PAL/NTSC demodulator system block., has to extract clock information from the input signal for the color subcarrier regeneration process. The frequency requirements for NTSC specify a tolerance of ±10 Hz (i.e., ±3 ppm) [6-5]. The central subcarrier frequency is 3,579,545 Hz (or exactly (63/88) × 5 MHz). The corresponding values for NTSC and other composite signals are summarized in Table 6.1. The requirements above define the precision of the oscillators for the modulator and, thus, the minimum locking range for the PLL at the receiver end.

There are also requirements for the short- and long-term frequency variations. The maximum allowed short-term frequency variation for an NTSC signal is 56 Hz within a line (or 1 ns/64 ms), whereas the corresponding value for a PAL signal is 69 Hz. This corresponds to a variation of the color frequency of 16 ppm/line in both cases [6-5]. If this requirement is satisfied, we can obtain a correct color representation for each line.

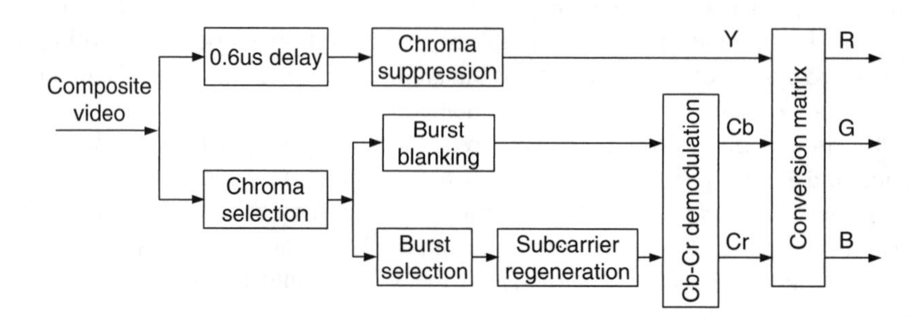

FIGURE 6.4 PAL/NTSC demodulator system block.

TABLE 6.1
Specifications for the Color Subcarrier of Various Video Formats

Standard	Frequency (Hz)	Tolerance (Hz)	Tolerance (ppm)
NTSC	3579545.4	±10	±3
PAL (M)	3575611.49	±10	±3
PAL (B, D, G, H, N)	4433618.75	±5	±1

The maximum long-term frequency variation (clock drift) that the composite NTSC or PAL signal must meet is 0.1 Hz/s. The drift may result because of temperature changes at the signal generator and can be determined in an averaging manner over different time-window sizes. In fact, the actual requirement for the color subcarrier frequency (3.57 MHz ± 10 Hz for NTSC) in broadcasting applications is an average value that can be measured over any reasonable time period. Averaging intervals in the range from 0.1 s to several seconds are common [6.5].

6.2.2 MPEG-2 Systems Timing Model

In MPEG-2 systems, the source timing is sent to the decoder by means of time stamps. The timing model for compressed video, shown in the Figure 6.5, is an expanded version of the one in MPEG-2 systems standard ISO/IEC 13818-1, Annex D[6-4].

Video data flows at a constant (or specified) rate at the input to the compression engine (Point A in Figure 6.5); then at a variable rate at the compression engine output before the encoder compressed data buffer (CDB) (Point B); then at a constant rate out of the encoder CDB (Point C), through the channel, and into the decoder CDB (point D); then at a variable rate into the decompression engine (Point E); and finally at a constant (or specified) rate output from the decompression engine (Point F).

The system target decoder (STD) model, specified in MPEG-2 Systems [6-4], is assumed to have zero delay in the decompression engine, and the compression engine and (transmission) channel are also assumed to have zero delay; that is, the

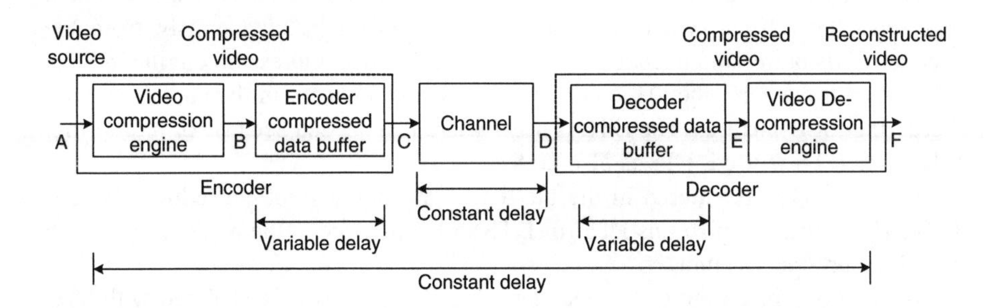

FIGURE 6.5 Constant delay timing model.

buffers represent the only system delay. It will be seen that the zero-delay elements can also have a constant delay without disturbing the operation.

The decoder is synchronized with the encoder (or the transcoder) by time stamps, which are introduced as follows. The encoder contains a master oscillator and counter, i.e., STC, the master time reference of an encoder (or transcoder) or decoder, consisting of a 27-MHz oscillator and a 33-bit counter. The STC uses a 27-MHz oscillator with a worst-case tolerance of 810 Hz. The STC first divides the 27 MHz by 300, giving a 90-kHz clock that is counted by a 33-bit counter to give the base STC value. The remainder is taken as a 9-bit value that may be used as an STC extension. Note that this is a looser tolerance than that of the NTSC color subcarrier. The ATSC standard [6-1] therefore contains a tighter constraint on the transmitted data rate so that it may be used if desired to regenerate the NTSC color carrier in equipment intended to feed an NTSC receiver. The data and STC frequencies are independent. There is a means provided to signal to the receiver the actual accuracy of the system time clock if it is better than the minimum allowed.

The STC belongs to a particular program and is the master clock of the video and audio compression engines for that program. The case where there are multiple programs, each with its own STC, will be discussed later on multiplexed programs. Note that it is valid within the MPEG-2 systems for a particular program component to have no time stamps, but there is no requirement to synchronize such unstamped component with other components.

At the input of the compression engine, Point A, the time of occurrence of an input video picture (and of the appearance of its coded version at the zero-delay compression engine output) is noted by sampling the STC. A constant quantity equal to the sum of encoder and decoder CDB delays is added, creating a PTS, which indicates the instant at which an access unit should be removed from the receiver buffer, instantaneously decoded, and presented for display. Any delay in a practical decoding or display process can be compensated for if it is fixed. Therefore, if a practical decoder has a variable delay, it must include buffering to restore a constant delay at the output. Note that the display itself may include a delay, for example, the time for the vertical scanning of a CRT to reach the middle of the picture. The PTS is inserted in the first of the packet(s) representing that picture, at Point B in Figure 6.5. In systems that do not send PTSs with every access unit (picture), the PTS value can be derived from a previous PTS by assuming a constant presentation rate.

Also inserted into the bit stream under certain conditions is a DTS, which represents the time at which the data should be taken instantaneously from the decoder CDB buffer and decoded. Since the delay of decompression engine in STD is assumed to be zero, the DTS and PTS are identical except in the case of picture reordering for B-pictures. The DTS is used where it is needed because of reordering. Whenever DTS is used, PTS is also coded.

PTS (or DTS) is entered in the bit stream at intervals not exceeding 700 ms. ATSC [6-1] further constrains PTS (or DTS) to be inserted at the beginning of each coded picture (access unit).

In addition, the output of the encoder CDB (Point C) is time stamped with STC values as the clock reference for decoder. In a program stream, the clock reference is called the *system clock reference* (SCR). The system clock reference time stamps are required to occur at a maximum of 700-ms intervals. In a transport stream, the clock reference is the PCR. The PCR time stamps are required to occur at a maximum of 100-ms intervals. The SCR and/or the PCR are used to synchronize the decoder STC with the encoder STC.

All video and audio streams included in a program must get their time stamps from a common STC so that synchronization of the video and audio decoders with each other can be accomplished. The important characteristic of the system of SCR/PCR time stamps is that the data rate and packet rate on the channel (at the multiplexer output) may be completely asynchronous with the STC, and the STC can still be synchronized at the decoder. This also means that different programs that may have different STCs can be carried in a multiplex with other programs while allowing recovery of the STC for each program.

Operation depends on there being a constant delay in the CDB buffers and (transmission) channel (in general, for both the video and the audio). This will naturally be the case if there is no buffer underflow or overflow. Data is neither generated nor destroyed by the transmission system, and the encoder input and decoder output run at equal and constant rates. Thus there is also a fixed end-to-end delay (from input to the encoder to output from the decoder).

In video systems where exact synchronization is not required, the decoder clock can be free running, with video pictures repeated or skipped as necessary to prevent buffer underflow or overflow, respectively. This type of decoder is not capable of lip sync.

Multiplexed Programs: An MPEG data stream may multiplex streams from several video and audio sources, each having a slight difference in clock rate from the others. In general, it will not be possible to synchronize all these sources to the transport multiplexer.

In order to multiplex several incoming MPEG streams into one, the individual streams must have some null packets that can be removed if necessary to adjust the overall multiplex data rate. Null packets are specially defined and do not carry any program data. The null packet removal, as well as the interleaving of packets from different sources, violates the assumption of a constant-delay transmission channel, and introduces jitter into the packet timing.

The MPEG time stamping process provides sufficient information to carry synchronization through such a multiplexer to a decoder, but the MPEG-2 System does not specifically provide methods to take care of packet jitter effects. In general, the resulting packet jitter may exceed the MPEG tolerance, and then methods must be developed to correct this in order to feed a standard decoder. Some commercial equipment avoids this problem by physically co-locating all encoders that feed a given multiplex (for example a satellite transponder), and driving them all from a common clock and time reference.

6.2.3 Decoder STC Synchronization

Decoder STC synchronization is the process of setting both the frequency of the STC 27-MHz oscillator and the value of the STC counter so that the counter matches the PCRs (or SCRs) in the incoming data stream. This effectively matches the local 27-MHz clock to the encoder's 27-MHz clock.

In MPEG applications, a standard PLL, as shown in Figure 6.3, is often used to recover the clock from the PCR time stamps transmitted within the stream. The PLL works as follows: Initially, the PLL waits for the reception of the first PCR value for use as the time base. This value is loaded in the local STC counter and the PLL starts operating in a closed-loop fashion. When a new PCR sample is received at the decoder, its value is compared with the value of the local STC. The difference gives an error term. This error term is then sent to an LPF. The output of the LPF controls the instantaneous frequency of a VCO whose output provides the decoder's system clock frequency. A detailed analysis on such digital PLL can be found in [6-3].

Synchronization Loop Model: The basic synchronizer consists of a first-order feedback loop. Each incoming PCR (or SCR) is subtracted from the receiver STC counter, and the filtered difference (times a proportionality constant) is the control voltage for a crystal VCO. This loop stabilizes with the correct frequency, but with an offset in STC that is proportional to the offset in frequency between the encoder 27-MHz oscillator and decoder 27-MHz oscillator free-running frequency. This implies that the decoder should have a slightly larger buffer to absorb the offset timing.

Difference signal filtering is needed to reduce unnecessary responses to PCR (or SCR) arrival time jitter. Jitter reduction must be traded for settling rate. MPEG sets a tolerance on input PCR (or SCR) jitter. The output jitter that is tolerable depends on the application, for example, the use of the 27-MHz clock to generate display pixel, horizontal, and vertical clocks. If the recovered 27-MHz clock is used to generate NTSC color subcarrier, more stringent requirements with respect to both jitter and settling rate may need to be met.

The feedback loop in the receiver can be modeled as a sampled phase locked loop, as shown in Figure 6.6 [6-3]. The loop compares each value of a local STC counter sequence LSTC(n), driven by a voltage-controlled crystal oscillator (VCXO) in the receiver, to a sequence of received samples of the encoder's counter PCR(n), driven by the encoder STC, where for both counters one can interpret

$$\theta(t) \equiv \text{counter value} \tag{6.1}$$

$$\frac{\Delta\theta(t)}{\Delta t} = \text{counter rate} = \text{frequency} \tag{6.2}$$

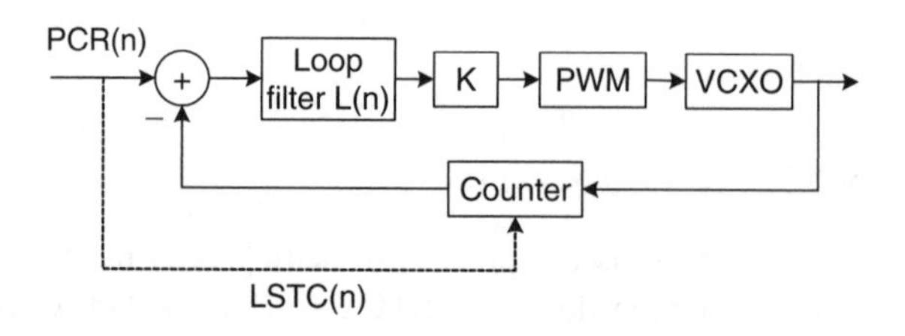

FIGURE 6.6 A PLL MODEL.

The sequence PCR(n) represents PCR samples that are assumed to arrive at a 10-Hz rate. In this model, the 27-MHz component of both the encoder and decoder clocks has been factored out of the sequences PCR(n) and LSTC(n). In other words, PCR(n) and LSTC(n) are referenced to 27 MHz. Therefore, the reception of an nonjittered 27-MHz encoder STC is represented by PCR(n). If LSTC(n) is frequency locked to an nonjittered PCR(n), then it will also be a sequence of a constant value. If LSTC(n) is locked to and in phase with PCR(n), then LSTC(n) = PCR(n). If LSTC(n) is not locked to PCR(n), then LSTC(n) can be modeled as a ramp sequence increasing (or decreasing) at a rate corresponding to the difference in frequency between the oscillators represented by PCR(n) and LSTC(n) [6-3].

Typically the operations of PCR(n) – LSTC(n), loop filter, and multiplication by the gain factor K can be performed in a microcontroller (e.g., 10 times per second), which provides the scaled filtered output to a pulse width modulator (PWM) circuit to drive an external VCXO, which in turn drives a counter. The VCXO block above is understood to include the PWM (often with a RC circuit), VCXO, and counter driven by the VCXO. The RC time constant may be ignored in this analysis if it is much less than the 100-ms spacing between PCR(n) samples.

If the *Z*-transform of the VCXO integrator block is modeled as 1/(*Z* – 1), then the *Z*- transform of the closed-loop transfer function for the above loop is given as

$$H(Z) = \frac{K \cdot L(Z)}{(Z-1) + K \cdot L(Z)} = \frac{LSTC(Z)}{PCR(Z)} \tag{6.3}$$

This model in the *Z*-domain is given in Figure 6.7.

Note that the 10-Hz PCR arrival rate sets an upper bound on loop bandwidth. Useful loop bandwidths for this application are typically much lower than 10 Hz.

Loop Characteristics: Frequency lock in the synchronization loop model occurs when

$$\frac{\Delta PCR(n)}{\Delta n} = \frac{\Delta LSTC(n)}{\Delta n} \tag{6.4}$$

A good choice for *L*(*Z*) is a unity gain moving window averaging filter having *N* taps each of value 1/*N*. If the loop filter *L*(*Z*) is an all zeros filter, then we have a first-order loop. Such a loop will achieve frequency lock with a nonzero phase

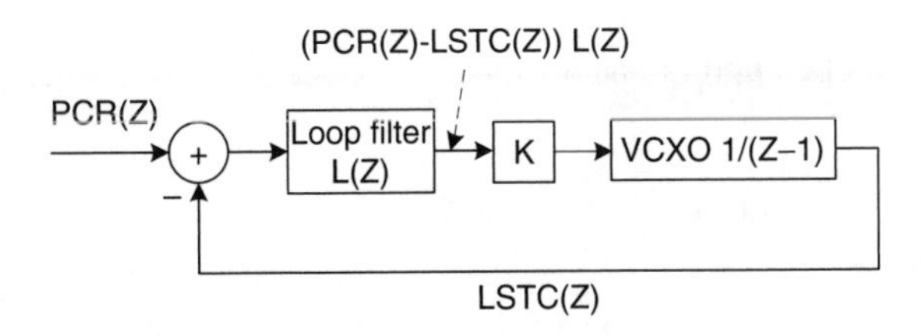

FIGURE 6.7 A Z-domain PLL model.

difference PCR(n) – LSTC(n). This fixed phase difference after frequency lock can be reduced to 0 if we use a second-order loop ($L(Z)$ with a pole at $Z = 1$). However, it is hard to maintain stability in such a loop. Small differences between the actual loop parameters and the design loop parameters can easily push the poles of $H(Z)$ outside the unit circle causing instability. A first-order loop is very stable and an all-zeros filter with many taps is easily implemented by an application-specific integrated circuit or a microprocessor. The fixed phase difference after frequency lock does increase the required receiver buffer size.

Unlike a first-order PLL, the first-order STC loop will pull in from any offset frequency to any transmitted frequency within the range of the VCXO, because of the integrating effect of the STC counter. The loop parameters do affect the final phase (STC value) error, the speed of pull-in, and the jitter response.

Phase Difference: A first-order loop will achieve frequency lock with a fixed phase difference between the encoder STC counter, i.e., PCR, and the decoder STC counter. The worst-case frequency difference between the encoder STC and the decoder VCXO center frequency is 1,620 Hz (either may be at 27 MHz ± 810 Hz). The decoder VCXO should be specified to have a midpoint frequency with this range. At a 10-Hz sampling rate,

$$|\Delta f| = \left| \frac{\Delta PCR(n)}{\Delta n} - \frac{\Delta LSTC(n)}{\Delta n} \right| = 162 \tag{6.5}$$

which counts per sample maximum between the encoder STC and the decoder VCXO center frequency. The largest fixed phase difference (between the received PCR values and the local VCXO counter) after lock-up for this first-order loop will be

$$\Delta\phi_{\max} = \frac{|\Delta f|}{K} = \frac{162}{K} \tag{6.6}$$

where K is the integer gain factor.

Lock Up Time: The closed-loop transfer function for the first-order loop [6-3] is given in Equation (6.3) and such transfer function can be expressed by

$$L(Z) = \frac{1}{1 - Z^{-1}} = 1 + Z^{-1} + Z^{-2} + \cdots\cdots \tag{6.7}$$

Equation (6.7) can be approximated by

$$L(Z) = \frac{1}{N}\left(1 + Z^{-1} + Z^{-2} + \cdots\cdots + Z^{-N}\right) \tag{6.8}$$

where the integer N is for normalizing the gain and, hence, this filter is also called a *unity gain* N*-tap averaging filter.*

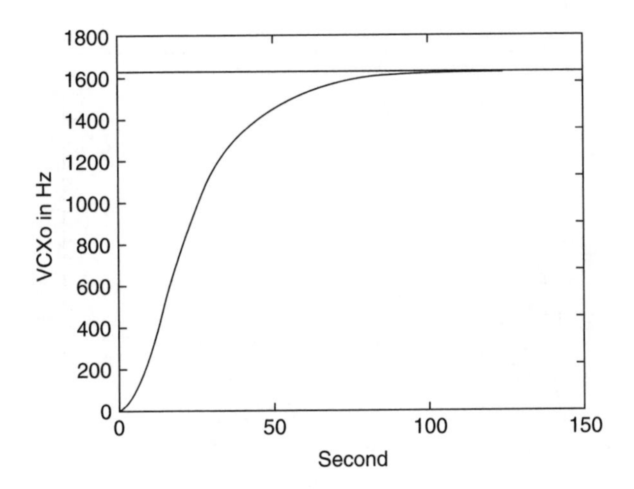

FIGURE 6.8 A VCXO frequency acquisition curve.

Example

(Lock-up time): For a unity gain 128-tap averaging filter ($N = 128$) and the loop transfer function with $K = 1/256$: to determine the frequency acquisition time for this loop, PCR(n) and LSTC(n) are initially zero, and PCR(n) is set to a ramp sequence increasing at a rate of 162 counts per sample. This represents the worst-case 1,620-Hz initial difference between the encoder and decoder clocks. With the transfer function in Equation (6.3), the sequence LSTC(n) can be calculated. Decoder instantaneous oscillator frequency in counts per second is $10 \cdot \frac{\Delta LSTC(n)}{\Delta n}$. This is plotted in Figure 6.8 below with respect to $n/10$, where n = number of samples, 10 samples per second. Figure 6.8 shows an example that lock occurs after about 130 s.

Next, the phase difference $PCR(n) - LSTC(n)$ is plotted in Figure 6.9 below with respect to $n/10$ (n = number of samples, 10 samples/s). It is seen that phase difference after lockup is about 41,000. This closely matches the predicted phase difference

$$\Delta\phi_{\max} = \frac{|\Delta f|}{K} = \frac{162}{K} = 162 \times 256 = 41472.$$

Jitter Response: We assume a locked steady-state condition to test the effect of jitter. The input sequence PCR(n) is set to be a random sequence with a uniform distribution from −108,000 to +108,000 representing a ±4-ms jitter. Then sequence LSTC(n) is calculated using the loop transfer function. The instantaneous VCXO frequency is calculated in counts per second as $10 \cdot \frac{\Delta LSTC(n)}{\Delta n}$ and plotted versus n (n = number of samples, 10 samples/s). Note that perfect jitter immunity would be represented as LSTC(n) equal to zero so that $\frac{\Delta LSTC(n)}{\Delta n}$ would also equal to zero. For more analysis on jitter and the decoder VCXO frequency stability, one can read the reference [6-3].

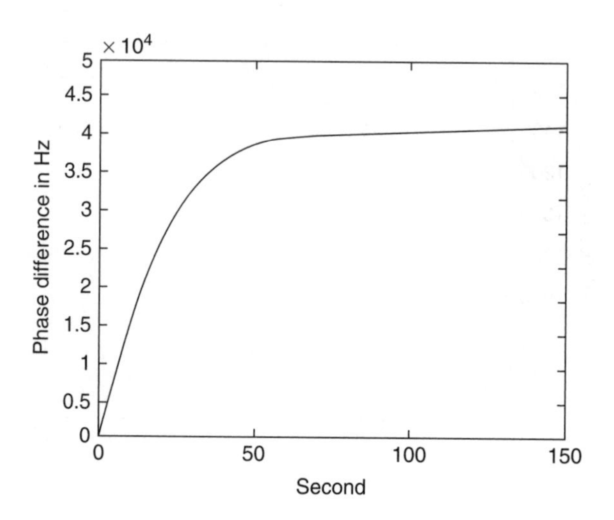

FIGURE 6.9 Phase difference curve in acquisition

6.2.4 Required Decoder Buffer Size for Video Synchronization

The MPEG 2 standard sets requirements for minimum decoder channel buffer size. The action of the 27 MHz clock recovery loop requires additional buffer in the decoder beyond the standard specified amount.

Buffer Requirements for Lock Up Time: The video for a newly acquired channel must be displayed immediately, long before the loop has achieved lock. So for some period of time (as long as a few minutes), the receiver will be decoding and displaying video, but its 27-MHz clock will be running faster or slower than the encoder's STC. Some additional buffer is needed to prevent both underflow and overflow for these cases.

Example

(Buffer requirement for lock-up time): Consider the case in which the encoder clock is 27 MHz + 810 Hz and that the receiver VCXO center frequency is 27MHz – 810 Hz (slow case) just prior to channel acquisition, and that a 19.393 Mbit/s (ATSC HDTV) [6-1] stream is received. For a 130-s lockup time, we can envision a worst-case (but unrealistic) scenario where the two clocks remain 1,620 Hz apart during the entire lockup process, and then at $t = 130$ s, they instantly lock at 27 MHz + 810 Hz. It means that for this period of time the buffer takes in excess data at $\frac{1620Hz}{27MHz} \cdot 19.393 Mbits/s = 1164$ bits/s. In 130 s, this becomes approximately 152 Kbits of extra buffer needed to prevent overflow.

This number can be substantially reduced if one accounts for the actual trajectory of the VCXO frequency as it approaches 27 MHz + 810 Hz. If this is done, one can multiply 152,000 by a factor equal to 1 minus the ratio of the area under the VCXO frequency acquisition curve in the loop characteristics subtopic and the area under the straight line at 1,620 in that same graph taken from 0 to 130 s.

Now consider the case in which the encoder clock is 27 MHz – 810 Hz and that the receiver VCXO runs at 27 MHz + 810 Hz (fast case) just prior to channel acquisition. In this case the receiver must wait for an additional 152 Kbits of data to arrive at 19.393 Mbits/s before it decodes in order to prevent channel buffer underflow. In other words the receiver must add 152,000/19,393,000 = 7.8 ms to all PTSs (or subtract that amount from all PCRs). Again this number can be substantially reduced if the actual VCXO trajectory is considered.

Assuming that the receiver does not know if it has an initially slow or initially fast VCXO, it must always add 7.8 ms to all PTSs. This has the effect of doubling the extra buffer requirement to 304 Kbits.

Buffer Requirements for Possible Jitter: To account for a possible ±4 ms of jitter, i.e., a 8-ms swing, extra buffer for 16 ms × 19.393 Mbits/s is needed, which is approximately 311 Kbits/s. This extra buffer is kept half full by the decoder adding 4 ms to each PTS (or subtract same from PCR's).

Buffer Requirements for Phase Difference After Lockup: Using a first-order loop that locks up with a fixed phase difference also requires additional channel buffer. The worst-case frequency difference between the encoder STC and the decoder VCXO center frequency is 1,620 Hz (162 counts per sample). If the VCXO is slower than the encoder clock, after lockup the VCXO phase will be 162/K counts behind that of the encoder clock. This will cause the decoder to present pictures late by the amount in seconds of

$$\text{Phase Difference} = \frac{\frac{162}{K}}{27 \times 10^6 \; counts\,/\,s}$$

This presentation delay requires additional buffer space of Phase Difference × 19.393 Mbits/s in order to prevent overflow.

If the receiver VCXO is initially fast with respect to the encoder clock, the receiver must delay presentation by adding some fixed value to all PTSs in order to prevent channel buffer underflow. This value is the same phase difference calculated above. Assuming that the receiver does not know if it has an initially slow or initially fast VCXO, it must always add phase difference to all PTSs. This has the effect of doubling the buffer requirement in the slow case to 2 × Phase Difference × 19.393 Mbits/s.

6.3 VIDEO DECODING AND PRESENTATION TIME STAMPS

6.3.1 Background

As discussed in previous sections, the encoder STC of a video program is used to create time stamps that indicate the presentation and decoding timing of video, as well as to create time stamps that indicate the sample values of the encoder STC itself at sampled intervals. The time stamps that indicate the presentation time of video are PTS, whereas those that indicate the decoding time are DTS. It is the presence of these time stamps and the correct use of the time stamps that provide the facility to synchronize properly the operation of the decoding.

In this section, methods for generating the DTS and PTS in the video encoder are discussed. In particular, the time stamping schemes for MPEG-2 video are introduced as examples. In MPEG-2, a compressed digital video elementary stream is assembled into a packetized elementary stream (PES). The PTS are carried in headers of the PES; DTS are also carried in PES headers that have the picture header of an I- or P-picture when bidirectional predictive coding is enabled. The DTS is not sent with a video PES stream that was generated with B-picture coding disabled. The PTS values (and DTS, if present) of the video stream are derived from the 90-kHz portion of the encoder STC (at 27 MHz) that is assigned to the program to which the video belongs.

Both PTS and DTS are determined in video encoder for coded pictures. If B-pictures are present in the video stream, coded pictures (sometime also called *video access units*) do not arrive at the decoder in presentation order. In this case, some pictures in the stream must be stored in a reorder buffer in the decoder after being decoded until their correct presentation time. In particular, I-pictures or P-pictures carried before B-pictures will be delayed in the reorder buffer after being decoded. Any I- or P-picture previously stored in the reorder buffer is presented before the next I- or P-picture is stored. While the I- or P-picture is stored in the reorder buffer, any subsequent B-picture(s) is(are) decoded and presented.

As discussed, the video DTS indicates the time when the associated video picture is to be decoded, whereas the video PTS indicates the time when the presentation unit decoded from the associated video picture is to be presented on the display. Times indicated by PTS and DTS are evaluated with respect to the current PCR value. For B-pictures, which are never reordered, PTS is always equal to DTS since these pictures are decoded and displayed instantaneously. For I- or P-pictures (if B-pictures are present), PTS and DTS differ by the time that the picture is delayed in the reorder buffer, which will always be a multiple of the nominal picture period, except in the film mode (which will be discussed later). If B-pictures are not present in the video stream, i.e., B-picture type is disabled, all I- and P-pictures arrive in presentation order at the decoder, and consequently their PTS and DTS values are identical. Note that if the PTS and DTS values are identical for a given access unit, only the PTS should be sent in the PES header.

The detailed video coding structures can be found in Chapter 2 or reference [6-30]. The most commonly operated MPEG video coding modes, termed m = 1, m = 2, or m = 3 by the MPEG-2 committee are described as follows. In m = 1 mode, no B-pictures are sent in the coded video stream, and therefore all pictures will arrive at the decoder in presentation order. In m = 2 mode, one B-picture is sent between each I- or P-picture, and pictures arrive at the decoder in the following decoding order (as the subindex):

$$I_1\ P_2\ B_3\ P_4\ B_5\ P_6\ B_7\ P_8\ B_9\ I_{10}\ B_{11}\ \ldots$$

These pictures will be reordered in the following presentation order:

$$I_1\ B_3\ P_2\ B_5\ P_4\ B_7\ P_6\ B_9\ P_8\ B_{11}\ I_{10}\ \ldots$$

In m = 3 mode, two B-pictures are sent between each I- or P-picture, i.e., pictures arrive at the decoder in the following decoding order:

$I_1 P_2 B_3 B_4 P_5 B_6 B_7 P_8 B_9 B_{10} I_{11} B_{12} B_{13} P_{14} B_{15} B_{16} \ldots$

These pictures will be reordered in the following presentation order:

$I_1 B_3 B_4 P_2 B_6 B_7 P_5 B_9 B_{10} P_8 B_{12} B_{13} I_{11} B_{15} B_{16} P_{14} \ldots$

Each time that the picture sync is active, the following picture information is usually required for time stamping of the picture in MPEG-2 video:

- Picture type: I-, P-, or B-picture.
- Temporal reference: A 10-bit count of pictures in the presentation order.
- Picture encoding time stamp (PETS): A 33-bit value of the 90-kHz clock of the encoder STC that was latched by the picture sync.

In the normal video mode, the DTS for a given picture is calculated by adding a fixed delay time, T_d, to the PETS. For some pictures in the film mode, the DTS is generated by (T_d + PETS – a field time) (this is detailed later in this section). Time T_d is nominally the delay from the input of the MPEG-2 video encoder to the output of the MPEG-2 video decoder. This is also called the *end-to-end delay* as shown in Figure 6.5. In real applications, the exact value of T_d is most likely determined during system integration testing.

The position of the current picture in the final display order is determined by using the picture type (I, P, or B). The number of pictures (if any) for which the current picture is delayed before presentation is used to calculate the PTS from the DTS. If the current picture is a B-picture or if it is an I- or P-picture in m = 1 mode, then it is not delayed in the reorder buffer and the PTS and DTS are identical. In this case, the PTS is sent usually in the PES header that precedes the start of the picture. If the current picture is instead an I- or P-picture and the B-processing mode is m = 2 or m = 3, then the picture will be delayed in the reorder buffer by the total display time required by the subsequent B-picture(s).

In addition to considering picture reordering when B-pictures are present, it is needed to check if the current picture is in the film mode in order to correctly compute the PTS and DTS.

Film Mode. In countries that use 525-line interlaced display systems, such as the United States and Canada, television video signals are sampled and transmitted at approximately 59.94 fields per second (fps). For these countries, digital television video streams are generally encoded and transmitted by using MPEG-2 video at approximately 29.97 frames per second (FPS).

Hereinafter, an integral value of fps or an integral value of FPS may be an approximation including, within its scope, a range of equivalent values. Thus, for example, the expression 30 FPS may be used to refer to rates such as approximately 29.97 FPS or approximately 30 FPS. Furthermore, the expression 24 FPS may be used to refer to such rates as approximately 23.976 FPS or approximately 24 FPS. Similarly, the expression 60 fps may be used to refer to such rates as approximately 59.95 fps or approximately 60 fps.

Film material produced at 24 FPS is routinely converted to 60 fps in many applications. Broadcast networks usually encode and transmit movies that were originally filmed at 24 FPS and not at 60 fps. However, at the receiver, the decoded video at 24 FPS is often converted to 60 fps for interlaced display. A conventional process for converting 24 FPS to 60 fps sampling includes the Telecine process (named after the original type of machine used to perform the conversion from film to video). It is also known as the 3:2 pull-down process. The Telecine Process inserts repeated fields derived from the original film frames in such a way that 5 video frames (i.e., 10 fields) are produced for every 4 original film frames. Figure 6.10 illustrates one example of a process that performs a 3:2 pull-down. The original film sequence filmed at 24 FPS is converted to a video sequence at 30 FPS.

For film material that has been converted to video, it is often desirable to restore the film sequence to a 24 FPS form prior to compression by eliminating, for example, the repeated fields inserted by the Telecine Process. Such a process reduces the amount of data for compression, thereby improving the quality of video or reducing the bit rate for transmission. The process of eliminating the repeated fields is commonly known as the *inverse Telecine process* or the *inverse 3:2 pull-down process*. Figure 6.10 also illustrates one example of the process that performs an inverse 3:2 pull-down. The video sequence at 30 FPS is restored or converted into the film sequence at 24 FPS.

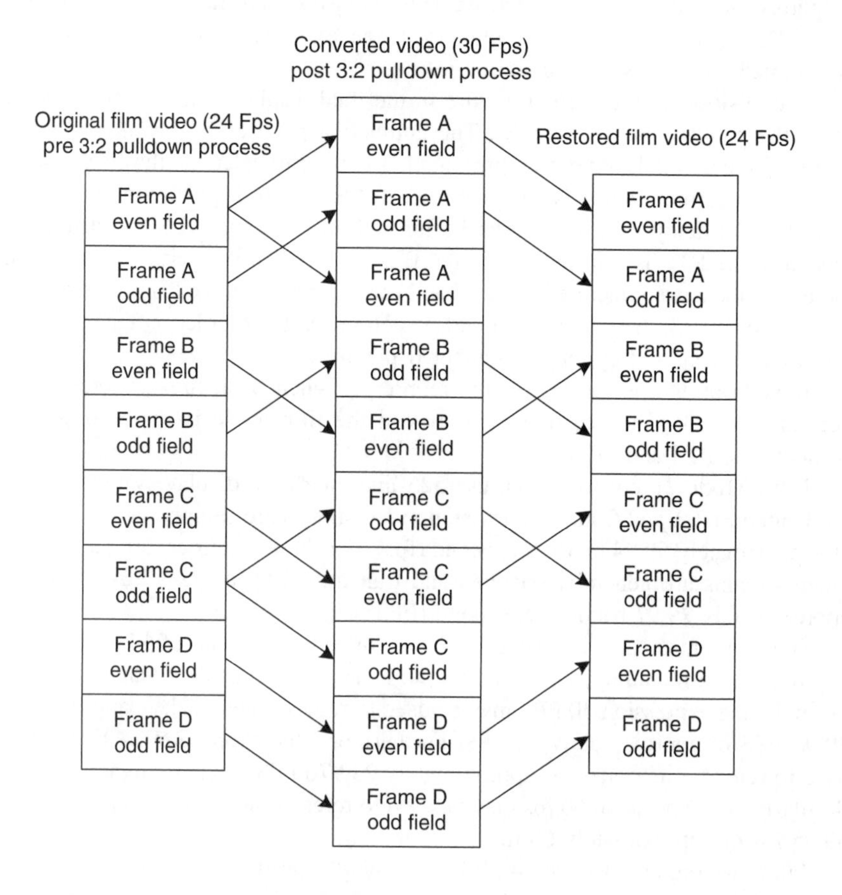

FIGURE 6.10 Mapping of the 3:2 pull-down process.

The mechanism for handling the 3:2 pull-down and/or the inverse 3:2 pull-down for film material in digital video systems is usually referred to as the film mode.

In the film mode, two repeated fields have been removed from each ten-field film sequence by the MPEG-2 video encoder. The PETS will not be stamped on the coded pictures arriving from the MPEG-2 video encoder at the nominal picture rate; two of every four pictures will be of a three-field duration (one and one-half times the nominal picture period), while the other two are of a two-field duration (the nominal picture period). Therefore, in the film mode, the time an I- or P-picture is delayed in the reordering buffer will not be merely the number of subsequent B-picture(s) multiplied by the nominal picture period, but instead will be the total display time of the B-picture(s). For example, if a P-picture is followed by two B-pictures, one of which will be displayed for a two-field duration and the other for a three-field duration, then the P-picture will be delayed for a total of five field times, or two and one-half picture times. The PTS for the picture then becomes the DTS plus two and one-half picture times. Note that for NTSC, one-half picture time cannot be expressed in an integral number of 90-kHz clock cycles, and must be either rounded up to 1,502 or rounded down to 1,501. A fixed set of rules is usually followed for governing rounding of the one-half picture time for the film mode. These rules are outlined later in this section.

6.3.2 Computation of MPEG-2 Video PTS and DTS

In this subsection, the following MPEG-2 video coding configurations for the frame-structured picture (i.e., a frame is a picture in this case) are discussed as examples for calculating video PTS and DTS.

Configuration 1: B-picture type disabled (m = 1) with low delay, without the film mode.
Configuration 2: B-picture type disabled (m = 1) with low delay and the film mode.
Configuration 3: Single B-picture (m= 2) without the film mode.
Configuration 4: Single B-picture (m = 2) with the film mode.
Configuration 5: Double B-picture (m = 3) without the film mode.
Configuration 6: Double B-picture (m = 3) with the film mode.

Example

(B-picture type disabled with low delay, without the film mode)**:** in this configuration, m = 1 and no B-pictures will be sent in the coded video stream. I-Pictures and P-pictures in the stream are sent in presentation order; therefore, no picture reordering is required at the MPEG-2 video decoder. Consequently, the PTS and the DTS are identical in this configuration.

For the i-th picture in the coded video stream, the PTS and DTS are computed as

$$DTS_i = PTS_i = PETS_i + T_d \tag{6.9}$$

where $DTS_i = DTS$ for the i-th picture; $PTS_i = PTS$ for the i-th picture; $PETS_i = PETS$ which tags the i-th picture; and T_d = nominal delay from the output of the encoder to the output of the decoder.

TABLE 6.2
Mapping of the 3:2 Pull-Down Flags

Pictures	Transmitted Fields	Repeat First Field Flag	Top Field First Flag	Displayed Field
A	A1	1	1	A1
	A2			A2
				A1
B	B1	0	0	B2
	B2			B1
C	C1	1	0	C2
	C2			C1
				C2
D	D1	0	1	D1
	D2			D2

If all pictures are processed without the film mode, then the difference F between PTS_i and PTS_{i-1} should be exactly equal to the nominal picture time in 90-kHz clock cycles (i.e., $F = \frac{90 \times 10^3}{29.97} = 3003$ for NTSC since the picture rate equals 29.97, and $F = \frac{90 \times 10^3}{25} = 3600$ for PAL since the picture rate equals 25).

In summary, the following rules are applied to the calculation of PTS and DTS for the pictures without the film mode with m = 1.

1. Verify that the difference between PTS_i and PTS_{i-1} should be exactly equal to the nominal picture time in 90-kHz clock cycles. If this is not true, the value of PTS_i has to be adjusted and an error should be reported.
2. Calculate PTS and DTS as $DTS_i = PTS_i = PETS_i + T_d$.
3. Send the PTS, but will not send the DTS in the PES header preceding the i-th picture.

Example

(B-picture type disabled with low delay and the film mode): Again, I- and P-pictures in a video stream processed without B-pictures (m = 1) are sent in presentation order, regardless of film mode. The PTS and the DTS are again identical.

In this case, for the i-th picture in the coded video stream processed in the film mode, the DTS and PTS are calculated by Equation (6.9).

In the film mode, two flags of MPEG-2 video in the coded picture header, top_field_first and repeat_first_field, are used to indicate the current film state. As shown in Table 6.2 and Figure 6.10, the four possible film mode states (represented as A, B, C, and D) are repeated in the same order every four pictures. Film mode processing should always commence with state A (or C) and exit with state D (or B). The decoder will display film state A and C pictures for three field times since they both contain a "repeated" field. The decoder redisplays the first field as the "repeated" field. This is because in the 3:2 pull-down algorithm, the first field is repeated every

TABLE 6.3
A film mode sequence

Encoded Pictures	Film States	Repeat First Field Flag	Top Field First Flag	Displayed Field
0		0	0	Frm0Bot
				Frm0Top
1	C	1	0	Frm1Bot
				Frm1Top
				Frm1Bot
2	D	0	1	Frm2Top
				Frm2Bot
3	A	1	1	Frm3Top
				Frm3Bot
				Frm3Top
4	B	0	0	Frm4Bot
				Frm4Top
5	C	1	0	Frm5Bot
				Frm5Top
				Frm5Bot
6	D	0	1	Frm6Top
				Frm6Bot
7	A	1	1	Frm7Top
				Frm7Bot
				Frm7Top
8	B	0	0	Frm8Bot
				Frm8Top
9		0	0	Frm9Bot
				Frm9Top
10		0	0	Frm10Bot
				Frm10Top

other picture to convert film material at 24 pictures/s to video mode at 30 pictures/s. Film state B and D pictures are displayed for only two field times. A film-mode sequence of four pictures will therefore be displayed as a total of 10 field times. In this way, the decoded video is displayed at the correct video picture rate.

Table 6.3 shows a sequence of eleven coded pictures that are output from the video encoder during which the film mode is enabled and then disabled. Picture 0 was not processed in the film mode. Picture 1 is the first picture to be processed in the film mode.

Unlike the case of m = 1 without the film mode, the difference between the PETS tagging successive pictures will not always be equal to the nominal picture time. As can be seen from Table 6.3, the time interval between a picture in film state A and the successive picture in film state B is three field times. Likewise, the time interval between a picture in film state C and the successive picture in film state D is also three field times. Note that for NTSC, three-field time cannot be expressed in an integral number of 90-kHz clock cycles, and must be either rounded up to 4,505 or rounded down to 4,504. As a convention, for example, the time interval

TABLE 6.4
PTS and DTS Calculation for the Film Mode (m = 1)

Encoded Pictures	Film States	$PETS_i$	$\Delta PETS_i$	$DTS_i = PTS_i$	Display Duration
I_0		$PETS_0$	F	$PETS_0 + T_d$	F
P_1	C	$PETS_1$	F	$PETS_1 + T_d$	$\lceil 1.5F \rceil$
P_2	D	$PETS_2$	$\lceil 1.5F \rceil$	$PETS_2 + T_d$	F
P_3	A	$PETS_3$	F	$PETS_3 + T_d$	$\lfloor 1.5F \rfloor$
P_4	B	$PETS_4$	$\lfloor 1.5F \rfloor$	$PETS_4 + T_d$	F
P_5	C	$PETS_5$	F	$PETS_5 + T_d$	$\lceil 1.5F \rceil$
P_6	D	$PETS_6$	$\lceil 1.5F \rceil$	$PETS_6 + T_d$	F
I_7	A	$PETS_7$	F	$PETS_7 + T_d$	$\lfloor 1.5F \rfloor$
P_8	B	$PETS_8$	$\lfloor 1.5F \rfloor$	$PETS_8 + T_d$	F
P_9		$PETS_9$	F	$PETS_9 + T_d$	F
P_{10}		$PETS_{10}$	F	$PETS_{10} + T_d$	F

between a state A picture and a state B picture may be rounded up to 4,505, and the interval between a state C picture and a state D picture may be rounded down to 4,504. Over the four-picture film mode sequence, the total time interval will be 15,015 90-kHz clock cycles for NTSC, or exactly five NTSC picture times. Table 6.4 summarizes the PTS and DTS calculations for a sequence of pictures in the film mode processed without B-pictures (m = 1).

In summary, the following general rules are applicable to the PTS and DTS for the *i*-th picture in the film mode with m = 1.

1. If picture i is in film state C and picture $i - 1$ is in nonfilm mode, then the difference between $PETS_i$ and $PETS_{i-1}$ is F, where F is the nominal picture period in 90-kHz clock cycles (3,003 for NTSC, 3,600 for PAL).
2. If picture i is in film state D, then the difference between $PETS_i$ and $PETS_{i-1}$ is $\lceil 1.5F \rceil$, where $\lceil 1.5F \rceil$ is the one and one-half nominal picture periods in 90-kHz clock cycles rounded up to the nearest integer (4,505 for NTSC, 5,400 for PAL).
3. If picture i is in film state A and picture $i - 1$ is in film state D, then the difference between $PETS_i$ and $PETS_{i-1}$ is F, where F is the nominal picture period in 90-kHz clock cycles.
4. If picture i is in film state B and picture $i - 1$ is in film state A, then the difference between $PETS_i$ and $PETS_{i-1}$ is $\lfloor 1.5F \rfloor$, where $\lfloor 1.5F \rfloor$ is the one and one-half nominal picture periods in 90-kHz clock cycles rounded down to the nearest integer (4,504 for NTSC, 5,400 for PAL).
5. If picture i is in nonfilm mode and picture $i-1$ is in film state B, then the difference between $PETS_i$ and $PETS_{i-1}$ is F, where F is the nominal picture period in 90-kHz clock cycles.
6. Compute DTS and PTS as $DTS_i = PTS_i = PETS_0 + T_d$, where T_d is the nominal delay from the output of the video encoder to the output of the decoder.
7. PTS is sent in the PES header preceding the *i*-th picture.

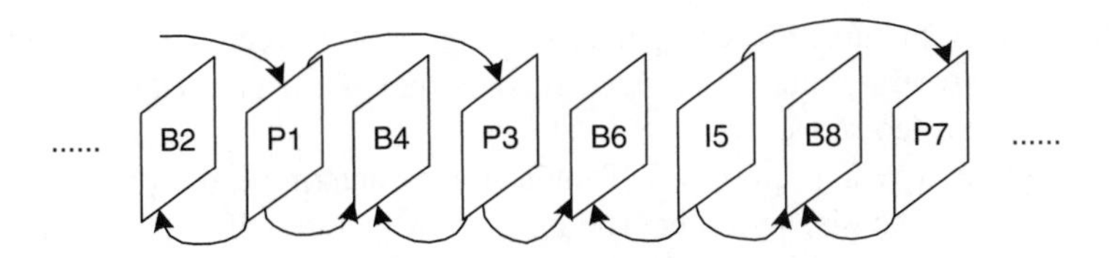

FIGURE 6.11 An open GOP sequence in display order.

EXAMPLE

(Single B-picture without the film mode): In this configuration (m = 2), a single B-picture will be sent between each anchor picture, i.e., each I- or P-picture. Pictures will arrive at the decoder in the following decoding order:

$I_1\ P_2\ B_3\ P_4\ B_5\ P_6\ B_7\ P_8\ B_9\ I_{10}\ B_{11}\ldots$

and will be reordered in the following presentation order:

$I_1\ B_3\ P_2\ B_5\ P_4\ B_7\ P_6\ B_9\ P_8\ B_{11}\ I_{10}\ \ldots$

The MPEG-2 video encoder may generate two types of I-pictures, an I-picture that follows the open group of pictures (GOP) or an I-picture that follows the closed GOP. An open GOP I-picture will begin a group of pictures to which motion vectors in the previous group of pictures point. For example, a portion of a video sequence is output from the video encoder in the decoding order as

$\ldots P_1B_2P_3B_4I_5B_6P_7B_8.\ldots$

and displayed in Figure 6.11.

The B-picture, B_6, has motion vectors that point to I_5, which is therefore an open I-picture. In m = 2 processing, the video encoder may generate an open GOP I-picture in any position within a video sequence which would normally be occupied by a P-picture.

A closed GOP I-picture will begin a group of pictures that are encoded without predictive vectors from the previous group of pictures. For example, a portion of a video sequence is output from the video encoder in the decoding order as

$\ldots\ \ldots P_1B_2P_3B_4I_5P_6B_7.\ldots..$

and displayed as in Figure 6.12.

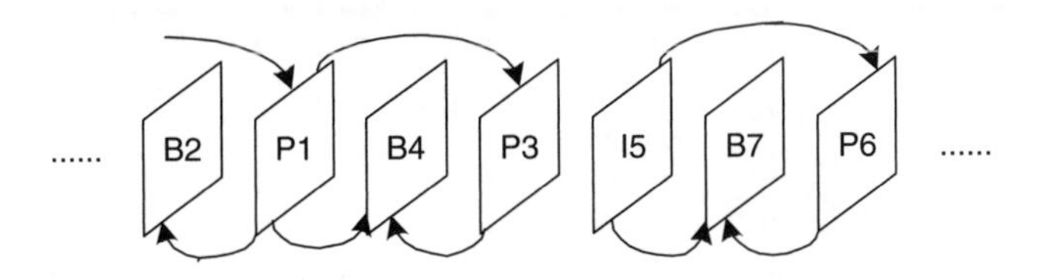

FIGURE 6.12 A closed GOP sequence in display order.

There are no pictures preceding I_5 that contain motion vectors pointing to it; I_5 is therefore a closed I-picture. The video encoder may place a closed GOP I-picture in any point in video sequence.

The picture type and closed GOP indicator to determine the position of the current picture in the display order. The number of pictures (if any) for which the current picture is delayed before presentation is used to calculate the PTS from the DTS as follows:

1. If the current picture is a B-picture, then the picture is not delayed in the reorder buffer and the PTS and DTS are identical.
2. If the current picture is either an open GOP I-picture or a P-picture that does not immediately precede a closed GOP I-picture, then the picture will be delayed in the reorder buffer by two picture periods while the subsequent B-picture is decoded and displayed. The PTS is equal to the DTS plus twice the picture period.
3. If the current picture is an open GOP I-picture or a P-picture that is immediately before a closed GOP I-picture, then the picture is delayed in the reorder buffer by only one picture period while a previously delayed open I-picture or P-picture is displayed. The PTS is equal to the DTS plus the picture period.
4. If the current picture is a closed GOP I-picture, then the picture is delayed one picture period while a previously delayed open GOP I-picture or P-picture is displayed. The PTS is equal to the DTS plus the picture period.

Table 6.5 summarizes the PTS and DTS calculations for a sequence of pictures without the film mode processed in m = 2 B-picture processing mode. Note that the subindices (i) are in decoding order.

TABLE 6.5
PTS and DTS Calculation for NonFilm Mode (m = 2)

Encoded Pictures	Display Order	$PETS_i$	$\Delta PETS_i$	DTS_i	PTS_i
I_0 (closed)		$PETS_0$	F	$PETS_0 + T_d$	$DTS_0 + F$
P_1	I_0	$PETS_1$	F	$PETS_1 + T_d$	$DTS_1 + 2F$
B_2	B_2	$PETS_2$	F	$PETS_2 + T_d$	DTS_2
I_3 (open)	P_1	$PETS_3$	F	$PETS_3 + T_d$	$DTS_3 + 2F$
B_4	B_4	$PETS_4$	F	$PETS_4 + T_d$	DTS_4
P_5	I_3	$PETS_5$	F	$PETS_5 + T_d$	$DTS_5 + 2F$
B_6	B_6	$PETS_6$	F	$PETS_6 + T_d$	DTS_6
P_7	P_5	$PETS_7$	F	$PETS_7 + T_d$	$DTS_7 + F$
I_8 (closed)	P_7	$PETS_8$	F	$PETS_8 + T_d$	$DTS_8 + F$
P_9	I_8	$PETS_9$	F	$PETS_9 + T_d$	$DTS_9 + 2F$
B_{10}	B_{10}	$PETS_{10}$	F	$PETS_{10} + T_d$	DTS_{10}

The rules used in computing the DTS and PTS for the *i*-th picture in nonfilm mode with m = 2 can be summarized as follows.

1. Verify to ensure that the difference between $PETS_i$ and $PETS_{i-1}$ is F, where F is the nominal picture period in 90-kHz clock cycles (3,003 for NTSC, 3,600 for PAL).
2. Calculate DTS as $DTS_i = PETS_i + T_d$, where T_d is the nominal delay from the output of the video encoder to the output of the decoder.
3. If picture *i* is a B-picture, then $PTS_i = DTS_i$.
4. If picture *i* is a P-picture or an open GOP I-picture and picture i + 1 is a closed GOP I-picture, then $PTS_i = DTS_i + F$, where F is the nominal picture period in 90-kHz clock cycles.
5. If picture *i* is a P-picture or an open I-picture and picture i + 1 is not a closed GOP I-picture, then $PTS_i = DTS_i + 2F$, where 2F is twice the nominal picture period in 90-kHz clock cycles (6,006 for NTSC, 7,200 for PAL).
6. If picture *i* is a closed GOP I-picture, then $PTS_i = DTS_i + F$, where F is the nominal picture period in 90-kHz clock cycles (3,003 for NTSC, 3,600 for PAL).
7. If $PTS_i = DTS_i$, then the PTS_i is sent, but the DTS_i will not be sent in the PES header preceding the *i*-th picture; otherwise, both the PTS_i and the DTS_i will be sent in the PES header preceding the *i*-th picture.

EXAMPLE

(Single B-picture with the film mode). In the case of m = 2, a sequence of coded pictures will arrive at the decoder in the same picture type order and be likewise reordered in an identical manner regardless of whether the film mode was active or inactive when the sequence was coded. The difference between the film mode and nonfilm mode is seen in how long each picture is displayed by the decoder.

As shown in Table 6.3, the display duration of a given picture processed in the film mode depends on which of the four possible film mode states (represented as A, B, C, and D) was active when the picture was processed. The video encoder usually needs to implement film mode processing, dropping two fields of redundant information in a sequence of five pictures, prior to predictive coding. A sequence of m = 2 coded pictures in the film mode will not output from the video encoder in the A, B, C, D order; the film state order will be rearranged by I, P, and B coding. However, after reordering by the decoder, the A, B, C, D order will be reestablished prior to display after decoding. The decoder will display film state A and C pictures for three field times since they both need to repeat a field. Film state B and D pictures are displayed for only two field times.

In m = 2 configuration, there are two different scenarios to be examined for developing an algorithm for calculating PTS and DTS in the film mode. The picture coding type/film state interaction will display two different patterns depending if the first picture to enter the film mode is a B-picture or a P-picture.

Tables 6.6 and 6.7, and Figure 6.13 provide examples of the PTS and DTS calculations for a series of pictures in film mode processed with B-pictures (m = 2).

TABLE 6.6
PTS and DTS Calculation, with the Film Mode, Entry at B-Picture, m = 2

Encoded Pictures (with Film State)	Display Pictures (with Film State)	$PETS_i$	$\Delta PETS_i$	DTS_i	PTS_i	Display Duration
I_0		$PETS_0$	F	$PETS_0 + T_d$	$DTS_0 + F$	F
P_1 (B)	I_0	$PETS_1$	F	$PETS_1 + T_d$	$DTS_1 + \lceil 2.5F \rceil$	F
B_2 (A)	B_2 (A)	$PETS_2$	F	$PETS_2 + T_d$	DTS_2	$\lceil 1.5F \rceil$
P_3 (D)	P_1 (B)	$PETS_3$	$\lceil 1.5F \rceil$	$PETS_3 + T_d$	$DTS_3 + \lfloor 2.5F \rfloor$	F
B_4 (C)	B_4 (C)	$PETS_4$	F	$PETS_4 + T_d$	DTS_4	$\lfloor 1.5F \rfloor$
I_5 (B)	P_3 (D)	$PETS_5$	$\lfloor 1.5F \rfloor$	$PETS_5 + T_d$	$DTS_5 + \lceil 2.5F \rceil$	F
B_6 (A)	B_6 (A)	$PETS_6$	F	$PETS_6 + T_d$	DTS_6	$\lceil 1.5F \rceil$
P_7 (D)	I_5 (B)	$PETS_7$	$\lceil 1.5F \rceil$	$PETS_7 + T_d$	$DTS_7 + \lfloor 2.5F \rfloor$	F
B_8 (C)	B_8 (C)	$PETS_8$	F	$PETS_8 + T_d$	DTS_8	$\lfloor 1.5F \rfloor$
P_9	P_7 (D)	$PETS_9$	$\lfloor 1.5F \rfloor$	$PETS_9 + T_d$	$DTS_9 + 2F$	F
B_{10}	B_{10}	$PETS_{10}$	F	$PETS_{10} + T_d$	DTS_{10}	F

The example shown in Table 6.6 is with a B-picture as the first picture to enter the film mode (film state A). Table 6.7 shows the case when a P-picture is the first picture to enter the film mode. Note that no closed GOP I-pictures are considered in here. This is because the encoders usually do not generate a closed I-picture in the film mode. The picture coding type/film state pattern in both cases shown win repeat every fourth picture. Again, the indices (i) are in decoding order.

For the case of m = 2, the following timing relationships for the *i*-th picture in the film mode need to be satisfied:

1. If picture *i* is in film state A and picture i – 1 is in nonfilm mode, then the difference between $PETS_i$ and $PETS_{i-1}$ is F, where F is the nominal picture period in 90-kHz clock cycles (3,003 for NTSC, 8,600 for PAL).
2. If picture i – 1 is in film state A, then the difference between $PETS_i$ and $PETS_{i-1}$ is $\lceil 1.5F \rceil$, where $\lceil 1.5F \rceil$ is the one and one-half nominal picture periods in 90-kHz clock cycles rounded up to the nearest integer (4,505 for NTSC, 5,400 for PAL).
3. If picture i – 1 is in film state B or D, then the difference between $PETS_i$ and $PETS_{i-1}$ is F.
4. If picture i – 1 is in film stats C, then the difference between $PETS_i$ and $PETS_{i-1}$ is $\lfloor 1.5F \rfloor$, where $\lfloor 1.5F \rfloor$ is the one and one-half nominal picture periods in 90-kHz clock cycles rounded down to the nearest integer (4,504 for NTSC, 5,400 for PAL).

The calculation for the PTS and DTS in the film mode with m = 2 is conditional to the current picture type and its film state and previous picture's

TABLE 6.7
PTS and DTS Calculation, with the Film Mode, Entry at P-Picture, m = 2

Encoded Pictures (with Film State)	Display Pictures (with Film State)	$\Delta PETS_i$	DTS_i	PTS_i	Display Duration
I_0		F	$PETS_0 + T_d$	$DTS_0 + F$	F
P_1 (A)	I_0	F	$PETS_1 + T_d$	$DTS_1 + 2F$	F
B_2	B_2	$\lceil 1.5F \rceil$	$PETS_2 + T_d - \lceil 0.5F \rceil$	DTS_2	F
P_3 (C)	P_1 (A)	F	$PETS_3 + T_d - \lfloor 0.5F \rfloor$	$DTS_3 + \lceil 2.5F \rceil$	$\lceil 1.5F \rceil$
B_4 (B)	B_4 (B)	$\lfloor 1.5F \rfloor$	$PETS_4 + T_d - \lfloor 0.5F \rfloor$	DTS_4	F
I_5 (A)	P_3 (C)	F	$PETS_5 + T_d - \lceil 0.5F \rceil$	$DTS_5 + \lfloor 2.5F \rfloor$	$\lfloor 1.5F \rfloor$
B_6 (D)	B_6 (D)	$\lceil 1.5F \rceil$	$PETS_6 + T_d - \lceil 0.5F \rceil$	DTS_6	F
P_7 (C)	I_5 (A)	F	$PETS_7 + T_d - \lfloor 0.5F \rfloor$	$DTS_7 + \lceil 2.5F \rceil$	$\lceil 1.5F \rceil$
B_8 (B)	B_8 (B)	$\lfloor 1.5F \rfloor$	$PETS_8 + T_d - \lfloor 0.5F \rfloor$	DTS_8	F
P_9	P_7 (C)	F	$PETS_9 + T_d - \lceil 0.5F \rceil$	$DTS_9 + \lfloor 2.5F \rfloor$	$\lfloor 1.5F \rfloor$
B_{10} (D)	B_{10} (D)	F	$PETS_{10} + T_d$	DTS_{10}	F

film state. One set of rules for video encoder and transcoder is summarized in Table 6.8.

EXAMPLE

(Double B-picture, without the film mode): In m = 3 configuration, two B-pictures are sent between each I- or P-picture. As described for the case of nonfilm, m = 2 mode, the normal I-P-B order will be altered when a closed I-picture is generated. The picture type and closed GOP indicator to determine the position of the current picture in the display order.

An example of the PTS and DTS calculations is given in Table 6.9 for a coded sequence without the film mode with m = 3.

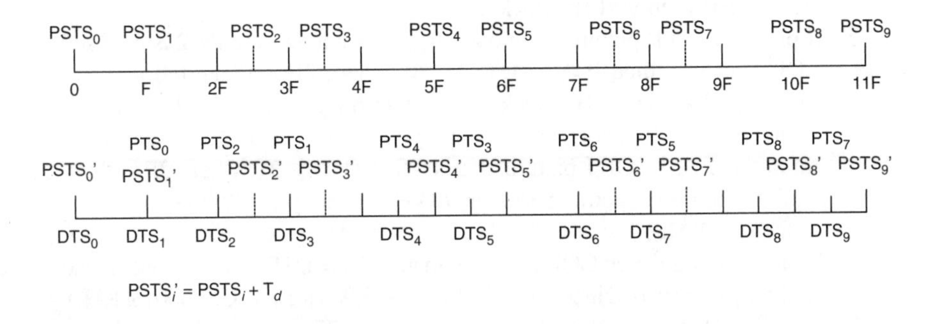

FIGURE 6.13 Illustration of time stamps relations for the example given in Table 6.7.

TABLE 6.8
Calculation of PTS and DTS in the Film Mode for m = 2

Picture Type	Film State for Picture i	Film State for Picture $i-1$	DTS_i	PTS_i
I (closed)	None	None	$PETS_i + T_d$	$DTS_i + F$
B	None	None	$PETS_i + T_d$	DTS_i
B	A	B		
B	C	D		
B	D	None		
P or I(open)	None	None	$PETS_i + T_d$	$DTS_i + 2F$
P or I(open)	A	None		
P or I(open)	C	None	$PETS_i + T_d - \lfloor 0.5F \rfloor$	$DTS_i + \lceil 0.5F \rceil$
P or I(open)	C	D		
P or I(open)	B	None	$PETS_i + T_d$	$DTS_i + \lceil 0.5F \rceil$
P or I(open)	B	C		
B	None	A	$PETS_i + T_d - \lceil 0.5F \rceil$	DTS_i
B	D	A		
B	B	C	$PETS_i + T_d - \lfloor 0.5F \rfloor$	DTS_i
P or I(open)	A	B	$PETS_i + T_d - \lceil 0.5F \rceil$	$DTS_i + \lfloor 0.5F \rfloor$
P or I(open)	D	A	$PETS_i + T_d$	$DTS_i + \lfloor 0.5F \rfloor$

The rules used in computing the DTS and PTS for the i-th picture in nonfilm mode with m = 3 are extension of those rules for m = 2. These can be summarized as follows.

1. Verify to ensure that the difference between $PETS_i$ and $PETS_{i-1}$ is F, where F is the nominal picture period in 90-kHz clock cycles (3,003 for NTSC, 3,600 for PAL).
2. Calculate DTS as $DTS_i = PETS_i + T_d$, where T_d is the nominal delay from the output of the video encoder to the output of the decoder.
3. If picture i is a B-picture, then $PTS_i = DTS_i$.
4. If picture i is a P-picture or an open GOP I-picture and picture i + 1 is a closed GOP I-picture, then $PTS_i = DTS_i + F$, where F is the nominal picture period in 90-kHz clock cycles.
5. If picture i is a P-picture or an open I-picture and picture i + 2 is a closed GOP I-picture, then $PTS_i = DTS_i + 2F$, where 2F is twice the nominal picture period in 90-kHz clock cycles (6,006 for NTSC, 7,200 for PAL).
6. If picture i is a P-picture or an open I-picture and pictures i + 1 and i + 2 are not a closed GOP I-picture, then $PTS_i = DTS_i + 3F$, where 3F is three times the nominal picture period in 90 kHz clock cycles (9,009 for NTSC, 10,800 for PAL).
7. If picture i is a closed GOP I-picture, then $PTS_i = DTS_i + F$, where F is the nominal picture period in 90-kHz clock cycles (3,003 for NTSC, 3,600 for PAL).
8. If $PTS_i = DTS_i$, then the PTS_i is sent, but the DTS_i will not be sent in the PES header preceding the i-th picture; Otherwise, both the PTS_i and the DTS_i will be sent in the PES header preceding the i-th picture.

TABLE 6.9
An Example for the PTS and DTS Calculations for m = 3 Without the Film Mode

Encoded Pictures (i)	Display Order	$PETS_i$	$\Delta PETS_i$	DTS_i	PTS_i
I_0 (closed)		$PETS_0$	F	$PETS_0 + T_d$	$DTS_0 + F$
P_3	I_0	$PETS_1$	F	$PETS_1 + T_d$	$DTS_1 + 3F$
B_1	B_1	$PETS_2$	F	$PETS_2 + T_d$	DTS_2
B_2	B_2	$PETS_3$	F	$PETS_3 + T_d$	DTS_3
I_6 (open)	P_3	$PETS_4$	F	$PETS_4 + T_d$	$DTS_4 + 3F$
B_4	B_4	$PETS_5$	F	$PETS_5 + T_d$	DTS_5
B_5	B_5	$PETS_6$	F	$PETS_6 + T_d$	DTS_6
P_7	I_6	$PETS_7$	F	$PETS_7 + T_d$	$DTS_7 + F$
I_8 (closed)	P_7	$PETS_8$	F	$PETS_8 + T_d$	$DTS_8 + F$
P_{11}	I_8	$PETS_9$	F	$PETS_9 + T_d$	$DTS_9 + 3F$
B_9	B_9	$PETS_{10}$	F	$PETS_{10} + T_d$	DTS_{10}
B_{10}	B_{10}	$PETS_{11}$	F	$PETS_{11} + T_d$	DTS_{11}
P_{13}	P_{11}	$PETS_{12}$	F	$PETS_{12} + T_d$	$DTS_{12} + 2F$
B_{12}	B_{12}	$PETS_{13}$	F	$PETS_{13} + T_d$	DTS_{13}
I_{14} (closed)	P_{13}	$PETS_{14}$	F	$PETS_{14} + T_d$	$DTS_{14} + F$

Note that in Table 6.9 the indices (i) are in the decoding order. However, the subindex of each picture is in the display order

EXAMPLE

(Double B-picture with the film mode): As in the case of m = 2, both the reordering caused by the presence of B-pictures and the difference in display duration for certain film states must be considered when calculating the PTS and DTS for m = 3 pictures in the film mode. An example of PTS and DTS calculation for m = 3 in the film mode is given next in Tables 6.10, 6.11, 6.12, and Figure 6.14.

The general rules for m = 3 in the film-mode can also be determined in a similar manner as that for m = 2 in the film mode. Interested readers can develop these rules as exercises.

Time Stamp Errors: As discussed in Section 6.2 System Clock Recovery, the clock-recovery process is designed to track the encoder timing and manage the absolute and relative system timing for video and other multimedia data during decoding operations. Specifically, the clock-recovery process monitors time stamps in the transport stream and updates the system clock in a multimedia program when necessary.

During MPEG-2 decoding, the clock-recovery process is programmed to monitor the PCRs in the transport stream. The clock-recovery process uses PCRs in the stream against its own system clock and indicates discontinuities every time an error is seen in a PCR that is larger than a programmable threshold. If a PCR discontinuity is detected in the incoming transport stream, the new PCR is used to update the

TABLE 6.10
An Example for the PTS and DTS Calculations in the Film Mode with m = 3 and Entering at the first B-Picture

Encoded Pictures (with Film State)	Display Pictures (with Film State)	$\Delta PETS_i$	DTS_i	PTS_i	Display Duration
I_0		F	$PETS_0 + T_d$	$DTS_0 + F$	F
P_1 (A)	I_0	F	$PETS_1 + T_d$	$DTS_1 + \lceil 3.5F \rceil$	F
B_2	B_2 (A)	F	$PETS_2 + T_d$	DTS_2	$\lceil 1.5F \rceil$
B_3	B_3 (B)	$\lceil 1.5F \rceil$	$PETS_3 + T_d$	DTS_3	F
P_4 (D)	P_1 (C)	F	$PETS_4 + T_d$	$DTS_4 + \lfloor 3.5F \rfloor$	$\lfloor 1.5F \rfloor$
B_5 (B)	B_5 (D)	$\lfloor 1.5F \rfloor$	$PETS_5 + T_d$	DTS_5	F
B_6 (C)	B_6	F	$PETS_6 + T_d$	DTS_6	F
P_7	P_4	F	$PETS_7 + T_d$	$DTS_7 + 3F$	F
B_8	B_8	F	$PETS_8 + T_d$	DTS_8	F
B_9	B_9	F	$PETS_9 + T_d$	DTS_9	F
P_{10}	P_7	F	$PETS_{10} + T_d$	—	F

video system clock counter (STC). After the video decoder STC is updated, the PLL begins to track PCR. The picture is decoded when DTS = STC.

However, the network jitters can cause time-stamp errors that, in turn, could cause decoder buffer over- or underflows. Therefore, at any moment, if the decoder

TABLE 6.11
An Example for the PTS and DTS Calculations in the Film Mode with m = 3 and Entering at the Second B-Picture

Encoded Pictures (with Film State)	Display Pictures (with Film State)	$\Delta PETS_i$	DTS_i	PTS_i	Display Duration
I_0		F	$PETS_0 + T_d$	$DTS_0 + F$	F
P_1 (A)	I_0	F	$PETS_1 + T_d$	$DTS_1 + \lceil 3.5F \rceil$	F
B_2	B_2 (A)	F	$PETS_2 + T_d$	DTS_2	F
B_3	B_3 (B)	F	$PETS_3 + T_d$	DTS_3	$\lceil 1.5F \rceil$
P_4 (D)	P_1 (C)	$\lceil 1.5F \rceil$	$PETS_4 + T_d$	$DTS_4 + \lfloor 3.5F \rfloor$	F
B_5 (B)	B_5 (D)	F	$PETS_5 + T_d$	DTS_5	$\lfloor 1.5F \rfloor$
B_6 (C)	B_6	$\lfloor 1.5F \rfloor$	$PETS_6 + T_d$	DTS_6	F
P_7	P_4	F	$PETS_7 + T_d$	$DTS_7 + 3F$	F
B_8	B_8	F	$PETS_8 + T_d$	DTS_8	F
B_9	B_9	F	$PETS_9 + T_d$	DTS_9	F
P_{10}	P_7	F	$PETS_{10} + T_d$	—	F

TABLE 6.12
An Example for the PTS and DTS Calculations in the Film Mode with m = 3 and Entering at P-Picture

Encoded Pictures (with Film State)	Display Pictures (with Film State)	$\Delta PETS_i$	DTS_i	PTS_i	Display Duration
I_0		F	$PETS_0 + T_d$	$DTS_0 + F$	F
P_1 (A)	I_0	F	$PETS_1 + T_d$	$DTS_1 + 3F$	F
B_2	B_2	$\lceil 1.5F \rceil$	$PETS_2 + T_d - \lceil 0.5F \rceil$	DTS_2	F
B_3	B_3	F	$PETS_3 + T_d - \lfloor 0.5F \rfloor$	DTS_3	F
P_4 (D)	P_1 (A)	$\lfloor 1.5F \rfloor$	$PETS_4 + T_d - F$	$DTS_4 + 4F$	$\lceil 1.5F \rceil$
B_5 (B)	B_5 (B)	F	$PETS_5 + T_d - \lceil 0.5F \rceil$	DTS_5	F
B_6 (C)	B_6 (C)	F	$PETS_6 + T_d - \lfloor 0.5F \rfloor$	DTS_6	$\lceil 1.5F \rceil$
P_7	P_4 (D)	F	$PETS_7 + T_d$	$DTS_7 + 3F$	F
B_8	B_8	F	$PETS_8 + T_d$	DTS_8	F
B_9	B_9	F	$PETS_9 + T_d$	DTS_9	F
P_{10}	P_7	F	$PETS_{10} + T_d$	—	F

buffer is overflowing, some coded pictures in the buffer will be dropped without decoding. If DTS = STC, but the decoder buffer is still underflow, the decoder may wait a certain amount of time for the current coded picture to completely enter the buffer for decoding. In these cases, error-concealment algorithms are usually required.

The above methods of calculating DTS and PTS for MPEG-2 video can be directly used in (or be generalized to) other compressed video, such as MPEG-4 (part 2) video [6-29], H.263 video [6-31], and MPEG-4 (part-10) Advanced Video Coding [6-33].

6.4 SUMMARY

In this chapter, we discussed techniques of video synchronization at the decoder through time stamps. As an example, we focused on MPEG-2 transport systems to illustrate the key function blocks of a video synchronization technique. In MPEG-2 systems, clock recovery is possible by transmitting time stamps called *program clock references* (PCRs) in the bit stream at the rate of at least 10 per second. The PCRs are generated at the encoder by sampling the system time clock (STC), which runs at 27 MHz ± 30 ppm. Since the decoder's free-running system clock frequency does not exactly match the encoder's STC, the reference time is reconstructed by means of a phase-locked loop (PLL) and the received PCRs.

The encoder STC of a video program is also used to create time stamps that indicate the presentation and decoding timing of video. The time stamps that

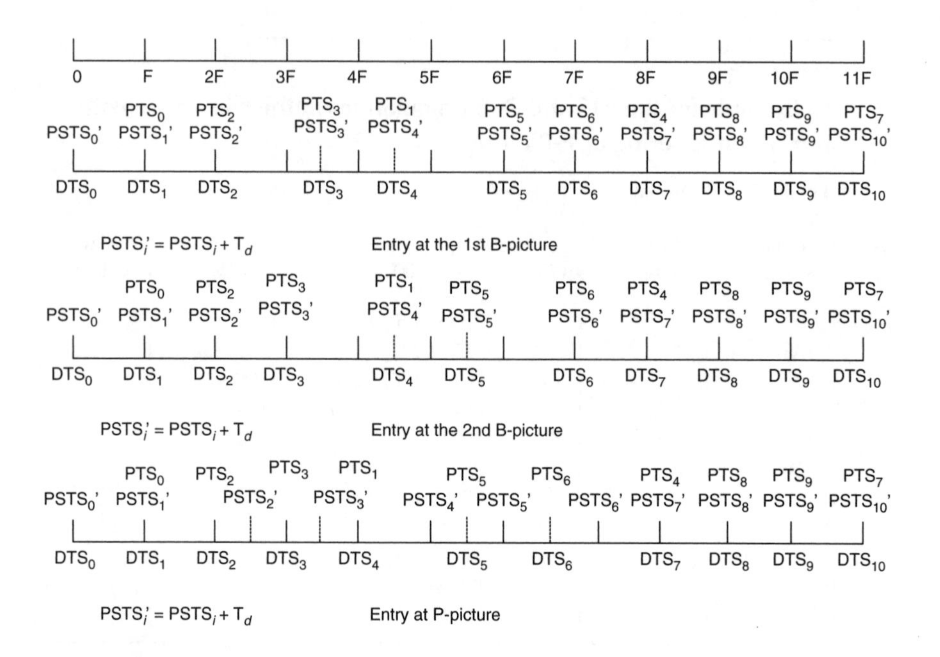

FIGURE 6.14 Time stamp relations of examples given in Tables 6.10, 6.11, and 6.12.

indicate the presentation time of video are called *presentation time stamps* (PTS) whereas those that indicate the decoding time are called *decoding time stamps* (DTS). It is the presence of these time stamps and the correct use of the time stamps that provide the facility to synchronize properly the operation of the decoding. Methods for generating the DTS and PTS in the video encoder are discussed in this chapter. In particular, the time stamping methods for MPEG-2 video are introduced as examples. These methods can be directly used in other video compression standards.

6.5 EXERCISES

6-1. Consider an MPEG-2 video transcoder that needs to extract the PTS/DTS in the incoming stream and perform decoding and then re-encoding. Draw a block diagram to show how the STC synchronization and PTS/DTS restamping should be done in such transcoder.

6-2 If a transcoder performs format conversion, e.g., from a stream of 720 × 480 30 frames/s compressed video to a stream of 352 × 480 30 frames/s compressed video, what are the impacts on its system timing information? Why ?

6-3. A diagram of a second-order digital PLL system is provided in the following figure.

In this diagram, basic building blocks are described:

- Low-pass (loop) filter: an IIR filter has been designed as the loop filter, $L(Z)$ is its transfer function

$$L(Z) = \frac{K_1 + K_2 - K_1 \cdot Z^{-1}}{1 - Z^{-1}}$$

where K_1 and K_2 are the gains of the IIR filter.

- A digitally-controlled VCO, or a discrete time oscillator, has the transfer function $D(Z)$

$$D(Z) = \frac{K_{DCO}}{1 - Z^{-1}}$$

where K_{DCO} is the gain of the discrete voltage controlled oscillator. With these building blocks of the digital PLL system, its closed-loop transfer function can be written as:

$$H(Z) = \frac{K_{PD} \cdot L(Z) \cdot D(Z) \cdot Z^{-1}}{1 + K_{PD} \cdot L(Z) \cdot D(Z) \cdot Z^{-1}}$$

where, K_{PD} is the gain of the phase detector. The format of this transfer function can be rewritten as:

$$H(Z) = \frac{(k_1 + k_2) \cdot Z - k_1}{Z^2 + (k_1 + k_2 - 2) \cdot Z + (1 - k_1)}$$

where $k_1 = K_{PD} \cdot K_{DCO} \cdot K_1$ and $k_2 = K_{PD} \cdot K_{DCO} \cdot K_2$. The denominator of the above equation is also called the characteristic equation of the system:

$$\Delta(Z) = Z^2 + (k_1 + k_2 - 2) \cdot Z + (1 - k_1)$$

By using the characteristic equation, k_1 and k_2 can be resolved [6-2]:

$$k_1 = 1 - e^{-2\lambda \cdot \omega \cdot T_s}$$

$$k_2 = 1 + e^{-2\lambda \cdot \omega \cdot T_s} - 2e^{-\lambda \cdot \omega \cdot T_s} \cos\left(\omega \cdot T_s \cdot \sqrt{1 - \lambda^2}\right)$$

A. One mandatory requirement for designing digital PLLs is that the system must be stable. Derive the stable conditions for the parameters k_2 and k_2 of this digital PLL architecture.

B. A steady-state error analysis of a Digital PLL is extremely important in the PLL design. The last paragraph describes the stable conditions of D-PLL system. The steady-state errors of phase and frequency of the D-PLL are studied here. Prove that both phase and frequency error of the D-PLL system given for H(z) will be zero when the system reaches steady state.

6-4. In PAL systems, video has a frame rate of 25 frames/s (or, say, 50 fields/s). Show an example on how to apply the "3:2 pull-down" process to the film mode in this case.

6-5. Try to fill up the blank DTS and PTS entries of Table 6.8.

6-6. Develop the general rules for the PTS and DTS calculations in the film mode with m = 3.

REFERENCES

[6-1] A54, Guide to the use of the ATSC digital television standard, Advanced Television Systems Committee, Oct. 19, 1995.

[6-2] D. Fibush. Subearrier Frequency, Drift and Jitter in NTSC Systems, ATM Forum, ATM940722, July 1994.

[6-3] X. Chen, *Transporting Compressed Digital Video,* Kluwer Academic Publishers, Boston, 2002.

[6-4] ITU-T Recommendation H.222.0 (1995) ISO/IEC 13818-1: 1996, Information technology—Generic coding of moving pictures and associated audio information: Systems.

[6-5] K. Jack, *Video Demystified,* HighText Interactive Inc., San Diego, 1996.

[6-6] H. Meyr and G. Ascheid. *Synchronization in Digital Communications,* John Wiley & Sons, New York, 1990.

[6-7] N. Ohta, *Packet Video,* Artech House Inc., Boston, 1994.

[6-8] B. G. Haskell, A. Puri, and A. N. Netravali, *Digital Video: An Introduction to MPEG-2,* Chapman & Hall, New York, 1997.

[6-9] The ATM Forum Technical Committee, Audiovisual Multimedia Services: Video on Demand Specification 1.0, December 1995.

[6-10] P. Hodgins and E. Itakura, The Issues of Transportation of MPEG over ATM, ATM Forum, ATM94-0570, July 1994.

[6-11] P. Hodgins and E. Itakura, VBR MPEC-2 over AAL5, ATM Forum, ATM94-1052, December 1994.

[6-12] R. P Singh, Sang-Hoon Lee, and Chong-Kwoon Kim, Jitter and clock recovery for periodic traffic in broadband packet networks, *IEEE Transactions on Communications,* 42, 2189, 1994.

[6-13] A. Leon Garcia, *Probability and Random Processes for Electrical Engineering,* Addison-Wesley Publishing Co., Reading, MA, 1994.

[6-14] B. C. Kuo, *Automatic Control Systems,* Prentice-Hall, Englewood Cliffs, NJ, 1995.

[6-15] A. V. Oppenheim and R. W. Schafer, *Discrete-Time Signal Processing,* Prentice-Hall, Englewood Cliffs, NJ, 1999.

[6-16] John L. Stensby, *Phase-Locked Loops: Theory and Applications,* CRC Press, Boca Raton, FL, 1997.

[6-17] C. Tryfonas and A. Varma, Time-stamping schemes for MPEG-2 systems layer and their effect on receiver clcok recovery, UCSC-CRL-98-2, University of California, Santa Cruz, 1998.

[6-18] M. De Prycker, *Asynchronous Transfer Mode : Solution for Broadband ISDN,* Ellis Horwood, 1993.

[6-19] S. Dixit and P. Skelly, MPEG-2 over ATM for video dial tone networks: issues and strategies, *IEEE Network,* 9, 30, 1995.

[6-20] Y. Kaiser, Synchronization and de-jittering of a TV decoder in ATM networks, In Proceedings of Packet Video '93, volume 1, 1993.

[6-21] M. Perkins and P. Skelly, A Hardware MPEG Clock Recovery Experiment in the Presence of ATM Jitter, ATM Forum, May 1994. ATM94-0434.

[6-22] J. Proakis and D. G. Manolakis. *Introduction to Digital Signal Processing,* Macmillan, New York, 1988.

[6-23] M. Schwartz and D. Beaumont, Quality of Service Requirements for Audio-Visual Multimedia Services, ATM Forum, July 1994. ATM94-0640.

[6-24] X. Chen, Synchronization of a stereoscopic video sequence," U.S. Patent Number 5886736, Assignee: General Instrument Corporation, March 23, 1999.

[6-25] X. Chen and R. O. Eifrig, Video rate buffer for use with push data flow, U.S. Patent Number 6289129, Assignee: Motorola Inc. and General Instrument Corporation, Sept. 11, 2001.

[6-26] WO9966734, X. Chen, F. Lin, and A. Luthra, Video encoder and encoding method with buffer control, 2000.

[6-27] X. Chen, Rate control for stereoscopic digital video encoding, U.S. Patent Number 6072831, Assignee: General Instrument Corporation, June 6, 2000.

[6-28] A. Puri and T. H. Chen, *Multimedia Standards and Systems,* Chapman & Hall, New York, 1999.

[6-29] ISO/IEC 14496-2:1998, Information technology: generic coding of audio-visual objects—Part 2: Visual.

[6-30] Test model editing committee, Test Model 5, MPEG93/457, ISO/IEC JTC1/SC29/WG11, April 1993.

[6-31] ITU-T Experts Group on Very Low Bitrate Visual Telephony, ITU-T Recommendation H.263: Video coding for low bitrate communication, Dec. 1995.

[6-32] J. Whitaker, *DTV Handbook,* McGraw-Hill, New York, 2001.

[6-33] A. Puri, X. Chen, and A. Luthra, "Video coding using the H.264/MPEG-4 AVC video compression standard," *Signal Processing: Image Communication Journal,* Vol. 19, Issue 10, 2004.

7 Transcoder Video Buffer and Hypothetical Reference Decoder

In various video compression standards, such as MPEG-1, MPEG-2, MPEG-4 video (Part 2), H.261, H.263, and H.264/MPEG-4 Advanced Video Coding (Part 10), a hypothetical reference decoder (HRD), or *video buffer verifier* (VBV), is specified for modeling the timing of the flow of compressed video data and decompressed pictures through a decoder. It serves to impose constraints on the variations in bit rate over time in a compressed bit stream. It may also serve as a timing and buffering model for a real decoder implementation or for a multiplexer. In this chapter, we will discuss basic concepts of video buffer management and specifications of HRD/VBV, especially for transcoders.

7.1 VIDEO BUFFER MANAGEMENT

The video buffer management in an encoder or a transcoder provides a protocol to prevent decoder buffer under- and/or overflows. With such a protocol, adaptive quantization is applied in the encoder or the transcoder along with rate control to ensure the required video quality and to satisfy the buffer regulation. In this chapter, we will derive the buffer dynamics and will determine general conditions for preventing both encoder and decoder buffers under- and/or overflow. In Chapter 6, we have discussed the time stamps for decoding and presentation. In this section, we will investigate conditions for preventing decoder buffer under-/overflows by using the picture encoding time, the picture decoding time and dynamics of encoded-picture size. We will also study some principles on video rate-buffer management of video encoders or transcoder.

Television broadcast applications require the video encoder, transcoder, and decoder to have the same clock frequency as well as the same frame rate, and to operate synchronously. For example, in an MPEG-2 enabled system, decoder to encoder (or transcoder) synchronization is maintained through the utilization of a program clock reference (PCR) and decoding time stamp (DTS) (or presentation time stamp [PTS]) carried in the stream. In an MPEG-2 transport stream, the adaptation field supplies a program clock reference (PCR). The packetized elementary stream (PES) packet carries DTS and PTS. Since compressed pictures usually have variable sizes, DTSs (and PTSs) are related to encoder and decoder buffer (FIFO) fullness at certain points. Figure 7.1 shows the video encoder/transcoder and decoder buffer model. In this figure, T is the picture duration of the original

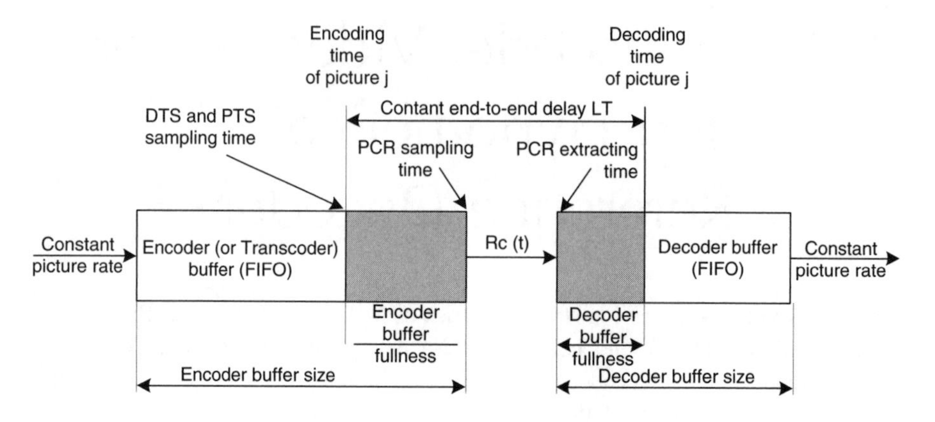

FIGURE 7.1 Video encoder/transcoder and decoder buffer model.

uncompressed video and L is a positive integer. Thus, after a picture being (instantaneously) compressed, it waits $L \cdot T$ before starting to decode.

The decoder buffer under- and/or overflows are usually caused by channel jitter and/or video encoder/transcoder buffer over- and/or underflows. If the decoder buffer underflows, the buffer is being emptied faster than it is being filled. Compressed data resided in the decoder buffer are removed completely by the decoder at some point and some data required by decoder are not yet received from the (assuming jitter-free) transmission channel. Consequently, too many data are being generated in the encoder/transcoder and then at some point, the encoder/transcoder buffer overflows. To prevent this, the following procedures are often used in the encoder/transcoder:

- Increase the quantization level.
- Adjust bit budget allocation for pictures.
- Discard high-frequency DCT coefficients.
- Skip pictures.

If the decoder buffer overflows, it is being filled faster than it is being emptied. Too many data are being transmitted and too few data are being removed by the decoder such that the buffer is full. Consequently, too few data are being generated in the encoder/transcoder at some point; i.e., the encoder/transcoder buffer underflows. To prevent this, the following procedures are often used in the encoder/transcoder at certain point:

- Decrease the quantization level.
- Adjust bit budget allocation for pictures.
- Stuff bits.

As shown in Chapter 2, the adjustments on quantization-level and bit budget allocation are usually accomplished by using the rate-control algorithms along with an adaptive quantizer.

Rate-control and adaptive quantizer are important function blocks for achieving good compression performance in video encoder/transcoder [7-2][7-3]. For this reason, every MPEG-2 encoding/transcoder system on the market has its own intelligent rate-control and quantization algorithms. For example, an encoder may have an optimized, and often complicated, bit-allocation algorithm to assign the number of bits for each type of pictures (I-, P-, and B-pictures). Such a bit budget allocation algorithm usually takes into account the prior knowledge of video characters (e.g., scene changes, fade) and coding types (e.g., picture types) for a group of pictures (GOP). Adaptive quantization is applied in the encoder along with rate control to ensure the required video quality and to satisfy the buffer constraints.

7.2 CONDITIONS FOR PREVENTING DECODER BUFFER UNDERFLOW AND OVERFLOW

The primary task of video rate-buffer management for an encoder is to control its output bit stream to comply with the buffer requirements, e.g., the VBV specified in MPEG-2 video (ISO/IEC 13818-2)[7-1] and MPEG-4 video (ISO/IEC 14496-2) [7-2] and the HRD specified in H.264/MPEG-4 AVC (ISO/IEC 14496-10) [7-12]. To accomplish such a task, rate-control algorithms are introduced in Chapter 2.

One of the most important goals for rate-control algorithms is to prevent video buffer under- and/or overflows. For constant bit-rate (CBR) applications, by a use of the rate-control, bit-count-per-second must precisely converge to the target bit rate with good video quality. For variable bit-rate (VBR) applications, the rate-control achieves the goal of maximizing the perceived quality of decoded video sequence with the maintained output-bit rate within permitted bounds.

In the following discussion, the encoder/transcoder buffer is characterized by the following new set of parameters (note that transcoder buffer here is a buffer in the transcoder for the re-encoding process):

- B_E^j denotes the encoder/transcoder buffer bit-level right before encoding/transcoding of the j-th picture.
- B_D^j denotes the decoder buffer bit-level right before encoding of the j-th picture.
- P_E^j denotes the bit-count of the j-th coded picture.
- $B_D^{\max}$ denotes the decoder buffer size, e.g., MPEG-2 VBV buffer size coded in the sequence header and sequence extension if present.
- $B_E^{\max}$ denotes the size of the video encoder/transcoder buffer.

Assume the encoding/transcoding time of j-th picture is $t_{e,j}$ and decoding time of j-th picture is $t_{d,j}$, i.e., DTS for the j-th picture. Then, in order to avoid decoder buffer underflow, it requires that all the coded data up to and including picture j must be completely transmitted to the decoder buffer before time $t_{d,j}$, i.e.,

$$B_E^j + P_E^j \leq \int_{t_{e,j}}^{t_{d,j}} R_C(t)dt \tag{7.1}$$

where $R_C(t)$ is the bit-rate function and the integral $\int R_C(t)dt$ represents the total of bits transmitted for the video service from time $t_{e,j}$ to $t_{d,j}$.

In order to avoid decoder buffer overflow, it requires that the decoder buffer fullness at time $t_{d,j}$ (before picture j is decoded) be less than $B_D^{\max}$. From time $t_{e,j}$ to $t_{d,j}$, the number of bits arriving at the decoder buffer will be $\int R_C(t)dt$ and the number of bits being removed from the decoder buffer will be all the coded video data in both encoder and decoder buffers at time $t_{e,j}$. Thus the decoder buffer fullness at time $t_{d,j}$ satisfies:

$$B_D^j + \int_{t_{e,j}}^{t_{d,j}} R_C(t)dt - (B_D^j + B_E^j) \leq B_D^{\max} \tag{7.2}$$

This inequality can be simplified to

$$\int_{t_{e,j}}^{t_{d,j}} R_C(t)dt - B_E^j \leq B_D^{\max} \tag{7.3}$$

By applying inequality (7.3) to the $(j + 1)$-th picture, one has

$$\int_{t_{e,j+1}}^{t_{d,j+1}} R_C(t)dt - B_E^{j+1} \leq B_D^{\max} \tag{7.4}$$

where $t_{e,j+1}$ and $t_{d,j+1}$ denote the encoding/transcoding and decoding time for picture $j + 1$, respectively.

The encoder/transcoder buffer fullness also satisfies the following recursive equation:

$$B_E^{j+1} = B_E^j + P_E^j - \int_{t_{e,j}}^{t_{e,j+1}} R_C(t)dt \tag{7.5}$$

Thus, inequalities (7.4) and (7.5) yield

$$\int_{t_{e,j}}^{t_{d,j+1}} R_C(t)dt \leq B_E^j + P_E^j + B_D^{\max} \tag{7.6}$$

Inequalities (7.1) and (7.6) are necessary and sufficient conditions for preventing buffer under- and/or overflows if they are held for all pictures. By combining the two inequalities (7.1) and (7.6), one obtains upper and lower bounds on the size of picture j:

$$-B_D^{\max} + \int_{t_{e,j}}^{t_{d,j+1}} R_C(t)dt \leq B_E^j + P_E^j \leq \int_{t_{e,j}}^{t_{d,j}} R_C(t)dt \tag{7.7}$$

The above upper and lower bounds imply

$$-B_D^{\max} + \int_{t_{e,j}}^{t_{d,j+1}} R_C(t)dt \le \int_{t_{e,j}}^{t_{d,j}} R_C(t)dt \tag{7.8}$$

This inequality (7.8) imposes a constraint on the transmission rate $R_C(t)$.

$$\int_{t_{d,j}}^{t_{d,j+1}} R_C(t)dt \le B_D^{\max} \tag{7.9}$$

Also, from inequality (7.6), one has

$$\int_{t_{e,j}}^{t_{d,j+1}} R_C(t)dt - B_D^{\max} \le B_E^{\max} \tag{7.10}$$

This inequality provides a lower bound on the encoder buffer size $B_E^{\max}$. Note that such a lower bound is determined by end-to-end (buffer) delay, transmission rate, and the decoder buffer size $B_D^{\max}$.

Example

(Encoder buffer size): In a MPEG-2 video transmission system, for an end-to-end (buffer) delay ($t_{d,j} - t_{e,j} = LT$, see Figure 7.1) of 0.6 s, the lag from $t_{e,j}$ to $t_{d,j+1}$ can be at most 0.6 s plus three field time (0.05 s) in the case of a video which has a 720 × 480 resolution and 29.97 frames/s. Therefore, from inequalities (7.1) and (7.10), one has

$$B_E^{\max} \le 0.6 * R_C^{\max}$$

$$0.65 * R_C^{\max} - B_D^{\max} \le B_E^{\max}$$

where $B_E^{\max}$ = vbv_buffer_size. Thus, for $B_C^{\max}$ = 15,000,000 bits/s, the minimum required the buffer size of the encoder (or the encoder in a transcoder) is 1.125 Mbytes.

The rate-buffer management protocol is an algorithm for checking a bit stream to verify that the amount of rate-buffer memory required in the decoder is bounded by $B_D^{\max}$, e.g., vbv_buffer_size in MPEG-2 video. The rate-control algorithm will be guided by the rate-buffer management protocol to ensure the bit stream satisfying the buffer regulation with good video quality.

One of the key steps in the rate-buffer management and rate-control process is to determine the bit budget for each picture. The condition given by inequality (7.1) on preventing the decoder buffer underflow provides an upper bound on the bit budget for each picture. The reason is that, at the decoding time, the current picture should be small enough so that it is contained entirely inside the decoder buffer. The condition given by inequality (7.6) on avoiding the decoder buffer overflow provides a lower bound on the bit budget for each picture. These conditions can also be directly applied to both MPEG-2 Test Model and MPEG-4 Verification Model rate-control algorithms [7-14].

7.3 HYPOTHETICAL REFERENCE DECODER

7.3.1 Background

An HRD provides a model to describe how the bit rate is controlled in the compression process. The HRD contains a decoder buffer (or VBV Buffer) through which compressed data flows with a precisely specified arrival and removal timing, as shown in Figure 7.2. An HRD may be designed for variable or constant bit rate operation, and for low-delay or delay-tolerant behavior.

Compressed data representing a sequence of coded pictures flows into the predecoder buffer according to a specified arrival schedule. All compressed bits associated with a given coded picture are removed from the decoder buffer by the instantaneous decoder at the specified removal time of the picture.

The decoder buffer overflows if the buffer becomes full and more bits are arriving. The buffer underflows if the removal time for a picture occurs before all compressed bits representing the picture have arrived. HRDs differ in the means to specify the arrival schedule and removal times, and the rules regarding overflow and underflow of the buffer.

7.3.2 H.261 and H.263 HRDs

The HRDs in H.263 [7-7] [7-8] and H.261 were designed for low-delay operation. They operate by removal of all bits associated with a picture at the first time the buffer is examined, rather than at a time explicitly transmitted in the bit stream. They do not specify precisely when data arrives in the buffer. This HRD does not require for precisely timed removal of bits from the buffer and this makes it hard to design systems that display pictures with precise timing based on this model.

7.3.3 MPEG-2 Video Buffering Verifier (VBV)

The VBV in MPEG-2 video [7-1] can operate in either variable or constant bit rate mode and also has a low-delay mode. The MPEG-2 VBV has two modes of

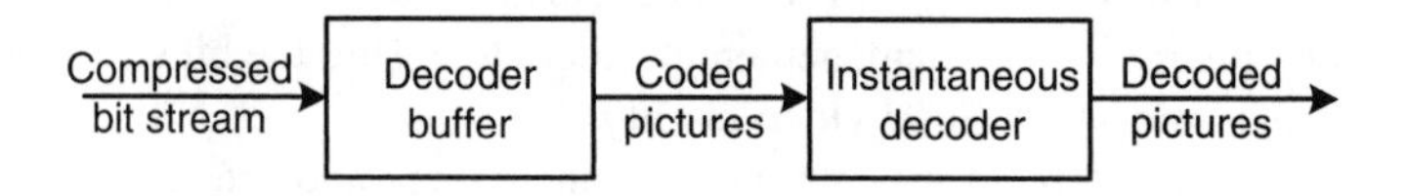

FIGURE 7.2 A hypothetical reference decoder.

operation based on whether a removal delay (the vbv_delay in MPEG-2 video syntax) is transmitted or not. In the first mode, when so-called vbv_delays are transmitted, the rate of arrival into the VBV buffer of each picture is computed based on picture sizes, vbv_delays, and additional removal time increments (which are based on the picture rate and some additional flags in the bit stream). This mode can be used by an encoder to create both variable and constant bit rate streams. However, without scanning through the entire stream, a decoder will not be able to determine whether or not a given stream is constant bit rate. The first mode also has an ambiguity at the beginning of the sequence that prevents the initial rate from being determined. Thus, technically it is impossible to know whether the stream is CBR. In this first mode, both underflow and overflow must be prevented by the encoder.

In the second mode (also known as the *leaky bucket* model), no explicit removal delays are transmitted; i.e., the encoder does not specify the vbv_delay (= 0xFFFF). In this mode, the arrival rate is constant unless the buffer is full. Thus the mode has no ambiguity about the initial rate. However, it has an arrival schedule that may not be causal with respect to the real production of bits. This limits its use as guidance for a multiplexer.

The term *leaky bucket* arises from the analogy of the encoder as a system that "dumps" water in discrete chunks into a bucket that has a hole in it. The departure of bits from the encoder buffer corresponds to water leaking out of the bucket. Here, the decoder buffer is described, which has an inverse behavior where bits flow *in* at a constant rate, and are removed in chunks.

In both modes, the removal times are based on a fixed frame rate. Neither of these MPEG-2 VBV modes can handle variable frame rate, except for the one special case of film content captured as video (3:2 pull-down). In this case, the removal time of certain pictures is delayed by one field period, based on the value of a syntax bit, repeat_first_field, in the picture header of that or a previous picture.

Further description of the nature of the noncausality in the second mode is warranted [7-13]. In the second mode, compressed data arrives in the VBV buffer at the peak rate of the buffer until the buffer is full, at which point it stops. The initial removal time is the exact point in time when the buffer becomes full. Subsequent removal times are delayed with respect to the first by fixed frame or field periods. Suppose now that an encoder produces a long sequence of very small pictures (meaning that few bits were used to compress them) at the start of the sequence. It would be possible in practice to produce 900 small pictures that all fit in the VBV buffer. The first of these pictures would be removed from the buffer after less than 1 s in a typical scenario. The last of these 900 pictures would sit in the buffer for 899 picture periods, or roughly 30 s for the NTSC video (29.97FPS). This would represent a noncausal situation unless the transmission of the initial picture was delayed by the same amount of time. However, in real-time broadcast applications, it is not possible to insert a 30-s delay at the encoder. What a real-time encoder actually does in a circumstance like this, rather than insert the 30-sdelay, is to wait until each picture is actually produced by the encoder before transmitting its bits associated with the small pictures. So clearly, a real-time encoder cannot mimic the buffer arrival timing of VBV of the second mode. This is the so-called noncausality of this mode.

In the leaky bucket mode, it is impossible for the buffer to overflow, as data stops entering when the buffer becomes full. However, an encoder must prevent underflow.

In the low-delay mode, the buffer may underflow occasionally and there are precise rules, involving skipping pictures, that define how the VBV is to recover.

Because of the number of modes of operation, and the arcane method of handling the one special case of variable frame rate that is handled, the MPEG-2 VBV is somewhat complex. It also suffers from the initial rate ambiguity of the first mode and the possible noncausality in the second mode.

Details of MPEG-2 VBV: MPEG-2 VBV are specified in Annex C of reference [7-1]. The VBV is conceptually connected to the output of an encoder. Its input buffer is the VBV buffer. Coded data is placed in the buffer and is removed instantaneously at certain examination time from the buffer as defined in C.3, C.5, C.6, and C.7 of reference [7-1]. The time intervals between successive examination points of the VBV buffer are specified in C.9, C.10, C.11, and C.12. The VBV occupancy is shown in Figure 7.3. It is required that a bit stream does not cause the VBV buffer to overflow. When there is no "skipped" picture, i.e., low_delay equals zero in MPEG-2 spec, the bit stream should not cause the VBV buffer to underflow.

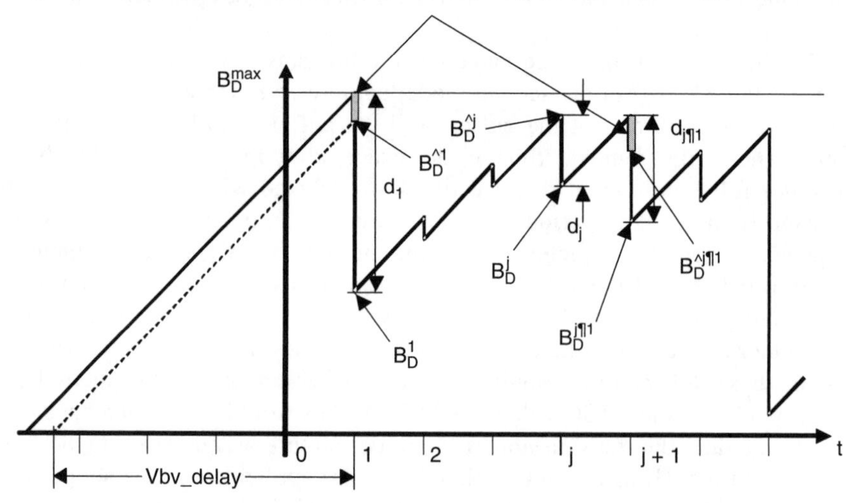

FIGURE 7.3 MPEG-2 VBV buffer occupancy.

Thus, the condition for preventing the VBV buffer to overflow is

$$\hat{B}_D^j \leq B_D^{\max}$$

And, the condition for preventing the VBV buffer to underflow is

$$0 \leq B_D^j$$

where $B_D^{\max}$ is the VBV buffer size. $\hat{B}_D^j$ is VBV occupancy, measured in bits, immediately before removing picture n from the buffer but after removing any header(s), user data and stuffing that immediately precedes the data elements of picture j. The term B_D^j is VBV occupancy, measured in bits, immediately after removing picture j from the buffer. Note that $\hat{B}_D^j - B_D^j$ is the size of the coded picture j; i.e., $\hat{B}_D^j - B_D^j = P_E^j = d_j$.

In the vbv_delay mode, for the constant bit-rate (CBR) case, B_D^j may be calculated by vbv_delay from the equation as follows:

$$\text{vbv\_delay}_j = \frac{90000\hat{B}_D^j}{R_C} \tag{7.11}$$

that is,

$$\hat{B}_D^j = \frac{\text{vbv\_delay}_j \cdot R_C}{90000} \tag{7.12}$$

where R_C denotes the actual bitrate (i.e. to full accuracy rather than the quantized value given by bit rate in the sequence header). An approach to calculate the piecewise constant rate $R_C(j \cdot t)$ from a coded stream is specified in C.3.1 of reference [7-1].

Note that the encoder is capable of knowing the delay experienced by the relevant picture start code in the encoder buffer and the total end-to-end delay. Thus, the value encoded in vbv_delay (the decoder buffer delay of the picture start code) is calculated as the total end-to-end delay subtract the delay of the corresponding picture start code in the encoder buffer measured in periods of a 90-kHz clock derived from the 27-MHz system clock. Therefore, the encoder knows how to generate a bit stream that does not violate the VBV constraints.

Initially, the VBV buffer is empty. The data input continues at the piecewise constant rate $R_C(j \cdot t)$. After filling the VBV buffer with all the data that precedes the first picture start code of the sequence and the picture start code itself, the VBV buffer is filled from the bit stream for the time specified by the vbv_delay field in the picture header. At this time decoding begins. By following this process (without looking the system's DTS), the decoder buffer will not over- and/or underflow for VBV compliant streams.

Note that the ambiguity can happen at the first picture and the end of a sequence since input bit-rate cannot be determined from the bit stream. The ambiguity may become a problem when the video bit stream is remultiplexed and delivered at a rate different from the intended piecewise constant rate $R_C(j \cdot t)$.

For the CBR channel, if the initial vbv_delay$_0$ can be obtained, the decoding time can be determined from the vbv_delay$_0$ and picture (frame) rate T. For example, the decoding time for non-film mode video can be determined as follows:

$$t_{d,j} = \frac{90000 \cdot j}{T} + vbv\_delay_0 \tag{7.13}$$

In the leaky-bucket mode, for the variable bit-rate (VBR) case, i.e., vbv_delay is coded with the value hexadecimal FFFF, data enters the VBV buffer as specified as follows:

- Initially, the VBV buffer is empty.
- If the VBV buffer is not full, data enters the buffer at R_C^{max} where R_C^{max} is the maximum bit-rate specified in the bit rate field of the sequence header.
- If the VBV buffer becomes full after filling at R_C^{max} for some time, no more data enters the buffer until some data is removed from the buffer.

This is a reversed leaky-bucket buffer model since the video encoder for VBR transmission can be simply modeled as a leaky-bucket buffer, as detailed in Section 3.3.2 in [7-14]. In this case, the bucket size satisfies $b_{max} = B_D^{max}$ and the leaky-bucket output rate $C_{output} = R_C^{max}$.

When there are skipped pictures, i.e., low_delay = 1, decoding a picture at the normally expected time might cause the VBV buffer to underflow. If this is the case, the picture is not decoded and the VBV buffer is reexamined at a sequence of later times specified in C.7 and C.8 of reference [7-1] until it is all present in the VBV buffer.

The VBV constraints ensure encoder buffer never over- and/or underflow. A decoder that is built on a basis of VBV can always decompress the VBV compliant video streams without over- and/or underflow the decoder buffer.

7.3.4 MPEG-4 Video Buffering Verifier

As discussed in the previous section, a video rate buffer model is required in order to bound the memory requirements for the bit stream buffer needed by a video decoder. With a rate buffer model, the video encoder can be constrained to make bit streams that are decodable with a predetermined buffer memory size.

The MPEG-4 (ISO/IEC 14496-2) [7-2][7-10] video buffering verifier (VBV) is an algorithm for checking a bit stream with its delivery rate function, $R_C(t)$, to verify that the amount of rate buffer memory required in a decoder is less than the stated buffer size. If a visual bit stream is composed of multiple video objects (VO) and each VO is with one or more video object layers (VOL), the rate buffer model is applied independently to each VOL (using buffer size and rate functions particular to that VOL). The concepts of VO, VOL, and video object plane (VOP) of MPEG-4 video are specified in [7-2].

In MPEG-4, the coded video bit stream is constrained to comply with the requirements of the VBV defined as follows:

- When the vbv_buffer_size and vbv_occupancy parameters are specified by systems-level configuration information, the bit stream shall be constrained according to the specified values. When the vbv_buffer_size and vbv_occupancy parameters are not specified (except for the short video header case for H.263 as described below), this indicates that the bit stream should be constrained according to the default values of vbv_buffer_size and vbv_occupancy. The default value of vbv_buffer_size is the maximum value of vbv_buffer_size allowed within the profile and level. The default value of vbv_occupancy is 170 × vbv_buffer_size, where vbv_occupancy is in 64-bit units and vbv_buffer_size is in 16,384-bit units. This corresponds to an initial occupancy of approximately two-thirds of the full buffer size.
- The VBV buffer size is specified by the vbv_buffer_size field in the VOL header in units of 16,384 bits. A vbv_buffer_size of 0 is forbidden. Define $B_D^{\max} = 16{,}384 \times$ vbv_buffer_size to be the VBV buffer size in bits.
- The instantaneous video object layer channel bit rate seen by the encoder is denoted by $R_C(t)$ in bits per second. If the bit rate field in the VOL header is present, it defines a peak rate (in units of 400 bits per second; a value of 0 is forbidden) such that $R_C(t) \leq R_C^{\max} = 400 \times$ bit_rate.
- The VBV buffer is initially empty. The vbv_occupancy field specifies the initial occupancy of the VBV buffer in 64-bit units before decoding the initial VOP. The first bit in the VBV buffer is the first bit of the elementary stream, except for basic sprite sequences.
- Define d_j to be size in bits of the j-th VOP plus any immediately preceding group of VOP (GOV) header, where j is the VOP index which increments by 1 in decoding order. A VOP includes any trailing stuffing code words before the next start code and the size d_j of a coded VOP is always a multiple of 8 bits due to start code alignment.
- Let $t_{d,j}$ be the decoding time associated with VOP j in decoding order. All bits d_j of VOP j are removed from the VBV buffer instantaneously at $t_{d,j}$. This instantaneous removal property distinguishes the VBV buffer model from a real rate buffer.
- The method of determining the value of $t_{d,j}$ is specified below. Assume $t_{p,j}$ is the composition time (or presentation time in a no-compositor decoder) of VOP j. For a VOP, $t_{p,j}$ is defined by vop_time_increment (in units of 1/vop_time_increment_resolution seconds) plus the cumulative number of whole seconds specified by module_time_base In the case of interlaced video, a VOP consists of lines from two fields and $t_{p,j}$ is the composition time of the first field. For example, the relationship between the composition time and the decoding time for a VOP is given by:

$$t_{d,j} = \begin{cases} t_{p,j}, & \text{if the } j-\text{th VOP is a B}-\text{VOP.} \\ t_{p,j} - m_j, & \text{otherwise.} \end{cases}$$

TABLE 7.1
An Example That Demonstrates m_j is Determined

J	$t_{p,j}$	$t_{d,j}$	m_j
0	0	0 − 1 = − 1	1
1	1	1 − 1 = 0	1
2	2	2 − 1 = 1	1
3	4	4 − 2 = 2	2
4	3	3	2
5	6	6 − 2 = 4	2
6	5	5	2
7	9	9 − 3 = 6	3
8	7	7	3
9	8	8	3
10	12	12 − 3 = 9	3
11	10	10	3
12	11	11	3

In the normal decoding, the composition time of I and P VOPs is delayed until all immediately temporally-previous B-VOPs have been composed. This delay period is $m_j = t_{p,j} - t_{p,k}$, where k is the index of the nearest temporally-previous non-B VOP relative to VOP j.

In order to initialize the model decoder when m_j is needed for the first VOP, it is necessary to define an initial decoding time $t_{d,0}$ for the first VOP (since the timing structure is locked to the B-VOP times and the first decoded VOP would not be a B-VOP). This defined decoding timing shall be that $t_{d,0} = 2 \cdot t_{d,1} - t_{d,2}$ (i.e., assuming that $t_{d,1} - t_{d,0} = t_{d,2}\ t_{d,1}$), since the initial $t_{p,k}$ is not defined in the case.

The example given in Table 7.1 demonstrates how m_j is determined for a sequence with variable numbers of consecutive B-VOPs:

Decoding order: $I_0P_1P_2P_3B_4P_5B_6P_7B_8B_9P_{10}B_{11}B_{12}$

Presentation order: $I_0P_1P_2B_4P_3B_6P_5B_8B_9P_7B_{11}B_{12}P_{10}$

In this example, assume that *vop_time_increment* = 1 and *modulo_time_base* = 0. The subindex j is in decoding order.

Define B_D^j as the buffer occupancy in bits immediately following the removal of VOP j from the rate buffer. Using the above definitions, B_D^j can be iteratively defined as

$$B_D^0 = 64 \cdot vbv\_occupancy - d_0$$

$$B_D^{j+1} = B_D^j + \int_{t_{d,j}}^{t_{d,j+1}} R_C(t) \cdot dt - d_{j+1} \text{ for } j \geq 0 \tag{7.14}$$

The rate buffer model requires that the VBV buffer never overflow or underflow, that is

$$0 \le B_D^j \text{ and } B_D^j + d_j \le B_D^{\max} \text{ for all } j$$

Also, a coded VOP size must always be less than the VBV buffer size; i.e., $d_j \le B_D^{\max}$ for all j. The MPEG-4 VBV buffer occupancy is shown in Figure 7.4.

If the short video header is in use (i.e., for H.263 baseline video [7-7][7-8]), then the parameter vbv_buffer_size is not present and the following conditions are required for VBV operation. The buffer is initially empty at the start of encoder operation (i.e., $t = 0$ being at the time of the generation of the first video plane with short header), and its fullness is subsequently checked after each time interval of 1,001/30,000 s (i.e., at $t = 1{,}001/30{,}000$, $2{,}002/30{,}000$, etc.). If a complete video plane with short header is in the buffer at the checking time, it is removed. The buffer fullness after the removal of a VOP, B_D^j, shall be greater than or equal to zero and less than $(4 \cdot R_C^{\max} \cdot 1{,}001)/30{,}000$ bits, where $R_C^{\max}$ is the maximum bit rate in bits per second allowed within the profile and level. The number of bits used for coding any single VOP, d_j, shall not exceed $k \cdot 16384$ bits, where $k = 4$ for QCIF and Sub-QCIF, $k = 16$ for CIF, $k = 32$ for 4CIF, and $k = 64$ for 16CIF, unless a larger value of k is specified in the profile and level definition. Furthermore, the total buffer fullness at any time shall not exceed a value of $B_D^{\max} = k \cdot 16384 + (4 \cdot R_C^{\max} \cdot 1{,}001)/30{,}000$.

It is a requirement on the encoder to produce a bit stream that does not overflow or underflow the VBV buffer. This means the encoder must be designed to provide correct VBV operation for the range of values of $R_C(t)$ over which the system will operate for delivery of the bit stream. A channel has constant delay if the encoder bit rate at time t when particular bit enters the channel, the bit will be received at $t + LT$ and L is constant. In the case of constant delay channels, the encoder can use its locally estimated $R_C(t)$ to simulate the VBV occupancy and control the number of bits per VOP, $D_j = P_E^j$, in order to prevent overflow or underflow.

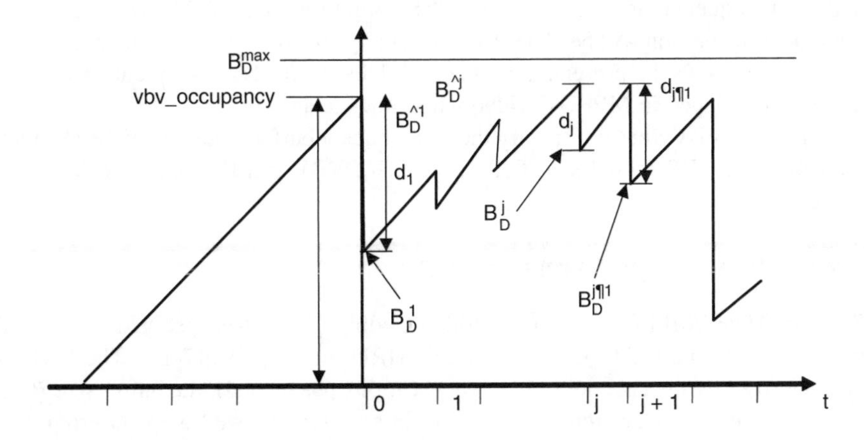

FIGURE 7.4 MPEG-4 VBV buffer occupancy.

MPEG-4 VBV model assumes a constant delay channel. This allows the encoder to produce an elementary bit stream that does not overflow or underflow the buffer using $R_C(t)$.

7.3.5 Comparison between MPEG-2 VBV and MPEG-4 VBV

Both MPEG-2 and MPEG-4 VBV models [7-1][7-2][7-10] specify that the rate buffer may not overflow or underflow and that coded pictures (VOPs) are removed from the buffer instantaneously. In both models a coded picture/VOP is defined to include all higher-level syntax immediately preceding the picture/VOP.

MPEG-2 video has a constant frame period (although the bit stream can contain both frame and field pictures and frame pictures can use explicit 3:2 pull-down via the repeat_first_field flag). In MPEG-4 terms, this frame rate would be the output of the compositor (the MPEG-2 terminology is the output of the display process that is not defined normatively by MPEG-2). This output frame rate together with the MPEG-2 picture_structure and repeat_first_field flag precisely defines the time intervals between consecutive decoded picture (either frames or fields) passed between the decoding process and the display process.

In general, the MPEG-2 bit stream contains B-pictures (assume that low_delay = 0). This means the coding order and display order of pictures is different (since both reference pictures used by a B-picture must precede the B-picture in coding order). The MPEG-2 video VBV specifies that a B-picture is decoded and presented (instantaneously) at the same time and the anchor pictures are re-ordered to make this possible. This is the same reordering model specified in MPEG-4 video.

A MPEG-4 model decoder using its VBV buffer model can emulate a MPEG-2 model decoder using the MPEG-2 VBV buffer model if the VOP time stamps given by vop_time_increment and the cumulative modulo_time_base agree with the sequence of MPEG-2 picture presentation times. Assume here that both coded picture/VOPs use the common subset of both standards (frame structured pictures and no 3:2 pulldown on the decoder, i.e., repeat_first_field = 0). For example, if the MPEG-4 sequence is coded at the NTSC picture rate 29.97 Hz, vop_time_increment_resolution will be 30,000 and the change in vop_ time_increment between consecutive VOPs in presentation order will be 1,001 because pictures are not allowed to skipped in MPEG-2 video when low_delay = 0.

MPEG-4 VBV does not specify the leaky bucket buffer model for VBR channel. However, the VBR model specified in MPEG-2 VBV can be applied to MPEG-4 video.

7.3.6 HRD in H.264/MPEG-4 AVC

Standard H.264/MPEG-4 Advanced Video Coding (AVC) has specified a so-called causal arrival time leaky bucket (CAT-LB) HRD model [7-12][7-13]. The HRD is characterized by the channel's peak rate R (in bits per second), the buffer size B (in bits), the initial decoder removal delay T (in seconds), as well as picture removal delays for each picture. The first three of these parameters $\{R, B, T\}$ represent levels of resources (transmission capacity, buffer capacity, and delay) used to decode a bit stream.

The leaky bucket in H.264/MPEG-4 AVC is called a *causal arrival time leaky bucket* because the arrival times of all pictures after the first are constrained to arrive at the buffer input no earlier than the difference in hypothetical encoder processing times between that picture and the first picture. In other words, if a picture is encoded exactly 5 s after the first picture was encoded, then its bits are guaranteed not to start arriving in the buffer prior to 5 s after the bits of the first picture started arriving. It is possible to know this encoding time difference because it is sent in the bit stream as the picture removal delay.

7.3.6.1 Operation of the CAT-LB HRD

The HRD input buffer has capacity B bits. Initially, the buffer begins empty. The lifetime in the buffer of the coded bits associated with picture n is characterized by the arrival interval $\{t_{ai}(n), t_{af}(n)\}$, known as the initial arrival time and the final arrival time, and the removal time $t_r(n)$.

At time $t_{ai}(0) = 0$, the buffer begins to receive bits at the rate R. The removal time $t_r(0)$ for the first picture is computed from the decoder removal delay associated with the buffer by the following:

$$t_r(0) = 90{,}000 \times \text{initial decoder removal delay}$$

Removal times $t_r(1)$, $t_r(2)$, $t_r(3)$, …, for subsequent pictures (in transmitted order) are computed with respect to $t_r(0)$, as follows. Let t_c denote the clock tick, which is a time interval no larger than the shortest possible inter-picture capture interval in seconds.

In the picture layer syntax for each picture, there is a syntax element representing decoder removal delay. This indicates the number of clock ticks to delay the removal of picture n after removing picture $n - 1$. Thus, the removal time is simply

$$t_r(n) = t_r(n-1) + t_c \times \text{decoder removal delay } (n)$$

Note that this recursion can be used to show that

$$t_r(n) = t_r(0) \times t_c \times \sum_{l=1}^{n} \text{decoder removal delay } (l)$$

The calculation of arrival times is more complex, because of the causality constraint. The initial arrival time of picture n is equal to the final arrival time of picture $n - 1$, unless that time precedes the earliest arrival time, computed by

$$t_{\text{ai,earliest}}(n) = t_c \times \sum_{l=1}^{n} \text{decoder removal delay } (l)$$

Let $b(n)$ be the number of bits associated with picture n. The duration of the picture arrival interval is always the time-equivalent of the picture size $P(n)$ in bits, at the rate R.

$$t_{af}(n) - t_{ai}(n) = P(n)/R$$

For each picture, the initial arrival time of picture n is the later of $t_{af}(n-1)$ and the sum of all preceding decoder removal delay times, as indicated in the following,

$$t_{ai}(n) = \max\{t_{af}(n-1),\quad t_c \times \sum_{i=1}^{n} \text{decoder removal delay } (l)\}.$$

7.3.6.2 Low-Delay Operation

Low-delay operation is obtained by selecting a low value for the initial pre-decoder removal delay. This results in true low delay through the buffer because, under normal operation, no removal delay $(t_r(n) - t_{ai}(n))$ can exceed the initial removal delay $t_r(0)$.

7.3.6.3 Stream Constraints

The buffer must not be allowed to underflow or overflow. Furthermore, all pictures except the isolated big pictures must be completely in the buffer before their computed removal times. Isolated big pictures are allowed to arrive later than their computed removal times, but must still obey the overflow constraint. In CBR mode, there must be no gaps in bit arrival.

7.3.6.4 Underflow

The underflow constraint, Buffer fullness $(t) \geq 0$ for all t, is satisfied if the final arrival time of each picture precedes its removal time.

$$t_{af}(n) \leq t_r(n)$$

This puts an upper bound on the size of picture n. The picture size $P(n)$ can be no larger than the bit-equivalent *be* of the time interval from the start of arrival to the removal time.

$$P(n) \leq be[t_r(n) - t_{ai}(n)]$$

Since the initial arrival time $t_{ai}(n)$ is in general a function of the sizes and removal delays of previous pictures, the constraint on $b(n)$ will vary over time as well.

7.3.6.5 Overflow

is prevented by providing the buffer fullness curve Buffer fullness (t) never exceeds the buffer size B. The constraints that the initial pre-decoder removal delay must be no larger than the time-equivalent *te* of the buffer size, $t_r(0) \le te(B)$, and that under normal operation no removal delay can exceed the initial one guarantee that no overflow occurs in normal operation. To avoid overflow of an isolated big picture, the picture size is constrained by

$$P(n) \le be[B - t_{ai}(n)]$$

7.4 BUFFER ANALYSIS OF VIDEO TRANSCODERS

7.4.1 BACKGROUND

Digital video compression algorithms specified in the MPEG and H.263 standards [7-1] [7-2] [7-3] [7-9] [7-10] have already enabled many video services such as, video on demand (VOD), digital terrestrial television broadcasting (DTTB), cable television (CATV) distribution, and Internet video streaming. Due to the variety of different networks making up the present communication infrastructure, a connection from the video source to the end user may be established through links of different characteristics and bandwidth.

In the scenario where only one user is connected to the source, or independent transmission paths exist for different users, the bandwidth required by the compressed video should be adjusted by the source in order to match the available bandwidth of the most stringent link used in the connection. For uncompressed video, this can be achieved in video encoding systems by adjusting coding parameters, such as quantization steps, whereas for precompressed video, such a task is performed by applying video transcoders [7-17][7-22].

In the scenario in which many users are simultaneously connected to the source and receiving the same coded video, as happens in VOD, CATV services, and Internet video, the existence of links with different capacities poses a serious problem. In order to deliver the same compressed video to all users, the source has to comply with the subnetwork that has the lowest available capacity. This unfairly penalizes those users that have wider bandwidth in their own access links. By using transcoders in communication links, this problem can be resolved. For a video network with transcoders in its subnets, one can ensure that users receiving lower quality video are those having lower bandwidth in their transmission paths. An example of this scenario is in CATV services where a satellite link is used to transmit compressed video from the source to a ground station, which in turn distributes the received video to several destinations through networks of different capacity.

In the scenario where the compressed video programs need to be reassembled and retransmitted, the bit rates of the coded video are often reduced in order to fit in the available bandwidth of the channel. For example, cable head-ends can reassemble programs from different video sources. Some programs from broadcast television and others from video servers. In order to ensure that the reassembled programs can match the available bandwidth, video transcoders are often used.

When a transcoder is introduced between an encoder and the corresponding decoder, the following issues should be considered for the system [7-14] [7-17] [7-21]:

- Buffer and delay
- Video decoding and re-encoding
- Timing recovery and synchronization

In Chapter 2, many video compression technologies are discussed. As an extension, two basic video transcoder architectures are overviewed here. Transcoding is an operation of converting a precompressed bit stream into another bit stream at a different rate. For example, a straightforward architecture of transcoder for MPEG bit stream can simply be a cascaded MPEG decoder/encoder, as shown in Figure 7.5. In the cascaded-based transcoder, the precompressed MPEG bit stream is first decompressed by the cascaded decoder, which includes a variable length decoder (VLD), inversed quantization (Q_1^{-1}), inverse discrete cosine transform (IDCT), and motion compensation (MC), and the resulting reconstructed video sequence is then re-encoded by the cascaded encoder, which includes additional functional blocks, such as motion estimation (ME) and variable length encoder (VLE), and generates a new bit stream. The desired rate of the new bit stream can often be achieved by adjusting quantization level, Q_2, in the cascaded encoder. The main concern with the cascaded-based transcoder is its implementation cost: one full MPEG decoder and one full MPEG encoder.

Recent studies showed that a transcoder consisting of a cascaded decoder/encoder can be significantly simplified if the picture types in precompressed bit stream can remain unchanged during transcoding [7-14] [7-17]; that is, a decoded I picture is again coded as an I-picture, a decoded P-picture is again coded as a picture and a decoded B-picture is again coded as a B-picture. In fact, by maintaining

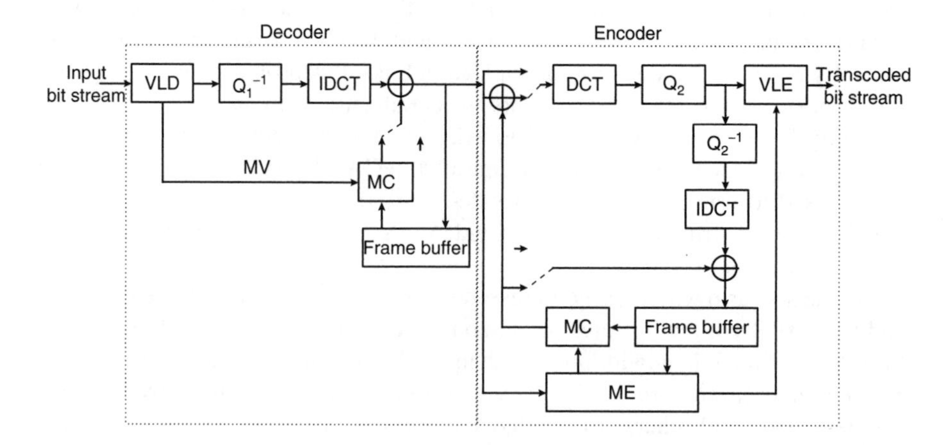

FIGURE 7.5 Cascaded MPEG decoder/encoder.

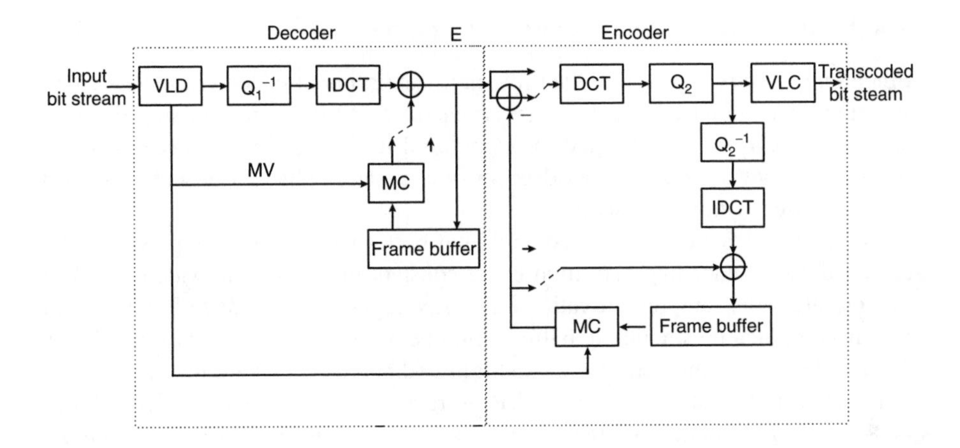

FIGURE 7.6 Cascaded MPEG decoder/encoder without ME where MV required for MC in the cascaded encoder is obtained from the cascaded decoder.

the picture types, one can possibly remove motion estimation (ME) (the most expensive operation) in the cascaded-based transcoder (Figure 7.6) because of the fact that there is a strong similarity between the original and the reconstructed video sequences. Hence, a motion vector (MV) field that is good for an original picture should be reasonably good for the corresponding reconstructed picture.

Figure 7.6 shows a cascaded-based transcoder without ME where the MV fields required for MC in the cascaded encoder are now obtained from the cascaded decoder. However, it should also be pointed out that although the MV fields obtained from the cascaded decoder can be reasonably good for motion compensation (MC) in the cascaded encoder in the transcoder (Figure 7.6), they may not be the best because they were estimated by the original encoder. For example, the half-pixel positions of reused MVs could be inaccurate.

Many other transcoder architectures [7-18] [7-19] can be derived or extended from the two basic architectures given in Figure 7.5 and Figure 7.6. For example, a transcoder with picture resolution change is developed in [7-19] [7-23].

In the remaining of the chapter, the discussion will be focused on analyzing buffer, timing recovery, and synchronization for video transcoder. The buffering implications of the video transcoder within the transmission path are analyzed. For transcoders with either fixed or variable compression ratio, it is shown that the encoder buffer size can be maintained as if no transcoder existed while the decoder has to modify its own buffer size according to both the bit rate conversion ratio and transcoder buffer size. The buffer conditions of both the encoder and transcoder are derived for preventing the final decoder buffer from underflowing or overflowing. It is also shown that the total buffering delay of a transcoded bit stream can be made less than or equal to its "encoded-only" counterpart.

7.4.2 Buffer Dynamics of Video Transcoders

Smoothing buffers play a important role in transmission of coded video. Therefore, if a transcoder is introduced between an encoder and the corresponding decoder, some modifications are expected to be required in the existing buffering arrangements of a conventional encoder-decoder only system, which is primarily defined for being used without transcoders.

It is known that encoders need an output buffer because the compression ratio achieved by the encoding algorithm is not constant throughout the video signal. If the instantaneous compression ratio of a transcoder could be made to follow that of the encoder, then no smoothing buffer would be necessary at the transcoder [7-21]. For a CBR system this requires a fixed-transcoding compression ratio equal to the ratio between the output and input CBR bit rates of the transcoder. In general, this is not possible to obtain in practice and a small buffer is necessary to smooth out the difference.

In the following analysis, the general case of buffer dynamics is first presented and then, in order to clarify the effect of adding a transcoder in the transmission path, the cases of fixed-transcoding compression ratio without buffering, and variable transcoding compression ratio with buffering, are analyzed. The concept of video data unit is usually defined as the amount of coded data that represents an elementary portion of the input video signal such as block, macroblock, slice, or picture (a frame or a field). In the following analysis, the video (data) unit is assumed to be a picture and the processing delay of a video unit is assumed to be constant in either the encoder, transcoder, or decoder and is much smaller than the buffering delays involved. Thus, it can be neglected in the analysis model. For the same reasons, the transmission channel delay is also neglected in the analysis model. A video unit is characterized by the instant of its occurrence in the input video signal, as well as by the bit rate of the corresponding video data unit. Since the processing time is ignored, video units are instantly encoded into video data units, and these then instantly decoded into video units. Although video data units are periodically generated by the encoder, their periodicity is not maintained during transmission (after leaving the encoder buffer) because, due to the variable compression ratio of the coding algorithm, each one comprises a variable number of bits. However, for real-time display, the decoder must recover the original periodicity of the video units through a synchronized clock.

7.4.3 Buffer dynamics of the encoder-decoder only system

Before analyzing buffering issues in transcoders, let us look again at the relationship between encoder and decoder buffers in a codec without transcoding, such as the general case of transmission depicted in Figure 7.1, where the total delay $L \cdot T$ from the encoder input (camera) to the decoder output (display) is the same for all video units (T is the picture duration of the original uncompressed video as described in Section 0). Therefore, since processing and transmission delays are constant, a video data unit entering into the encoder buffer at time t will leave the decoder buffer at $t + L \cdot T$ where $L \cdot T$ is constant. Since the acquisition and display rates of corresponding video units are equal, the output bit rate of the decoder buffer at time $t + L \cdot T$ is exactly the same as that of the input of the encoder buffer at time t. Thus, in Figure 7.1, assume that $E_0(t)$ represents the bit rate of a video data unit encoded at time t and the coded video data is transmitted at a rate $R_C(t)$. If underflow or overflow never occurs in either the encoder or decoder buffers, then the encoder buffer fullness $B_E(t)$ is given by Equation (7.15), while that of the decoder buffer $B_D(t)$ is given by Equation (7.16)

$$B_E(t) = \int_0^t E_0(\tau) \cdot d\tau - \int_0^t R_C(\tau) \cdot d\tau \tag{7.15}$$

$$B_D(t) = \begin{cases} \int_0^t R_C(\tau) \cdot d\tau, & t < L \cdot T \\ \int_0^t R_C(\tau) \cdot d\tau - \int_0^{t-L \cdot T} E_0(\tau) \cdot d\tau, & t \geq L \cdot T \end{cases} \tag{7.16}$$

Note that in general, it is possible for encoder buffer underflow to occur if transmission starts at the same time as the encoder puts the first bit into the buffer, as implied by Equation (7.15). In practice, this is prevented by starting the transmission only after a certain initial delay such that the total system delay is given by the sum of the encoder and decoder initial delays, L_e and L_d, respectively. For simplicity, one can assume that encoder buffer underflow does not occur and then these two delays are included in the total initial decoding delay, i.e., $L \cdot T = L_e \cdot T + L_d \cdot T$. From Equation (7.16), it can be seen that during the initial period $0 \leq t < L \cdot T$ the decoder buffer is filled at the channel rate up to the maximum $\int_0^{L \cdot T} R_C(\tau) \cdot d\tau$, hence decoding of the first picture only starts at $t = L \cdot T$. Combining Equation (7.15) and Equation(7.16) yields that the sum of the encoder and decoder buffer occupancies at times t and $t + L \cdot T$, respectively, is bounded and equal to the buffer size required for the system; i.e.,

$$\begin{aligned} & B_E(t) + B_D(t + L \cdot T) \\ & = \int_0^t E_0(\tau) \cdot d\tau - \int_0^t R_C(\tau) \cdot d\tau + \int_0^{t+L \cdot T} R_C(\tau) \cdot d\tau - \int_0^t E_0(\tau) \cdot d\tau \\ & = \int_t^{t+L \cdot T} R_C(\tau) \cdot d\tau \leq L \cdot T \cdot R_C^{\max}, \quad t \geq L \cdot T \end{aligned} \tag{7.17}$$

For a VBR channel, the sum of buffer occupancies of both encoder and decoder is the total amount of bits that have been transmitted from $(t, t + LT)$. Thus, in this case, the above equation shows that, for a constant delay channel, the buffer size required for the system is $L \cdot T \cdot R_C^{\max}$, where $R_C^{\max}$ is the maximum channel rate. For a CBR channel, one has $R_C^{\max} = R_C^{\min}$ = constant where $R_C^{\min}$ is the minimum channel rate. Then, the above equation also shows that the total number of bits stored in both the encoder and decoder buffers at any times t and $t + L \cdot T$, respectively, is always the same; i.e., $B_E(t) + B_D(t + L \cdot T) = R \cdot L \cdot T$. Thus, if these bits "travel" at a CBR, the delay between encoder and decoder is maintained constant for all video units while the sum of buffer occupancies of both encoder and decoder is a constant for all video units. The principal requirement for the encoder is that it must control its buffer occupancy such that decoder buffer overflow or underflow never occurs. Decoder buffer overflow implies loss of data whenever its occupancy reaches beyond the required buffer size $B_D^{\max}$. On the other hand, underflow occurs when the decoder buffer occupancy is zero at display time of a video unit that is not fully decoded yet (display time is externally imposed by the display clock).

Equation (7.17) relates encoder and decoder buffer occupancies at time t and $t + L \cdot T$, respectively. This buffer fullness equation provides the conditions for preventing encoder and decoder buffers being over- or underflow. Decoder buffer underflow is prevented if $B_D(t + L \cdot T) \geq 0$ is ensured at all times. Thus, using Equation (7.19), at time t the encoder buffer fullness should be

$$B_E(t) \leq \int_t^{t+L \cdot T} R_C(\tau) \cdot d\tau \tag{7.18}$$

On the other hand, decoder buffer overflow does not occur if

$$B_D(t + L \cdot T) \leq \int_t^{t+LT} R_C(\tau) \cdot d\tau \tag{7.19}$$

holds all the time, which requires that the encoder buffer occupancy at time meets the condition: $B_E(t) \geq 0$.

Therefore, it can be seen that decoder buffer underflow and overflow can be prevented by simply controlling the encoder buffer occupancy such that $0 \leq B_E(t) \leq \int_t^{t+L \cdot T} R_C(\tau) \cdot d\tau$ at any time t. By preventing the encoder buffer from overflowing, its decoder counterpart never underflows while preventing encoder buffer underflow ensures that the decoder buffer never overflows. More buffering requirements can be found in [7-14].

Inequalities (7.18) and (7.19) also imply that the maximum needed encoder and decoder buffer sizes satisfy: $B_E^{\max} = B_D^{\max} = L \cdot T \cdot R_C^{\max}$. This means that the specified buffer size for either encoder or decoder needs no more than $L \cdot T \cdot R_C^{\max}$.

The MPEG-2 standard defines both VBR and CBR transmission. The sequence header of an MPEG-2 bit stream includes the decoder buffer size and the maximum bit rate that can be used. Also, in each picture header is included the time e.g., vbv_delay, that the decoder should wait after receiving the picture header until start decoding the picture. For a CBR transmission in the vbv delay mode, vbv_delay is generated in such a way that the encoder and decoder buffer dynamics are as explained above and the total delay is kept constant. It is calculated by the encoder

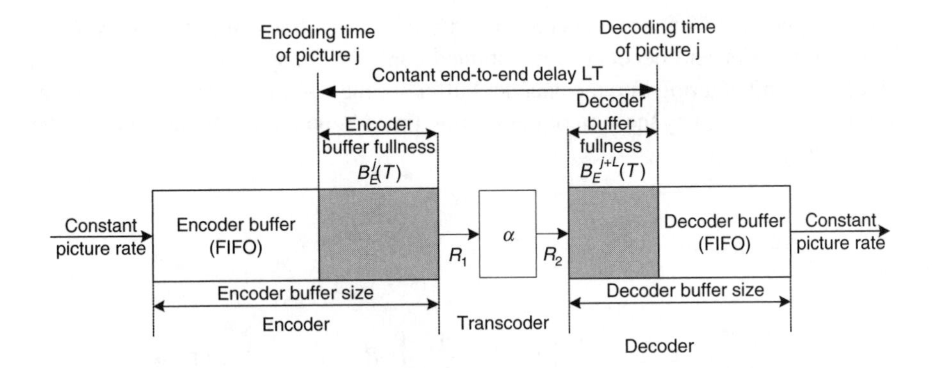

FIGURE 7.7 A transcoder model with fixed-transcoding compression ratio.

as the difference (in number of periods of the system clock) between the total delay and the delay that each picture header undergoes in the encoder buffer.

7.4.4 Transcoder with a Fixed Compression Ratio

Next, consider the CBR transmission case. Assume that a hypothetical transcoder, capable of achieving a fixed-transcoding compression ratio of $\alpha = \frac{R_2}{R_1}$ such that $0 < \alpha \leq 1$ is inserted in the transmission path as shown in Figure 7.7. The terms R_1 and R_2 are the input and output CBRs of the transcoder, respectively. The bitrate that enters into the encoder buffer is reduced through the factor α such that the output of the decoder buffer is a delayed and scaled version of $E_0(t)$, given by $\alpha \cdot E_0(t - L \cdot T)$. Because of the lower channel rate at the decoder side, if the total delay is to be kept the same as if no transcoder was used, then the decoder buffer fullness level is usually lower than that of the decoder without transcoder being used in the system. The encoder assumes a normal CBR transmission without transcoders in the network; thus, if any of the system parameters encoded in the original bit stream, such as bit rate, buffer size, and vbv_delay in the headers of MPEG-2 video bit streams need to be updated, the transcoder has to perform the task in a transparent manner with respect to both the encoder and decoder. The initial delay is set up by the decoder as waiting time before decoding the first picture. Hence, if neither buffer underflows nor overflows, the encoder and decoder buffer occupancies are given by

$$B_E(t) = \int_0^t E_0(\tau) \cdot d\tau - R_1 \cdot t \tag{7.20}$$

$$B_D(t) = \begin{cases} R_2 \cdot t, & t < L \cdot T \\ R_2 \cdot t - \int_0^{t - L \cdot T} \alpha \cdot E_0(\tau) \cdot d\tau, & t \geq L \cdot T \end{cases} \tag{7.21}$$

Using Equation (7.20) and Equation (7.21), it can be shown that the delay $L \cdot T$ between encoder and decoder is maintained constant at any time t. In this case, a video data unit entering into the encoder buffer at time t will leave the decoder buffer at time $t + L \cdot T$; hence, the sum of the waiting times in both the encoder and decoder buffers is given by

$$
\begin{aligned}
&\frac{1}{R_1} \cdot B_E(t) + \frac{1}{R_2} \cdot B_D(t + L \cdot T) \\
&= \frac{1}{R_1} \int_0^t E_0(\tau) \cdot d\tau - \frac{1}{R_1} \int_0^t R_1 \cdot d\tau + \frac{1}{R_2} \int_0^{t+L\cdot T} R_2 \cdot d\tau - \frac{\alpha}{R_2} \int_0^t E_0(\tau) \cdot d\tau \\
&= \frac{1}{R_1} \int_0^t E_0(\tau) \cdot d\tau - t + t + L \cdot T - \frac{1}{R_2} \int_0^t \alpha \cdot E_0(\tau) \cdot d\tau \\
&= L \cdot T
\end{aligned}
\tag{7.22}
$$

Since $\alpha = \frac{R_2}{R_1}$, the above equation shows that the total delay from encoder to decoder is still the same constant regardless of the transcoder being inserted along the transmission path. However, because the encoder and decoder buffers work at different CBRs, the sum of the buffer occupancies $B_E(t) + B_D(t + L \cdot T)$ is no longer constant as was in the previous case of transmission without the transcoder. Since the input bit rate of the decoder buffer is lower than the output bit rate of the encoder buffer, for the given end-to-end delay $L \cdot T$, the maximum needed encoder and decoder buffer sizes, $B_E^{\max}$ and $B_D^{\max}$, respectively, can be derived as follows. Then, from Equation (7.24), one has

$$
\begin{aligned}
&\frac{B_E(t)}{R_1} \le L \cdot T \le \frac{B_E(t)}{R_1} + \frac{B_D^{\max}}{R_2}, \\
&\frac{B_D(t)}{R_2} \le L \cdot T \le \frac{B_E^{\max}}{R_1} + \frac{B_D(t)}{R_2}
\end{aligned}
\tag{7.23}
$$

Thus,

$$L \cdot T \cdot R_1 \ge B_E(t) \ge L \cdot T \cdot R_1 - \frac{R_2}{R_1} \cdot B_D^{\max},$$

$$L \cdot T \cdot R_2 \ge B_D(t) \ge L \cdot T \cdot R_2 - \frac{R_2}{R_1} \cdot B_D^{\max},$$

Therefore, the maximum needed buffer sizes satisfy

$$L \cdot T = \frac{B_E^{\max}}{R_1} = \frac{B_D^{\max}}{R_2} \tag{7.24}$$

that is,

$$B_D^{\max} = \frac{R_2}{R_1} \cdot B_E^{\max} = \alpha \cdot B_E^{\max} \tag{7.25}$$

Equation (7.25) shows that by using a smaller decoder buffer with size $B_D^{\max} = \alpha \cdot B_E^{\max}$, the same total delay can be maintained as if no transcoder existed.

Let us now analyze the implications of the small decoder buffer size on the encoder buffer constraints needed to prevent decoder buffer underflow and overflow. Assuming that the encoder is not given any information about the transcoder then, recalling the previous case of CBR transmission without transcoders, the encoder prevents decoder buffer overflow and underflow by always keeping its own buffer occupancy within the limits $0 \leq B(t) \leq B_E^{\max}$. With a similar approach to Equation (7.22), the system delay is

$$\frac{1}{R_1} \cdot B_E(t) + \frac{1}{R_2} \cdot B_D(t + L \cdot T) = L \cdot T \tag{7.26}$$

where it can be seen that decoder buffer underflow never occurs if at display time $t + L \cdot T$ all the bits of the corresponding video data unit are received; i.e., $B_D(t + L \cdot T) \geq 0$ after removing all its bits from the buffer. Hence, using Equation (7.24) and Equation (7.26)

$$B_D(t + L \cdot T) = B_D^{\max} - \frac{R_2}{R_1} \cdot B_E(t) \geq 0 \tag{7.27}$$

and the condition for preventing decoder buffer underflow is given by

$$B_E(t) \leq \frac{R_1}{R_2} \cdot B_D^{\max} = B_E^{\max} \tag{7.28}$$

On the other hand, decoder buffer does not overflow if its fullness is less than the buffer size immediately before removing all the bits of any video data unit; i.e., hence, using againEquation (7.24) and Equation (7.26)

$$\frac{R_2}{R_1} \cdot B_E(t) \geq 0 \tag{7.29}$$

since $\frac{R_2}{R_1} \geq 0$, then decoder buffer overflow is prevented, providing that

$$B_E(t) \geq 0 \tag{7.30}$$

Inequalities (7.28) and (7.30) show that no extra modification is needed at the encoder for preventing decoder buffer underflow or overflow. By controlling the occupancy of its own buffer of size $B_E^{\max}$ such that overflow and underflow never occurs, the encoder is automatically preventing the smaller decoder buffer from underflowing and overflowing. This means that, in this case, the presence of the transcoder can be simply ignored by the encoder without adding any extra buffer restrictions on the decoder. In this case, an MPEG-2 transcoder would have to modify the buffer size $B_E^{\max}$ specified in its incoming bit stream to a new value $\frac{R_2}{R_1} . B_E^{\max}$ while the delay parameter in picture headers should not be changed because the buffering delay at the decoder is exactly the same as in the case where no transcoder is used.

However, a transcoder with a fixed compression ratio as was assumed in this case is almost impossible to obtain in practice, mainly because of the nature of the video-coding algorithms and the compressed bit streams they produce. Such a transcoder would have to output exactly $\alpha \cdot N$ bits for each incoming bits N. Since each video data unit consists of a variable number of bits and the quantized DCT blocks cannot be finely encoded such that a given number of bits is exactly obtained, a perfectly fixed compression ratio transcoder cannot be implemented in practice. Moreover, a transcoder with variable compression ratio may be even desirable if the objective is, for instance, to enforce a given variable transcoding function. The above analysis of a fixed compression ratio transcoder provides relevant insight into the more practical case to be described next.

Transcoder with a Variable Compression Ratio

As was pointed out before, a transcoder with variable compression ratio must incorporate a smoothing buffer in order to accommodate the rate change of the coded stream. The conceptual model of a CBR transmission system including such a transcoder with a local buffer of size $B_{TC}^{\max}$ is illustrated in Figure 7.8. The encoder buffer size $B_E^{\max}$ is maintained as in the previous cases while that of the decoder should be given by $B_D^{\max} = \alpha \cdot B_E^{\max} + B_{TC}^{\max}$ as shall be explained later. Here, transcoding is modeled as a scaling function $\beta(t)$ which multiplied by $r(t) = E_0(t)$ produces the transcoded VBR $r'(t)$, i.e.,

$$r'(t) = \beta(t) \cdot r(t) \tag{7.31}$$

The effect of multiplying $\beta(t)$ by $r(t)$ can be seen as equivalent to reducing the number of bits used in the video data unit encoded at time t. The output of the decoder buffer consists of a delayed version of $r'(t)$. In the system of Figure 7.8, transcoding is performed on the CBR R_1 which consists of the video data units of $r(t)$ after the encoder buffering delay $L_e(t)$, defined as the delay that a video data unit encoded at time waits in the encoder buffer before being transmitted.

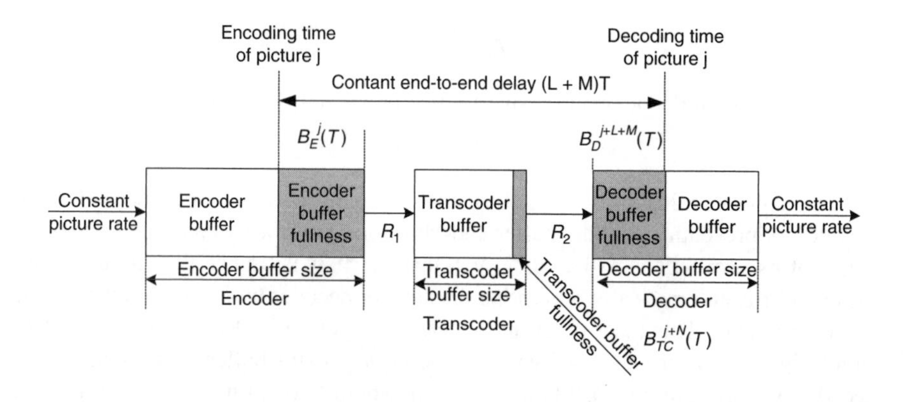

FIGURE 7.8 A transcoder model with variable-transcoding compression ratio.

Let us now verify that under normal conditions, where neither of the three buffers underflows or overflows, the total delay between encoder and decoder is still a constant $M \cdot T + L \cdot T$ where $M \cdot T$ is the extra delay introduced by the transcoder. A video data unit entering the encoder buffer at time t will arrive at the transcoder at $t + L_e(t)$ and will be decoded at $t + M \cdot T + L \cdot T$, respectively. Since the processing delay in the transcoder is neglected, $t + L_e(t)$ is also the time at which the video data unit is transcoded and put in the transcoder buffer. Therefore, in order to calculate the total delay of the system, the encoder, transcoder and decoder buffers should be analyzed at instants t, $t + L_e(t)$, and $t + M \cdot T + L \cdot T$, respectively. The following expressions, Equation (7.32) through Equation (7.34)provide the encoder buffer occupancy at time t, the transcoder buffer occupancy at time $t + L_e(t)$, and the decoder buffer occupancy at time $t + M \cdot T + L \cdot T$, respectively.

$$B_E(t) = \int_0^t E_0(\tau) \cdot d\tau - R_1 \cdot t \tag{7.32}$$

$$B_{TC}(t + L_e(t)) = \int_0^t E_0(\tau) \cdot d\tau - (t + L_e(t)) \cdot R_2 \tag{7.33}$$

$$B_D(t + (M + L) \cdot T + L_e(t)) = (t + (M + L) \cdot T) \cdot R_2 - \int_0^t r'(\tau) \cdot d\tau \tag{7.34}$$

A video data unit entering the encoder buffer at time t has to wait for $L_e(t) = \frac{B_E(t)}{R_1}$ seconds before leaving this buffer, plus $\frac{B_{TC}(t+Le(t))}{R_2}B$ in the transcoder buffer before being transmitted to the decoder buffer from which it is finally removed at $t + M \cdot T + L \cdot T$. Using the above equations for the buffer occupancies, the total buffering delay from L_{TOTAL} encoder to decoder is given by

$$\begin{aligned} L_{TOTAL} &= \frac{1}{R_1} \cdot B_E(t) + \frac{1}{R_2} \cdot B_{TC}(t + L_e(t)) + \frac{1}{R_2} \cdot B_D(t + (M + L) \cdot T) \\ &= \frac{1}{R_1} \int_0^t E_0(\tau) \cdot d\tau - t \\ &+ \frac{1}{R_2} \int_0^t r'(\tau) \cdot d\tau - (t - L_e(t)) \\ &+ \frac{1}{R_2} \left((t + (M + L) \cdot T) \cdot R_2 - \int_0^t r'(\tau) \cdot d\tau \right). \end{aligned} \tag{7.35}$$

where $L_e(t) = \frac{1}{R_1} \int_0^t E_0(\tau) \cdot d\tau - t$.

By simplifying the above expression, one has

$$
\begin{aligned}
L_{TOTAL} &= L_e(t) + \frac{1}{R_2}\int_0^t r'(\tau)\cdot d\tau - t - L_e(t) \\
&\quad + t + (M+L)\cdot T - \frac{1}{R_2}\int_0^t r'(\tau)\cdot d\tau \qquad (7.36) \\
&= (M+L)\cdot T
\end{aligned}
$$

It can be seen that the total delay is constant as given by the initial decoding delay. Note that, similar to the case of a transcoder with fixed compression ratio, the sum of the occupancies of the three buffers is not constant because of the different CBRs involved.

Since the encoder is assuming a decoder buffer of maximum size $L \cdot T \cdot R_C^{\max}$ (from Equation (7.16), its own buffer occupancy is kept within the limits $0 \leq B(t) \leq B_E^{\max}$, as was shown earlier, is necessary for preventing decoder buffer overflow and underflow. However, since the actual required size of the decoder buffer is $B_D^{\max}$, the constraints that the transcoder buffer should meet in order to prevent decoder buffer underflow and overflow are derived from the system delay Equation (7.37),

$$
\begin{aligned}
\frac{B_D^{\max}}{R_2} &= \frac{1}{R_1}\cdot B_E(t) + \frac{1}{R_2}\cdot B_{TC}(t + L_e(t)) + \frac{1}{R_2}\cdot B_D(t + (M+L)\cdot T) \\
&= L_{TOTAL}
\end{aligned} \qquad (7.37)
$$

and substituting $B_D^{\max} = \alpha \cdot B_E^{\max} + B_{TC}^{\max}$ where $\alpha = \frac{R_2}{R_1}$, the decoder buffer occupancy is given by

$$
\begin{aligned}
B_D(t + (M+L)\cdot T) &= \alpha \cdot B_E^{\max} + B_{TC}^{\max} \\
&\quad - \alpha \cdot B_E(t) - B_{TC}(t + L_e(t))
\end{aligned} \qquad (7.38)
$$

To ensure the decoder buffer not underflowing one has that

$$B_D(t + (M+L)\cdot T) \geq 0,$$

that is,

$$\alpha \cdot B_E^{\max} + B_{TC}^{\max} - \alpha \cdot B_E(t) - B_{TC}(t + L_e(t)) \geq 0 \qquad (7.39)$$

This is equivalent to constrain the transcoder buffer occupancy such that

$$B_{TC}(t + L_e(t)) \leq \alpha \cdot B_E^{\max} + B_{TC}^{\max} - \alpha \cdot B_E(t) \qquad (7.40)$$

Since $B_E(t) \le B_E^{max}$, the decoder buffer never underflows if the transcoder buffer fullness is constrained such that

$$B_{TC}(t) \le B_{TC}^{max} \tag{7.41}$$

Similarly, $B_D(t+(M+L)\cdot T) \le \alpha \cdot B_E^{max} + B_{TC}^{max}$ is the condition that the decoder buffer should meet for not overflowing. Thus, using Equation (7.38), one obtains

$$\alpha \cdot B_E^{max} + B_{TC}^{max} - \alpha \cdot B_E(t) - B_{TC}(t + L_e(t)) \le \alpha \cdot B_E^{max} + B_{TC}^{max} \tag{7.42}$$

which is equivalent to constrain the transcoder buffer occupancy, such that

$$-\alpha \cdot B_E(t) \le B_{TC}(t + L_e(t)) \tag{7.43}$$

Hence, in order to prevent the decoder buffer from overflowing, it is sufficient that the condition of (7.44) holds all the time

$$B_{TC}(t) \ge 0 \tag{7.44}$$

To summarize, if both the encoder and transcoder buffers never underflow or overflow, then decoder buffer overflow or underflow will never occur either. The basic idea is that by increasing the total delay between encoder and decoder, a buffer corresponding to this extra delay can be used in the transcoder, which in turn is responsible for preventing it from overflowing and underflowing. Therefore, the encoder "sees" a decoder buffer of size and the decoder "is told" that the encoder is using a buffer size $B_D^{max} = \alpha \cdot B_E^{min} + B_{TC}^{min}$. Between them, the transcoder performs the necessary adaptation, such that the process is transparent for both the encoder and decoder. Using MPEG-2 coded video, the transcoder is responsible for updating the buffer size specified in the sequence header, as well as the delay parameter of each picture header.

7.5 REGENERATING TIME STAMPS IN TRANSCODER

As studied in section 0, a transcoder involves decoding and re-encoding processes. Thus, the idea method for re-generating PCR, PTS and DTS is to use a phase-lock loop for the video transport stream. Figure 7.9 shows a model of re-generating time stamps in transcoder.

In this model, assume that the encoding time of the transcoder can be ignored. PCR_E denotes PCRs being inserted in the encoder while PCR_T, PTS_T, and DTS_T denote time stamps being inserted in the transcoder. STC_T is the new time-base (clock) for the transcoder.

For many applications, it is too expensive to have a PLL for each video transcoder, especially for multiple channel transcoder. Instead, the transcoder can use one free running system clock (assuming that the clock is accurate, e.g., exactly 27 MHz) and perform PCR correction. An example of the PCR correction is shown in Figure 7.10.

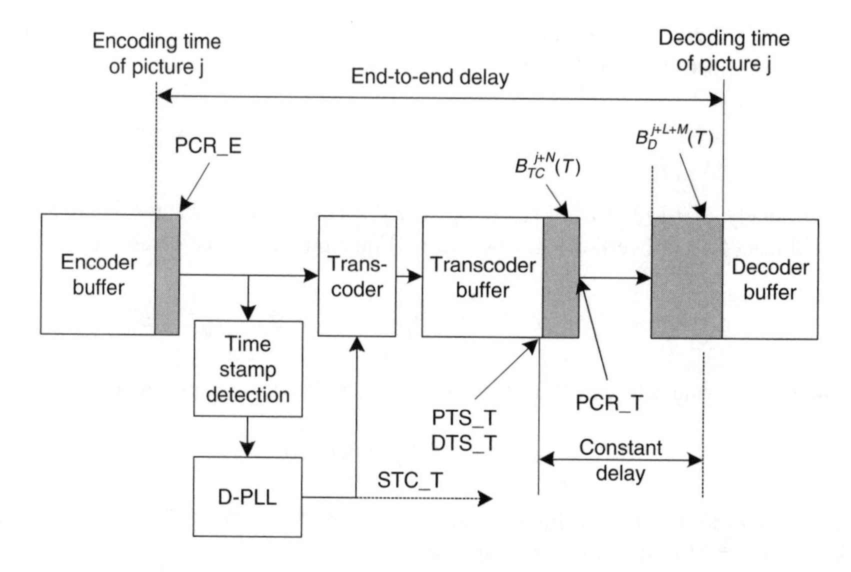

FIGURE 7.9 A model of regenerating time stamps in transcoder.

In Figure 7.10, one free-running system time clock, STC_{free_runing} is used (for all channels). When a TS packet with a PCR (PCR_E) arrives, the snapshot value S_{PCR_in} of STC_{free_runing} is taken and the difference is computed, PCR_{offset}. Then, the instantaneous system time clock for the channel is

$$STC = STC_{free\_runing} + PCR_{offset} \tag{7.45}$$

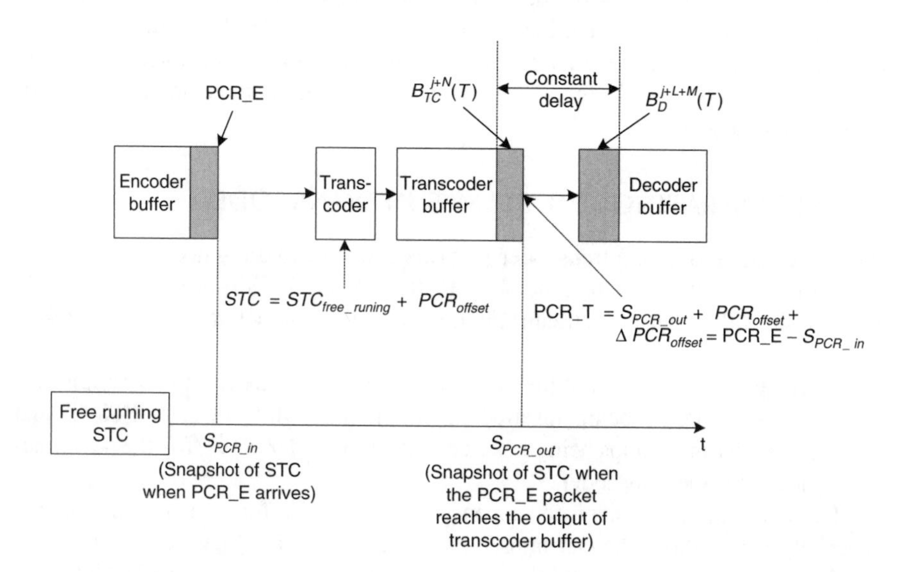

FIGURE 7.10 A model of the PCR correction.

The snapshot of the same PCR packet when it reaches the output of the transcoder buffer can also be taken as S_{PCR_out}. Then, the new PCR value for the transcoder output can be generated by

$$\text{PCR\_T} = S_{PCR\_out} + PCR_{offset} + \Delta \tag{7.46}$$

where Δ is an estimated error due to small difference between the transcoder free-running clock counter and the video encoder STC counter.

Both PTS and DTS values for the transcoder can be generated in a similar manner. One can also keep the original PTS and DTS values, but only adjust PCR_T by subtracting the delay between the transcoder decoding time and the final decoder decoding time.

7.6 SUMMARY

In this chapter, constraints on video compression/decompression buffers and the bit rate of a compressed video bit stream have been discussed. The transmission channels impose these constraints to video encoder/transcoder buffers. First, we have introduced concepts of compressed video buffers. Then, conditions that prevent the video encoder/transcoder and decoder buffer overflow or underflow are derived for the channel that can transmit a variable bit rate video. Next, the discussion has focused on analyzing buffer, timing recovery, and synchronization for video transcoder. The buffering implications of the video transcoder within the transmission path have also been analyzed.

7.7 EXERCISES

7-1. Rate control and buffer regulation are an important for both VBR and CBR applications. In the case of VBR encoding, the rate controller attempts to achieve optimum quality for a given target rate. In the case of CBR encoding and real-time application, the rate control scheme has to satisfy the low-latency requirement. Both CBR and VBV rate control schemes have to satisfy buffer constraints. Provide a rate-control algorithm, e.g., MPEG-2 Test Model 5 (TM5) rate control, to show how these buffer constraints are imposed.

7-2. Let $R_C(t)$ be the channel bit rate and $B_E(t)$ be the fullness of the encoder/transcoder buffer at time t. This buffer has a maximum size, $B_E^{\max}$. Define $S(iT)$ ($i = 1, 2, \ldots$) to be the number of units (bits, bytes or packets) output by the encoder/transcoder in the interval $[(i-1)T, iT]$, where T is the picture duration of the original uncompressed video, e.g., $T = 1/29.97$ s for digitized NTSC video. Also, let $C(iT)$ be the number of bits that are transmitted during the i-th picture period. Prove that conditions for preventing encoder/transcoder buffer overflow and underflow are:

$$0 \le B_E^i(T) = B_E^{i-1}(T) + S(iT) - C(iT) \le B_E^{\max}$$

$$C(iT) - B_E^{i-1}(T) \le S(iT) \le C(iT) + B_E^{\max} - B_E^{i-1}(T)$$

which is a constraint on the number of bits per coded picture for a given channel rate.

7-3. Let $R_C(t)$ be the channel bit rate and $B_D(t)$ be the fullness of the decoder buffer at time t. This buffer has a maximum size, $B_D^{\max}$. Define $S(iT)$ ($i = 1, 2, \ldots$) to be the number of units in the interval $[(i-1)T, iT]$, where T is the picture duration of the original uncompressed video. Also, let $C(iT)$ be the number of bits that are transmitted during the i-th picture period. Prove that the conditions for preventing decoder buffer overflow and underflow are,

$$0 \le C((i+L)\cdot T) + B_D^{i-1}(T) - S(i\cdot T) \le B_D^{\max}$$

$$C((i+L)\cdot T) + B_D^{i-1}(T) - B_D^{\max} \le S(i\cdot T) \le C((i+L)\cdot T) + B_D^{i-1}(T)$$

which is a constraint on the encoder bit rate for a given channel rate.

7-4. Assume that the system delay is LT, show that the system must store L pictures worth of data; i.e.,

$$0 \le \sum_{j=i-L+1}^{i} S(iT) \le B_D^{\max} + B_D^{\max}.$$

(These bounds arise because of the finite memory of the video system. The system can store no more than $B_E^{\max} + B_D^{\max}$ bits at any given time, but it must always store L pictures of data).

REFERENCES

[7-1] ITU-T Recommendation H.262 | ISO/IEC 13818-2: 1995. Information technology: generic coding of moving pictures and associated audio information: Video.

[7-2] ISO/IEC 14496-2:1998, Information Technology: generic coding of audio-visual objects—Part 2: Visual.

[7-3] Test model editing committee, Test Model 5, MPEG93/457, ISO/IEC JTC1/SC29/WG11, April 1993.

[7-4] X. Chen and R. O. Eifrig, "Video rate buffer for use with push data flow," U.S. Patent Number 6289129, Assignee: Motorola Inc. and General Instrument Corporation, Sept. 11, 2001.

[7-5] A. Puri and T. H. Chen, *Multimedia Standards and Systems,* Chapman & Hall, New York, 1999.

[7-6] T. Sikora, "The MPEG-4 Video Standard Verification Model," IEEE Transactions on Circuits and Systems for Video Technology, 7, 19–31, 1997.

[7-7] ITU-T Experts Group on Very Low Bitrate Visual Telephony, ITU-T Recommendation H.263 Version 2: Video Coding for Low Bitrate Communication, Jan. 1998.

[7-8] ITU-T Experts Group on Very Low Bitrate Visual Telephony, ITU-T Recommendation H.263: Video Coding for Low Bitrate Communication, Dec. 1995.

[7-9] B. G. Haskell, A. Puri, and A. N. Netravali, *Digital Video: An Introduction to MPEG-2,* Chapman & Hall, New York, 1997.

[7-10] X.Chen and B. Eifrig, Video rate buffer, ISO/IEC JTC1/SC29/WG11, M3596, July 1998.

[7-11] X. Chen and A. Luthra, A brief report on core experiment Q2—improved rate control, ISO/IEC JTC1/SC29/WG11, M1422 Maceio, Brizal, Nov. 1996.

[7-12] ITU-T Rec. H.264 | ISO/IEC 14496-10 AVC: 2003 Information technology—Generic coding of moving pictures and associated audio information: Advanced Video Coding.

[7-13] E. Viscito and D. T. Hoang, Proposal for Hypothetical Reference Decoder—Overview, JVT-D131, July 2002.

[7-14] X. Chen, *Transporting Compressed Digital Video,* Kluwer Academic Publishers, Boston, 2001.

[7-15] X. Chen, B. Eifrig. and A. Luthra, Rate control for multiple higher resolution VOs: a report on CE Q2, ISO/IEC JTC1/SC29/WG11, M1657, Seville, Spain, Feb. 1997.

[7-16] R. Schafer and T. Sikora, Digital video coding standards and their role in video communications, *Proceedings of the IEEE,* 83, 907, 1995.

[7-17] G. Keesman, R. Hellinghuizen, F. Hoeksema, and G. Heideman, Transcoding of MPEG bitstreams, *Signal Processing: Image Communication,* 8, 481, 1996.

[7-18] X. Chen and F. Ling, Implementation architectures of a multi-channel MPEG-2 video transcoder using multiple programmable processors, U.S. Patent No. 6275536B1, Aug. 14, 2001.

[7-19] X. Chen, L. Wang, A. Luthra, and R. Eifrig, Method of architecture for converting MPEG-2 4:2:2-profile bitstreams into main-profile bitstreams, U.S. Patent No. 6259741B1, July 10, 2001.

[7-20] X. Chen, F. Lin, and A. Luthra, Video rate-buffer management scheme for MPEG transcoder, WO0046997, 2000.

[7-21] P. A. A. Assuncao and M. Ghanbari, Buffer analysis and control in CBR video transcoding, *IEEE Transcations on Circuits and Systems for Video Technology,* 10, 83–92, 2000.

[7-22] L. Wang, A. Luthra, and B. Eifrig, Rate-control for MPEG transcoder," *IEEE Transactions on Circuit and Systems for Video Technology,* 11, 222–234, Feb. 2001.

[7-23] K. Panusopone and X. Chen, Video size conversion and transcoding from MPEG-2 to MPEG-4, U.S. Patent, U.S. 6647061B1, Nov. 11, 2003.

[7-24] X. Chen, K. Panusopone, and F. Ling, Digital transcoder with logo insertion, U.S. Patent, U.S. 6658057B1, Dec. 2, 2003.

8 Cryptography and Conditional Access for Video Transport Systems

To provide digital video services, there are technologies on how best to compress, transmit, store, and retrieve video data. Among these technologies, no one need be reminded of the importance, not only of the reliable transmission and efficient compression, but also of the security of the video delivery process. In traditional video communication systems, such as analog cable television networks, it is important to secure transmission channels so that only authorized customers can view the transmitted programs. For MPEG-enabled digital video transmission systems, it is more important to provide a higher level of security for the digital information that is transmitted. As public communication networks are increasingly used for commercial digital video transmission, the need to provide security becomes critical. In digital cable and satellite systems, high-value movies and video programs ("content") must be protected from piracy by conditional access and copy protection systems. Video content protection is now an integral part of digital video services. Secure communications service prevents imposters from impersonating legitimate users. Achieving video security in an electronic society requires a vast array of technical and legal skills. There is, however, no guarantee that all of the video security objectives deemed necessary can be adequately met. The technical means of security is provided through cryptography.

Cryptography is the study of mathematical techniques related to aspects of information security such as confidentiality, data integrity, entity authentication, and data origin authentication. Cryptography is not the only means of providing video security, but rather one set of techniques [8-1]-[8-8]. Cryptography is about the prevention and detection of cheating and other malicious activities. This chapter describes a number of basic cryptographic algorithms used to provide video security, especially to compressed video [8-10][8-10][8-15]. Examples of these algorithms include the data encryption standard (DES) algorithm [8-17] and the Rivest, Shamir, and Adleman (RSA) encryption algorithm [8-29][8-30]. This chapter also discusses conditional access and copy protection schemes [8-14][8-22][8-23][8-24] for video transport systems [8-12].

8.1 BASIC TERMINOLOGY AND CONCEPTS

While this chapter is not a treatise on abstract mathematics, a familiarity with basic mathematical concepts will be useful [8-8][8-11][8-13]. One concept that is absolutely fundamental to cryptography is that of a function in the mathematical sense. A function is alternately referred to as a *mapping*, or a *transformation*.

8.1.1 Functions (One-to-one, One-way, Trapdoor One-way)

A set consists of distinct objects that are called *elements of the set*. For example, a set X might consist of the digits 0 and 1, and this is denoted X = {0, 1}.

Function: A function is defined by two sets X and Y and a rule f that assigns to each element in X precisely one element in Y. The set X is called the *domain of the function* and Y the *range*. If x is an element of X (usually written $x \in X$) the image of x is the element in Y which the rule f associates with x; the image y of x is denoted by y = f(x). Standard notation for a function f from set X to set Y is $f: X \rightarrow Y$. If $y \in Y$, then a pre-image of y is an element $x \in X$ for which f(x) = y. The set of all elements in Y that have at least one pre-image is called *the image of f*, denoted Im(f).

One-to-one functions: A function is one-to-one if each element in the range Y is the image of at most one element in the domain X.

Onto: A function (or transformation) is onto if each element in the range Y is the image of at least one element in the domain. Equivalently, a function $f: X \rightarrow Y$ is onto if Im(f) = Y.

Bijection: If a function $f : X \rightarrow Y$ is one-to-one and Im(f) = Y, then f is called a *bijection*.

Inverse Function: If f is a bijection from X to Y, then it is a simple matter to define a bijection g from Y to X as follows: for each $y \in Y$ define g(y) = x, where $x \in X$ and f(x) = y. This function g obtained from f is called the *inverse function of f* and is denoted by $g = f^{-1}$.

Note that if f is a bijection, then so is f^{-1}. In cryptography, bijections are used as the tool for encrypting messages and the inverse transformations are used to decrypt. This will be made clearer in Section 8.1.2 when some basic terminology is introduced. Notice that if the transformations were not bijections then it would not be possible to always decrypt to a unique message.

There are certain types of functions which play significant roles in cryptography. At the expense of rigor, an intuitive definition of a one-way function is given.

One-way functions: A function f from a set X to a set Y is called a one-way function if f(x) is "easy" to compute for all $x \in X$ but for "essentially all" elements $y \in Im(f)$ it is "computationally infeasible" to find any $x \in X$ such that f(x) = y.

Trapdoor one-way functions: A trapdoor one-way function is a one-way function $f : X \rightarrow Y$ with the additional property that given some extra information (called the *trapdoor information*) it becomes feasible to find for any given $y \in Im(f)$, an $x \in X$ such that f(x) = y.

Permutations are functions that are often used in various cryptographic constructs.

Permutations: Let S be a finite set of elements. A permutation p on S is a bijection from S to itself (i.e., p: S → S).

Example

(Permutation): Let S = {1, 2, 3, 4}. A permutation p: S → S is defined as follows: p(1) = 2, p(2) = 3, p(3) = 4, p(4) = 1.

Another type of function commonly used in cryptography is an involution. Involutions have the property that they are their own inverses.

Involutions: Let S be a finite set and let f be a bijection from S to S (i.e., f : S → S). The function f is called an involution if $f = f^{-1}$. An equivalent way of stating this is f(f(x)) = x for all x ∈ S.

Example

(Involution): Let S = {1, 2, 3, 4}. An involution f: S → S is defined as follows: f(1) = 3, f(2) = 4, f(3) = 1, f(4) = 2.

8.1.2 Basic Concepts of Encryption and Decryption

The scientific study of any discipline must be built upon rigorous definitions arising from fundamental concepts. What follows is a list of terms and basic concepts used throughout this chapter. Where appropriate, rigor has been sacrificed for the sake of clarity.

Sender and receiver: Suppose a sender A wants to send a message to a receiver B. Moreover, the message needs to be sent securely such that an attacker cannot read the message.

Messages and encryption: A message **m** is an element in the plaintext (sometimes called *cleartext*) set **M**; i.e., **m** ∈ **M.** The *message space* is represented by **M**. The process of disguising a message in such a way as to hide its substance is encryption. An encrypted message **c** is an element in the ciphertext set **C**; i.e., **c** ∈ **C**. **C** is called the *ciphertext space*. The process of turning ciphertext back into plaintext is decryption.

Encryption and decryption transformations: Let **K** denotes a set called the *key space.* An element **e** of **K** is called a *key.* Each element **e** ∈ **K** uniquely determines a bijection from **M** to **C**, denoted by **Ee**, and **Ee** is called an *encryption function*, or an *encryption transformation.* Note that **Ee** must be a bijection if the process is to be reversed and a unique plaintext message recovered for each distinct ciphertext. For each **d** ∈ **K**, **Dd** denotes a bijection from **C** to **M** (i.e., **Dd : C** → **M**). The term **Dd** is called a *decryption function*, or *decryption transformation.*

The process of applying the transformation **Ee** to a message **m** $\in$ **M** is usually referred to as *encrypting **m***, or *the encryption of **m***, whereas the process of applying the transformation **Dd** to a ciphertext **c** is usually referred to as *decrypting **c***, or *the decryption of **c***.

An encryption scheme consists of a set {**Ee** : **e** $\in$ **K**} of encryption transformations and a corresponding set {**Dd** : **d** $\in$ **K**} of decryption transformations with the property that for each **e** $\in$ **K** there is a unique key **d** $\in$ **K** such that $D_d = E_e^{-1}$; that is, **Dd**(**Ee**(**m**)) = **m** for all **m** $\in$ **M**. An encryption scheme is sometimes referred to as a *cipher.*

The keys **e** and **d** in the preceding definition are referred to as a *key pair* and sometimes denoted by (**e**, **d**). Note that **e** and **d** could be the same.

Achieving confidentiality: An encryption scheme may be used as follows for the purpose of achieving confidentiality. Two parties A and B first secretly choose or secretly exchange a key pair (**e**, **d**). At a subsequent point in time, A sends a message **m** $\in$ **M** to B by computing **c** = **Ee**(**m**) and transmitting **c** to B. Upon receiving **c**, B recovers the original message **m** by computing **Dd**(**c**) = **m**.

The question arises as to why keys are necessary. (Why not just choose one encryption function and its corresponding decryption function?) Having transformations that are very similar but characterized by keys means that if some particular encryption/decryption transformation is revealed then one does not have to redesign the entire scheme but simply change the key. It is sound cryptographic practice to change the key (encryption/decryption transformation) frequently. As a physical analogue, consider an ordinary resettable combination lock. The structure of the lock is available to anyone who wishes to purchase one but the combination is chosen and set by the owner. If the owner suspects that the combination has been revealed he can easily reset it without replacing the physical mechanism. Figure 8.1 provides a simple model of a two-party communication using encryption.

Communication participants: Referring to Figure 8.1, the following terminology is defined:

- An entity or party is someone or something that sends, receives, or manipulates information. An entity may be a set top box, a video transmission device, etc.
- A sender is an entity in a two-party communication that is the legitimate transmitter of information. In Figure 8.1, the sender is A.
- A receiver is an entity in a two-party communication that is the intended recipient of information. In Figure 8.1, the receiver is B.
- An attacker is an entity in a two-party communication that is neither the sender nor receiver, and that tries to defeat the information security service being provided between the sender and receiver. Various other names are synonymous with attacker such as *enemy, adversary opponent, tapper, eavesdropper intruder* and *interloper.* An attacker will often attempt to play the role of either the legitimate sender or the legitimate receiver.

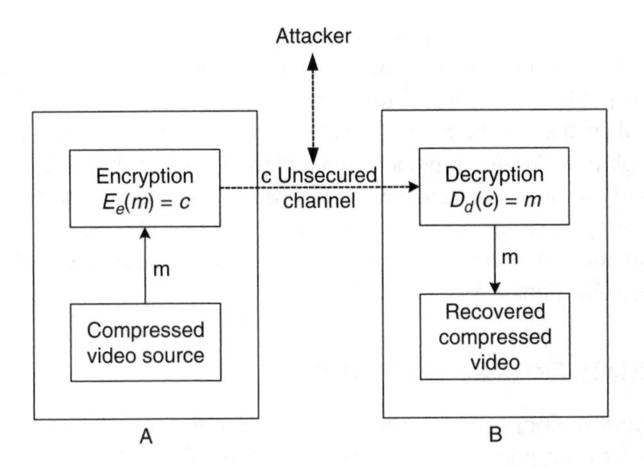

FIGURE 8.1 Schematic of a two-party communication using encryption.

Channels: A channel is a means of conveying information from one entity to another:

- A physically secure channel is one that is not physically accessible to the attacker.
- An unsecured channel is one from which parties other than those for which the information is intended can reorder, delete, insert, or read.
- A secured channel is one from which an attacker does not have the ability to reorder, delete, insert, or read.

Note that the subtle difference between a physically secure channel and a secured channel: a secured channel may be secured by physical or cryptographic techniques, the latter being the topic of this chapter. Certain channels are assumed to be physically secure. These include trusted networks, secure data paths, and a dedicated communication link, to name a few.

A *fundamental premise in cryptography* is that the sets **M, C, K, {Ee : e ∈ K}, {Dd : d ∈ K}** are public knowledge. When two parties wish to communicate securely using an encryption scheme, the only thing that they keep secret is the particular key pair (**e, d**) that they are using, and that they must select. One can gain additional security by keeping the class of encryption and decryption transformations secret but one should not base the security of the entire scheme on this approach. History has shown that maintaining the secrecy of the transformations is very difficult indeed.

Breakable system: An encryption system is said to be breakable if a third party, without prior knowledge of the key pair (**e, d**), can systematically recover plaintext from corresponding ciphertext within some appropriate time frame.

An appropriate time frame will be a function of the useful lifespan of the data being protected by a given key. For example, an instruction to buy a pay-per-view (PPV) movie may only need to be kept secret for a few minutes whereas the encrypted movie needs to be secured in its entire delivery process.

An encryption scheme can be broken by trying all possible keys to see which one the communicating parties are using (assuming that the class of encryption functions is public knowledge). This is called an *exhaustive search of the key space.* It follows then that the number of keys (i.e., the size of the key space) should be large enough to make this approach computationally infeasible. It is the objective of a designer of an encryption scheme that this be the best approach to break the system.

Cryptographic techniques are typically divided into two generic types: symmetric-key and public-key ciphers. Encryption methods of these types will be discussed separately in Sections 8.2, 8.3, and 8.6.

8.2 SYMMETRIC-KEY CIPHERS

Symmetric-Key Encryption Method: Consider an encryption scheme consisting of the sets of encryption and decryption transformations {**Ee** : **e** ∈ **K**} and {**Dd** : **d** ∈ **K**}, respectively, where **K** is the key space. The encryption scheme is said to be symmetric-key if for each associated encryption/decryption key pair (**e**, **d**), it is computationally "easy" to determine **d** knowing only **e**, and to determine **e** from **d**.

Since **e** = **d** in most practical symmetric-key encryption schemes, the term *symmetric-key* becomes appropriate. Other terms used in the literature are *single-key, one-key*, and *private key* encryption. Figure 8.2 illustrates the idea of symmetric-key encryption.

A two-party communication using symmetric-key encryption can be described by the block diagram of Figure 8.2, which is Figure 8.1 with the addition of the secure (both confidential and authentic) channel. One of the major issues with symmetric-key systems is to find an efficient method to agree upon and exchange keys securely. This problem is often referred to as the *key distribution problem.*

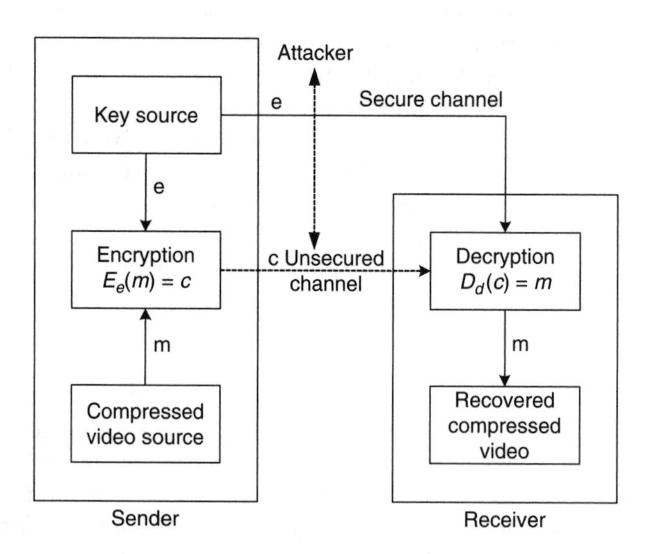

FIGURE 8.2 Two-party communication using encryption, with a secure channel for key exchange. The decryption key **d** can be efficiently computed from the encryption key **e**.

It is assumed that all parties know the set of encryption/decryption transformations (i.e., they all know the encryption scheme). As has been emphasized several times, the only information that should be required to be kept secret is the key **d**. However, in symmetric-key encryption, this means that the key **e** must also be kept secret, as **d** can be deduced from **e**. In the encryption key, **e** is transported from one entity to the other with the understanding that both can construct the decryption key **d**.

There are two classes of symmetric-key encryption schemes that are commonly distinguished: block ciphers and stream ciphers.

Block Cipher: A block cipher is an encryption scheme that breaks up the plaintext messages to be transmitted into strings (called *blocks*) of a fixed length n over an alphabet **A**, and encrypts one block at a time.

Stream Cipher: A stream cipher, in one sense, is a very simple block cipher having block length equal to one.

Most well-known symmetric-key encryption techniques for digital video services are block and stream ciphers. A number of examples of these will be given in this chapter. Two important classes of block ciphers are substitution ciphers and permutation ciphers (Section 8.2.1). Product ciphers (Section 8.2.2) combine these. Stream ciphers and the key space are considered in Section 8.2.3.

8.2.1 Substitution and Permutation Ciphers

Substitution ciphers are block ciphers thatreplace symbols (or groups of symbols) by other symbols or groups of symbols.

Simple substitution ciphers: Let **A** be an alphabet of q symbols and **M** be the set of all strings of length n over **A**. Let **K** be the set of all permutations (also called *S-Box*) on the set **A**. Define for each **e** $\in$ **K** an encryption transformation **Ee** as:

$$E_e(m) = (e(m_1)e(m_2)\cdots e(m_n)) = (c_1c_2\cdots c_n) = \mathbf{c},$$

where $\mathbf{m} = (m_1m_2\cdots m_n) \in \mathbf{M}$. In other words, for each symbol in a n-tuple, replace (substitute) it by another symbol from **A** according to some fixed permutation **e**. To decrypt $\mathbf{c} = (c_1c_2\cdots c_n)$ compute the inverse permutation $\mathbf{d} = \mathbf{e}^{-1}$ and

$$D_d(c) = (d(c_1)d(c_2)\cdots d(c_n)) = (m_1m_2\cdots m_n) = \mathbf{m}$$

Ee is called a simple substitution cipher.

Example

(Simple substitution cipher): Consider $\mathbf{A} = \{0, 1, 2, \ldots, p-1\}$, $e(m_i) = m_i + \left\lfloor \frac{p}{2} \right\rfloor \bmod p$ and $m_i \in \mathbf{A}$. For $p = 5$, if $\mathbf{m} = (4,2,1)$, then $\mathbf{c} = (1,4,3)$.

The number of distinct substitution ciphers is q! and is independent of the block size in the cipher. Simple substitution ciphers over small block sizes provide inadequate security even when the key space is extremely large. If the alphabet set is the set of 26 English alphabets, then the size of the key space is $26! \approx 4 \times 10^{26}$, yet the key being used can be determined quite easily by examining a modest amount of ciphertext. This follows from the simple observation that the distribution of letter frequencies is preserved in the ciphertext. For example, the letter E occurs more frequently than the other letters in ordinary English text. Hence the letter occurring most frequently in a sequence of ciphertext blocks is most likely to correspond to the letter E in the plaintext. By observing a modest quantity of ciphertext blocks, a cryptanalyst can determine the key.

Homophonic substitution ciphers: To each symbol $\mathbf{a} \in \mathbf{A}$, associate a set $\mathbf{H(a)}$ (S-Box) of strings of n symbols, with the restriction that the sets $\mathbf{H(a)}$, $\mathbf{a} \in \mathbf{A}$, be pairwise disjoint. A homophonic substitution cipher replaces each symbol a in a plaintext message block with a randomly chosen string from $\mathbf{H(a)}$. To decrypt a string $\mathbf{c}$ of n symbols, one must determine an $\mathbf{a} \in \mathbf{A}$ such that $\mathbf{c} \in \mathbf{H(a)}$. The key for the cipher consists of the sets $\mathbf{H(a)}$.

Example

(Homophonic substitution cipher): Consider $\mathbf{A} = \{\mathbf{u}, \mathbf{v}\}$, $\mathbf{H(u)} = \{11, 10\}$, and $\mathbf{H(v)} = \{01, 00\}$. The plaintext message block **uv** encrypts to one of the following: 1101,1100, 1001, 1000. Observe that the range of the encryption function (for messages of length two) consists of the following pairwise disjoint sets of four-element bit-strings: $\mathbf{uu} \in \{1111, 1110, 1011, 1010\}$, $\mathbf{uv} \in \{1101, 1100, 1001, 1000\}$, $\mathbf{vu} \in \{0111, 0011, 0110, 0010\}$, $\mathbf{vv} \in \{0101, 0100, 0001, 0000\}$. Any 4-bitstring uniquely identifies a range element, and hence a plaintext message.

Often the symbols do not occur with equal frequency in plaintext messages. With a simple substitution cipher this nonuniform frequency property is reflected in the ciphertext. A homophonic cipher can be used to make the frequency of occurrence of ciphertext symbols more uniform, at the expense of data expansion. Decryption is not as easily performed as it is for simple substitution ciphers.

Polyalphabetic substitution ciphers: A polyalphabetic substitution cipher is a block cipher with block length n over an alphabet A having the following properties:

- The key space $\mathbf{K}$ (S-Box) consists of all ordered sets of n permutations $(p_1, p_2, \ldots, p_n)$, where each permutation p_i is defined on the set $\mathbf{A}$.
- Encryption of the message $\mathbf{m} = (m_1, m_2, \ldots, m_n)$ under the key $\mathbf{e} = (p_1, p_2, \ldots, p_n)$ is given by $\mathbf{Ee(m)} = (p_1(m_1)p_2(m_2)\ldots p_n(m_n))$.
- The decryption key associated with $\mathbf{e} = (p_1, p_2, \ldots, p_n)$ is $\mathbf{d} = \left(p_1^{-1}, p_2^{-1}, \ldots, p_n^{-1}\right)$.

Example

(Polyalphabetic substitution cipher): Let $\mathbf{A} = \{a, b, c, \ldots, x, y, z\}$ and n = 2. Choose $\mathbf{e} = (p_1, p_2)$, where p_1 maps each letter to the letter two positions to its right in the alphabet and p_2 to the one five positions to its right. If

$$\mathbf{m} = (\text{TR AN SC OD IN GV ID EO})$$

then

$$\mathbf{c} = \mathbf{Ee(m)} = (\text{VW CS UH QI KS IA KI GT})$$

Polyalphabetic ciphers have the advantage over simple substitution ciphers that symbol frequencies are not preserved. In the example above, the letter E is encrypted to both O and L. However, polyalphabetic ciphers are not significantly more difficult for attacker to analyze ciphertext, the approach being similar to the simple substitution cipher. In fact, once the block length n is determined, the ciphertext letters can be divided into n groups (where each group i consists of those ciphertext letters derived using permutation p_i), and a frequency analysis can be done on each group.

Permutation ciphers: Another class of symmetric-key ciphers is the simple permutation (also called *transposition*) cipher, which simply permutes the symbols in a block. Consider a symmetric-key block encryption scheme with block length n. Let **K** be the set of all permutations (also called *P-Box*) on the set {1,2,…,n}. For each **e** $\in$ **K** define the encryption function

$$\mathbf{Ee(m)} = \left(m_{e(1)}, m_{e(2)}, \ldots, m_{e(n)}\right)$$

where $\mathbf{m} = (m_1, m_2, \ldots, m_n) \in \mathbf{M}$, the message space. The set of all such transformations is called a *simple permutation cipher*. The decryption key corresponding to **e** is the inverse permutation $\mathbf{d} = \mathbf{e}^{-1}$. To decrypt **c**, compute $\mathbf{Dd(c)} = \left(c_{d(1)}, c_{d(2)}, \ldots, c_{d(n)}\right)$. A simple permutation cipher preserves the number of symbols of a given type within a block, and thus is easily cryptanalyzed.

8.2.2 Product Cipher System

In order to describe product ciphers, the concept of composition of functions is introduced. Compositions are a convenient way of constructing more complicated functions from simpler ones.

Composition of functions: Let **U**, **V**, and **W** be finite sets and let $f : \mathbf{U} \to \mathbf{V}$ and $g : \mathbf{V} \to \mathbf{W}$ be functions. The composition of g with f, denoted $g \circ f$ (or simply gf), is a function from **U** to **W** and defined by $(g \circ f)(\mathbf{x}) = g(f(\mathbf{x}))$ for all $\mathbf{x} \in \mathbf{U}$. Composition can be easily extended to more than two functions. For functions $f_1, f_2, \ldots, f_n$, one can define $f_n \circ f_{n-1} \circ \ldots \circ f_2 \circ f_1$, provided that the domain of f_n equals the range of f_{n-1} and so on.

Compositions and involutions: Involutions were introduced in Section 8.1 as a simple class of functions with an interesting property that **Ek**(**Ek**(x)) = x for all x in the domain of **Ek**. That is, **Ek** $\circ$ **Ek** is the identity function.

Note that the composition of two involutions is not necessarily an involution. However, involutions may be composed to get somewhat more complicated functions whose inverses are easy to find. This is an important feature for decryption. For example, if $Ek_1, Ek_2, \ldots, Ek_n$ are involutions, then the inverse of $Ek = Ek_1 Ek_2 \ldots Ek_n$ is $Ek^{-1} = Ek_n \ldots Ek_2\, Ek_1$, the composition of the involutions in the reverse order.

Product ciphers: Simple substitution and permutation ciphers individually do not provide a very high level of security. However, by combining these transformations it is possible to obtain strong ciphers. Shannon [8-1] suggested using a product chipper or a combination of substitution (S-Box) and permutation (P-Box) transformations, which together could yield a cipher system more secure than either one alone. As will be seen in next two sections some of the most practical and effective symmetric-key systems are product ciphers. One example of a product cipher is a composition of $n \geq 2$ transformations $Ek_1Ek_2 \ldots Ek_n$ where each Ek_i, $1 \leq i \leq n$, is either a substitution or a permutation cipher. For the purpose of this introduction, let the composition of a substitution and a transposition be called a *round.*

Example

(Product cipher): Let $\mathbf{M} = \mathbf{C} = \mathbf{K}$ be the set of all binary strings of length five. The number of elements in $\mathbf{M}$ is $2^5 = 32$. Let $\mathbf{m} = (m_1m_2m_3m_4m_5)$ and define

$$\mathbf{Ek}_1 (\mathbf{m}) = \mathbf{m} \oplus \mathbf{k}_1, \text{ where } \mathbf{k}_1 \in \mathbf{K}, \text{ (S-Box)}$$

$$\mathbf{Ek}_2 \text{ (m)} = (m_3m_4m_5m_1m_2). \text{ (P-Box)}$$

Here, $\oplus$ is the exclusive-OR (XOR) operation defined as follows: $0 \oplus 0 = 0$, $0 \oplus 1 = 1$, $1 \oplus 0 = 1$, $1 \oplus 1 = 0$. $\mathbf{Ek}_1$ $(\mathbf{m})$ is a polyalphabetic substitution cipher and $\mathbf{Ek}_2$ $(\mathbf{m})$ is a permutation cipher (not involving the key). The product $\mathbf{Ek}_1$ $(\mathbf{Ek}_2\ (\mathbf{m}))$ is a round. Figure 8.3 shows two rounds of the product cipher for this example.

Confusion and diffusion: A substitution in a round is said to add confusion to the encryption process whereas a permutation is said to add diffusion. Confusion is intended to make the relationship between the key and ciphertext as complex as possible. Diffusion refers to rearranging or spreading out the bits in the message so that any redundancy in the plaintext is spread out over the ciphertext. A round then can be said to add both confusion and diffusion to the encryption. Most modern block cipher systems apply a number of rounds in succession to encrypt plaintext.

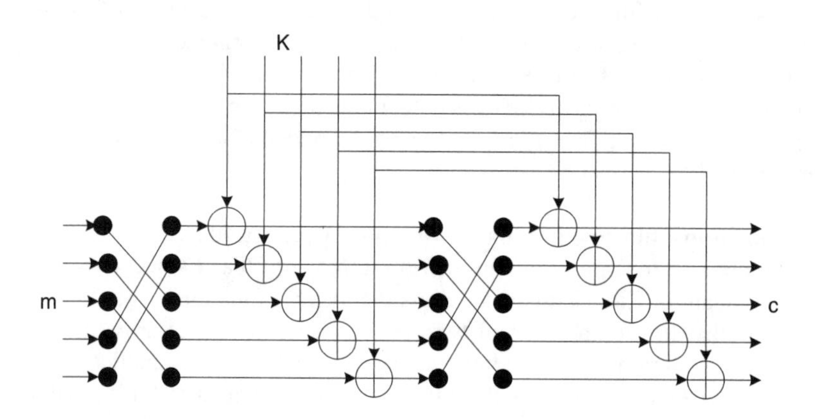

FIGURE 8.3 A product cipher with 2 rounds.

8.2.3 Stream Cipher and the Key Space

Stream ciphers form an important class of symmetric-key encryption schemes. What makes them useful is the fact that the encryption transformation can change for each symbol of plaintext being encrypted. In situations where transmission errors are highly probable, stream ciphers are advantageous because they have no error propagation. They can also be used when the data must be processed one symbol at a time (e.g., if the equipment has no memory or buffering of data is limited).

Key Stream: Let **K** be the key space for a set of encryption transformations. A sequence of symbols $\{\mathbf{e}_1,\mathbf{e}_2,\mathbf{e}_3 \ldots\ldots\}$, $\mathbf{e}_i \in \mathbf{K}$, is called a *keystream.*

Stream Cipher: Let **A** be a set of alphabets of q symbols and let **Ee** be a simple substitution cipher with block length 1 where $\mathbf{e} \in \mathbf{K}$. Let $\{\mathbf{m}_1,\mathbf{m}_2,\mathbf{m}_3\ldots\ldots\}$ be a plaintext string and let $\{\mathbf{e}_1,\mathbf{e}_2,\mathbf{e}_3 \ldots\ldots\}$ be a key stream from **K**. A stream cipher takes the plaintext string and produces a ciphertext string $\{\mathbf{c}_1,\mathbf{c}_2,\mathbf{c}_3 \ldots\ldots\}$ where $\mathbf{c}_i = \mathbf{Ee}_i\,(\mathbf{m}_i)$. If $\mathbf{d}_i$ denotes the inverse of $\mathbf{e}_i$, then $\mathbf{Dd}_i(\mathbf{c}_i) = \mathbf{m}_i$ decrypts the ciphertext string $\{\mathbf{c}_1,\mathbf{c}_2,\mathbf{c}_3 \ldots\ldots\ldots\}$.

A stream cipher applies simple encryption transformations according to the key stream being used. The key stream could be generated at random, or by an algorithm that generates the key stream from an initial small key stream (called a *seed*), or from a seed and previous ciphertext symbols. Such an algorithm is called a *key stream generator.*

Example

(The Vernam stream cipher [8-2]): The Vernam stream cipher is defined on the alphabet $\mathbf{A} = \{0,1\}$. A binary message $m_1m_2 \ldots\ldots m_n$ is operated on by a binary key string $e_1e_2 \ldots\ldots e_n$ of the same length to produce a ciphertext string $c_1c_2 \ldots\ldots c_n$ where

$$c_i = m_i \oplus e_i,\ 1 \le i \le n.$$

If the key string is randomly chosen and never used again, the Vernam stream cipher is called a one-time system or a one-time pad.

A motivating factor for the Vernam stream cipher was its simplicity and ease of implementation. It is observed in the Vernam stream cipher that there are precisely two substitution ciphers on the set **A**. One is simply the identity map E_0 that sends 0 to 0 and 1 to 1; the other E_1 sends 0 to 1 and 1 to 0. When the keystream contains a 0, apply E_0 to the corresponding plaintext symbol; otherwise, apply E_1.

The one-time pad can be shown to be theoretically unbreakable. That is, if an attacker has a ciphertext string $\{\mathbf{c}_1,\mathbf{c}_2,\mathbf{c}_3 \ldots\ldots\ldots\}$ encrypted using a random key string that has been used only once, the attacker can do no better than guess at the plaintext being any binary string of length n (i.e., n-bit binary strings are equally likely as plaintext). It has been proven that to realize an unbreakable system requires a random key of the same length as the message. This reduces the practicality of the system in all but a few specialized situations.

The key space: The size of the key space is the number of encryption/decryption key pairs that are available in the cipher system. A key is typically a compact way to specify the encryption transformation (from the set of all encryption transformations) to be used. For example, a permutation cipher of block length n has $n! = n \cdot (n-1) \cdots 2 \cdot 1$ encryption functions from which to select. Each can be simply described by a permutation which is called the *key.* It is a great temptation to relate the security of the encryption scheme to the size of the key space. The following fact is very important to remember: a necessary, but usually not sufficient, condition for an encryption scheme to be secure is that the key space be large enough to preclude exhaustive search.

EXAMPLE

(The key space): The example on simple substitution cipher (in Section 8.2.1) has a key space of size $26! \approx 4 \times 10^{26}$. The other example on polyalphabetic substitution cipher (also in Section 8.2.1) has a key space of size $(26!)^2 \approx 16 \times 10^{52}$. Exhaustive search of either key space is completely infeasible; however, both ciphers are relatively weak and provide little security. In Section 8.3, we will discuss some strong ciphers with large key spaces.

8.3 DATA ENCRYPTION STANDARD

No block cipher is ideally suited for all applications, even one offering a high level of security. This is a result of inevitable tradeoffs required in practical applications, including those arising from, for example, real-time requirements and memory limitations (e.g., code size, data size, cache memory), other constraints imposed by implementation methods (e.g., hardware, software, chip cards), and differing tolerances of applications to properties of various modes of operation. In addition, efficiency must typically be traded off against security. Thus it is beneficial to have a number of candidate ciphers from which to draw. Of the many block ciphers currently available, the data encryption standard (DES) is chosen as the standard scrambler in many MPEG-enabled digital video networks.

This section explains the various steps involved in DES-encryption [8-21], illustrating each step by means of examples. Since the creation of DES, many other algorithms (recipes for changing data) have emerged that are based on design principles similar to DES. Once you understand the basic transformations that take place in DES, you will find it easy to follow the steps involved in these more recent algorithms.

Iterated block cipher. An iterated block cipher is a block cipher involving the sequential repetition of an internal function called a *round function*. Parameters include the number of rounds r, the block size n in bits, and the input key **e** with size k in bits. The r sub-keys $\mathbf{e}_i$ (round keys) are derived from **e**. For invertibility (allowing unique decryption), for each value $\mathbf{e}_i$ the round function is a bijection on the round input.

Feistel cipher. A Feistel cipher is an iterated cipher mapping a 2t-bit plaintext (L_0,R_0), for t-bit blocks L_0 and R_0, to a ciphertext L_r,R_r, through an r-round process where $r \geq 1$. For $1 \leq i \leq r$, round i maps $\left(L_{i-1}, R_{i-1}\right)^{e_i} \rightarrow \left(L_i, R_i\right)$ as follows: $L_i = R_{i-1}$, $R_i = L_{i-1} \oplus f\left(R_{i-1}, e_i\right)$ where each sub-key e_i is derived from the cipher key e.

Typically in a Feistel cipher, $r \geq 3$ and often is even. The Feistel structure specifically orders the ciphertext output as (L_r, R_r) rather than (R_r, L_r); the blocks are exchanged from their usual order after the last round. Decryption is thereby achieved using the same r-round process but with sub-keys used in reverse order, e_i through e_1; for example, the last round is undone by simply repeating it. The f function of the Feistel cipher may be a product cipher, though f itself need not be inver-tible to allow inversion of the Feistel cipher. Note the round function $R_i = L_{i-1} \oplus f\left(R_{i-1}, e_i\right)$ may also be rewritten to eliminate $L_i : R_i = R_{i-2} \oplus f\left(R_{i-1}, e_i\right)$. In this case, the final ciphertext output is $\left(R_r, R_{r-1}\right)$, with input labeled $\left(R_{-1}, R_0\right)$.

The data encryption standard is the most well-known symmetric-key Feistel block cipher in the world. It sets a precedent in the middle of 1970s as the first commercial-grade modern encryption algorithm.The DES encrypts plaintext blocks of n = 64 bits, and decrypts ciphertext blocks of n = 64 bits, using a 64-bit key with an effective key length k = 56 bits. It takes a 64-bit block of input data and outputs a 64-bit block of encrypted data. Since it always operates on blocks of equal size and it uses both permutations and substitutions in the algorithm, DES is both a block cipher and a product cipher.

The core DES algorithm has 16 round of iterations, meaning the algorithm is repeated 16 times to produce the encrypted data. It has been found that the number of rounds is exponentially proportional to the amount of time required to find a key using a brute-force attack. So as the number of rounds increases, the security of the algorithm increases exponentially.

8.3.1 Key Scheduling

A key consists of 64 binary digits (0s or 1s) of which 56 bits are randomly generated and used directly by the DES algorithm. For each byte of the 64 binary digits, the least significant (rightmost) bit is a parity bit for error detection. Such a bit should be set so that there are always an odd number of 1s in every byte. Sometimes keys are generated in an encrypted form. A random 64-bit number is generated and defined to be the cipher formed by the encryption of a key using a key-encrypting key. In this case the parity bits of the encrypted key cannot be set until after the key is decrypted. For the DES algorithm, these parity bits are ignored and only the seven most significant bits of each byte are used (resulting in a key length of 56 bits).

The first step is to pass the 64-bit key through a permutation called *Permuted Choice 1* in Table 8.1. Note that in all subsequent descriptions of bit numbers, 1 is the leftmost bit in the number, and n is the rightmost bit. The parity bits 8, 16, 24, 32, 40, 48, 56, and 64 of the original key are not included as input entries in the table. Thus, these unused parity bits are discarded when the final 56-bit key is created.

Example

(Permutation choice 1): We can use Table 8.1 to figure out how bit 23 of the original 64-bit key transforms to a bit in the new 56-bit key. Find the number 23 in the table,

TABLE 8.1
Permuted Choice 1

Permutation Outputs	0	1	2	3	4	5	6
1	57	49	41	33	25	17	9
8	1	58	50	42	34	26	18
15	10	2	59	51	48	35	27
22	19	11	3	60	52	44	36
29	63	55	47	39	31	23	15
36	7	62	54	46	38	30	22
43	14	6	61	53	45	37	29
50	21	13	5	28	20	12	4

and notice that it belongs to the column labeled 5 and the row labeled 29. Add up the value of the row and column to find the new position of the bit within the key. For bit 23, 29 + 5 = 34, so bit 23 becomes bit 34 of the new 56-bit key.

Now that we have the 56-bit key, the next step is to use this key to generate 16 48-bit sub-keys, called *K[1]-K[16]*, which are used in the 16 iterations of DES for encryption and decryption. The procedure for generating the sub-keys, known as *key scheduling*, is quite simple:

Step 1: Set the initial iteration number i = 1.
Step 2: Split the current 56-bit key, K, up into two 28-bit blocks, L (the left-hand half) and R (the right-hand half).
Step 3: Rotate L left by the number of bits specified in the sub-key rotation table, and rotate R left by the same number of bits as well.
Step 4: Merge L and R together to get the new K.
Step 5: Apply Permuted Choice 2 to K to get the final K[i], where i is the iteration number.
Step 6: If i < 16, set i = i + 1 and go to Step 2; Otherwise, output sub-keys K[1]-K[16].

The tables involved in these operations are Table 8.2 and Table 8.3.

TABLE 8.2
Sub-key Rotation

Iteration Number	1	2	3	4	5	6	7	8	9	10	11	12	13	14	15	16
Number of bits to rotate	1	1	2	2	2	2	2	2	1	2	2	2	2	2	2	1

TABLE 8.3
Permuted Choice 2

Permutation Output	0	1	2	3	4	5
1	14	17	11	24	1	5
7	3	28	15	6	21	10
13	23	19	12	4	26	8
19	16	7	27	20	13	2
25	41	52	31	37	47	55
31	30	40	51	45	33	48
37	44	49	39	56	34	53
43	46	42	50	36	29	32

8.3.2 Input Data Preparation

Once the key scheduling has been performed, the next step is to prepare the input data for the actual encryption. The first step is to pass a 64-bit input data block through a permutation called the *initial permutation* (IP) in Table 8.4. Table 8.4 also has an inverse, called the *inverse initial permutation*, or IP^{-1}, given in Table 8.5. Sometimes IP^{-1} is also called the *final permutation.*

Table 8.4 and Table 8.5 are used just like Table 8.1 and Table 8.3 were for the key scheduling.

Example

(Use of permutation Table 8.4 and Table 8.5): Let's examine how bit 60 is transformed under IP. In the table, bit 60 is located at the intersection of the column labeled 0 and the row labeled 9. So this bit becomes bit 9 of the 64-bit block after the permutation. Now let's apply IP^{-1}. In IP^{-1}, bit 9 is located at the intersection of the column labeled 3 and the row labeled 57. So this bit becomes bit 60 after the permutation. And this

TABLE 8.4
Initial Permutation

Permutation Output	0	1	2	3	4	5	6	7
1	58	50	42	34	26	18	10	2
9	60	52	44	36	28	20	12	4
17	62	54	46	38	30	22	14	6
25	64	56	48	40	32	24	16	8
33	57	49	41	33	25	17	9	1
41	59	51	43	35	27	19	11	3
49	61	53	45	37	29	21	13	5
57	63	55	47	39	31	23	15	7

TABLE 8.5
Inverse Initial Permutation

Permutation Output	0	1	2	3	4	5	6	7
1	40	8	48	16	56	24	64	32
9	39	7	47	15	55	23	63	31
17	38	6	46	14	54	22	62	30
25	37	5	45	13	53	21	61	29
33	36	4	44	12	52	20	60	28
41	35	3	43	11	51	19	59	27
49	34	2	42	10	50	18	58	26
57	33	1	41	9	49	17	57	25

is the bit position that we started with before the permutation IP. So IP^{-1} really is the inverse of IP. It does the exact opposite of IP. If one runs a block of input data through IP and then pass the resulting block through IP^{-1}, one will end up with the original block.

8.3.3 The Core DES Function

Once the key scheduling and input data preparation have been completed, the core DES algorithm performs the actual encryption or decryption. The 64-bit block of input data is first split into two halves, L and R; L is the leftmost 32 bits, and R is the rightmost 32 bits. The following process is repeated 16 times, making up the 16 iterations of standard DES. Denote the 16 sets of halves L[0]-L[15] and R[0]-R[15].

Step 1: Set the initial iteration number $i = 1$.
Step 2: R[i – 1] is fed into the E-Bit Selection Table (Table 8.6), which is like a permutation, except that some of the bits are used more than once. This expands the number R[i – 1] from 32 to 48 bits to prepare for Step 3.
Step 3: The 48-bit R[i – 1] is XORed with K[i] and stored in a temporary buffer so that R[i – 1] is not modified.

TABLE 8.6
E-Bit Selection Table

Permutation Output	0	1	2	3	4	5
1	32	1	2	3	4	5
7	4	5	6	7	8	9
13	8	9	10	11	12	13
19	12	13	14	15	16	17
25	16	17	18	19	20	21
31	20	21	22	23	24	25
37	24	25	26	27	28	29
43	28	29	30	31	32	1

Step 4: The result from Step 3 is now split into eight segments of 6 bits each. The leftmost 6 bits are B[1], and the rightmost 6 bits are B[8]. These blocks form the index into the S-boxes, which are used in the next step. The Substitution boxes, known as S-boxes, are a set of eight two-dimensional arrays, each with 4 rows and 16 columns. The numbers in the boxes are always 4 bits in length, so their values range from 0 to 15. The S-boxes are numbered S[1]-S[8] (Table 8.7).

Step 5: Starting with B[1], the first and last bits of the 6-bit block are taken and used as an index into the row number of S[1], which can range from 0 to 3, and the middle four bits are used as an index into the column number, which can range from 0 to 15. The number from this position in the S-box is retrieved and stored away. This is repeated with B[2] and S[2], B[3] and S[3], and the others up to B[8] and S[8]. Eight 4-bit numbers are obtained at this point. When strung together one after the other in the order of retrieval, these numbers give a 32-bit result.

Step 6: The result from the previous stage is now passed into the P Permutation (Table 8.8).

Step 7: This number is now XORed with $L[i-1]$, and moved into $R[i]$. $R[i-1]$ is moved into $L[i]$.

Step 8: If $i < 16$, set $i = i + 1$ and go to Step 2; Otherwise output L[16] and R[16].

When L[16] and R[16] have been obtained, they are joined back together in the same fashion in which they were split apart (L[16] is the left-hand half, R[16] is the right-hand half), and the resultant 64-bit number is called the *pre-output.*

Figure 8.4 shows one round of the DES iteration.

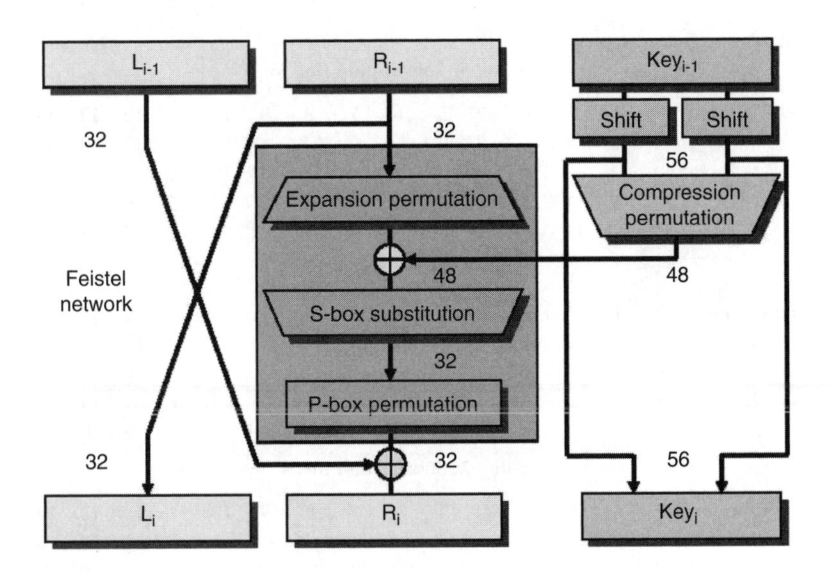

FIGURE 8.4 One round of DES.

TABLE 8.7
S-Boxes

S-Box 1: Substitution Box 1

Row/Column	0	1	2	3	4	5	6	7	8	9	10	11	12	13	14	15
0	14	4	13	1	2	15	11	8	3	10	6	12	5	9	0	7
1	0	15	7	4	14	2	13	1	10	6	12	11	9	5	3	8
2	4	1	14	8	13	6	2	11	15	12	9	7	3	10	5	0
3	15	12	8	2	4	9	1	7	5	11	3	14	10	0	6	13

S-Box 2: Substitution Box 2

Row/Column	0	1	2	3	4	5	6	7	8	9	10	11	12	13	14	15
0	15	1	8	14	6	11	3	4	9	7	2	13	12	0	5	10
1	3	13	4	7	15	2	8	14	12	0	1	10	6	9	11	5
2	0	14	7	11	10	4	13	1	5	8	12	6	9	3	2	15
3	13	8	10	1	3	15	4	2	11	6	7	12	0	5	14	9

S-Box 3: Substitution Box 3

Row/Column	0	1	2	3	4	5	6	7	8	9	10	11	12	13	14	15
0	10	0	9	14	6	3	15	5	1	13	12	7	11	4	2	8
1	13	7	0	9	3	4	6	10	2	8	5	14	12	11	15	1
2	13	6	4	9	8	15	3	0	11	1	2	12	5	10	14	7
3	1	10	13	0	6	9	8	7	4	15	14	3	11	5	2	12

S-Box 4: Substitution Box 4

Row/Column	0	1	2	3	4	5	6	7	8	9	10	11	12	13	14	15
0	7	13	14	3	0	6	9	10	1	2	8	5	11	12	4	15
1	13	8	11	5	6	15	0	3	4	7	2	12	1	10	14	9
2	10	6	9	0	12	11	7	13	15	1	3	14	5	2	8	4
3	3	15	0	6	10	1	13	8	9	4	5	11	12	7	2	14

S-Box 5: Substitution Box 5

Row/Column	0	1	2	3	4	5	6	7	8	9	10	11	12	13	14	15
0	2	12	4	1	7	10	11	6	8	5	3	15	13	0	14	9
1	14	11	2	12	4	7	13	1	5	0	15	10	3	9	8	6
2	4	2	1	11	10	13	7	8	15	9	12	5	6	3	0	14
3	11	8	12	7	1	14	2	13	6	15	0	9	10	4	5	3

S-Box 6: Substitution Box 6

Row/Column	0	1	2	3	4	5	6	7	8	9	10	11	12	13	14	15
0	12	1	10	15	9	2	6	8	0	13	3	4	14	7	5	11
1	10	15	4	2	7	12	9	5	6	1	13	14	0	11	3	8
2	9	14	15	5	2	8	12	3	7	0	4	10	1	13	11	6
3	4	3	2	12	9	5	15	10	11	14	1	7	6	0	8	13

S-Box 7: Substitution Box 7

Row/Column	0	1	2	3	4	5	6	7	8	9	10	11	12	13	14	15
0	4	11	2	14	15	0	8	13	3	12	9	7	5	10	6	1
1	13	0	11	7	4	9	1	10	14	3	5	12	2	15	8	6
2	1	4	11	13	12	3	7	14	10	15	6	8	0	5	9	2
3	6	11	13	8	1	4	10	7	9	5	0	15	14	2	3	12

(Continued)

TABLE 8.7
S-Boxes (*Continued*)

S-Box 8: Substitution Box 8

Row/Column	0	1	2	3	4	5	6	7	8	9	10	11	12	13	14	15
0	13	2	8	4	6	15	11	1	10	9	3	14	5	0	12	7
1	1	15	13	8	10	3	7	4	12	5	6	11	0	14	9	2
2	7	11	4	1	9	12	14	2	0	6	10	13	15	3	5	8
3	2	1	14	7	4	10	8	13	15	12	9	0	3	5	6	11

Example

(Use of the S-boxes): The purpose of this example is to clarify how the S-boxes work. Suppose we have the following 48-bit binary number:

011101000101110101000111101000011100101101011101

In order to pass this through Steps 3 and 4 of the core function as outlined above, the number is split up into eight 6-bit blocks, labeled B[1] to B[8] from left to right:

011101 000101 110101 000111 101000 011100 101101 011101

Now, eight numbers are extracted from the S-boxes—one from each box:

B[1] = S[1](01, 1110) = S[1][1][14] = 3 = 0011
B[2] = S[2](01, 0010) = S[2][1][2] = 4 = 0100
B[3] = S[3](11, 1010) = S[3][3][10] = 14 = 1110
B[4] = S[4](01, 0011) = S[4][1][3] = 5 = 0101
B[5] = S[5](10, 0100) = S[5][2][4] = 10 = 1010
B[6] = S[6](00, 1110) = S[6][0][14] = 5 = 0101
B[7] = S[7](11, 0110) = S[7][3][6] = 10 = 1010
B[8] = S[8](01, 1110) = S[8][1][14] = 9 = 1001

In each case of S[n][row][column], the first and last bits of the current B[n] are used as the rowindex, and the middle four bits as the column index.The results are now

TABLE 8.8
P-Permutation

Permutation Output	0	1	2	3
1	16	7	20	21
5	29	12	28	17
9	1	15	23	26
13	5	18	31	10
17	2	8	24	14
21	32	27	3	9
25	19	13	30	6
29	22	11	4	25

joined together to form a 32-bit number, which serves as the input to stage 5 of the core function (the P Permutation):

00110100111001011010010110101001

Preparation of Encrypted Data: The final step is to apply the permutation IP^{-1} to the pre-output. The result is the completely encrypted data.

Encryption and Decryption: The same algorithm can be used for either encryption or decryption. The method described above will encrypt a block of plaintext and return a block of encrypted data. In order to decrypt the ciphertext and get the original plaintext again, the procedure is simply repeated but the sub-keys are applied in reverse order, from K[16]-K[1]. That is, stage 2 of the core function as outlined above changes from $R[i-1]$ XOR $K[i]$ to $R[i-1]$ XOR $K[17-i]$. Other than that, decryption is performed exactly the same as encryption.

8.4 MODES OF OPERATION

A block cipher encrypts plaintext in fixed-size n-bit blocks (e.g., n = 64). For messages exceeding n bits, the simplest approach is to partition the message into n-bit blocks and encrypt each separately. This electronic-codebook (ECB) mode has disadvantages in most applications, motivating other methods of employing block ciphers (modes of operation) on larger messages. Four modes of operation are specified in FIPS PUB 81 [8-18].

ECB (Electronic code book) mode of operation: For a q-bit key **k**, n-bit plaintext blocks $\mathbf{m}_1, \mathbf{m}_2, \ldots\ldots, \mathbf{m}_L$ and their corresponding n-bit ciphertext blocks $\mathbf{c}_1, \mathbf{c}_2, \ldots\ldots, \mathbf{c}_L$ are related by:

- Encryption: $\mathbf{c}_j = \mathbf{Ek}(\mathbf{m}_j)$, $1 \leq j \leq L$.
- Decryption: $\mathbf{m}_j = \mathbf{Dk}(\mathbf{c}_j)$, $1 \leq j \leq L$.

The ECB mode of operation has the following important properties:

- Identical plaintext blocks (under the same key) result in identical ciphertext.
- Separate encryptions with different blocks are totally independent of each other. This means that if data is transmitted over a network or phone line, transmission errors will only affect the block containing the error. It also means, however, that the blocks can be rearranged, thus scrambling a file beyond recognition, and this action would go undetected. Furthermore, block ciphers do not hide data patterns—identical ciphertext blocks imply identical plaintext blocks. For this reason, the ECB mode is not recommended for messages longer than one block, or if keys are reused for more than a single one-block message.
- If one or more bit errors occur in a single ciphertext block, these errors only affect decipherment of that block. There is no error propagation.

Electronic code book mode is the regular DES algorithm, exactly as described above. Data is divided into 64-bit blocks and each block is encrypted one at a time; ECB is the fastest and easiest to implement, making it the most common mode of DES seen in commercial applications. However, ECB is the weakest of the various modes because no additional security measures are implemented besides the basic DES algorithm.

CBC (Cipher block chaining) mode of operation: For a q-bit key **k** and an n-bit initialization vector, denoted IV, n-bit plaintext blocks m1, m2, ……, mL and their corresponding n-bit ciphertext blocks c1,c2,……, cL are related by

- Encryption: cj = Ek(cj − 1 ⊕ mj), where $1 \leq j \leq L$ and c0 = IV.
- Decryption: mj = cj − 1 ⊕ Dk(cj), where $1 \leq j \leq L$ and c0 = IV.

The CBC mode of operation has the following important properties:

- Identical ciphertext blocks result when the same plaintext is enciphered under the same key and IV. Changing the IV, key, or first plaintext block results in different ciphertext. While the IV in the CBC mode need not be secret, its integrity should be protected, since malicious modification thereof allows an attacker to make predictable bit changes to the first plaintext block recovered. Using a secret IV is one method for preventing this. In this mode, each block of ECB encrypted ciphertext is XORed with the next plaintext block to be encrypted, thus making all the blocks dependent on all the previous blocks. This means that in order to find the plaintext of a particular block, you need to know the ciphertext, the key, and the ciphertext for the previous block. The first block to be encrypted has no previous ciphertext, so the plaintext is XORed with an IV.
- Data chaining causes ciphertext cj to depend on mj and all preceding plaintext blocks (the entire dependency on preceding blocks is, however, contained in the value of the previous ciphertext block). Consequently, rearranging the order of ciphertext blocks affects decryption. Proper decryption of a correct ciphertext block requires a correct preceding ciphertext block. So if data is transmitted over a network or phone line and there is a transmission error, the error will be carried forward to all subsequent blocks since each block is dependent upon the last. That is, a single bit error in ciphertext block cj affects decipherment of blocks cj and cj + 1 since mj depends on cj and cj − 1.
- The CBC mode is self-synchronizing in the sense that if an error (including loss of one or more entire blocks) occurs in block cj but not cj + 1, cj + 2 is correctly decrypted to mj + 2.
  - Although CBC mode decryption recovers from errors in ciphertext blocks, modifications to a plaintext block mj during encryption alter all subsequent ciphertext blocks. This impacts the usability of chaining modes for applications requiring random read/write access to encrypted data. The ECB mode is an alternative.

  - Although self-synchronizing in the sense of recovery from bit errors, recovery from "lost" bits causing errors in block boundaries (framing integrity errors) is not possible in the CBC or other modes.
- The CBC mode of operation is more secure than ECB because the extra XOR step adds one more layer to the encryption process.

While the CBC mode processes plaintext n bits at a time (using an n-bit block cipher), some applications require that m-bit plaintext units be encrypted and transmitted without delay, for some fixed m < n (e.g., m = 1 or m = 8 while n = 64). In this case, the cipher feedback (CFB) mode may be used, as specified below.

CFB (Cipher feedback) mode of operation: For a q-bit key **k** and two n-bit registers $\mathbf{R}_1$ and $\mathbf{R}_2$, m-bit plaintext blocks $\mathbf{m}_1, \mathbf{m}_2, \ldots\ldots, \mathbf{m}_L$ where $1 \leq m \leq n$ and their corresponding m-bit ciphertext blocks $\mathbf{c}_1, \mathbf{c}_2, \ldots\ldots, \mathbf{c}_L$ are related by:

- Encryption: The register $\mathbf{R}_1$ is initialized by the so-called the initialization vector IV, i.e., $\mathbf{R}_1$ (1) = IV. For $1 \leq j \leq L$:
  - $\mathbf{R}_2(\mathrm{j}) = \mathbf{Ek}(\mathbf{R}_1(\mathrm{j}))$
  - $\mathbf{c}_j = \mathbf{m}_j \oplus$ (left most m bits of $\mathbf{R}_2(\mathrm{j})$)
  - $\mathbf{R}_1(\mathrm{j}+1) = 2^m \cdot \mathbf{R}_1(\mathrm{j}) + \mathbf{c}_{j,}$ for $j < L$
- Decryption: $\mathbf{R}_1$ (1) = IV. For $1 \leq j \leq L$:
  - $\mathbf{R}_2(\mathrm{j}) = \mathbf{Ek}(\mathbf{R}_1(\mathrm{j}))$
  - $\mathbf{m}_j = \mathbf{c}_j \oplus$ (left most m bits of $\mathbf{R}_2(\mathrm{j})$)
  - $\mathbf{R}_1(\mathrm{j}+1) = 2^m \cdot \mathbf{R}_1(\mathrm{j}) + \mathbf{c}_{j,}$ for $j < L$

Properties of the CFB mode of operation:

- As per the CBC mode, changing the IV results in the same plaintext input being enciphered to a different output. The IV need not be secret (although an unpredictable IV may be desired in some applications).
- Similar to the CBC mode, the chaining mechanism causes ciphertext block $\mathbf{c}_j$ to depend on both $\mathbf{m}_j$ and preceding plaintext blocks; consequently, reordering ciphertext blocks affects decryption. Proper decryption of a correct ciphertext block requires a number of preceding ciphertext blocks to be correct (so that the shift register contains the proper value).
- One or more bit errors in any single m-bit ciphertext block $\mathbf{c}_j$ affect the decipherment of that and the next several ciphertext blocks (i.e., until n bits of ciphertext are processed, after which the error block $\mathbf{c}_j$ has shifted entirely out of the shift register) in error. Thus an attacker may cause predictable bit changes in $\mathbf{m}_j$ by altering corresponding bits of $\mathbf{c}_j$.
- Similar to the CBC mode, the CFB mode is self-synchronizing, but requires a number of ciphertext blocks to recover.
- Compared with the CBC mode, the throughput of the CFB mode is decreased by a factor of n/m in that each execution of **Ek** yields only m bits of ciphertext output.
- The encryption function **Ek** is used for both CFB encryption and decryption. Therefore, the block cipher **Ek** must be a strong one-way function.

For the DES algorithm in the CFB mode, blocks of plaintext that are less than 64 bits long can be encrypted. Normally, special processing has to be used to handle files whose size is not a perfect multiple of 8 bytes, but this mode removes that necessity. The plaintext itself is not actually passed through the DES algorithm, but merely XORed with an output block from it, in the following manner. A 64-bit block called the *shift register* is used as the input plaintext to DES. This is initially set to some arbitrary value, and encrypted with the DES algorithm. The ciphertext is then passed through an extra component called the *M-box*, which simply selects the leftmost m bits of the ciphertext, where m is the number of bits in the block we wish to encrypt. This value is XORed with the real plaintext, and the output of that is the final ciphertext. Finally, the ciphertext is fed back into the shift register, and used as the plaintext seed for the next block to be encrypted. As with CBC mode, an error in one block affects all subsequent blocks during data transmission. This mode of operation is similar to CBC and is very secure, but it is slower than ECB due to the added complexity.

The output feedback (OFB) mode of operation may be used for applications in which all error propagation must be avoided. It is similar to CFB, and allows encryption of various block sizes (characters), but differs in that the output of the encryption block function **Ek** (rather than the ciphertext) serves as the feedback. Two versions of OFB are discussed next.

OFB (Output feedback) mode of operation: For a q-bit key **k** and two n-bit registers R1 and R2, m-bit plaintext blocks m1, m2, ……, mL where $1 \le m \le n$ and their corresponding m-bit ciphertext blocks c1,c2,……, cL are related by:

- Encryption: The register R1 is initialized by the so-called the initialization vector, IV; i.e., R1 (1) = IV. For $1 \le j \le L$:
  - R2 (j) = Ek(R1 (j))
  - cj = mj ⊕ (leftmost m bits of R2 (j)):
    - (per ISO/IEC 10116 [8-34]) R1 (j + 1) = R2 (j), for j < L.
    - (per FIPS 81 [8-18]) R1 (j + 1) = $2^m \cdot$ R1 (j) + (leftmost m bits of R2 (j)), for j < L.
- Decryption: R1 (1) = IV. For $1 \le j \le L$:
  - R2 (j) = Ek(R1 (j))
  - mj = cj ⊕ (leftmost m bits of R2 (j)) :
    - (per ISO/IEC 10116 [8-34]) R1 (j + 1) = R2 (j), for j < L.
    - (per FIPS 81 [8-18]) R1 (j + 1) = $2^m \cdot$ R1 (j) + (leftmost m bits of R2 (j)), for j < L.

Properties of the OFB mode of operation:

- As per CBC and CFB modes, changing the IV results in the same plaintext being enciphered to a different output.
- The key stream is plaintext-independent. Thus, the IV, which need not be secret, must be changed if an OFB key K is reused. Otherwise an identical key stream results, and by XORing corresponding ciphertexts an attacker

may reduce cryptanalysis to that of a running-key cipher with one plaintext as the running key.

- One or more bit errors in any ciphertext character cj affects the decipherment of only that character, in the precise bit position(s) cj is in error, causing the corresponding recovered plaintext bit(s) to be complemented.
- The OFB mode recovers from ciphertext bit errors, but cannot self-synchronize after loss of ciphertext bits, which destroys alignment of the decrypting key stream (in which case explicit re-synchronization is required).
- For $r < n$, throughput is decreased as per the CFB mode. However, in all cases, since the key stream is independent of plaintext or ciphertext, it may be precomputed (given the key and IV).
- The encryption function Ek is used for both OFB encryption and decryption. Therefore, the block cipher Ek must be a strong one-way function.

For the DES algorithm, this is similar to CFB mode, except that the ciphertext output of DES is fed back into the shift register, rather than the actual final ciphertext. The shift register is set to an arbitrary initial value, and passed through the DES algorithm. The output from DES is passed through the M-box and then fed back into the shift register to prepare for the next block. This value is then XORed with the real plaintext (which may be less than 64 bits in length, like CFB mode), and the result is the final ciphertext. Note that unlike CFB and CBC, a transmission error in one block will not affect subsequent blocks because once the recipient has the initial shift register value, it will continue to generate new shift register plaintext inputs without any further data input. However, this mode of operation is less secure than CFB mode because only the real ciphertext and DES ciphertext output is needed to find the plaintext of the most recent block. Knowledge of the key is not required.

EXAMPLE

(Modes for stream ciphers): It is clear that the OFB mode with full feedback (per ISO 10116) employs a block cipher as a key stream generator for a stream cipher. Similarly, the CFB mode encrypts a character stream using the block cipher as a (plaintext-dependent) key stream generator. The CBC mode may also be considered a stream cipher with n-bit blocks playing the role of very large characters. Thus, modes of operation allow one to define stream ciphers from block ciphers. Figure 8.5 shows an example of block cipher and stream cipher constructed from different modes of operation.

8.5 CASCADE CIPHER AND MULTIPLE ENCRYPTION

The size of the key defines an upper bound on the security of a block cipher. If a block cipher is inadequate on security for its applications due to key length, encryption of the same message block more than once may increase security. Many such techniques for multiple encryption of n-bit messages can be defined. Once defined, they may be extended to messages exceeding one block by using standard modes of operation, with **E** denoting multiple rather than single encryption.

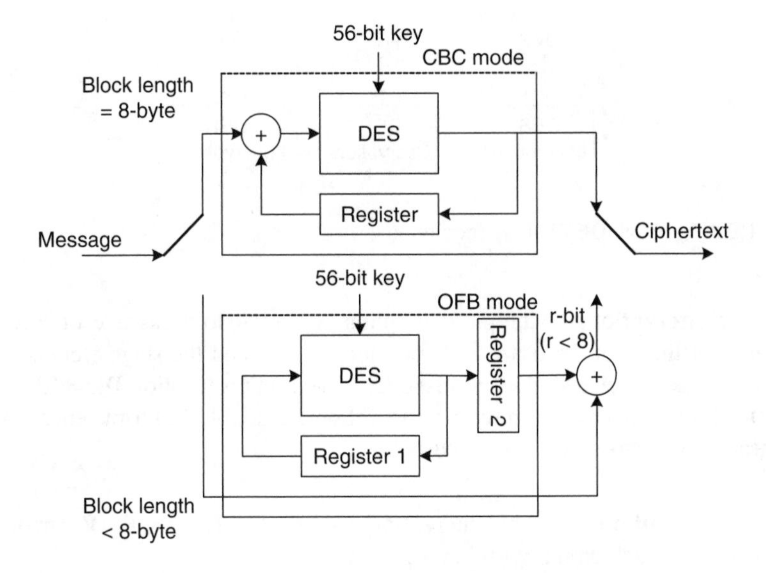

FIGURE 8.5 Block and stream ciphers constructed from DES CBC and OFB modes.

Cascade cipher: A cascade cipher is the concatenation of $v \geq 2$ block ciphers (called *stages*), each with independent keys. Plaintext is input to first stage; the output of stage i is input to stage $i + 1$; and the output of stage v is the cascade's ciphertext output. A cascade of v (independently keyed) ciphers is at least as strong as the strongest component cipher.

In the simplest case, all stages in a cascade cipher have k-bit keys, and the stage inputs and outputs are all n-bit quantities. The stage ciphers may differ (general cascade of ciphers), or may all be identical (cascade of identical ciphers).

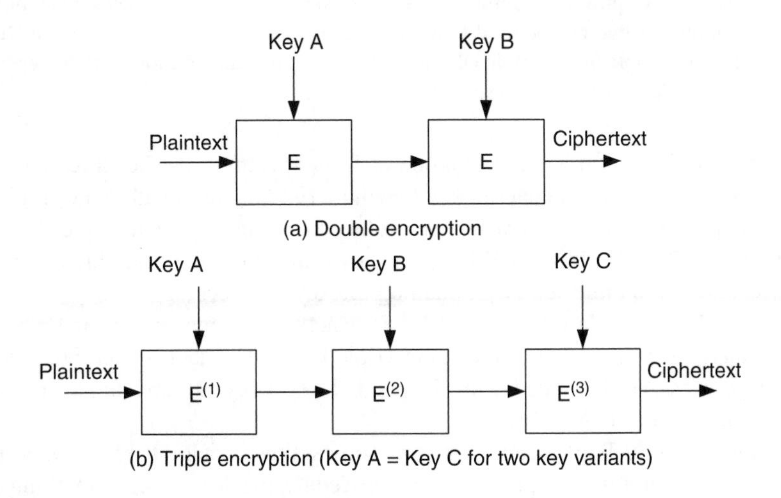

FIGURE 8.6 Two cases of multiple encryption.

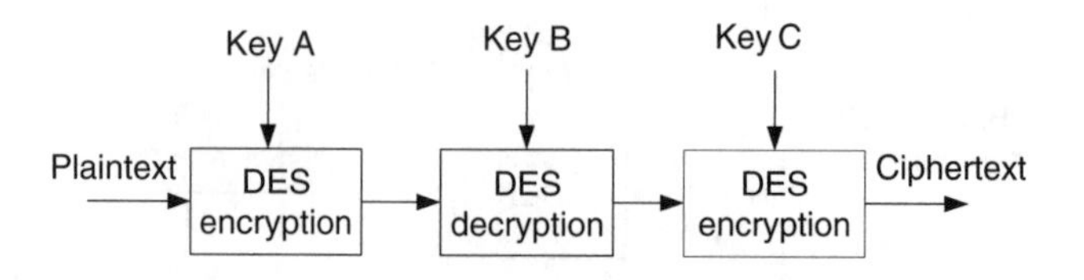

FIGURE 8.7 Triple DES E-D-E encryption.

Multiple encryption: Multiple encryption is similar to a cascade of v identical ciphers, but the stage keys need not be independent, and the stage ciphers may be either a block cipher **Ek** or its corresponding decryption function $\mathbf{Dk} = \mathbf{Ek}^{-1}$.

Two important cases of multiple encryption are double and triple encryption, as illustrated in Figure 8.6 and defined below.

Double encryption: Double encryption is defined as $\mathbf{E(m)} = \mathbf{E}_{k2}(\mathbf{E}_{k1}(\mathbf{m}))$, where $\mathbf{E}_{kj}$ denotes a block cipher with key **kj**.

Triple encryption: Triple encryption is defined as $\mathrm{E(m)} = \mathrm{E}_{k3}^{(3)}\left(\mathrm{E}_{k2}^{(2)}\left(\mathrm{E}_{k1}^{(1)}(\mathrm{m})\right)\right)$, where $\mathrm{E}_{kj}^{(j)}$ denotes a block cipher either E_{kj} or $\mathrm{D}_{kj} = \mathrm{E}_{kj}^{-1}$. The case = is called *E-D-E triple-encryption* and the subcase k3 = k1 is often referred to as *two-key triple encryption.*

Example

(Triple DES [8-32]): Triple DES is a minor variation of DES. It is three times slower than regular DES but can be billions of times more secure if used properly. Triple DES was an answer to many of the shortcomings of DES. Since it is based on the DES algorithm, it is very easy to modify existing software to use Triple DES. It also has the advantage of proven reliability and a longer key length that eliminates many of the shortcut attacks that can be used to reduce the amount of time it takes to break DES. For the foreseeable future Triple DES is an excellent and reliable choice for the security needs of highly sensitive information.

Triple DES is simply another mode of DES operation. It takes three 64-bit (56 effective bits) keys, for an overall key length of 192 bits. In Stealth, you simply type in the entire 192-bit (24 character) key rather than entering each of the three keys individually. The Triple DES E-D-E, as shown in Figure 8.7, then breaks the user provided key into three sub-keys, padding the keys if necessary so they are each 64 bits long. The procedure for encryption is exactly the same as regular DES, but it is repeated three times (hence the name Triple DES). The data is encrypted with the first key (Key A), decrypted with the second key (Key B), and finally encrypted again with the third key (Key C).

Consequently, Triple DES runs three times slower than standard DES, but is much more secure if used properly. The procedure for decrypting something is the same as the procedure for encryption, except it is executed in reverse. Like DES, data is encrypted and decrypted in 64-bit chunks. Unfortunately, there are some weak keys that one should be aware of: if all three keys, the first and second keys, or the second and third keys are the same, then the encryption procedure is essentially the same as standard DES. This situation is to be avoided because it is the same as using a really slow version of regular DES.

Note that although the input key for DES is 64 bits long, the actual key used by DES is only 56 bits in length. The least-significant (rightmost) bit in each byte is a parity bit, and should be set so that there are always an odd number of 1s in every byte. These parity bits are ignored, so only the seven most significant bits of each byte are used, resulting in a key length of 56 bits. This means that the effective key strength for Triple DES is actually 168 bits because each of the three keys contains 8 parity bits that are not used during the encryption process.

Multiple-encryption modes of operation: In contrast to the single modes of operation in Section 8.4, multiple modes are variants of multiple encryption constructed by concatenating selected single modes. For example, the combination of three single-mode CBC operations provides triple-CBC while the combination of three single-ECB operations generates triple-ECB. With replicated hardware, multiple modes such as triple-CBC and triple-ECB may be pipelined allowing computational performance comparable to single encryption.

Security of multiple encryption modes: Many multiple modes of operation are weaker than the corresponding multiple-ECB mode (i.e., multiple encryption operating as a black box with only outer feedbacks), and in some cases multiple modes (e.g., ECB-CBC-CBC) are not significantly stronger than single encryption.

Counterintuitively, it is possible to devise examples whereby cascading of ciphers actually reduces security. However, under a wide variety of attack models and meaningful definitions of "breaking," a cascade of *v* (independently keyed) ciphers, in general, is at least as strong as the strongest component cipher.

EXAMPLE

(Triple DES ECB): This variant of Triple DES works exactly the same way as the ECB mode of DES. Triple ECB is the type of encryption used in video broadcasting applications. This is the most commonly used mode of operation.

EXAMPLE

(Triple DES CBC): This method is very similar to the standard DES CBC mode. As with Triple ECB, the effective key length is 168 bits and keys are used in the same manner, as described above, but the chaining features of CBC mode are also employed. The first 64-bit key acts as the initialization vector to DES. Triple ECB is then executed for a single 64-bit block of plaintext. The resulting ciphertext is then XORed with the next plaintext block to be encrypted, and the procedure is repeated. The CBC mode adds an extra layer of security to Triple DES and is therefore more secure than Triple DES ECB, although it is not used as widely as Triple DES ECB.

8.6 PUBLIC-KEY CIPHERS

In symmetric-key ciphers the encryption algorithm can be revealed since the security of the cipher depends on a safeguarded key. For most of cases, the same key is used for both encryption and decryption. Public-key ciphers, also referred to as *asymmetric-key ciphers*, utilize two different keys, one for encryption and the other for decryption. In public-key ciphers, not only the encryption algorithm, but also the encryption key can be publicly revealed without compromising the security of the cipher. Only the decryption key is kept secret. Figure 8.8 shows two-party communication using public-key cipher.

Public-key encryption method: The encryption algorithm **Ee** and the decryption algorithm **Dd** are invertible functions on the plaintext **m**, or the ciphertext **c**, defined by the keys **e** for the encryption **c** = **Ee**(**m**) and **d** for the decryption **m** = **Dd**(**c**). For a given **e** (or **d**), **Ee** (or **Dd**) is easy to compute. More important, it is computationally "infeasible" to determine **d** knowing only **e**; i.e., for each e, the computation of **Dd** from **Ee** is computationally "infeasible."

Public key ciphers are based on the concept of trapdoor one-way function, as described in 8.1.1. Public-key cipher would enable secure communication between subscribers who have never met or communicated before.

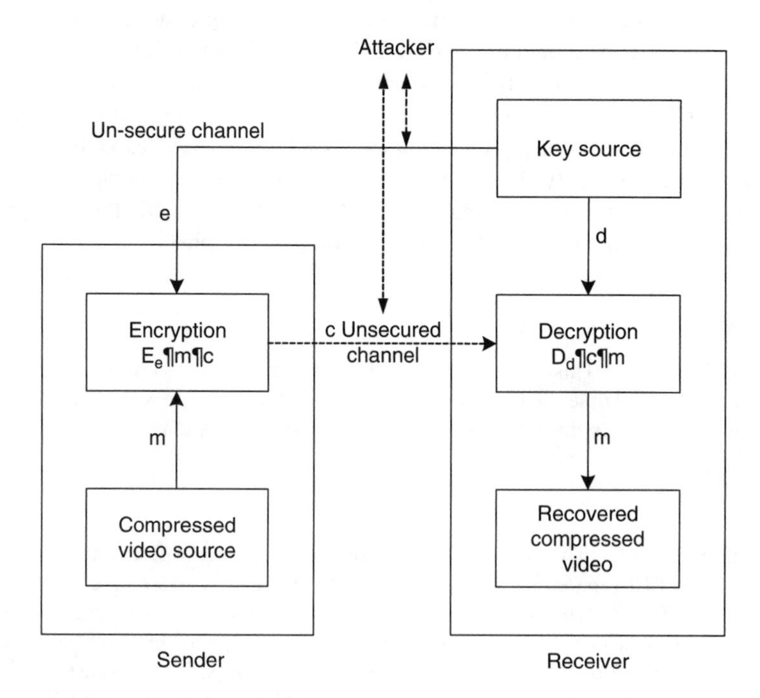

FIGURE 8.8 Two-party communication using public-key cipher.

EXAMPLE

(Secure communication using a public-key cipher): Assume that a public directory, much like a telephone directory, is envisioned, which contains the encryption keys of all the subscribers for a service. Subscriber A can send a message, **m**, to subscriber B by looking up B's public (encryption) key $\mathbf{e}_b$ in the directory and applying the encryption algorithm, $\mathbf{Ee}_b$, to obtain the ciphertext $\mathbf{c} = \mathbf{Ee}_b(\mathrm{m})$, which is being transmitted on a public channel. Subscriber B is the only party who can decrypt **c** by applying the secret key $\mathbf{d}_b$ to the decryption algorithm, $\mathbf{Dd}_b$, to obtain $\mathbf{m} = \mathbf{Dd}_b(\mathbf{c})$.

8.6.1 RSA PUBLIC-KEY ENCRYPTION

The RSA cipher, named after its inventors, R. Rivest, A. Shamir, and L. Adleman, is the most widely used public-key cipher. It may be used to provide both secrecy and digital signatures and its security is based on the intractability of the integer factorization problem [8-29][8-30][8-35]. This subsection describes the RSA encryption scheme, its security, and some implementation issues; the RSA signature scheme will be covered in Subsection 8.6.1.

RSA public-key algorithm: In the RSA scheme, messages m are first represented as integers in the range $(0, n - 1)$. Each entity chooses a value of n and another pair of positive integers e and d, in a manner to be described below. The public encryption key consists of the number pair (n, e), whereas the decryption key consists of the number pair (n, d), of which d is kept secret. Encryption of a message m and decryption of a ciphertext c are defined as follows:

- Encryption: $c = \mathbf{Ee}(m) = (m)^{\mathrm{e}} \pmod{n}$
- Decryption: $m = \mathbf{Dd}(c) = (c)^{\mathrm{d}} \pmod{n}$

These are each easy to compute and the results of each operation are integers in the range $(0, n - 1)$. The integers e and d in the RSA scheme are called the *encryption exponent* and the *decryption exponent*, respectively, while n is called the modulus. In the RSA scheme, the entity creates an RSA public key and a corresponding private key as follows:

- Generate two large random (and distinct) primes p and q, each roughly the same size.
- Calculate $n = pq$ where n is made public, p and q are kept hidden (trapdoors).
- Compute Euler's totient function $\phi(n) = (p - 1)(q - 1)$.
- Select a random integer e, $1 < \mathrm{e} < \phi(n)$, such that $\gcd(\mathrm{e}, \phi(n)) = 1$, where gcd means "great common divisor."
- The integer d is computed, $1 < \mathrm{d} < \phi(n)$, such that $\mathrm{ed} \equiv 1 \pmod{\phi(n)}$.
- The entity's public key is (n, e). The entity's private key is d.

Note that since ed ≡ 1 (mod $\phi(n)$), there exists an integer v such that ed = 1 + $v\phi(n)$. Now, if gcd(m, p) = 1 then by Fermat's theorem [8-8],

$$m^{p-1} \equiv 1 \pmod{p}$$

Raising both sides of the congruence to the power $v(q - 1)$ yields the following congruence,

$$m \equiv m^{v\phi(n)+1} \pmod{p}, \text{ i.e., } m \equiv m^{ed} \pmod{p}$$

On the other hand, if gcd(m, p) = p, then this congruence is again valid since each side is congruent to 0 module p. Hence, in both cases

$$m \equiv m^{ed} \pmod{p}$$

By the same argument,

$$m \equiv m^{ed} \pmod{q}$$

Finally, since p and q are distinct primes, it follows that

$$m \equiv m^{ed} \pmod{n}$$

Therefore,

$$\mathbf{Ee}(\mathbf{Dd}(m)) = \mathbf{Dd}(\mathbf{Ee}(m)) = m$$

and decryption works correctly.

Example

(RSA encryption with artificially small parameters): Let primes p = 47, q = 59. Therefore, $n = pq$ = 1773 and computes $\phi(n) = (p - 1)(q - 1) = 2668$. The encryption exponent e, 1 < e < $\phi(n)$, is chosen such that gcd(e, $\phi(n)$) = 1. For example, choose e = 17. Next, the value of d is computed as follows:

$$\text{ed} \equiv 1 \pmod{f(n)}$$

$$17\text{d} \equiv 1 \pmod{2668}$$

Therefore, d = 157. Consider an example of plaintext transcoding video. By replacing each letter with a two-digit number in the range (01, 26) corresponding to its position in the alphabet, i.e.,

A	B	C	D	E	F	G	......	T	U	V	W	X	Y	Z
01	02	03	04	05	06	07	......	20	21	22	23	24	25	26

and encoding a blank as 00, the plaintext message can be written as

2018 0114 1903 1504 0914 0700 2209 0405 1500

Each message needs to be expressed as an integer in the range (0, n – 1); therefore, for this example, encryption can be performed on blocks of four digits at a time since this is the maximum number of digits that will always yield a number less than $n - 1 = 2772$. The first four digits (2018) of the plaintext are encrypted as follows:

$$c = (m)^e \pmod n = (2018)^{17} \bmod 2773 = 0985.$$

Continuing this process for the remaining plaintext digits, we get

$$c = 0985\ 1684\ 1628\ 2444\ 2503\ 0761\ 1504\ 0702\ 2417$$

The plaintext is returned by applying the decryption key, as follows:

$$m = (c)^{157} \pmod{2773}.$$

Property and security of RSA:

- The Euclidean algorithm [8-11] for calculating the gcd of $\phi(n)$ and e can be used to compute d (see Exercise 8.4).
- Given an encryption key (n, e), one way that an attack might attempt to break the cipher is to factor n into p and q, compute $\phi(n) = (p - 1)(q - 1)$, and compute d by using the Euclidean algorithm. This is all straightforward except for the factoring of n.
- The problem of computing the RSA decryption exponent d from the public key (n, e) and the problem of factoring n are computationally equivalent. When generating RSA keys, it is imperative that the primes p and q be selected in such a way that factoring $n = pq$ is computationally infeasible.

RSA encryption in practice: There are numerous ways of speeding up RSA encryption and decryption in software and hardware implementations. Some of these techniques, for example, are covered in [8-8], including fast modular multiplication, fast modular exponentiation, and the use of the Chinese remainder theorem for faster decryption. Even with these improvements, RSA encryption/decryption is substantially slower than the commonly used symmetric-key encryption algorithms, such as DES. In practice, RSA encryption is most commonly used for the protecting keys of symmetric-key algorithms and for the encryption of small data items.

8.6.2 Diffie-Hellman Key Agreement

Diffie-Hellman (DH) public key agreement algorithm provides a method for multiple entities to compute a shared secret that is used in the encryption/decryption key generation. The DH protocol provides the system with a cryptographic property known as *perfect forward secrecy* [8-26][8-27][8-31][8-38].

Diffie-Hellman Key Exchange algorithm: Two entities A and B each send the other one message over an open channel. The DH algorithm generates a shared secret key K known to both parties A and B as follows:

- One-time setup. An appropriate prime p and generator g such that $0 < g \leq p - 2$ are selected and published.
- Algorithm steps. Performing the following steps each time a shared key is required:

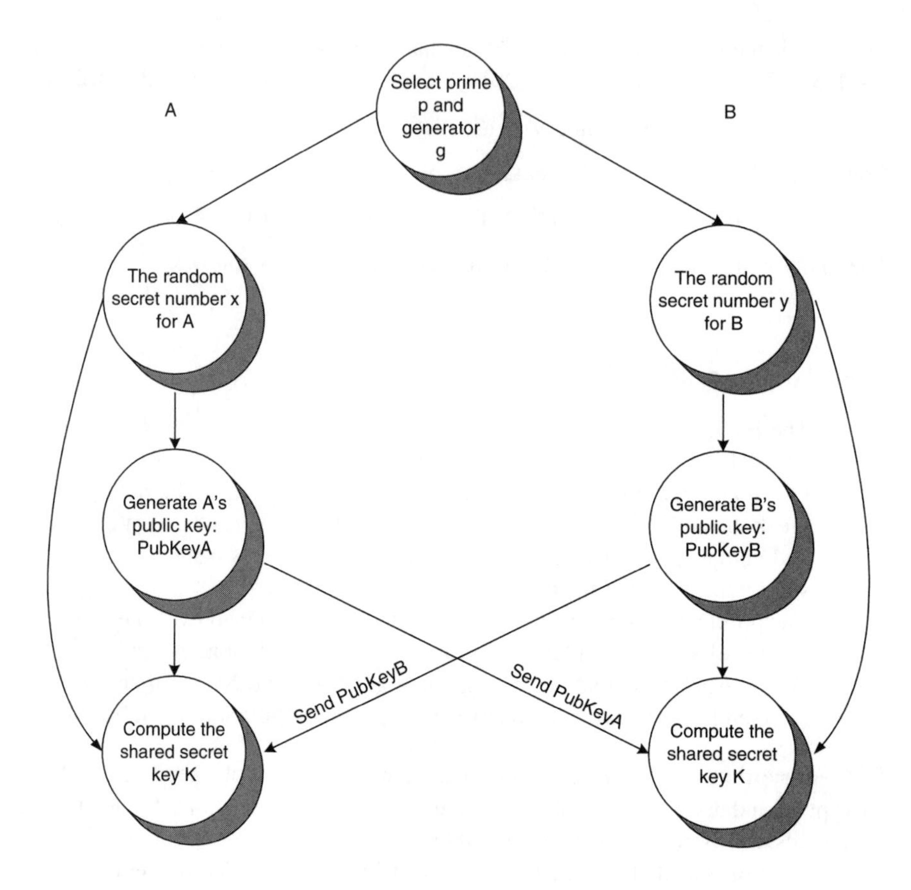

FIGURE 8.9 DH Key agreement protocol between two entities.

- A chooses a random secret number x, $0 \le x \le p-2$, and sends B message (A′ public key): $A \to B$: PubKeyA $= g^x \pmod{p}$.
- B chooses a random secret number y, $0 \le y \le p-2$, and sends A message (B′ public key): $A \leftarrow B$: PubKeyB $= g^y \pmod{p}$.
- B receives g^x and computes the shared key as $K = \text{Pubkey}A^y = \left(g^x\right)^y$ (mod p).
- A receives g^y and computes the shared key as $K = \text{Pubkey}B^x = \left(g^y\right)^x$ (mod p).

Figure 8.9 illustrates the steps of DH algorithm operations conducted between two entities.

Security of DH algorithm: The DH algorithm is based on the difficulty of the discrete logarithm problem for finite field [8-1]. It assumes that it is computationally infeasible to compute g^{xy} knowing only g^x and g^y. The choice of g and p can have a substantial impact on the security of DH algorithm. The number $(p - 1)/2$ should also be a prime [8-8]. And most important, p should be large: the security of the system is based on the difficulty of factoring numbers the same size as p. One can choose any g, such that g is primitive mod p. The DH algorithm also gets its security from the difficulty of calculating discrete logarithms in a finite field, as compared with the ease of calculating exponentiation in the same field.

DH key agreement protocol in practice: The DH algorithm can be used for key distribution, e.g., A and B can use this algorithm to generate a shared secret key for symmetric key scrambler/descrambler. However, the DH algorithm itself cannot be used to encrypt and decrypt messages. However, a variation of the Diffie-Hellman algorithm, often called *Elgamal variation* [8-13], can be used for message encryption.

8.6.3 Authentication

Authentication is a term that is used in a very broad sense. By itself it has little meaning other than to convey the idea that some means has been provided to guarantee that entities are who they claim to be, or that information has not been manipulated by unauthorized parties [8-2]. Authentication is specific to the security objective that one is trying to achieve. Examples of specific objectives for digital video security include access control, entity authentication, message authentication, and key authentication.

Example

(Key authentication problem): When the broadcasting service provider wants to send an encrypted key to a subscriber over an insecure channel, one or both parties might want the ability to monitor such communication. Both parties, however, would like to be assured of the identity of each other, and of the integrity and origin of the key information they send and receive.

Digital signatures: A digital signature is fundamental in authentication. The purpose of a digital signature is to provide a means for an entity to bind its identity to a piece of information. The process of signing entails transforming the message and some secret information held by the entity into a tag called a *signature*. A generic description follows.

Signing procedure: Entity A (the signer) creates a signature by using signing function SA for a message m by doing the following:

- Compute s = SA(m).
- Transmit the pair (m, s). Term s is called the *signature* for message m.

Verification procedure: To verify that a signature s on a message m was created by A, an entity B (the verifier) performs the following steps:

- Obtain the verification function VA of A.
- Compute u = VA(m, s).
- Accept the signature as having been created by A if u = true, and reject the signature if u = false.

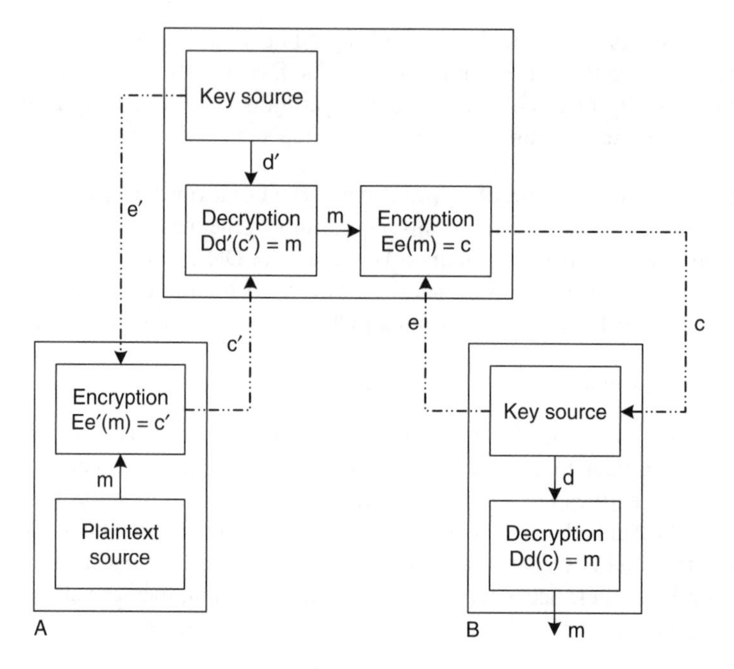

FIGURE 8.10 Impersonation attack to the public-key system.

Properties required for signing and verification functions: There are several properties that the signing and verification functions must satisfy:

- s is a valid signature of A on message m if and only if VA(m, s) = true.
- It is computationally infeasible for any entity other than A to find, for any m, an s such that VA(m, s) = true.

Signature authentication in public-key systems: It would appear that public-key cryptography is an ideal system, not requiring a secure channel to pass the encryption key. This would imply that two entities could communicate over an unsecured channel without ever having met to exchange keys. Unfortunately, this is not the case. Figure 8.10 illustrates how an active attack can defeat the system (decrypt messages intended for a second entity) without breaking the encryption system. This is a type of impersonation. In this scenario the attacker impersonates entity B by sending entity A a public key e, which A assumes (incorrectly) to be the public key of B. The attacker intercepts encrypted messages from A to B, decrypts with its own private key d, re-encrypts the message under B's public key e, and sends it on to B. This highlights the necessity to authenticate public keys to achieve data origin authentication of the public keys themselves. Party A must be convinced that the encryption is using the legitimate public key of party B. Fortunately, public-key techniques also allow an elegant solution to this problem.

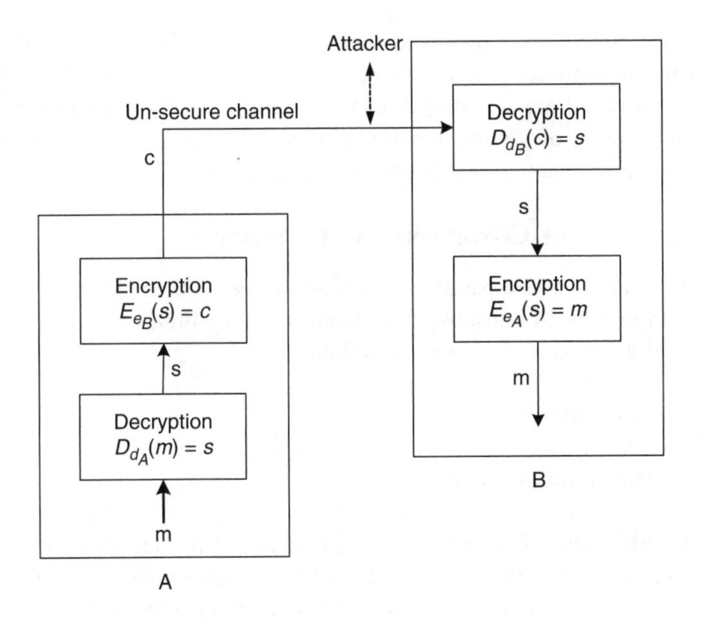

FIGURE 8.11 A public-key system for signature authentication.

EXAMPLE

(Signature authentication): An example in Figure 8.11 illustrates the use of a public-key system for signature authentication. Entity A "signs" his message by first applying its decryption algorithm $D_{d_A}(m)$, to the message m, yielding s. Next, he uses the encryption algorithm $E_{e_B}(s)$, of entity B to encrypt s, yielding c, which he transmits on a public channel. When entity B receives c, he first decrypts it using his private decryption algorithm $D_{d_B}(c)$, yielding s. Then he applies the encryption algorithm $E_{e_A}(s)$ of entity A to produce m.

If the result is an intelligible message, it must have been initiated by entity A, since no one else could have known A's secret decryption key d_A to form s. Notice that c is both message dependent and signer dependent, which means that while B can be sure that the received message indeed came from A, at the same time A can be sure that no one can attribute any false message to him.

8.7 CONDITIONAL ACCESS

Conditional access (CA) is defined as protection against unauthorized use of resources, including protection against the use of resources in an unauthorized manner. Digital broadcast television systems have to provide conditional access facilities to be economically viable. The sooner these facilities are taken into account in the definition, specification, and implementation of the systems, the earlier their deployment is. Conditional access is a means of selectively controlling the ability of system users to interpret information contained in widely available data. In general, conditional access relates to environments in which data is broadcast. The mechanisms of conditional access for digital broadcast signals (compressed audio/video or data) are introduced in this section. The discussion focuses on the general architecture of a conditional access system. The architecture described in here is applicable to any digital conditional access system.

8.7.1 Functions of Conditional Access System

The implementation of fee-based video broadcasting requires the conditional access (CA) system to prevent nonsubscribers from receiving broadcasts. A complete CA system usually includes three main functions:

- Scrambling/descrambling
- Entitlement control
- Entitlement management

Scrambling/descrambling function: The scrambling/descrambling function aims at making the program incomprehensible for unauthorized receivers. Scrambling can be applied commonly or separately to the different elementary streams of a program (e.g., video, audio, and data for a TV program) to make these streams unintelligible. Descrambling can be achieved by any receiver that holds a secret key, called *control words* (CW), used for a scrambling algorithm. Scrambling and descrambling operations do not cause any impairment on the quality of the signals.

The commonly used algorithms for scrambling digital data in CA systems are symmetric-key ciphers. Table 8.9 summarizes various standardized symmetric-key algorithms for CA systems.

The CW (or, say, the key) is a secret parameter known only by the scrambler and the authorized descramblers. It has to be changed frequently to avoid any exhaustive search. For example, a CA system with the transport rate of 34 to 45 Mbits/s may use a single DES algorithm with 56-bit CW that is renewed around every 10 s. The period counter, sometimes, may be a public parameter that enables the fast synchronization of the CW generators or the scrambler and of the descramblers. No changes in the access conditions are allowed during a period.

Entitlement Control Function: The rights and associated keys needed to descramble a program are called *entitlements*. The entitlement control function

TABLE 8.9
Standard Symmetric-Key Algorithms for CA Ssystems

Scrambling Algorithm	Standard	Applied Region
DES (triple-DES)	ATSC and SCTE [8-19][8-22]	North America
DVB Common Scrambling Algorithm	DVB [8-24]	Europe
MULTI2	ARIB [8-43]	Japan

provides the conditions required to access a scrambled program together with the encrypted secret parameters enabling the descrambling for the authorized receivers. These data are broadcast as conditional access messages, called *entitlement control messages* (ECMs), which carry an encrypted form of the control words or a means to recover the control words, together with access parameters, i.e., a identification of the service and of the conditions required for accessing this service.

Upon reception of an ECM, the receiver sends the encrypted control word and the access conditions to the security device (e.g., a smart card) which checks their origin and their integrity before decrypting the control word and sending it to the descrambler if the customer is authorized to watch the program.

Entitlement management function: The entitlement management function consists of distributing the entitlements to the receivers. There are several kinds of entitlements matching the different means of "buying" a video program. These entitlement data are also broadcast as conditional access messages, called *entitlement management messages* (EMMs), used to convey entitlements or keys to users, or to invalidate or delete entitlements or keys.

The control and management functions require the use of secret keys and cryptographic algorithms. For example, most of modern conditional access systems have chosen the smart card—the physical packaging standard conforming to ISO-7816 Part 1 and 2 [8-44][8-45]—to store the secret keys and to run the cryptographic algorithms safely.

8.7.2 Configuration of a Conditional Access System

A CA system operates on the principle of randomizing the transmitted data so that unauthorized decoders cannot decode the signal. Authorized decoders are delivered a "key" that initializes the circuit that inverts the bit randomization. In subsequent discussion, the term *scrambling* is used to mean the pseudo-random inversion of data bits based on a "key" that is valid for a short time. The term is used to mean the process of transforming the "key" into an encrypted key by a means that protects the key from unauthorized users. From a cryptographic point of the view, this transformation of the key is the only part of the system that protects the data from a highly motivated pirate. The scrambling portion of the process alone, in the absence of key encryption, can be defeated. *Conditional access* is a blanket term for the system that implements the key encryption and distribution. The general requirements that a scrambling and CA subsystem must meet for digital video delivery are:

- Robust protection to against piracy—it must be difficult for a third party to perform unauthorized reception, and scrambled signal content must not be understandable.
- Efficient scrambling of all kinds of signals (as in multimedia broadcasts) must be possible and quality must not deteriorate (perceptibly) on restoring the signal (quality restoration).

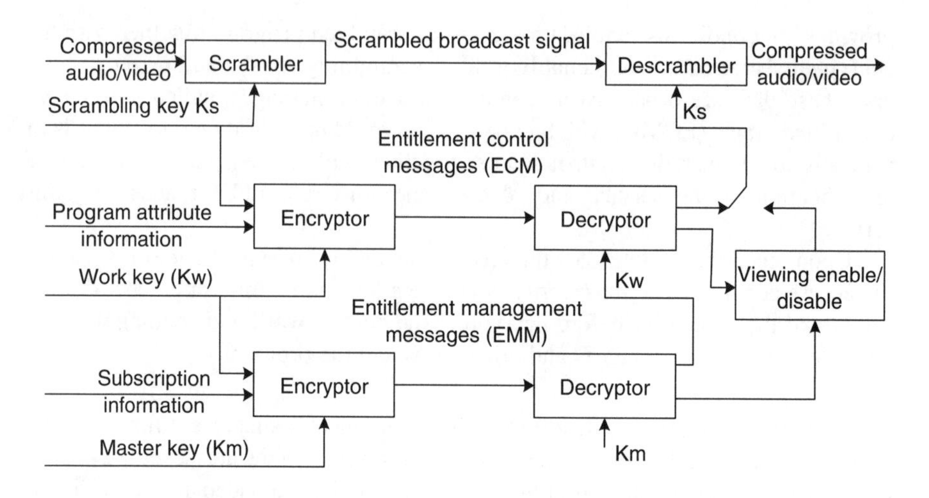

FIGURE 8.12 A basic configuration of CA system.

- Conditional access is also flexible in the sense that it can be exercised on an elementary stream-by-stream basis, including the ability to selectively scramble bit streams in a program if desired.
- Various business formats, such as multichannel services and billing schemes, must be supported with low operating costs. Private encryption system can be used by each program provider.
- The system can be implemented in standard consumer instruments (means no secrets in consumer equipment), to ensure cost-effective receivers.

A CA system can be achieved in various ways depending on types of services, required functions, and security. One basic configuration, shown in Figure 8.12, has been recommended by ITU-R [8-50]. First, the system should perform scrambling according to the properties of the data for transmission. Second, it should change the key regularly to maintain the security of the scrambling system, and transmit the key information to the receiver in a secure manner using a hierarchical encryption system. Third, for the purpose of operating fee-based broadcasting service, reception should be controlled according to the details of each user's subscription. The CA functions implemented in digital broadcasting are described as follows.

The signal scrambling system is selected in accordance with the properties of the transmitted data. In this regard, CA systems for television broadcasting have once applied to analog broadcasting, and degradation in the restored signal due to transmission-path characteristics could occur if advanced signal scrambling was performed. Under these circumstances, many systems have been developed and deployed according to the level of security, restoration quality, etc., required by the target system.

On the other hand, scrambling of digital television broadcast signals causes no degradation of the restored signal provided that the transmitted signal is correctly received. This makes it possible to use advanced scrambling techniques, such as DES.

Various kinds of encryption technologies for digital data can be applied to the scrambling of a digital signal. Broadcast scrambling, however, may also employ an "effect control" system to reduce the degree of concealment and allow a slight amount of content to be recognized to promote subscriptions. In this case, control techniques based on detailed properties of the signal will be required in addition to encryption technology for digital signals. Technology for encrypting digital signals can be broadly classified as stream ciphers and block ciphers as discussed in Section 8.2.

In digital broadcasting systems, signals such as video, audio, and various kinds of data with very different properties, are unified in the form of transport packets and multiplexed into a single stream for transmission. The most commonly used system for performing this multiplexing is the one standardized by MPEG-2. This system features two signal scrambling methods: the first method scrambles transport stream (TS) packets, and the second method scrambles packetized elementary streams (PES). In either case, a detailed scrambling method is not specified and is instead prescribed on the application system side. For example, the digital video broadcasting (DVB) system [8-23][8-24] in Europe has standardized a signal scrambling system that combines block and stream ciphers. A description of this scrambling system has not been officially released, though, and the system itself cannot be used outside Europe. As a result, different scrambling systems must be used when operating DVB-based systems outside of Europe. Such a situation has occurred in Japan, where some companies have implemented digital television broadcasting via communications satellite (CS) in which the signaling system conforms to DVB, but the scrambling system has been independently developed. As shown in Figure 8.5, this scrambling system combines a block cipher and a stream cipher (which is applied to those sections of data targeted for scrambling that exceed an integer multiple of the block length in the block cipher). A similar system that uses MULTI2 scrambler [8-43] (instead of DES) has also been introduced in Japanese digital broadcasting via Broadcasting Satellite (BS) and terrestrial waves.

Key configuration and transmission of key information: In signal scrambling, a CW (or, say, scrambling key; Ks) determines the scrambling pattern, and it is common to change this key at fixed intervals of time, such as every few seconds, to maintain a secure system. This Ks information must therefore be continuously transmitted to the subscriber's receiver, and this is done by encrypting and transmitting it within ECM together with the scrambled broadcast signal.

Entitlement control messages include program attribute information for determining whether a subscriber is entitled to view a program on the basis of his or her subscription. To prevent the ECM that includes the Ks from being understood by a third party, it is encrypted before transmission by using a work key (Kw) that is also updated typically on a monthly or yearly basis. This work key is sent to the receiver through EMM together with subscription contents that are sent with subscription updates.

Besides broadcast waves, other physical media like Internet, telephone lines, or smart cards may be used to transmit EMMs. On transmission, the EMM is also encrypted by a master key (Km) unique to each receiver. This means that security

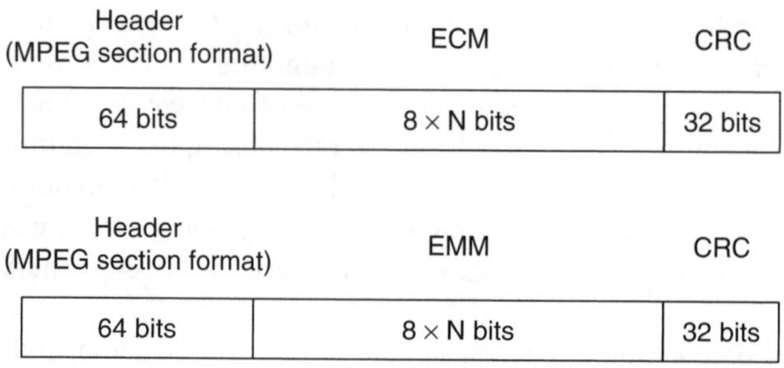

Header (MPEG section format)	ECM	CRC
64 bits	8 × N bits	32 bits

Header (MPEG section format)	EMM	CRC
64 bits	8 × N bits	32 bits

FIGURE 8.13 Example of signal packet configurations for sending ECMs and EMMs.

for master keys must be commonly managed among different broadcast operators that use the same type of receiver. This can normally be accomplished by setting up an organization for uniform key management.

A system using the MPEG-2 transport multiplexing system sends out ECMs and EMMs using signal configurations like the ones shown in Figure 8.13. (Note that packet identifier (PID) values of the packets used to send these ECM and EMM sections are included in the program map table (PMT) and conditional access table (CAT), respectively.) Here, specific ECM and EMM contents can be specified according to business format, billing system, etc., enabling a variety of conditional access systems to be implemented.

The processes are specified in some standards [8-22] for transmitting ECMs and EMMs and the information needed to receive and process these messages. These standards allow operators to decide on contents, encryption systems, etc. The conditional access system, moreover, includes a function for sending individual information to each receiver within a broadcast system that targets an unspecified number of the general public. This function can therefore be used to send different messages to different receivers.

The transport protocol implements functions useful for supporting conditional access. The functionality that is available is flexible and complete in the sense of supporting all transmission aspects of applicable key encryption and descrambling approaches that may be used. There are two features of the MPEG-2 transport stream that support conditional access. The first feature is the 2-bit transport_scrambling_control field that signals the decoder whether the transport packet was scrambled or not. In the case that it was scrambled, the field identifies which scrambling key was used. As will be shown shortly, the use of two bits in the field transport_scrambling_control to define the descrambling process is a necessary and sufficient bound for the key distribution function. The second feature is the ability to insert "private" data at several places in the CA transport stream. These include entirely private streams and private fields in the adaptation header of the transport bit stream being scrambled. These private fields can be used to transmit the encrypted scrambling key to the decoding device.

The key distribution and usage process is clarified in Figure 8.14. Basically, when the bit stream is scrambled, one descrambling key needs to be in use while

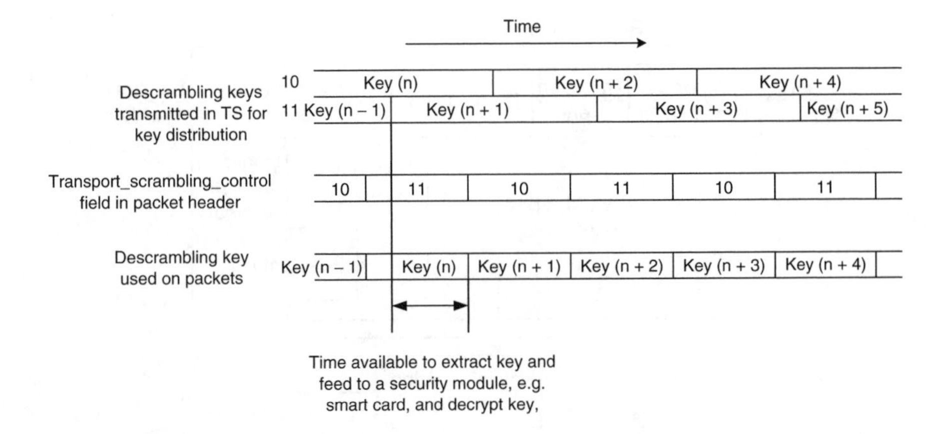

FIGURE 8.14 Key distribution and usage process.

the other is being received and decrypted. Two keys are transmitted at any time, with the keys being linked to a transport_scrambling_control value as shown in the figure. The transmission of the key should begin before it is going to be used, to allow time to decrypt it. Note that this function does not place a bound on the total number of keys that may be used during an entire transmission session.

Configuration of a conditional access receiver: A conditional access receiver consists of a subsystem that descrambles the scrambled signal and a subsystem that processes related control information like ECMs and EMMs. Figure 8.15, Figure 8.16, and Figure 8.17 show three typical configurations of the conditional access receiver. Configuration A makes both the signal-descrambling subsystem and

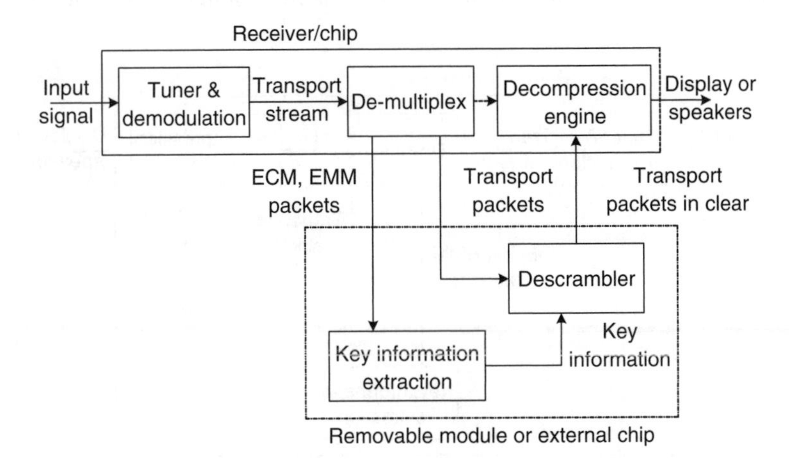

FIGURE 8.15 Configuration A makes both the signal-descrambling subsystem and key information extraction subsystem removable (or external).

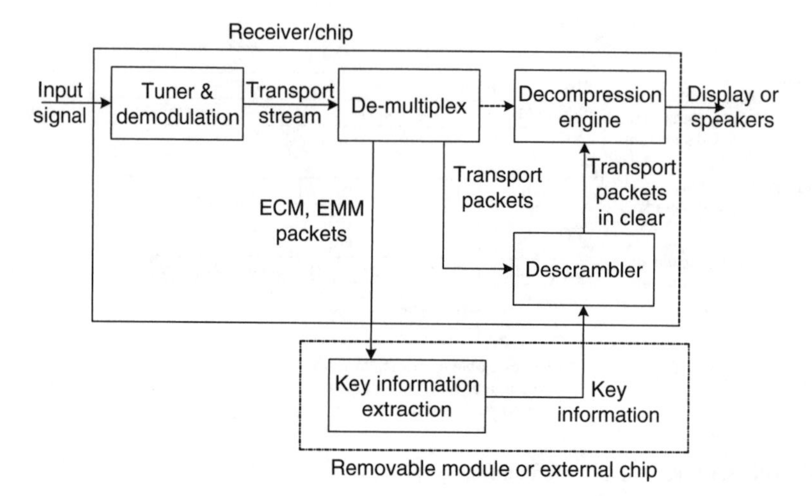

FIGURE 8.16 Configuration B includes the signal-descrambling subsystem in the receiver unit but implements the key information extraction subsystem in a removable security module.

related-information-processing subsystem removable. Configuration B, on the other hand, includes the signal-descrambling subsystem in the receiver unit but implements the related-information processing subsystem in a removable security module. Configuration C integrates the signal-descrambling subsystem and related-information processing subsystem in the receiver unit.

To deal with threats to security, changes to the billing system, and the like, configuration C requires that the entire receiver be repaired or replaced. Configuration B can require only the key information extraction subsystem being replaced, whereas configuration A allows even the descrambling algorithm to be updated, enabling system security to be improved all the more. Configuration B can deal with

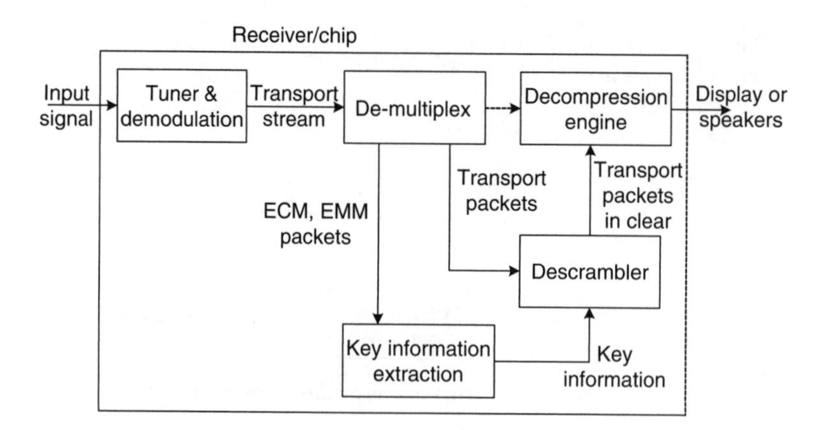

FIGURE 8.17 Configuration C integrates the signal-descrambling subsystem and key information extraction subsystem in the receiver unit.

security threats by simply replacing the key information extraction security module, which can be implemented in the form of a smart card [8-44]-[8-49] with a built-in processor. This smart card approach has recently been adopted in many systems, including some systems in Europe, DirecTV in the United States, and the CS and BS digital broadcasting systems in Japan. For example, Japanese CS digital broadcasting has also initiated a conditional access system in which unauthorized reception can be prevented by simply replacing the smart card. For CS digital broadcasting in which the receiver simply receives signals, a system has been standardized that allows a CA processing program to be downloaded by using a radio signal sent to the receiver. As for Japanese BS digital broadcasting, configuration B was also standardized for the launch of this form of broadcasting.

With regard to module interface signals for CA, configuration B requires them only for the subsystem concerned with related information, for which low-speed exchange of these signals would be acceptable. In contrast, configuration A requires, for example, an interface for an MPEG transport stream using high-speed data. The low-speed CA interface of configuration B has been standardized in Japanese CS digital television broadcasts. The interface of configuration A has been standardized in the European DVB system, since various security modules with different interfaces have already been deployed and a uniform interface under configuration B is impossible. The CA interface of configuration A features high security and extendibility, and studies continue on its use in future receivers providing diversified broadcast services.

As stated previously, the amount of data to be scrambled in a packet is variable depending on the length of the adaptation header. It should be noted that some padding of the adaptation field might be necessary for certain block mode algorithms.

Example

(Conditional access implementation): In this example, a simple implementation of a CA system is discussed. Consider the receiver architecture shown in Figure 8.18. The high-speed manipulations required to implement the descrambling are embedded in the transport demultiplexer, where they are shown as a DES block. The "key" that properly configures the descrambler achieves the data security. This element is delivered to the decoder through an ancillary data service, and is encrypted by the conditional access administrator. In the equipment at the customer's premises, the key is decrypted within the outboard smart card. The smart-card interface will conform to ISO standard ISO-7816 [8-44]-[8-49], which permits a variety of implementations and conditional access solutions.

The smart card maintains a short list of two keys commonly denoted as the "odd" key and the "even" key. The proper key to be used to descramble is signaled in the transport prefix in the transport_scrambling_control field. The transport_scrambling_control takes on one of the following four states:

Transport_Scrambling_Control	Description
00	unscrambled
01	reserved
10	even key
11	odd key

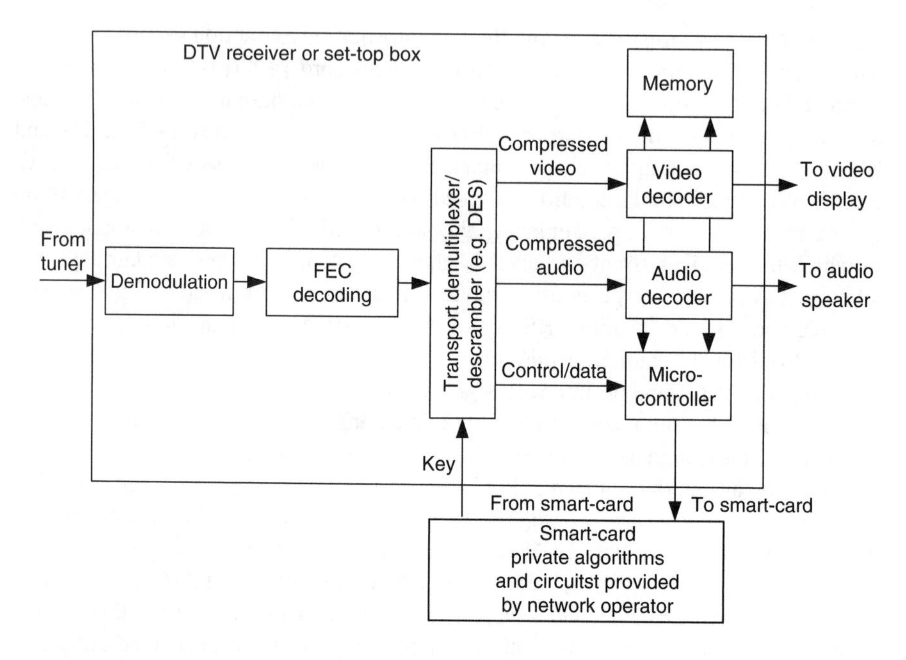

FIGURE 8.18 An example on CA implementation.

The scrambler in this example can be a block cipher using the ECB or CBC mode of the DES. The television electronics are "standard," while the smart card implementation is proprietary to the service provider.

8.7.3 Termination of Short Blocks in Block Cipher for Transport Packets

Transport packet often has a fixed length. For example, the MPEG-2 transport stream (TS) packet has a length of 188 bytes. However, video payload data carried in the transport packet can have variable length. It may not necessarily be a multiple of the size of the block cipher. For example, video payload data in an MPEG-2 TS packet might not be a multiple of 64-bit when it uses the DES algorithm for scrambling. Therefore, the remaining short block needs to be processed by following certain rules. This is called *termination of short blocks*. Figure 8.19 shows an example of terminating short blocks (residual termination blocks) for a scrambler, e.g., DES, operated in the CBC mode for the MPEG-2 TS packet [8-19][8-22].

8.7.4 Multi-Hop Encryption

There is significant flexibility for multi-hop encryption and nesting of encryption systems to ensure data security at every point in the transmission chain. Figure 8.20 illustrates nested encryption systems, where the service provider gives authorization, keys and the scrambled data to the end user through Encryption System A. During transit, System B is employed by the carrier to protect the data while in the

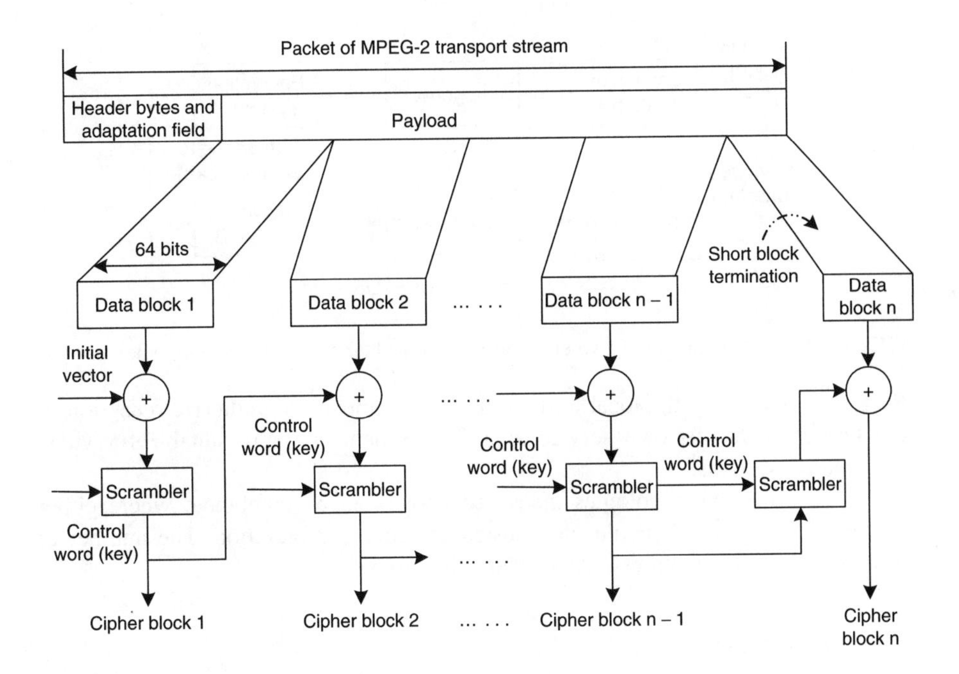

FIGURE 8.19 An example of termination of short blocks.

communication network. In fact, there are two implementation choices even for this segment of the delivery system. The provider can use both a second layer of scrambling in conjunction with a different authorization and key distribution. (This system need not comply with the transmission stand method.) Alternatively, System B could simply encrypt or scramble the encrypted keys distributed by system A, without the requirement of scrambling the actual service data. The main appeal of this method is that it is a low-bandwidth/complexity solution, whereaas its drawback is that it requires some knowledge of the System A key distribution method.

In Figure 8.21, two encryption systems are connected in series. System B is again used to protect the integrity of the data while in the communications network;

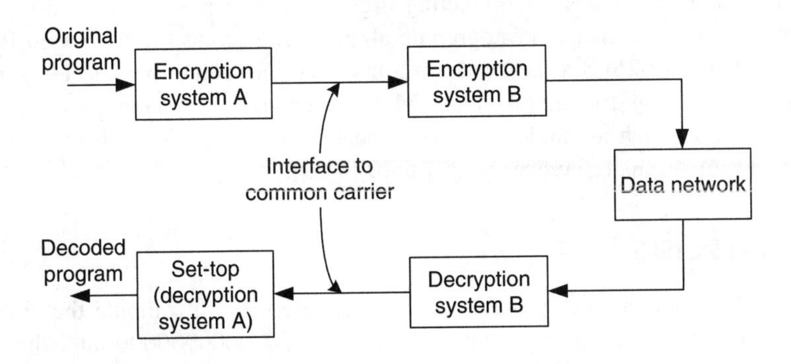

FIGURE 8.20 A nested encryption system.

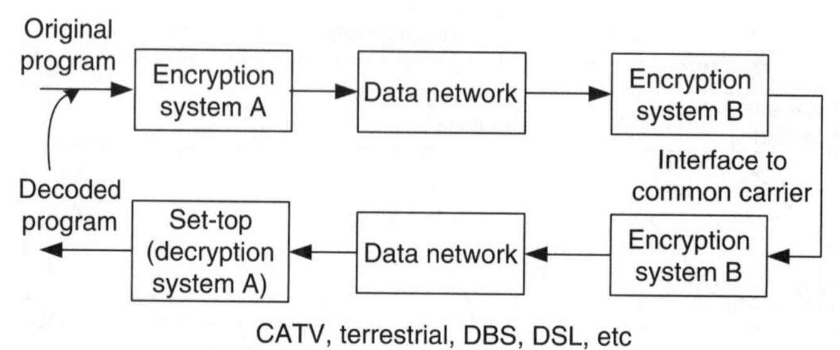

FIGURE 8.21 An example of two encryption systems being concatenated together.

System A is used by the local affiliate or cable company to authorize reception of the service within its own service area. This scheme is more suitable for video transcoding systems.

In fact, the two topologies discussed above can be combined, where either System A or System B could be a nested or series configuration. The number of nesting/series combinations can be arbitrarily large.

8.8 SUMMARY

In this chapter we have presented the basic security concepts, algorithms, and models for the transporting of compressed digital video. We discussed a number of important cryptographic algorithms that are used to protect high-value video content and have reviewed several configurations of conditional access systems for digital video transport systems. In particular, we detailsed the DES algorithm. We introduced four modes of operation and their applications in both block cipher and stream cipher. We also looked at the area of public-key cryptosystems and examined two algorithms, the RSA encryption algorithm, based on the product of two large prime numbers, and the DH key exchange algorithm, based on the discrete logarithm problem for finite field. We have also reviewed three configurations of conditional access schemes. Finally, we have introduced some basic ideas of the multi-hop encryption systems.

Although this chapter is intended to summarize many important and applicable cryptological techniques for protecting digital video, it is still far from complete. Recently, some advanced cryptographic algorithms, such as the Advanced Encryption Standard [8-25][8-41], have been introduced into conditional access and copy protection systems for digital video. Many other useful cryptographic algorithms and protocols, such as random number generators and secure hush functions, can be found in various references [8-37][8-40][8-42].

8.9 EXERCISES

8-1. Write a computer program of the DES algorithm. Compute the ciphertext c for the given plaintext m = 0x0000000000000000, and the key k = 0x00000000000000.

8-2. In a cipher, a key $\boldsymbol{k}$ is called *self-dual* if $E_k(E_k(m)) = m$. Try to find some self-dual keys in the DES algorithm. Explain why self-dual keys are weak for the security.
8-3. Describe the steps used for message decryption employed by the DES algorithm. How different is the operation when using triple-DES?
8-4. Try to describe advantages and disadvantages of block cipher and stream cipher in terms of security and processing complexity. Which one is more suitable for digital television applications?
8-5. A variation of the Euclidean algorithm for computing the gcd of $\phi(n)$ and d is used to compute e in the RSA algorithm. First, compute a series x_0, $x_1, x_2, \ldots$, where $x_0 = \phi(n)$, $x_1 = d$, and $x_{i+1} = x_{i-1} \pmod{x_i}$, until an $x_k = 0$ is found. Then the $\gcd(x_0, x_1) = x_{k-1}$. For each x_i compute numbers a_i and b_i such that $x_i = a_i\, x_0 + b_i\, x_1$. If $x_{k-1} = 1$, then b_{k-1} is multiplicative inverse of $x_1 \bmod x_0$. If b_{k-1} is a negative number, the solution is $b_{k-1} + \phi(n)$. For $p = 47$, $q = 59$, and $d = 157$, use the Euclidean algorithm to verify that $e = 17$.
8-6. Use the RSA scheme with the prime $p = 5$ and $q = 7$ to encrypt the message $m = 3$. Choose the decryption key, d, to be 11, and calculate the value of the encryption key e.
8-7. What are major requirements for a conditional access system?
8-8. Write a program to encrypt a 184-byte block using DES CBC mode. Do you need termination of short blocks? Why? If the block size is 185-byte, add the termination described in Figure 8.19 in your program.
8-9. Explain the major difference between symmetric-key ciphers and public-key ciphers.

REFERENCES

[8-1] C. E. Shannon, Communication theory of secrecy systems, Bell System Technical Journal, 28, 656, 1949.
[8-2] R. Demillo, *Applied Cryptography, Cryptographic Protocols, and Computer Security Models,* American Mathematical Society, Providence, RI, 1983.
[8-3] B. Schneier, *Applied Cryptography: Protocols, Algorithms, and Source Code in C,* John Wiley & Sons, New York, 1995.
[8-4] G. J. Simmons, *Contemporary Cryptology: The Science of Information Integrity,* IEEE Press, Piscataway, NJ, 1992.
[8-5] D. Denning, *Cryptography and Data Security,* Addison-Wesley Publishing Co., Inc., Reading, MA, 1982.
[8-6] C. Meyer, *Cryptography: A New Dimension in Computer Data Security,* John Wiley & Sons, New York, 1982.
[8-7] D. Stinson, *Cryptography: Theory and Practice,* CRC Press, Boca Raton, FL, 1995,
[8-8] P. C. Van Oorschot, S. A. Vanstone, and A. Menezes. *Handbook of Applied Cryptography,* CRC Press, Boca Raton, FL 1996.
[8-9] W. Friedman, *Military Cryptanalysis,* Aegean Park Press, Laguna Hills, CA, 1984.
[8-10] ITU-T Recommendation H.262 (1995)|ISO/IEC 13818-1: 2000, Information technology—Generic coding of moving pictures and associated audio information: Systems.

[8-11] ITU-T Recommendation H.264 (2003)|ISO/IEC 14496-10: 2003, Information technology—Generic coding of audio-visual objects—Part 10: Advanced Video Coding.

[8-12] I. S. Reed and X. Chen, *Error-Control Coding for Data Networks,* Kluwer Academic Publishers, Boston, 2001.

[8-13] X.Chen, *Transporting Compressed Digital Video*, Kluwer Academic Publishers, Boston, 2002.

[8-14] B Sklar, *Digital Communications: Fundamentals and Applications,* Prentice Hall, Englewood Cliffs, NJ, 2001.

[8-15] S. Namba, Scrambling (Conditional Access System), Technologies and Services on Digital Broadcasting (6), Corona Publishing Co., Ltd, Tokyo, 2002.

[8-16] ISO/IEC 14496-2:1998, Information Technology—Generic coding of audio-visual objects—Part 2: Visual.

[8-17] FIPS-PUB 46-2 Data Encryption Standard (DES), December 30, 1993.

[8-18] FIPS-PUB 46-3 Data Encryption Standard (DES), October 25, 1999.

[8-19] FIPS-PUB 81 DES Modes of Operation, December 2, 1980.

[8-20] DES CBC Packet Encryption, SCTE DVS 042, Society of Cable and Television Engineers, 25 October 1996.

[8-21] FIPS PUB 74, Guidelines for Implementing and Using NBS DES, National Institute of Standards and Technology, 1981.

[8-22] NBS PUB 500-20, Validating the Correctness of Hardware Implementations of the NBS DES, National Bureau of Standards, 1980.

[8-23] ATSC A70, Conditional Access System for Terrestrial Broadcast, July 1999.

[8-24] DVB—Head-end Implementation of DVB Simulcrypt, Draft ETSI TS103 197 V1.3.1 (02-06) TM2117R3.

[8-25] ETR 289 ed.1 (1996-10), Digital Video Broadcasting (DVB); Support for use of scrambling and Conditional Access (CA) within digital broadcast systems, European Telecommunications Standards Institute (Informative).

[8-26] FIPS-PUB 197 Advanced Encryption Standard (AES), November 26, 2001.

[8-27] SCTE 41 2001 (Formerly DVS 301) POD Copy Protection Standard, Society of Cable Telecommunications Engineers.

[8-28] OC-SP-SEC-I01-021126, OpenCable System Security Specification, November 26, 2002.

[8-29] EIA 679-B, National Renewable Security Standard (NRSS), Electronic Industry Association, 1999.

[8-30] RSA1, PKCS #1: RSA Encryption Standard, Version 1.5, RSA Laboratories, November 1993.

[8-31] RSA2, PKCS #1 v2.0: RSA Encryption Standard, Version 2.0, RSA Laboratories, October 1, 1999.

[8-32] PKCS #3: Diffie-Hellman Key-Agreement Standard, Version 1.4, November 1, 1993.

[8-33] ANSI X9.52, Triple Data Encryption Algorithm Modes of Operation.

[8-34] ITU-T Recommendation X.509, Information technology—Open Systems Interconnection —The Directory: Public-key and attribute certificate frameworks, March 2000.

[8-35] ISO/IEC 10116, Information processing—Modes of operation for an n-bit block cipher algorithm, International Organization for Standardization, Geneva, Switzerland, 1991.

[8-36] FIPS PUB 186-1 (May 18, 1994), FIPS PUB 186-2 (January 27, 2000), Digital Signature Standard Federal Information Processing Standards Publication (FIPS PUB).

[8-37] FIPS PUB 140-1 (January 11, 1994), FIPS PUB 140-2 (May 25, 2001), Security Requirements for Cryptographic Modules.

[8-38] FIPS-PUB 180-1, Secure Hash Standard Federal Information Processing Standards Publication (FIPS PUB), January 27, 2000.

[8-39] OC-SP-PODCP-IF-I08-021126: OpenCable POD Copy Protection System Specification, November 26, 2002, Cable Television Laboratories, Inc.

[8-40] IETF RFC 1750, Randomness Recommendations for Security, (Donald Eastlake, Stephen Crocker and Jeff Schiller), December 1994.

[8-41] IETF RFC 2104, HMAC: Keyed-Hashing for Message Authentication, (Krawczyk, Bellare, and Canetti), March 1996.

[8-42] IETF RFC 3268, Advanced Encryption Standard (AES) Ciphersuites for Transport Layer Security (TLS)), June 2002.

[8-43] H. Beker, and F. Piper. *Cipher Systems,* John Wiley & Sons, Inc., New York, 1982.

[8-44] ISO/IEC 9979 Register of Cryptographic Algorithms, 2nd ed. 1999.

[8-45] ISO/IEC 7816-1, Identification cards–Integrated circuit(s) cards with contacts–Part 1: Physical characteristics.

[8-46] ISO/IEC 7816-2, Identification cards–Integrated circuit(s) cards with contacts–Part 2: Dimensions and location of the contacts.

[8-47] ISO/IEC 7816-3, Identification cards–Integrated circuit(s) cards with contacts–Part 3: Electronic signals and transmission protocols.

[8-48] ISO/IEC 7816-4, Identification cards–Integrated circuit(s) cards with contacts–Part 4: Inter-industry commands for interchange.

[8-49] ISO/IEC 7816-5, Identification cards–Integrated circuit(s) cards with contacts–Part 5: Numbering system and registration procedure for application identifiers.

[8-50] ISO/IEC 7816-6, Identification cards–Integrated circuit(s) cards with contacts–Part 6: Inter-industry data elements.

[8-51] ITU-R Rec. BT.810, Conditional-access broadcasting systems, Sept. 1992.

9 Application and Implementation of Video Transcoders

Recently, video transcoding has found many applications. We present several applications in this chapter. The first application is the transcoding of MPEG-2 to MPEG-4 bit streams. Since the MPEG-4 video coding standard has been used for video streaming and mobile terminals, but some content in the video servers is encoded with MPEG-2 video coding, we need convert the MPEG-2 to MPEG-4 for those users with the MPEG-4 decoder to receive the MPEG-2 encoded content. The second application is the video transcoding for IP or wireless networks. In these kinds of transmission applications, the function of error resilience is very important. We will introduce several methods of error resilience transcoding. Finally, the topic of object-based transcoding technique for MPEG-4 will be presented.

9.1 MPEG-2 TO MPEG-4 TRANSCODER

The transcoding methods themselves can be applied within the same syntax format or between different syntax formats. Since the MPEG-4 simple profile has been adopted as the solution for mobile multimedia communications and a large amount of MPEG-1/2 content is available, we focus our discussion on MPEG-2 to MPEG-4 transcoding. The transcoding from MPEG-2 to MPEG-4 is necessary and useful for allowing mobile or PDA terminals, with their limited display size, to receive MPEG-2 compressed content. In this section, we describe the principles and techniques used in the MPEG-2 to MPEG-4 video transcoder. The discussion includes techniques for bit rate reduction, spatial resolution down-sampling, temporal resolution down-scaling and picture type change. The main difficulty of MPEG-2 to MPEG-4 transcoding is to perform transcoding on both bit rate reduction and spatial resolution reduction at the same time. The simultaneous transcoding for bit rate reduction and spatial resolution reduction causes serious error drift due to the change of predictive references. To address this problem, several issues have been investigated. First, an analysis of drift errors is provided to identify the sources of quality degradation when transcoding to a lower spatial resolution. Two types of drift error are considered: a reference picture error and error due to the noncommutative property of motion compensation and down-sampling. To overcome these sources of error, several novel transcoding architectures are then presented. One architecture attempts to compensate for the reference picture error in the reduced resolution, and another architecture attempts to do the same in the original resolution. We then present a third architecture that attempts to eliminate the second type of drift error and a final architecture that relies on an intra block refresh method to compensate all types of errors. In all these architectures, a variety of macroblock-level conversions are required, such as motion vector mapping and texture down-sampling. These conversions are discussed in detail. Another important issue for the transcoder is rate control. This is especially important for the intra refresh architecture since it must find a balance between number of intra blocks used to compensate errors and the associated rate-distortion characteristics of the low-resolution signal. The complexity and quality of the architectures are compared. Based on the results, we find that the intra refresh architecture offers the best trade-off between quality and complexity, and is also the most flexible.

9.1.1 Introduction

In many applications, we need scalable video coding techniques for robustness of the selection of bit rate, spatial resolution, and temporal resolution. Therefore, several scalable coding techniques have been proposed in the video coding standards. These techniques include spatial scalability, temporal scalability, and SNR scalability in MPEG-2 and MPEG-4. In order to meet the requirements of compressed video over IP or wireless networks, recently, the MPEG-4 encoding method has been developed with fine granular scalability (FGS), in which the enhancement layer bit stream can be truncated at any point to achieve different levels of video quality for the given bandwidth requirement. Although the scalable coding can provide robustness and quality of service (QoS) for the video transmission, it has several problems in application. First, the layered bit stream generated by the scalable coding provides a certain level of flexibility, but it sacrifices coding performance at the full bit rate compared with single-layer coding. Also, the scalable coding increases the complexity of decoders. Moreover, the main problem is that the content providers for DTV and DVD applications have already adopted the one-layer MPEG-2 coding as their default format; a large number of MPEG-2 coded video contents already exist. To access this existing MPEG-2 content with decoders of limited capability or using other kinds of decoders such as MPEG-4 decoders, we need transcoding to convert these files to the appropriate format.

In the context of coding and transmission, there is an increasing need to perform many types of conversions to accommodate terminal constraints, network limitations, or user preferences. On one side there exists a rich set of content that is rapidly growing by the day, and on the other we have terminals with varying capabilities. Furthermore, these two entities may be connected through a variety of different network configurations. It should be clear that the transmission and playback of content in such a diverse and dynamic environment may require many types of different conversions. MPEG-2 video can be seen as the most successful digital video coding standard so far. It is extensively employed in many applications including digital television (DTV) broadcasting, digital video disc (DVD), and direct satellite television broadcasting, etc. Therefore, a huge amount of MPEG-2 coded video content has been created for these applications. MPEG-4 video is the relatively new video coding standard from MPEG that mainly targets multimedia applications

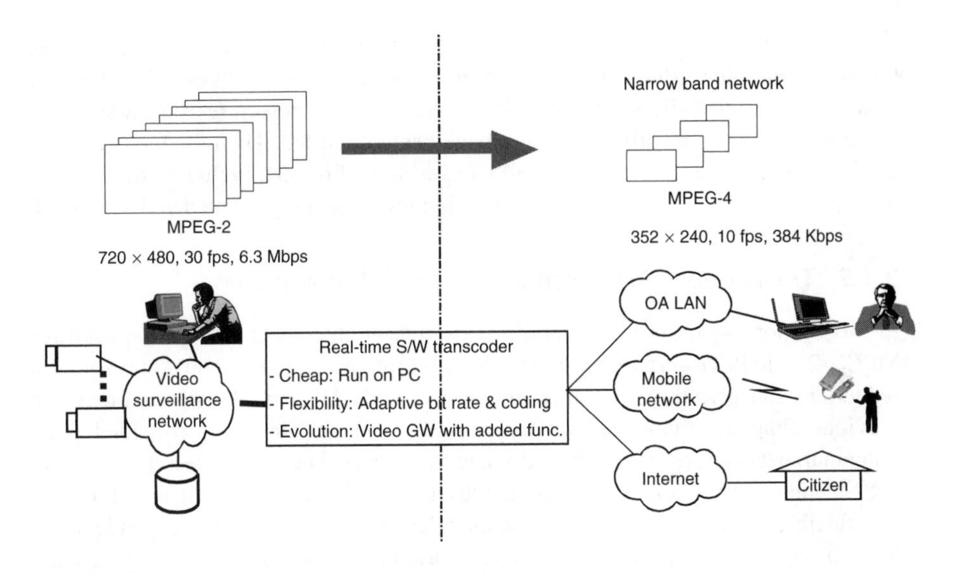

FIGURE 9.1 An application example of an MPEG-2 to MPEG-4 transcoder.

such as video camcorder, streaming video, and other low bit rate video applications. One motivation to investigate MPEG-2 to MPEG-4 transcoding is to enable broadcast-quality video streams through narrow bandwidth networks to be decoded and displayed on different devices such as the mobile devices as shown in Figure 9.1.

In this application, transcoding from MPEG-2 to MPEG-4 is needed. This conversion implies a reduction in bit rate from, for instance, approximately 6 Mbps to 384 kbps or lower, as well as a reduction in spatial resolution from 720 × 480 interlace to 352 × 240 progressive. In the following, we focus our attention on the problem of MPEG-2 to MPEG-4 transcoding with both reduced bit rate and reduced spatial resolution. Specifically, we consider techniques and architectures that convert a compressed video bit stream being encoded at one spatial resolution to an output bit stream with half that spatial resolution. In this section, we provide a different approach to the problem of spatial resolution transcoding by providing a detailed analytical analysis of the overall architecture. In doing so, we give significant attention to the degradation of reconstructed macroblocks in reduced-resolution pictures. Based on this analysis, several architectures for compressed domain transcoding are developed and analyzed. We also address the issue of constructing a reduced-resolution macroblock from macroblocks in the original bit stream that have different coding modes. We refer to this problem as the *mixed block problem* and present several ways to address it. First, we focus on the development of drift compensation architectures with transcoding performed in the compressed domain. Since B-frames do not introduce any drift propagation, our discussion is focused on P-frames only. Before presenting the various architectures, an analysis of the drift problem for reduced-resolution transcoding is given. Here, we introduce a reference architecture and compare it to a simple open-loop transcoding architecture. The open-loop architecture is used to identify the sources of drift error with respect to the reference. In the following, the various types of macroblock-level conversions that are needed for reduced resolution transcoding are then discussed. Based on the results of our analysis, we present four transcoding architectures that attempt to overcome the types of drift errors that have been identified. Then the technique for motion refinement is presented. Finally, experimental results, including quality and complexity analysis, as well as our concluding remarks, are presented and discussed.

9.1.2 Transcoding Architecture and Drift Error Analysis

In this section we introduce several transcoding architectures that can be used for MPEG-2 to MPEG-4 transcoding with both bit rate reduction and spatial resolution reduction, and give the error drift analysis for each of them. As we mentioned in previous chapters, there are two major transcoder architectures: cascaded pixel domain transcoder and compressed domain transcoder. The former decodes the input coded video to pixel domain and re-encodes the decoded video into the target format. It is flexible since its decoder loop and encoder loop can be totally independent and they can operate at different bit rates, temporal resolutions, spatial resolutions, and even different standards. Also this architecture can be drift-free if a full re-encoding process is performed. The big problem of this architecture is the high complexity. The compressed domain transcoder directly processes discrete cosine transform (DCT) coefficients instead of decoded pixels. However the compressed domain transcoder lacks the flexibility of the pixel domain transcoder. Generally, it cannot handle temporal or spatial resolution changes without causing drift errors. Therefore, the major efforts of video transcoding research focus on techniques that can minimize the drift error and at the same time can avoid high computational complexity.

In video coding, drift error refers to the continuous decrease in picture quality when a group of motion compensated (MC) inter-frame pictures are decoded. In general, it is due to an accumulation of error in the decoder's MC loop; however, many different factors cause drift. The problem of drift error has been studied for full-resolution transcoding [9-1], and for multilayer coding [9-2], but no explicit analysis has been done for reduced-spatial-resolution transcoding. In the following, we will analyze drift errors by comparing a cascaded closed-loop architecture that is drift-free with an open-loop architecture. The open-loop architecture is the simplest and lowest-complexity architecture that one can consider and it is characterized by severe drift errors due to many sources. In Section 9.1.3, we consider new drift-compensating architectures based on the drift error analysis. These new architectures attempt to simplify the drift-free architecture, while improving the quality of the open-loop architecture.

With regard to notation, lowercase variables indicate spatial domain signals, and uppercase variables represent the equivalent signal in the DCT domain. The subscript on the variable indicates time. A superscript of one denotes an input signal and a superscript of two denotes an output signal. D refers to down-sampling and U to up-sampling. M_f denotes full-resolution motion compensation and M_r denotes reduced-resolution motion compensation. x indicates a full-resolution image signal, y indicates a reduced-resolution image signal, e indicates a full-resolution residual signal, and g indicates a reduced-resolution residual signal. mv_f denotes the set of full-resolution motion vectors and mv_r denotes the set of reduced-resolution motion vectors.

9.1.2.1 Reference Architecture

Figure 9.2 shows a cascaded closed-loop transcoding architecture, which is referred to as the *Reference* architecture. This architecture is the most complex and costly, but it is assumed to have no drift errors and provides a basis for drift error analysis and further simplification. It should be noted that there is no motion estimation process in the encoder loop; the motion vectors are obtained with the MV mapping processor using the motion vectors in the original compressed bit stream. In this sense we should note that the *Reference* architecture does not perform a full re-encoding process, and this is considered in the draft analysis. Since there is no motion-compensated prediction for I-frames, $x_n^1 = e_n^1$. The reconstructed signal is then down-sampled,

$$y_n^1 = D(x_n^1) \tag{9.1}$$

Then, in the encoder input, $g_n^2 = y_n^1$.

In the case of P-frames, the identity

$$x_n^1 = e_n^1 + M_f(x_{n-1}^1) \tag{9.2}$$

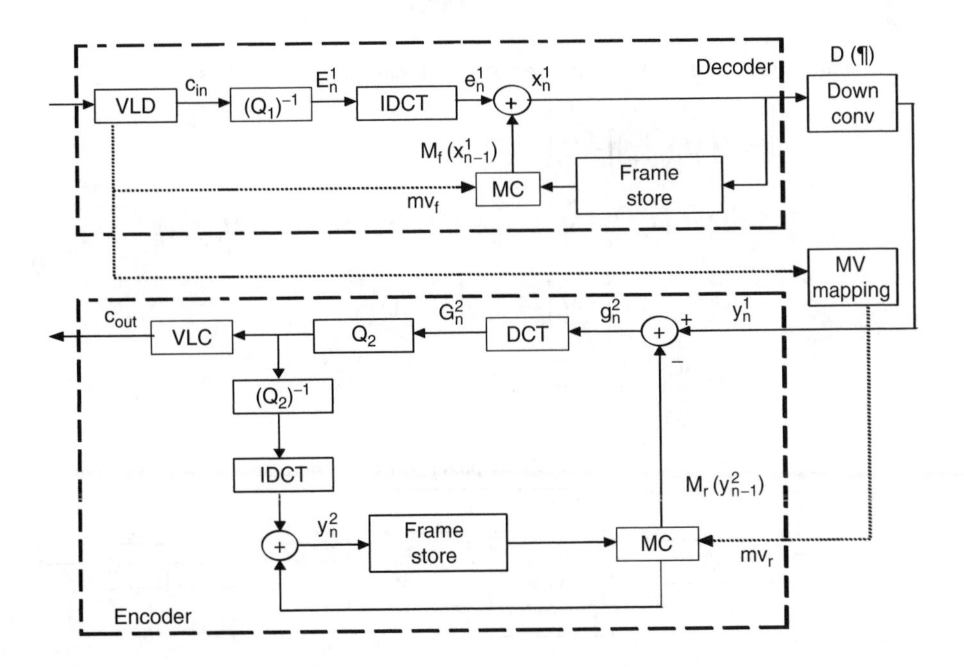

FIGURE 9.2 *Reference* architecture.

yields the reconstructed full-resolution picture. As with the I-frame, this signal is then down-sampled via Equation (9.1). Then, the reduced-resolution residual is generated according to

$$g_n^2 = y_n^1 - M_r\left(y_{n-1}^2\right) \tag{9.3}$$

which is equivalently expressed as,

$$g_n^2 = D\left(e_n^1\right) + D\left(M_f\left(x_{n-1}^1\right)\right) - M_r\left(y_{n-1}^2\right) \tag{9.4}$$

The signal given by Equation (9.4) represents the reference signal that is free of drift errors. Based on this equation, we can analyze drift errors.

9.1.2.2 Drift Error Analysis of Open-Loop Architecture

Figure 9.3 shows the open-loop architecture for a transcoder, referred as *Open-Loop*. The objectives and functions of certain blocks, such as the mixed-block processor, motion vector (MV) mapping, and downs-sampling will be described later in Section 9.1.3.2. At this time, we are mainly concerned with the analysis of drift errors caused by reduced-resolution transcoding. With this *Open-Loop* architecture, the reduced-resolution residual is given by

$$g_n^2 = D\left(e_n^1\right) \tag{9.5}$$

Compared to Equation (9.4), the drift error, d, can be expressed as

$$\begin{aligned} d &= D\left[M_f\left(x_{n-1}^1\right)\right] - M_r\left(y_{n-1}^2\right) \\ &= \left|D\left[M_f\left(x_{n-1}^1\right)\right] - M_r\left(y_{n-1}^1\right)\right| + \left|M_r\left(y_{n-1}^1\right) - M_r\left(y_{n-1}^1\right) - M_r\left(y_{n-1}^2\right)\right| \\ &= \left|D\left[M_f\left(x_{n-1}^1\right)\right] - M_r\left[D\left(x_{n-1}^1\right)\right]\right| + \left|M_r\left(y_{n-1}^1 - y_{n-1}^2\right)\right| \\ &= d_r + d_q \end{aligned} \tag{9.6}$$

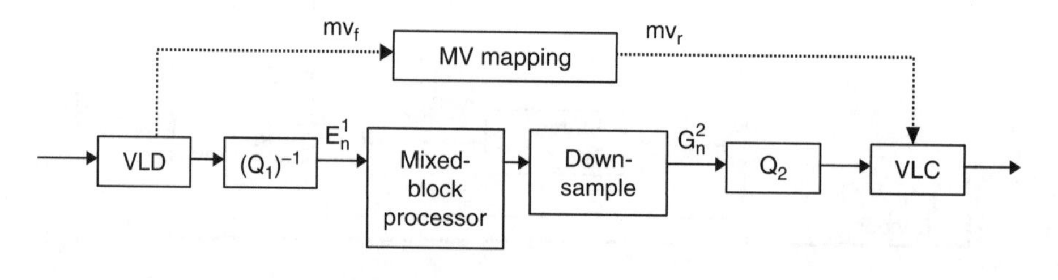

FIGURE 9.3 *Open-Loop* architecture.

where

$$d_q = M_r\left(y_{n-1}^1 - y_{n-1}^2\right) \tag{9.7}$$

and

$$d_r = D\left(M_f\left(x_{n-1}^1\right)\right) - M_r\left(D\left(x_{n-1}^1\right)\right) \tag{9.8}$$

In Equation (9.6), the drift error has been decomposed into two categories. The first component, d_q, represents an error in the reference picture that is used for MC. This error may be caused by requantization, elimination of some nonzero DCT coefficients, or arithmetic error caused by integer truncation. This is a common drift error that has been observed in other transcoding works [9-1]. As a result of this drift error, the pictures originally used as references by the transcoder will be different from their counterparts in the decoder, thus creating a mismatch between predictive and residual components. The second component, d_r, is due to the noncommutative property of motion compensation and down-sampling, which is unique to reduced-resolution transcoding. There are two main factors contributing to the impact of d_r: motion vector (MV) mapping and down-sampling. In mapping MVs from the original resolution to a reduced resolution, a truncation of the MV is experienced due to the limited coding precision. In down-sampling to a lower spatial resolution in the compressed domain, block constraints are often observed to avoid filters that overlap between blocks. Due to this effort to reduce complexity, the quality of the down-sampling process must be compromised and some errors are typically introduced. Regardless of the magnitude of these errors for a single frame, the combination of these two transformations generally creates a further mismatch between the predictive and residual components that will increase with every successively predicted picture. To illustrate this mismatch between predictive and residual components due to the noncommutative property of motion compensation and down-sampling, we consider an example with 1-D signals and neglect any error due to requantization (or d_q). Let b denote the reconstructed block, a denote the reference block, and e denote the error (residual) block, all at the original resolution. Furthermore, let h_v denote a full-resolution motion compensation filter and $h_{v/2}$ denote a reduced-resolution motion compensation filter. Then, the reconstructed block in the original resolution is given by

$$b = h_v a + e \tag{9.9}$$

If we apply a down-conversion process to both sides, we have

$$D(b) = D(h_v a) + D(e), \tag{9.10}$$

The quality produced by the above expression would not be subject to the drift errors included in d_r. However, this is not the signal that is produced by the reduced-resolution transcoder. The actual reconstructed signal is given by

$$\tilde{D}(b) = h_{v/2} D(a) + D(e), \tag{9.11}$$

Since $D(h_v a)$ does not usually equal $h_{v/a}D(a)$, there is a mismatch between the reduced-resolution predictive and residual components. To achieve the quality produced by Equation (9.10), either or both of the predictive and residual components would need to be modified to match each other. In the *Reference* architecture, this mismatch is eliminated with the second (encoder) loop that computes a new reduced-resolution residual. With this second loop, the predictive and residual components are realigned. Our objective in the following section is to consider alternative ways to compensate for drift with reduced complexity. Before that, we will first explain several issues related to reduced-resolution transcoding.

9.1.3 Transcoding at Macroblock Layer

In this section, we will discuss some issues related to reduced-resolution transcoding that correspond to three functional units: mixed block processing, MV mapping, and texture down-sampling. These blocks are presented in the *Open-Loop* architecture, which is shown in Figure 9.3, and will also be used in the new drift-compensating architectures.

9.1.3.1 Mixed Block Processing

In transcoding the compressed video to a lower spatial resolution, say one quarter of the original size, a group of four macroblocks in the original video corresponds to one macroblock in the transcoded video. So, the purpose of the mixed block processor is to preprocess selected groups of macroblocks to ensure that the down-sampling process will not generate an output macroblock in which its sub-blocks have different coding modes, e.g., both inter and intra sub-blocks within a single macroblock. Mixed coding modes within a macroblock are not supported by any known video coding standards. The function of the mixed-block processor includes producing motion vectors and corresponding residues for inter-frames. Because inter-frames contain both intra and inter macroblocks, we need first to decide the macroblock (MB) mode in the output bit stream from the MB modes of the four corresponding macroblocks in the original bit stream. If the four input MB modes are intra, the output MB mode is intra. Similarly, if the four input MB modes are inter, the output MB mode will also be inter. If neither of these cases is true, we encounter a group of *mixed blocks*, i.e., a group of macroblocks that contain both intra and inter coded macroblocks. In this case, we need to decide on a consistent mode and modify the associated motion vectors and DCT coefficients accordingly so that a non-mixed block is produced in the output. We propose three methods to accomplish this mixed block processing. In the first method, *Zero-Out*, the MB modes of the mixed macroblocks are all modified to inter mode. The MV's for the intra macroblocks are reset to zero and so are corresponding DCT coefficients. In this way, the input macroblocks that have been converted are replicated with data from corresponding blocks in the reference frame.

The second method is called *Intra-Inter*. In this method, the MB modes are also modified to be inter mode, but the motion vectors for the intra macroblocks are predicted. The prediction is based on the data in neighboring blocks, which can include both texture and motion data. In our experiments, the following estimate of the new motion vector is used,

$$mv_{\text{pred}} = \sum w_i \bullet mv_{ni},$$

$$\sum w_i = 1, \quad i = 1,2,... \tag{9.12}$$

where mv_{pred} is the predicted motion vector for the intra macroblock that needs to be converted, mv_{ni} represents the neighboring inter macroblocks, and w_i is the corresponding weighting factor based on the DCT energy of the inter macroblocks. As an alternative, we can simply set the motion vector to zero, depending on which produces less residual. In an encoder, the mean absolute difference of the residual blocks is typically used for mode decision. The same principles can be applied here. Based on the predicted motion vector, a new residual for the modified macroblock is calculated.

In the third method, *Inter-Intra*, the MB modes are all modified to intra mode. In this case, there is no motion information associated with the reduced-resolution macroblock, therefore all associated motion vector data is reset to zero. This step must be performed in the transcoder because the motion vectors of neighboring blocks are predicted from the motion vectors of this block. To ensure proper reconstruction in the decoder, the motion vectors for the particular group of macroblocks must be reset to zero in the transcoder. The intra DCT coefficients are generated to replace the inter DCT coefficients. It should be noted that to implement the second and third methods, we need a decoding loop to reconstruct a full-resolution picture. The reconstructed data is used as a reference to convert the DCT coefficients from intra to inter, or from inter to intra. For a sequence of frames with a small amount of motion and a low level of detail, the low-complexity strategy of *Zero-Out* can be used. Otherwise, either *Intra-Inter* or *Inter-Intra* should be used. The performance of *Inter-Intra* is a little better than that of *Intra-Inter*, because *Inter-Intra* can stop drift propagation by transforming inter blocks to intra blocks.

9.1.3.2 Motion Vector Mapping

When down-sampling four macroblocks to one macroblock, the associated motion vectors have to be mapped. Several methods suitable for frame-based motion vector mapping have been described in past work [9-3], [9-4]. To map from four frame-based motion vectors, i.e., one for each macroblock in a group, to one motion vector for the newly formed macroblock, a weighted average or median filters can be applied. This is referred to as a 4:1 mapping. However, with certain compression standards, such as MPEG-4 and H.263, there is a support in the syntax for advanced prediction modes that allow one motion vector per block, i.e., sub-block motion. In this case, each motion vector is mapped from a 16 × 16 macroblock in the original resolution to an 8 × 8 block in the reduced resolution macroblock with appropriate scaling by 2. This is referred as a 1:1 mapping. An illustration of 4:1 and 1:1 motion

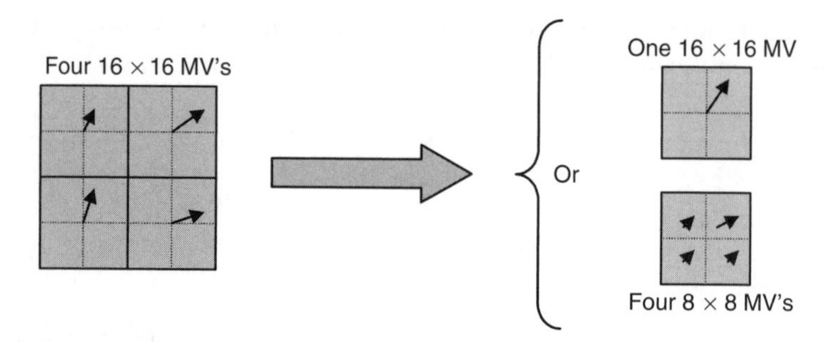

FIGURE 9.4 Illustration of 4:1 and 1:1 motion vector mapping.

vector mapping is shown in Figure 9.4. The 1:1 mapping usually performs better than the 4:1 mapping if no motion compensation is performed [9-4]. However, it is inefficient to always use the 1:1 mapping, because more bits are needed to code four motion vectors. We propose a mixed mapping method based on the variance of the four input motion vectors. If the variance of these motion vectors is greater than a threshold, then use the 1:1 mapping method presented in [9-4], otherwise use 4:1 mapping. Because MPEG-2 supports interlaced video, we also need to consider field-based MV mapping. We first convert the field-based MVs to a frame-based MV by using the top-field MV as the default [9-5]. However, if the bottom-field MV is used to predict the top field, the top-field and bottom-field motion vectors are averaged. Then, we divide the horizontal MV by 2 (the vertical MV is already halved by using the top-field MV). It should be pointed out that there is a relation between the methods used for texture down-sampling (discussed in the next section) and MV mapping. This is especially true when converting from an interlace source to a progressive output. We have tried various combinations of down-conversion filters and MV mappings, and have verified experimentally that this particular MV mapping method works slightly better than if the top-field MV is always used.

9.1.3.3 Texture Down-Sampling

For the purpose of down-sampling texture data, this section mainly relies on the filtering concepts that we developed a few years ago for memory saving decoders. A review of this work can be found in [9-6] and references therein. For the transcoder, there are some new aspects that must be considered. However, since the derivation of new filters is beyond the scope of this chapter, we only provide a brief explanation of this process and expressions for particular filters used in our experiments.

In our earlier work [9-48], the concept of frequency synthesis has been proposed for the transformation of an input DCT macroblock consisting of four 8 × 8 DCT blocks into a single 8 × 8 DCT block. The computations are actually performed on the rows and columns of the macroblock using separable 1-D filters. Let $\underline{A}$ and $\underline{B}$ denoted the input vectors of size N. Then, the output data, $\underline{E}$, also of size N, is computed according to

$$\underline{E} = f_1 \cdot \underline{A} + f_2 \cdot \underline{B} \tag{9.13}$$

where

$$f_1(k,p) = \sum_{i=0}^{N-1} \psi_p^N(i) \cdot \psi_k^{2N}(2i)$$

$$f_2(k,p) = \sum_{i-0}^{N-1} \psi_p^N(i) \cdot \psi_k^{2N}(2i+N) \tag{9.14}$$

and

$$\psi_k^N(i) = \sqrt{\frac{2}{N}}\alpha(k)\cos\left(\frac{2i+1}{2N}k\pi\right) \tag{9.15}$$

and $\alpha(k) = 1/\sqrt{2}$ for $k = 0$, and 1 for $k \neq 0$. The down-conversion filters given by Equation (9.14) can be applied in both the horizontal and vertical directions, and to both frame-DCT and field-DCT blocks. In our experiments, we aim at producing a progressive (MPEG-4) output. Therefore, if the input is a frame-DCT block, we apply the down-conversion filters of Equation (9.14) to both fields. On the other hand, if the input is a field-DCT block, a different set of filters is used to simultaneously perform a field-to-frame conversion and down-conversion in the DCT domain. These filters are given by

$$f_1(k,p) = \sum_{i=0}^{N-1} \psi_p^N(i) \cdot \psi_k^{2N}(2i) \tag{9.16}$$

$$f_2(k,p) = \sum_{i-0}^{N-1} \psi_p^N(i) \cdot \psi_k^{2N}\left(2i+\frac{N}{2}\right)$$

We would also like to point out that all of the above DCT filters could easily be transformed to mathematically equivalent spatial domain filters [9-6]. In fact, such filters are used in our *Reference* architecture to obtain a pure comparison of the architecture and avoid any differences due to down-sampling.

9.1.4 Architectures for Drift Compensation

In the previous section, the drift errors related to reduced-resolution transcoding were analyzed by comparing the signals produced by the *Reference* and *Open-Loop* architectures. The *Open-Loop* architecture is the simplest, but the drift error can be very severe. On the other hand, the *Reference* architecture is completely drift-free, but the complexity is high. Therefore, it is desirable to approximate the quality of the *Reference* architecture with a method that achieves significant complexity reduction. In this section, we will present four new architectures that provide different means of drift error compensation. The first two can be viewed as extensions of the full-resolution transcoding architectures presented in [9-1]. The signal given by Equation (9.4) represents the reference signal that the architectures described here approximate. Since I-frames do not cause drift and the processing of I-frames is simple and mature [9-3], we will focus our discussion on P-frames only. We note that several methods have been proposed to perform motion compensation in the frequency domain, e.g., [9-1]. In this way, some DCT/IDCT computation can be avoided, but these involve matrix multiplications that may be computationally complex. In the architectures presented in this section, we adopt spatial-domain motion compensation for simplicity. However, corresponding frequency-domain approaches can easily be applied.

9.1.4.1 Drift Compensation in Reduced Resolution

Figure 9.5 shows the first closed-loop architecture that performs drift compensation in the reduced resolution. This architecture will be referred to as *Drift-Low*. The reduced-resolution residual signal in Figure 9.5 is expressed as

$$g_n^2 = D\left(e_n^1\right) + M_r\left(y_{n-1}^1 - y_{n-1}^2\right) \tag{9.17}$$

which can be generated from Equation (9.4), with the approximation

$$D\left(M_f\left(x_{n-1}^1\right)\right) \approx M_r\left(D\left(x_{n-1}^1\right)\right) = M_r\left(y_{n-1}^1\right) \tag{9.18}$$

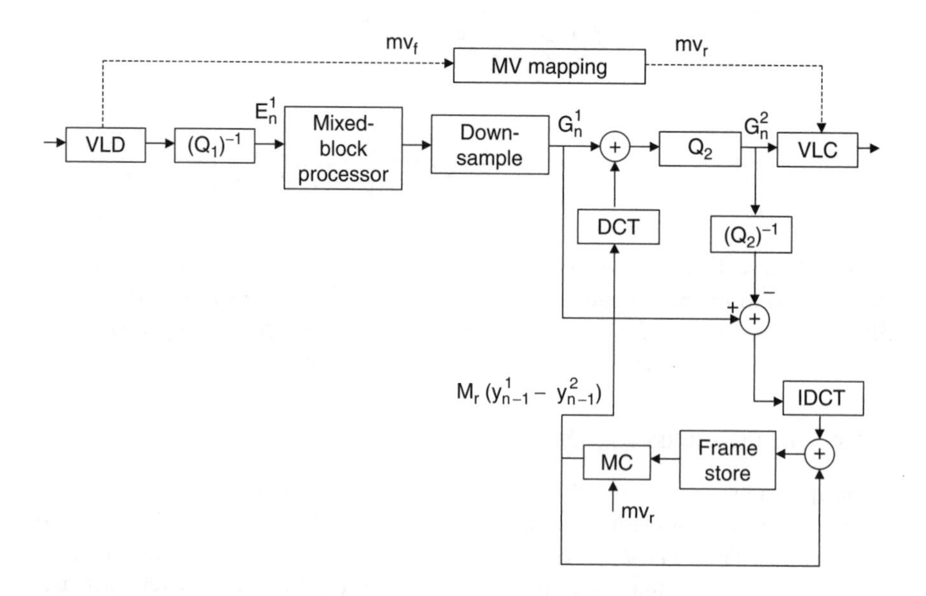

FIGURE 9.5 Drift compensation in reduced spatial resolution.

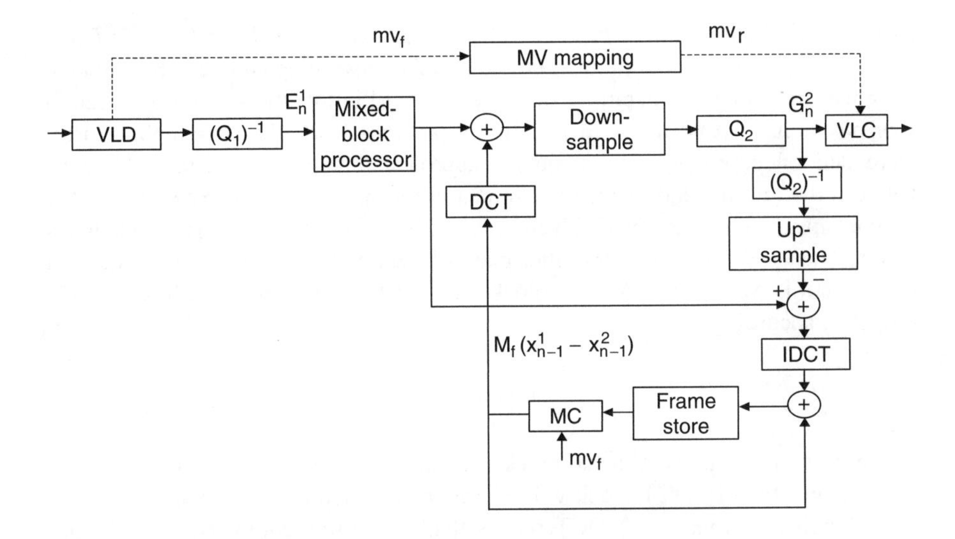

FIGURE 9.6 Drift compensation in original resolution.

The drift compensation is performed on a macroblock-by-macroblock basis using the set of reduced-resolution motion vectors. This architecture attempts to eliminate drift errors caused by d_q. It should be noted that the mixed block processor would be implemented as an MC loop with frame memory if the *Intra-Inter* or *Inter-Intra* configurations discussed in Section 9.1.3.1 are chosen. Only the *Zero-Out* configuration would avoid this additional loop.

9.1.4.2 Drift Compensation in Original Resolution

Figure 9.5 presents the second closed-loop architecture that performs drift compensation in the original resolution. This architecture will be referred to as *Drift-Full*. The reduced-resolution residual signal in Figure 9.6 is expressed as

$$g_n^2 = D\left(e_n^1 + M_f\left(x_{n-1}^1 - x_{n-1}^2\right)\right) \tag{9.19}$$

which can be generated from equation (9.4), with the approximation

$$M_r\left(y_{n-1}^2\right) \approx D\left(M_f\left(U\left(y_{n-1}^2\right)\right)\right) = D\left(M_f\left(x_{n-1}^2\right)\right) \tag{20}$$

This architecture requires up-sampling in the DCT domain, which is briefly discussed later. In this architecture, the drift compensation is performed on a macroblock-by-macroblock basis using the set of original-resolution motion vectors. This architecture attempts to eliminate drift errors caused by d_q, as well as drift errors caused by d_r. As with the *Drift-Low* architecture, an additional MC loop with frame memory may be substituted for the mixed block processor depending on the particular configuration that has been chosen, i.e., *Zero-Out, Intra-Inter,* or *Inter-Intra*. The up-sampling operation required by this transcoding architecture is unique from past up-sampling operations described in [9-6] in that the subdivision back to four blocks has not been considered. In our previous work, the up-sampling filters were derived based on a least-squares approach that minimized the difference between the original and reconstructed data in the spatial domain. Specifically, the optimal upsampling filter, X^+, is given by $X^+ = X^T(XX^T)^{-1}$, where $X = [f_1, f_2]$ denotes a concatenated down-conversion filter matrix based on the filters given in Equation (9.14). In this way, an $N \times N$ DCT block, $\underline{E}$, can be up-sampled to a $2N \times 2N$ DCT block, $\underline{C}$, according to

$$\underline{C} = X^+ \cdot \underline{E} \tag{9.21}$$

However, with the above, the sub-block structure of a macroblock is not recovered, i.e., a single 16 × 16 DCT block will be obtained, which is clearly not suitable for the difference of four 8 × 8 DCT blocks that need to be computed after the up-sampling operation. Without proof, we claim that two reconstructed $N \times N$ DCT blocks, $\hat{\underline{A}}$ and $\hat{\underline{B}}$, can be obtained through the following expressions:

$$\hat{\underline{A}} = y_1 \cdot \underline{E}$$

$$\hat{\underline{B}} = y_2 \cdot \underline{E} \tag{9.22}$$

where

$$y_1(k,p) = X^+(k,p) \cdot N - 1 \sum_{i=0}^{N-1} \psi_p^{N-1}(i) \cdot \psi_k^N(i)$$

$$y_2(k,p) = X^+(k,p) \cdot N - 1 \sum_{i=0}^{N-1} \psi_p^{N-1}(i+N) \cdot \psi_k^N(i) \tag{9.23}$$

Applying the expression given in Equation (9.22) to rows and columns of the down-sampled DCT block allows us to reconstruct four 8 × 8 DCT blocks. These blocks can be then be subtracted from the DCT blocks prior to encoding to yield the difference signal used for compensating the drift in the original resolution.

9.1.4.3 Partial-Encode Architecture

Figure 9.7 shows the third closed-loop architecture that performs drift compensation in the reduced resolution. We shall refer to this architecture as *Partial-Encode*. The reduced-resolution residual signal in Figure 9.6 is expressed as

$$g_n^2 = D(e_n^1) + D\left(M_f\left(x_{n-1}^1\right)\right) - M_r\left(D\left(x_{n-1}^1\right)\right) \tag{9.24}$$

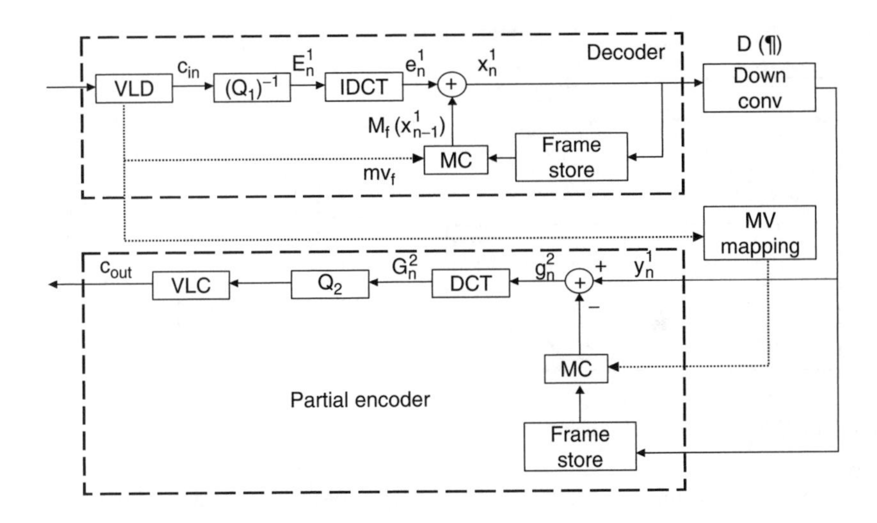

FIGURE 9.7 *Partial-Encode* architecture.

which can be generated from Equation (9.4), with the approximation

$$y_{n-1}^2 \approx y_{n-1}^1 = D\left(x_{n-1}^1\right) \tag{9.25}$$

The *Partial-Encode* architecture attempts to eliminate drift errors caused by d_r. It should be noted that since full-resolution decoding is performed with this architecture, there is no mixed block problem.

9.1.4.4 Intra Refresh Architecture

As presented in Section 9.1.3.1, the *Inter-Intra* and *Intra-Inter* configurations of the mixed block processor require an additional MC loop to reconstruct a full-resolution picture. In this section, we exploit this structure by considering methods to control or minimize the drift. We will present a new method using intra refresh to stop the drift, which can be applied in both open-loop and closed-loop architectures. In the mixed-block processor, the *Inter-Intra* method provides a way to convert an inter coded block to an intra-coded block. Because intra-coded blocks are not subject to drift, this kind of conversion stops the drift propagation, for both d_q and d_r. We refer to this technique of using *Inter-Intra* for drift compensation as *Intra-Refresh*. This method has successfully been applied to error resilience coding schemes [9-7], and we find the principle is also very useful for reducing drift in a transcoder. The *Intra-Refresh* architecture is illustrated in Figure 9.8. As seen from the figure, this architecture has only one motion-compensated loop, which is used to reconstruct a full-resolution picture to convert the DCT coefficients from inter to intra. As we will show, this architecture is very flexible in that it can adapt itself based on the input signal characteristics and transcoding condition. Besides providing an effective yet simple means to combat both types of drift errors, d_q and d_r, it is also capable of correcting any errors that may be caused by incorrect MV mapping. These properties make this architecture

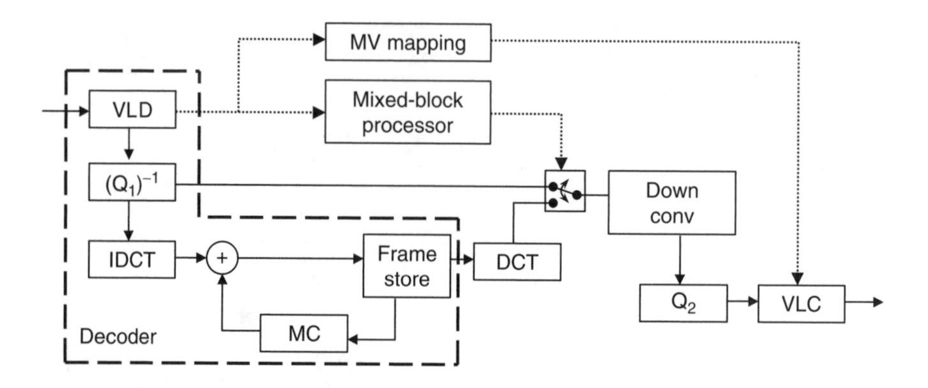

FIGURE 9.8 *Intra-Refresh* architecture.

very appealing. At this time, however, we focus our discussion on its ability to correct drift errors only and do not consider any other sources of error. In the following, we consider two steps for the intra refresh process. First, the amount of drift in the compressed bit stream is estimated. The drift can be estimated based on data pertaining to the amount of motion vector truncation, residual energy, motion activity, and requantization error. Second, the estimated value of drift is translated to an intra refresh rate, β, that is used as input to the mixed-block processor. β is the percentage of intra-coded macroblocks in one frame. One point to keep in mind about the intra refresh technique is that more bits are usually required for coding intra blocks. For this reason, the intra refresh process and the rate control must be considered jointly. The specific issues involved will be discussed in more detail in the following section.

9.1.4.5 Experimental Results

In this section, we present a comparative study of the quality and complexity of the architectures under consideration. We simulated all the proposed architectures in software transcoding MPEG-1 and MPEG-2 bit streams to quarter-resolution MPEG-4 bit streams. The machine we used to run in the simulations is an Ultra 60 workstation with processor speed 360 MHz, model UltraSPARC-II, with 4 MB L2 Cache and 512 MB memory, and the source code used in these experiments has not been optimized for speed. Table 9.1 shows the main blocks contributing to the complexity of each implementation, independent of the platform specifications. The table also provides a relative indication of the expected amount of drift error and the primary source of that error. With regard to the complexity of the architectures that employ inter-to-intra or intra-to-inter conversions, it should be noted that certain blocks involved in the conversion process are not invoked for every macroblock. For instance, the DCT block in the *Intra-Refresh* architecture is only used for blocks that require a conversion, which may be only a fraction of the total number of blocks in a given frame, e.g., 10%. This is especially important for software implementations that rely mainly on average processing times. Also, in our simulations, the *Inter-Intra* configuration for the mixed block processor has been adopted for the *Drift-Low* and *Drift-Full* architectures since this leads to the best quality. The numbers in Table 9.1 take

TABLE 9.1
Comparison of Transcoding Architectures

Transcoding Architecture	DCT/IDCT	MC Loop	Frame Buffer	Up-sample	Drift Source	Drift Amount
Reference	4	2	2	No	No	Low
Open-Loop	0	0	0	No	$d_r + d_q$	High
Drift-Low	4	2	2	No	d_r	Medium
Drift-Full	4	2	2	Yes	d_q	Medium
Partial-Encode	2	2	2	No	d_q	Low
Intra-Refresh	1	1	1	No	Partial	Low

this configuration into account, and would be lower if the *Zero-Out* configuration were used. The tests are conducted with various input sequences at different resolutions, formats, and encoding pattern (GOP structure). All the sequences are coded at 30 fps. In the *Intra-Refresh* simulations, we set initial thresholds: T_r = 1,000,000 and T_m = 6000. During transcoding, both thresholds are adaptively adjusted with the algorithm of adaptive intra refresh (AIR) technique, which will be discussed in Section 9.2.3. For comparison, we also simulate an *Optimal* architecture, which consists of an MPEG-2 decoder, followed by spatial-domain subsampling, and a full MPEG-4 encoder with motion re-estimation. This surely has the best quality, but has at least 37 times more computational cost than the *Reference* architecture [9-3].

We first conduct some experiments using input bit streams that have been encoded with periodic I-frames. The primary purpose of this experiment is to analyze the quality and complexity of all the presented architectures under common conditions that are somewhat tolerant to the accumulation of drift, but not completely forgiving. The original sequence is CIF resolution (352 × 288) and coded as an MPEG-1 bit stream with N = 15 and M = 3. In the first experiment, we consider the Akiyo sequence, which has low motion and a low level of detail. The source bit rate is 512 kbps and the target rates for the transcoder are 32, 64, and 96 kbps. The simulation shows that even the simplest open-loop architecture *Zero-Out* can achieve reasonably good quality, and higher quality can be reached by other architectures with higher complexity.

Table 9.2 shows processing time in seconds, as well as the mean and variance of the frame-based PSNR values (in dB) for all the architectures. We can see from the table that *Inter-Intra, Intra-Inter, Intra-Refresh, Partial-Encode*, and *Drift-Low* have comparable quality with the *Reference*. Also, for these architectures, we have recorded speed-up factors of 3.59, 3.59, 3.4, 1.68, and 1.36, respectively, in comparison to the *Reference*. For the second experiment, the *Foreman* sequence, which has a medium amount of motion and a medium level of detail, is considered. The original bit rate is 2 Mbps. The target rates for the transcoder are 128, 384, and 512 kbps. From simulation results, we observed that artifacts could be found in *Zero-Out, Intra-Inter, Inter-Intra*, and *Drift-Low*. We believe these artifacts are the result of d_r becoming more dominant in d with increased motion. However, such artifacts were not observed in *Intra-Refresh* and *Partial-Encode*. For the range of bit rates tested, these two architectures were able to achieve similar quality to the *Reference*, while decreasing

TABLE 9.2
Experimental Results for Akiyo Sequence

	Time	32 kbps		64 kbps		96 kbps	
		mean	var	mean	var	mean	var
Optimal	/	32.13	0.23	34.83	0.67	36.12	0.82
Reference	17.6	31.77	0.57	34.78	0.55	36.26	0.96
Intra-Refresh	5.2	31.30	2.84	34.31	1.87	35.34	2.28
Partial-Encode	10.5	31.83	2.25	34.40	1.79	35.66	1.92
Drift-Low	12.9	31.12	0.99	34.02	0.97	35.08	1.86
Drift-Full	21.7	31.52	0.69	34.73	0.44	36.09	0.88
Intra-Inter	4.9	31.46	3.20	34.16	2.89	35.06	2.81
Inter-Intra	4.9	31.46	3.46	34.22	2.51	35.11	2.71
Zero-Out	3.6	31.51	3.33	33.91	3.27	34.81	3.62

the time by a factor of 1.7 and 1.4, respectively. We note that the speed-up for *Intra-Refresh* has decreased in this simulation compared to the simulation with the Akiyo sequence. This is due to the increased need in transcoding the Foreman sequence for inter-to-intra conversion, which mainly requires an added DCT for each converted block. Summarized results for all the other architectures are provided in Table 9.3. To truly test the drift compensation capabilities of the *Intra-Refresh* and *Partial-Encode* architectures, we test two sequences that have been coded with $N = 100$ and $M = 1$. The first sequence is Foreman with CIF resolution and coded as MPEG-1 bit stream at a bit rate of 2 Mbps. The target rates for the transcoder are 512 kbps and 256 kbps. The second sequence is Football, which has fast motion and a high level of detail. The original sequence is a ITU-T 601 format (720480, interlace) and coded as an MPEG-2 bit stream at a rate of 6 Mbps. The target rates for the transcoder are 1.5 Mbps and 750 kbps. Figure 9.9 and Figure 9.10 provide frame-based PSNR plots of the transcoded results using selected architectures.

TABLE 9.3
Experimental Results for Foreman Sequence

	Time	128 kbps		384 kbps		512 kbps	
		mean	var	mean	var	mean	var
Optimal	/	30.91	0.59	32.17	1.37	32.36	1.63
Reference	20.8	29.76	1.56	32.02	1.09	32.32	1.29
Intra-Refresh	12.2	29.28	2.72	31.33	3.43	31.69	3.11
Partial-encode	14.7	28.25	2.69	30.90	1.51	31.54	1.64
Drift-Low	16.2	28.88	2.40	30.13	3.71	30.30	4.08
Drift-Full	26.9	29.61	1.53	31.03	2.44	31.25	2.78
Intra-Inter	7.7	28.00	3.69	29.56	3.85	29.85	4.18
Inter-Intra	7.9	28.34	3.26	29.86	3.98	30.14	4.40
Zero-Out	5.8	25.79	13.01	26.46	14.75	26.58	14.93

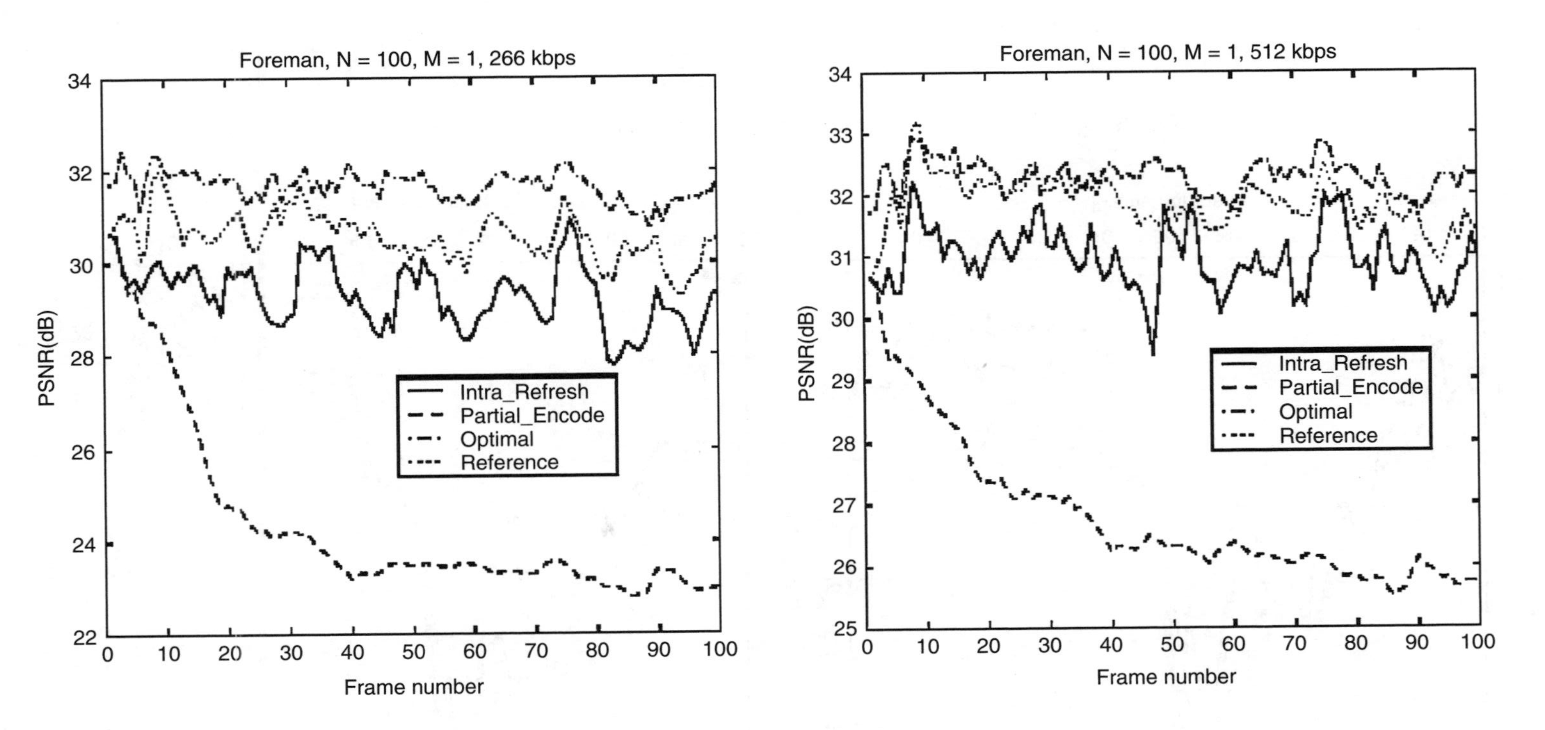

FIGURE 9.9 Experimental results for reduced-spatial-resolution transcoding (by a factor of 2) for Foreman sequence.

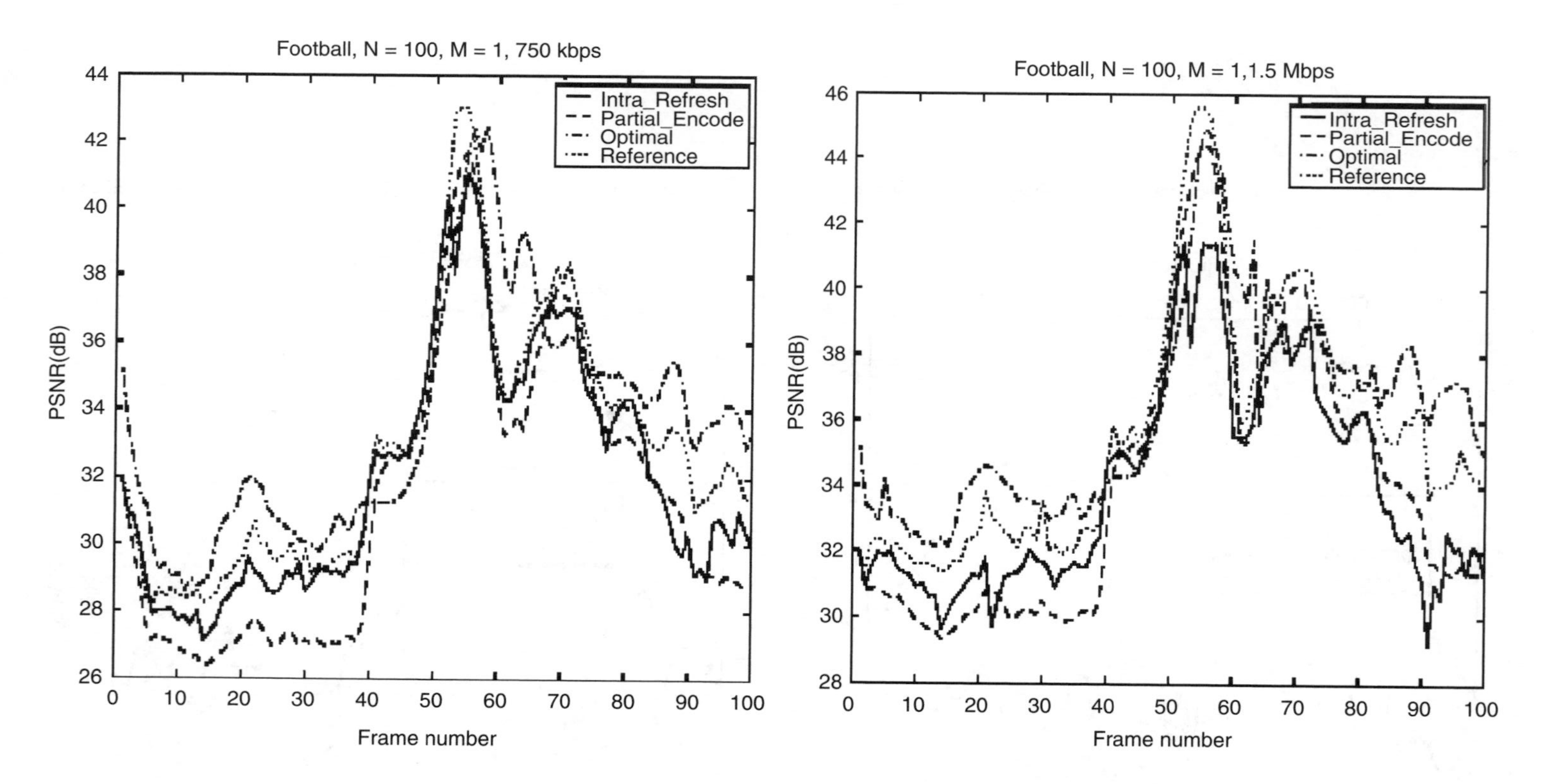

FIGURE 9.10 Experimental results for reduced-spatial-resolution transcoding (by a factor of 2) for Football sequence.

From these plots, we can see that *Intra-Refresh* compensates for drift quite well in both sequences, even with such a long series of successive predictions. However, the results are not so consistent for *Partial-Encode*, i.e., a strong decay over time is observed for Foreman and almost no decay for Football. There are two explanations for this occurrence. The first reason is that for Foreman, d_q plays a more critical role as N increases, whereas for Football, d_r is always more influential than d_q due to the high motion. The second reason is due to the large number of intra blocks in P-frames that were used to encode Football, which also relates to the high motion complexity in the sequence. This result indicates that drift propagates less in such sequences. From the experiments overall, we notice that *Drift-Full* with *Inter-Intra* is even more complex than the *Reference*. Therefore, this architecture is not recommended for any application. For simple sequences with low motion, low level of detail, low bit rate, short GOP, and small frame size, *Zero-Out* can be used to obtain reasonably good quality. If higher quality is required, other architectures, such as *Inter-Intra, Intra-Inter, Intra-Refresh, Partial- Encode*, and *Drift-Low* should be considered. For sequences with medium to fast motion, artifacts can be found in *Zero-Out*, *Intra-Inter*, *Inter-Intra*, and *Drift-Low*. Compared to the *Reference* architecture, only *Intra-Refresh* can achieve similar quality with less complexity. *Partial-Encode* can only be used in sequences with a short GOP, or sequences with a long GOP that have a high amount of intra-coded blocks. Through the simulations, we observe that the *Intra-Refresh* architecture can achieve quality comparable to *Reference*, especially for sequences with large motion, long GOP, and large frame size. As the scene complexity increases, we have observed that the speed-up over the *Reference* decreases, i.e., by a factor of 3.4 for Akiyo, 1.7 for Foreman, and 1.44 for Football. These results demonstrate the flexibility of the *Intra-Refresh* architecture and the balance that it provides in terms of quality and complexity.

9.1.5 Motion Vector Refinement

In all of the transcoding methods described so far a significant complexity is reduced by assuming that the motion vectors computed at the original bit rate are simply reused in the reduced-rate bit stream. It has been shown that reusing the motion vectors in this way leads to nonoptimal transcoding results due to the mismatch between prediction and residual components [9-9],[9-10]. To overcome this loss of quality without performing a full motion re-estimation, motion vector refinement schemes have been proposed [9-11],[9-12]. Typically, the search window used for motion vector refinement is relatively small compared to the original search window, e.g., [–2,+2]. This not only keeps the added complexity down, but it provides a significant amount of the achievable gains. Such schemes can easily be used with most bit-rate-reduction architectures for improved quality, as well as with the spatial- and temporal-resolution-reduction architectures. A comparison of results obtained with and without motion vector refinement is presented below in the context of spatial resolution reduction. The impact on the search window size is illustrated in Figure 9.11.

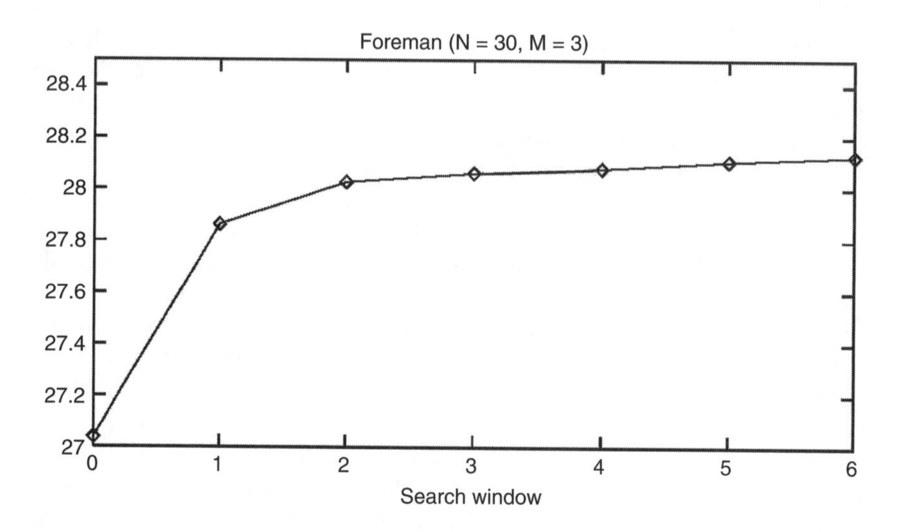

FIGURE 9.11 Average PSNR of the luminance component as a function of motion vector refinement search window.

The simulation results provided in Figure 9.11 and Figure 9.12 illustrate the impact of MV refinement techniques for spatial resolution reduction. The input bit stream we used is the CIF resolution Foreman sequence coded as an MPEG-1 video bit stream at 2 Mbps with a GOP structure of $N = 30$, $M = 3$. The QCIF resolution output is transcoded to an MPEG-4 visual format (simple profile) with a bit rate of 64 kbps and frame rate of 10 fps. It can be seen from the plot in Figure 9.11 that

FIGURE 9.12 Sample frames for comparison of transcoded quality with and without motion vector refinement. For the transcoded output that uses refinement, a search window size of 3 is used.

the average PSNR of the luminance component increases as a function of the search window size. However, a very small search window achieves the majority of the gain. This is due to the fact that that the majority of blocks find the best-matching motion vector (according to the specified criterion) within this range. Increasing the search window further allows more blocks to find their best match; however, since the number of blocks that will find a better match is fewer, the overall gain is less. It should be noted that finding a better match will decrease the residuals that need to be coded for each macroblock, hence allowing a finer quantization (better quality) under the same rate constraints.

In Figure 9.12 sample frames are displayed to compare the visual quality of transcoded frames with and without MV refinement. It is evident from these frames that the motion vector refinement process eliminates a significant amount of noise in the reconstructed output.

9.1.6 Motion Vector Re-Estimation and Residual Re-Estimation

The problems of MPEG-2 to MPEG-4 transcoding addressed so far include the bit rate reduction, spatial resolution reduction, and frame rate reduction. It should be noted that the frame rate reduction in the previously discussed transcoding architectures is achieved by dropping only B-frames but not any P-frames. If we want to achieve more flexibility on the frame skipping in the transcoding, we have to solve the problem of re-estimating a new motion vector from the current frame to a previous non-skipped frame when the original reference frame is skipped, as shown in Figure 9.13. As described in [9-10], [9-13], and [9-14], the problem of re-estimating a new motion vector from the current frame to a previous non-skipped frame can be solved by tracing the motion vectors back to the desired reference frame. However, the blocks in the current frame are predicted generally with multiple blocks in the skipped reference frame. Bilinear interpolation of the motion vectors in the previous skipped frame may be used, where the weighting of each input motion vector is proportional to the amount of overlap with the predicted

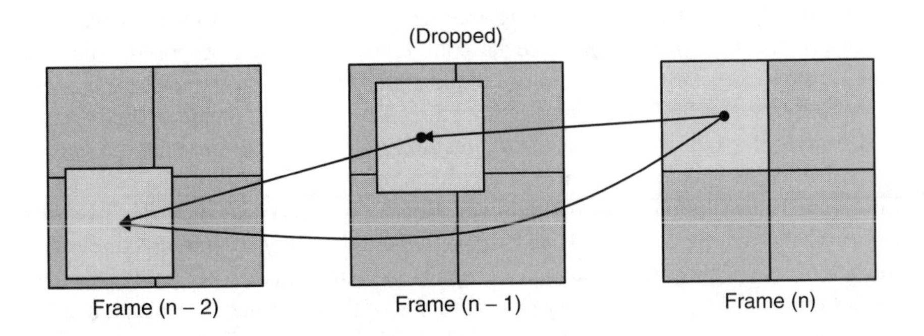

FIGURE 9.13 Motion vector re-estimation. Since frame $n - 1$ is dropped, a new motion vector to predict frame n from frame $n - 2$ is estimated.

block. In place of this bilinear interpolation, a dominant vector selection scheme as proposed in [9-10] and [9-15] may also be used, where the motion vector associated with the largest overlapping region is chosen. In order to trace back to the desired reference frame in the case of skipping multiple frames, the process can be repeated. It is suggested, however, that a refinement of the resulting motion vector be performed for better coding efficiency as discussed in the Section 9.1.5. In [9-14], an algorithm to determine an appropriate search range based on the motion vector magnitudes and the number of frames skipped has been proposed. To dynamically determine the number of skipped frames and maintain smooth playback, frame rate control based on characteristics of the video content have also been proposed [9-13][9-14].

When the new motion vector is reestimated, the residual data need to also be reestimated accordingly. The problem of estimating a new residual for temporal resolution reduction is primarily an issue for DCT-domain transcoding architectures. With pixel-domain architectures, the residual between the current frame and the new reference frame can be easily computed given the new motion vector estimates. For DCT-domain transcoding architectures, this calculation should be done directly using DCT-domain motion compensation techniques [9-16]. A novel architecture to compute this new residual in the DCT domain has been presented in [9-11] and [9-15]. In these works, the authors utilize direct addition of DCT coefficients for macroblocks without motion compensation, as well as an error-compensating feedback loop for motion-compensated macroblocks. The combination of these techniques has been shown to reduce requantization errors incurred during transcoding, and do so with less computational complexity.

9.1.7 Summary of MPEG-2 to MPEG-4 Transcoder

This section presents several architectures and methods for MPEG-2 to MPEG-4 with reduced resolution transcoding. Through the analysis of the drift error for this type of transcoding, we have identified two distinct sources of drift: one type due to errors in the reference picture, which have been studied in other works and arise mainly from requantization errors; and a second type that is due to the noncommutative property of motion compensation and down-sampling. With these sources of error in mind, four drift-compensating architectures that attempt to approximate the quality of the reference and reduce its complexity have been proposed. The first architecture, *Drift-Low*, attempted to compensate for the reference picture error in the reduced resolution. The second architecture, *Drift-Full*, attempted to do the same in the original resolution. The third architecture, *Partial-Encode* targeted the second type of drift error only and the final architecture, *Intra-Refresh*, relied on a novel intra-block refresh method to partially compensate for all types of errors. From the experiments we have conducted, we believe that the *Intra-Refresh* architecture offers the best trade-off between quality and complexity. In most cases, the quality provided by this transcoding method is very similar to that of the reference. It is also a flexible and adaptable architecture that can easily be scaled in terms of its complexity and quality to meet any number of system needs. With regard to future work, a better rate control model to combine the intra refresh process with the quantizer selection requires more investigation. Another interesting topic is the combination of temporal resolution reduction with spatial resolution reduction for video transcoding. Finally, we think there are some interesting studies to be done regarding the relation between MV mapping and texture down-sampling, especially within the context of interlace-to-progressive transcoding.

9.2 ERROR RESILIENCE VIDEO TRANSCODER

Transmitting video bit stream over wireless channels has to take into account the harsh conditions to which the transmitted video bit stream is subjected. In general, wireless channels have relatively low bandwidth and higher error rates than wired channels; therefore, the error resilience capability is important for the transmission of video bit streams over wireless networks. In this section we introduce the basic concept and techniques for error resilience video transcoding.

9.2.1 Basic Concept of Error Resilience Transcoding

In the video coding system many basic techniques have been used to remove the spatial and temporal redundancies for increasing the compression performance, as described in Chapter 2. The main techniques used include: predictive coding with motion compensation for removing temporal redundancy, DCT for decomposing the original video data to the unequal energy distribution, and adaptive quantization for removing some less important information. All of these techniques produce certain data: the motion vectors, the transformed/quantized coefficients, and the predictive residues encoded with the variable length coding (VLC) with look-up tables. The look-up tables are generated by way of statistical optimization (e.g., Huffman coding) on the extensive video training sequences. If the VLC is used, any bit errors can break the synchronization of the coded information and the following bits may undecodable until the next synchronization codeword appears. In order to protect the bit stream from errors, some methods use error-correction codes such as forward error correction (FEC) codes that add some redundancy to the bit stream. Adding error correction bits may cause two problems. First, it will reduce the coding efficiency due to adding redundant bits. Second, those error correction bits are often insufficient to detect or correct bit errors due to limitations in their capability. Ideally, the video system might use a layered or scalable coding method to increase its error resilience capability by adapting the encoding parameters, such as bit rate, frame rate, or others according to the varying conditions of the communication channels. However as we discussed previously, scalable coding may not be the solution for many applications. First, in the multicast case, the encoder can only provide a single-layer bit stream, although the decoders may have different capabilities and different channel conditions. Second, in the server application the video bit streams are pre-encoded and stored for delivery by request. In this case, the error resilience capability of the video bit stream cannot be changed. Third, the content providers usually try to provide the video bit stream at the highest coding performance for the given bit rate and might not consider sufficient robustness for error resilience since these bit streams may not be generated for transmission over wireless channels. Therefore, error resilience transcoding is necessary for delivering video streams over wireless channels or other noisy channels. Actually, during the process of transcoding, re-encoding or partial re-encoding is needed; therefore, video transcoding may be considered as a solution to protect the video against the channel errors by dynamically introducing error resilient features in the output bit stream. Furthermore, if the transcoder or transcoder proxy is located near the decoder, the video transcoder can also apply error concealment during the decoding or partial decoding process to alleviate the effects of errors encountered in the originating network before the transcoding operation is started. Another important issue of error resilience transcoding is rate control. Depending on the status of the output buffer, the output bit rate of the transcoder can be adaptively adjusted in compliance with the varying bandwidth conditions and channel error characteristics. Similarly, the amount of error protection to be added to the compressed video bit stream can also be controlled with monitoring of the output bit rate of the transcoder and the changing error conditions if feedback messages from the network are available. For the purpose of error resilience, the video transcoder can use all means in the error resilience encoding such as data partitioning, insertion of resynchronization markers into the incoming bit stream, and unequal error protection for various segments of video data. In this section, we will introduce several new techniques to increase the robustness of error resilience transcoding that may be applied instead of those existing tools.

In the MPEG-like video coding technique, the predictive coding schemes are frequently used in both the spatial and temporal domains. The decoding process depends on the previously decoded data in most cases except for the intra coding. Therefore, the error would be easily propagated in the decoding process. One task of error resilience is to prevent the error propagation. Based on this observation, several algorithms have been proposed to address the problem of error propagation [9-17],[9-18]. There are two basic ideas behind these algorithms. The first is to use short slice length and break the link between slices to prevent the error propagation within the frame, and the second is to convert a certain number of inter-coded blocks in the frame into intra-coded ones, which can provide temporal resilience between transcoded frames and prevent errors from propagating to the subsequent frames. In the following section, we give a more detailed introduction of the second technique.

9.2.2 Techniques for Spatial and Temporal Error Resilience Coding

Error resilience transcoding for video over wireless channels has been studied by many researchers [9-17],[9-19] and has been briefly introduced in Section 4.5. Here, we first introduce an error resilience transcoding technique that is referred to as spatial and temporal localization. The basic idea of spatial and temporal localization is to convert a certain number of inter-coded blocks to intra-coded blocks. Then the rate control strategy for determining the number of intra blocks will be presented.

The basic idea of spatial and temporal localization is to prevent errors caused by a bit error from propagating within a frame or to the subsequent frames due to loss of synchronization and predictive link. In the method proposed in [9-17],

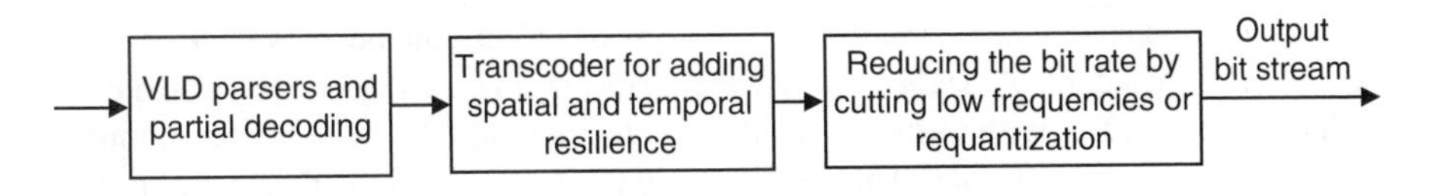

FIGURE 9.14 Architecture for spatial and temporal localization.

the system consists of three components as shown in Figure 9.14. First, a VLD parser and partial decoding is used to decode the input bit stream to the DCT coefficients. A transcoder is then used to inject spatial and temporal resilience into an encoded bit stream where the amount of resilience is tailored to the content of the video and the prevailing error conditions, as characterized by bit error rate. The transcoder increases the spatial resilience in two aspects: one is by reducing the number of blocks per slice and another is by increasing the proportion of intra blocks that are transmitted in each frame for increasing the temporal resilience. Since the bit rate will increase due to an increase in the overhead for short slices and adding intra-coded blocks for the error resilience, the transcoder has to remain at the output bit rate the same as at the input by dropping less significant coefficients. This task is implemented in the third component that performs the bit reduction by cutting low frequencies or performing requantization.

In order to optimally allocate the bit rate between spatial resilience, temporal resilience, and source rate, analytical models are derived. These models are used to characterize how the corruption propagates in a compressed video stream when it is subjected to bit errors. The analytical models are used to generate the resilience rate-distortion functions that are used to compute the optimal resilience. The optimal resilience is then applied to the bit stream by the transcoder. Simulation results have shown that overall video quality is improved by using a transcoder to optimally adjust the resilience in the presence of errors while maintaining the same input bit rate.

In [9-19], an error resilience video transcoding for inter-network communications is proposed by using a General Packet Radio Services (GPRS) mobile access network. The error resilience takes place in a proxy, which provides the necessary output rate with the required amount of robustness. It should be noted that the video proxy with the error resilience support allows a more rapid and dynamic method for handling error at the edge of different networks. In this system, two error resilience schemes are utilized: adaptive intra refresh (AIR) and feedback control signaling (FCS). The two schemes can work independently or in a combined way. Since both AIR and FCS increase the bit rate, a simple bit rate reduction mechanism is needed that adapts the quantization parameters accordingly. The system uses two primary control feedback mechanisms. In the first mechanism, feedback signal contains information related to the output channel conditions, such as bit error rate, delay, lost/received packets, etc. Based on this kind of received feedback, AIR and/or FCS can be used to insert the necessary robustness in the transcoded data. For example, if the bit error is increased, AIR is used as the major resilience scheme to stop the potential error accumulation effects. Especially, more inter-coded blocks in the high motion areas are transcoded to intra-coded blocks, which don't require motion compensation and, therefore, error propagation can be prevented. In the second feedback control mechanism, the feedback signal is originally from the output video frame buffer within the network-monitoring module, which continuously monitors the flow conditions. The feedback signal is then used to adaptively control the bit rate of transcoding. In case of overflow, the signal indicates to the transcoder that it should decrease the bit rate. This is a relatively straightforward rate-controlling scheme for congestion control. Experiments showed superior transcoding performance of these methods over the error-prone GPRS channels compared to the non-resilient video transcoding methods.

Another scheme of error resilience transcoding proposed in [9-20] is to convert the MPEG-2 video bit stream into a more resilient bit stream without increasing the bit rate. In this algorithm, each variable-length block starts at a known position in the transmitted bit stream. This is achieved by using a technique known as error resilient entropy coding (EREC). The bit stream is reordered without adding redundancy such that longer blocks fill up the spaces left by shorter blocks. However, in this scheme, the decoder at the receiver end also needs corresponding modifications, which may make it difficult to convince the receiver manufacturers to accept it.

9.2.3 Error Resilience Transcoding Using AIR

As we discussed in the previous section, the basic idea of spatial and temporal localization is to convert more inter-coded blocks to intra-coded blocks to prevent error propagation. Converting inter-coded blocks to intra-coded blocks can also be considered intra refreshment as presented in Section 9.1.4.4. Adding more intra-coded blocks causes an increase in the bit rate. For the output bit rate to remain the same as the input bit rate, we could do two things: the first option is to control the number of intra-coded blocks, and the second is to reduce the bit rate of the transcoded bit stream by either dropping the higher frequencies or requantizing the transformed coefficients with larger quantization steps [9-21]. In this section, we discuss the rate control issue for the scheme of intra refreshment.

Theoretically, the effect of intra refresh can be characterized by the operational rate-distortion (*R-D*) function $D(\beta, R)$, i.e., the average distortion D is expressed as a function of the average bit rate, R, and intra refresh rate, β. To illustrate the behavior of this function, consider the Foreman sequence at CIF resolution, encoded at 2 Mbps with GOP size $N = 30$ and no B-frame, i.e., $M = 1$. This bit stream is transcoded with a number of different fixed quantizer scales and various values of β. The *R-D* curves for each β are shown in Figure 9.15.

The intra refresh scheme used here for generating these results is similar to that of H.263 [9-7]. In this scheme, each macroblock is assigned a counter that is increased if the macroblock is encoded in inter-frame mode. If the counter reaches a threshold, $T = 1/\beta$, which denotes the update interval, the macroblock is encoded in intra mode and the counter is reset to zero. By assigning a different initial offsets to each macroblock, the updates of individual macroblocks can be spread over time. Figure 9.15 shows that overall quality is decreased at low bit rates when the intra refresh rate is high. The reason is that too many bits are consumed by the intra-coded blocks without a sufficient increase in quality. The opposite is observed at

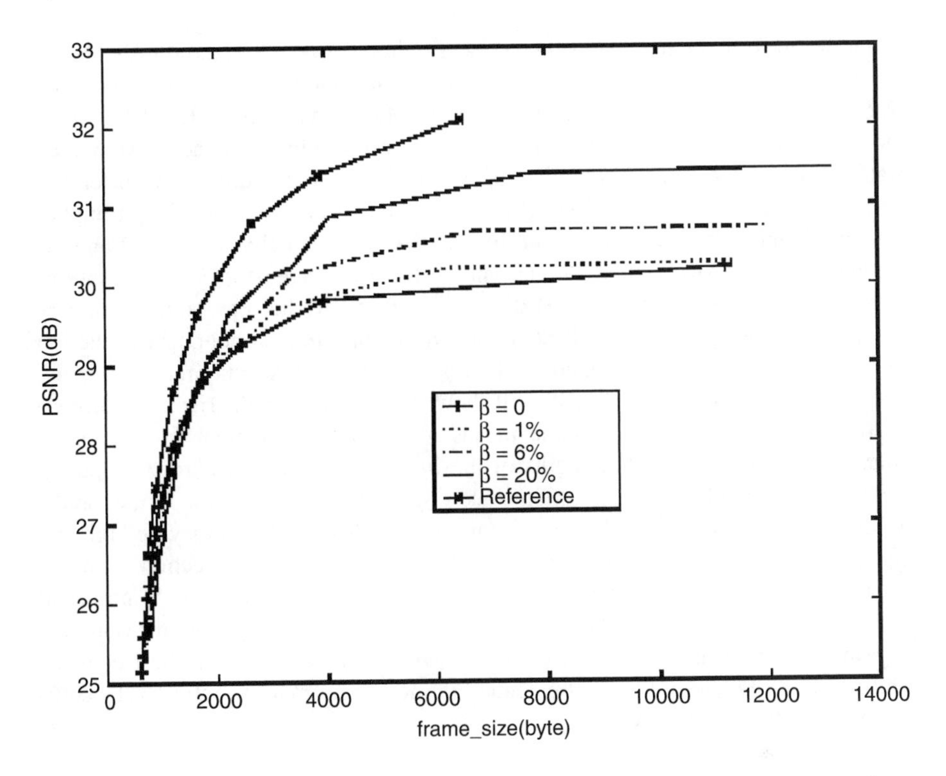

FIGURE 9.15 The *R-D* curve of the Foreman sequence for *Intra-Refresh* architecture.

higher bit rates. With a larger amount of bits that can be spent per frame, the overall quality is increased with more intra-coded blocks. Since the goal of the intra refresh techniques is to minimize the effect of drift, we should point out that at lower bit rates, d_q is likely to dominate the overall error, whereas at higher bit rates, the impact of d_q is significantly less and d_r is likely to be more dominant.

In the *Intra-Refresh* architecture, it is important that the intra refresh process be adaptive to account for these characteristics, and also that the outcome of the process, i.e., the number of intra blocks to be coded in a frame, be accounted for in certain aspects of the rate control, specifically, the quantizer selection and model parameter calculation. In the following, we give an example of error resilience transcoding algorithm using the AIR technique. In general, the objective of this technique is to dynamically determine the blocks to be converted from inter to intra. The particular scheme that we describe in this example is adaptive according to available bit rate and block attributes. After describing the technique itself, we explain how the outcome is accounted for by the rate control. As analyzed in Section 9.1.2, drift errors come from two mismatches, d_q and d_r. Through observation, we find that a large drift error always correlates to inter-coded blocks with large residual energy or motion activity. Consequently, AIR decides that a group of (four) macroblocks needs to be intra-coded if the sum of residual energy in this group of macroblocks is larger than a threshold, T_r, or if the sum of motion vector variance in this group of macroblocks is larger than a threshold, T_m. Initial values for the thresholds are determined experimentally through a simple linear relationship with the distortion (MSE). Specifically, the relations are given by MSE = αT_r and MSE = βT_m, where the parameters α and β are fitted based on several training sequences. After each frame is encoded, the thresholds are dynamically adjusted according to the difference between the target bit rate and actual bit rate. If the difference is positive, it implies that the target quality is higher than the bits that have actually been spent and we set the thresholds lower. On the other hand, if the difference is negative, the thresholds are set higher. Since this AIR decision may ignore the inter-coded boundary blocks of a moving object, we further expand the intra refresh boundary to the left or right to cover an object boundary. It should be noted that this procedure is not optimal, but as experimental results will show, it works quite well. The main purpose of this scheme is to provide some means of adaptive intra block conversion to illustrate the concepts and strengths of the *Intra-Refresh* architecture. An optimal scheme that considers rate-distortion trade-offs would be a topic for further study. With the intra refresh procedure, the total number of intra blocks may be high and must be accounted for in the rate control. For the quantizer selection, a single quantization parameter is selected for the frame and is applied to both intra- and inter-coded blocks. With this scheme, we consider a hybrid complexity measure that accounts for both inter and intra DCT coefficients. In other words, the frequency complexity in Equation (2.12) is extended to include normalized intra DCT coefficients as well. Specifically,

$$\tilde{S} = \frac{1}{M}(\sum_{k \in K} \sum_{i=1}^{63} \rho_1(i) \cdot |B_k(i)|^2 + \sum_{l \in L} \sum_{i=1}^{63} \rho_2(i) \cdot |B_l(i)|^2 \tag{9.26}$$

where $B_l(i)$ are the AC coefficients of an intra-coded block, l is a macroblock index in the set L of intra-coded blocks, M is the total number of non-skipped blocks in a frame, and $\rho_1(i)$ and $\rho_2(i)$ are frequency-dependent weights for inter- and intra-coded blocks, respectively. This hybrid complexity measure is used to calculate the updated model parameters after coding. With these small modifications to the rate control, we have found that a better fit between the rate-quantizer model and the actual data is achieved.

9.3 OBJECT-BASED TRANSCODING

This section introduces a new framework for video content delivery that is based on the transcoding of multiple video objects. The main material of this section is from our paper [9-22]. Generally speaking, transcoding can be defined as the manipulation or conversion of data into another more desirable format. Here, we consider manipulations of object-based video content, and more specifically, from one set of compressed bit streams to another. Given the object-based framework, we present a set of new algorithms that are responsible for manipulating the original set of video bit streams. Depending on the particular strategy that is adopted, the transcoder attempts to satisfy network conditions or user requirements in various ways. One of the main topics of this section is to discuss the degrees of freedom within an object-based transcoder and demonstrate the flexibility that it has in adapting the content. Two approaches are considered: a dynamic programming approach and a metadata–based approach. Simulations with these two approaches provide insight regarding the bit allocation among objects and illustrate the trade-offs that can be made in adapting the content. When certain metadata about the content are available, we show that bit allocation can be significantly improved, key objects can be identified, and varying the temporal resolution of objects can be considered.

9.3.1 Background

Recently, a new video coding standard, MPEG-4, has been developed. MPEG-4 coding allows arbitrary-shaped objects to be encoded and decoded as separate video object planes (VOPs) [9-23]. At the receiver, video objects are composed to form compound objects or scenes. In the context of video transmission, these compression standards are needed to reduce the amount of bandwidth that is required by the network. Since the delivery system must accommodate various transmission and load constraints, it is sometimes necessary to further convert the already compressed bit stream before transmission [9-1],[9-21]. Depending on these constraints, we know that the conventional transcoding techniques can be classified into three major categories: bit rate conversion or scaling, resolution conversion, and syntax conversion. A recent paper by Shanableh and Ghanbari [9-24] examines heterogeneous transcoding (e.g., from MPEG-1 to H.263), which simultaneously considers several of the specific conversions listed previously. In this section, we focus on bit rate scaling and look to exploit the object-level access of MPEG-4 in new and interesting ways. Until now, most research has focus on the delivery of frame-based video, which has been encoded using such standards as MPEG-1/2 and H.263. We build on this work by proposing a framework that considers an adaptive means of transcoding each of the objects in the scene based on available bandwidth and complexity of each object. The scheme is adaptive in that various techniques can be employed to reduce the rate depending on the ratio of incoming to outgoing rate, and since the goal is to provide the best overall quality for objects of varying complexity, the degradation of each object need not be the same. In fact, some objects may not even be transmitted. Within this object-based framework, various transcoding strategies can be employed. For instance, in a conservative approach, all objects would be delivered with varying spatial and temporal quality. However, in an aggressive approach and under severe network conditions, the transcoder may consider dropping less-relevant objects from the scene. These distinct approaches are two examples that will be used throughout this section to characterize the transcoder behavior. Not only do they differ in the actions taken to satisfy constraints on the network or user device, but they also differ in how quality is measured. These points will be elaborated on later. In this section, techniques that support both conservative and aggressive transcoding strategies within the object-based framework are presented. For the conservative strategy, the main objective is to consider the various spatio-temporal tradeoffs among all objects in a scene. To this end, a dynamic programming approach that jointly considers the quantization parameter (QP) selection for each object and frameskip is presented. The bit reallocation among objects is a critical process in the transcoder and is accomplished with the use of modified rate-quantizer models that were originally used for bit allocation in the encoder. The main difference is that the transcoder extracts information directly from the compressed domain, such as the residual energy of DCT blocks, hence certain parameters must be redefined. To further enhance the quality that can be achieved with this method, and also consider a set of tools that would support the aggressive transcoding strategy, we consider a new transcoding method that is based on the emerging MPEG-7 standard [9-25]. MPEG-7 aims at standardizing a set of descriptors and description schemes that can be used to describe multimedia content for a variety of applications, such as content navigation, content management, and fast search and retrieval. For video, the descriptors are low-level features of the content, such as color, texture, and motion [9-26]. These low-level features can be organized and referenced by higher-level description schemes. The description schemes may refer to various levels of the program, such as a segment or frame, and may also be used to facilitate video summarization, specify user preferences, and support the annotation of video segments. A more complete specification of the various description schemes that have been adopted by MPEG can be found in [9-27]. Besides using metadata for search and browsing applications, it is also of interest to consider metadata that would be suitable to guide the operation of a transcoder. Such metadata is referred to as a *transcoding hint* within the MPEG-7 community. In [9-28], such hints were introduced within the context of a universal multimedia access (UMA) framework. The specifics of this work and extensions of it have been investigated by the MPEG committee; a complete specification of the transcoding hints that have been adopted by the standard can be found in [9-27]. The specific extensions that we have proposed for object-based transcoding are discussed in [9-29]. The use of metadata for transcoding applications is most applicable in non–real-time scenarios, i.e., with stored content. For instance, before the content is transmitted or needs to be converted, the metadata can be efficiently extracted. When a request for the content is presented, any external conditions, e.g., network conditions, can be assessed and the transcoding can be appropriately executed. The advantages that one gains in using metadata are improved quality and reduced complexity. It should be noted that these metadata are not limited to descriptions related to the complexity of the scene, e.g., in terms of motion or texture, but may also describe (directly or indirectly) its subjective importance to the user. With this, algorithms have been developed to consider the distribution of available bit rate to different segments of the video sequence, and within those segments to the different regions or objects in the scene. More generally, we consider algorithms that are capable of adapting the quality of the scene in various ways. Several transcoding hints are proposed here to meet a number of objectives. The first of these objectives is to improve bit allocation among objects, the second is to identify key objects in the scene, and the third is to regulate the variable temporal resolution of objects in the scene.

9.3.2 Object-Based Transcoding Framework and Strategies

This section introduces the object-based transcoding framework and discusses two transcoding strategies: a conservative approach and an aggressive approach.

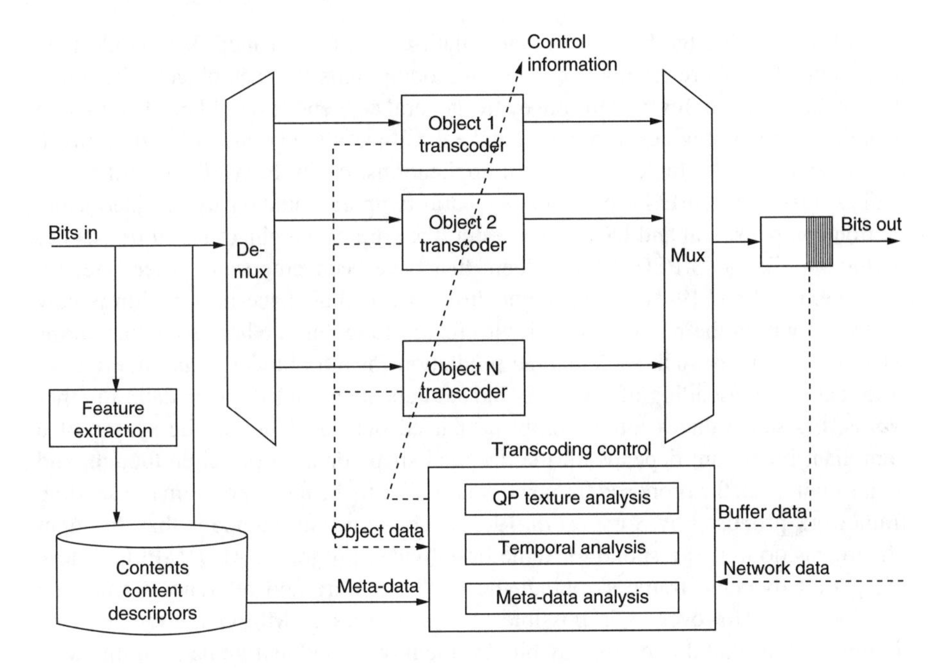

FIGURE 9.16 Object-based adaptive transcoding using metadata.

9.3.2.1 Object-Based Adaptive Transcoding System

Figure 9.16 shows a block diagram of an adaptive object-based transcoding system. In Figure 9.16, the input to the transcoder includes object-based bit streams, where the bit streams have a total bit rate R_{in}. The outputs of the system are the same bit streams with a reduced total rate $R_{out} < R_{in}$. In a non–real-time scenario, features are extracted from the bit stream before being processed by the transcoder. The features may include descriptors that can be used by a search engine or other types of metadata that describe the characteristics of the content. The content and associated metadata are stored on the server until a request for the content is made. In a real-time scenario, the content must be transmitted right away; therefore, any information extracted from the bit stream is of a causal nature. In this way, the feature extraction analysis is limited in that the metadata produced are instantaneous, i.e., they can only be associated with the current content or content that has already been sent. The transcoding control unit that we consider here includes three types of analysis units: one responsible for the selection of QPs, another to investigate temporal decisions such as frameskip, and a final one for metadata analysis.

The details of the metadata analysis will become clearer later in the context of specific transcoding hints, but we should mention that the outcome of metadata analysis would impact decisions made on the QP and frameskip. Also, the elements in the transcoding control unit are not limited to the ones mentioned here. In fact, as reported in [9-30], the control unit may include a resize analysis and shape analysis to consider reductions in texture and shape resolution, respectively. In any case, the control information that is passed to each transcoder is relayed from this transcoding control unit so that the bit stream manipulations can be executed. We should note that, in practice, there are not separate transcoding units for each object. The figure is only illustrated in this way to convey the general concept. It should be clear though that all of the new degrees of freedom make the object-based transcoding framework unique and very desirable for network applications. As in the MPEG-2 and H.263 coding standards, MPEG-4 exploits the spatio-temporal redundancy of video using motion compensation and DCT. As a result, the core of the object-based transcoder is quite similar to MPEG-2 transcoders that have been proposed earlier (see, for example, [9-1] and [9-21]). The major difference is that shape information is now contained within the bit stream, and with regard to texture coding, some new tools have been adopted, such as dc and ac prediction for intra blocks. It is important to note that the transcoding of texture is indeed dependent on the shape data. In other words, the shape data cannot simply be parsed out and ignored; the syntax of a compliant bit stream depends on the decoded shape data. In principle though, and at this high level, the proposed framework is similar to the multi-program transcoding framework proposed by Sorial *et al.* [9-31]. The major difference is that the input bit streams do not correspond to frame-based video programs. Also, MPEG-2 does not permit dynamic frame skipping; the GOP structure and reference frames are usually fixed. However, it is possible to skip frames in MPEG-1/2 by inserting dummy frames that do not use any bits for the motion and texture data. In this way, all of the macroblocks in a frame are skipped, where at least 1 bit per macroblock must be spent [9-32]. The main point worth noting is that the objects themselves need not be transcoded with equal quality. For example, the texture data of one object may be reduced, keeping its temporal rate intact, while the temporal rate of another object is reduced, keeping its texture information intact. Many other combinations can also be considered, including dropping frames or objects. In a news clip, for example, it would be possible to reduce the temporal rate along with the texture bits for the background, while keeping the information associated with a news reader intact.

9.3.2.2 Strategies of Object-Based Transcoding

Within the object-based transcoding framework, we consider two types of transcoding strategies. One is referred to as conservative, and another is referred to as aggressive. These distinct strategies are useful in characterizing the behavior of different transcoders and also in justifying the need for some of the transcoding hints proposed later on. Both strategies aim to maximize quality, but differ in the actions taken to do so, as well as the measures that indicate quality. In the conservative approach, quality should be measured objectively, i.e., by traditional PSNR-like measures, whereas in the aggressive approach, overall perceptual quality should be considered. With traditional PSNR-like measures, there is a limitation in what can be measured. In other words, given a scene with several objects, PSNR cannot indicate the relevance of each object to a user; it can only measure bit-to-bit differences. To measure the overall perceptual quality is difficult; it requires relevance feedback from the user. The best that a transcoder can do without this feedback is to attempt to identify the key objects in a scene through characteristics of the signal itself. Although this is not always possible, since less-relevant objects may display relevant object characteristics, it may be possible for certain applications in which the nature of the content is known. If we define the conservative approach to maximize PSNR quality, this implies that all objects in the scene must be transmitted. Of course, the objects will vary in spatial and temporal quality, but the PSNR between the original and reconstructed scene can be calculated. However, when varying the temporal quality of multiple objects, a major subtlety exists, that is, it is possible that different temporal rates for each object may produce undefined areas in the reconstructed scene. To illustrate, consider two objects, VO1 at a rate of 10 Hz and VO2 at a rate of 30 Hz. If the shape boundary of VO2 changes before the next VO1 is received, such undefined areas will appear in the composite image; we refer to this as a composition problem. To overcome this problem, we propose a measure that indicates the change in the shape boundaries over time, i.e., a shape hint. Such a measure is needed to allow for variable temporal quality of objects, but has some other uses as well, which are discussed later. If we define the aggressive approach to maximize perceptual quality, this implies that all objects in the scene are not necessarily transmitted, i.e., less-relevant objects may be dropped. When objects are dropped, traditional PSNR-like measures cannot be used, since the contents contained in the original and reconstructed scene have changed. To enable such a transcoding strategy, it is essential that key objects in the scene be identified. For this purpose, we will identify certain transcoding hints that can be used and demonstrate their utility.

9.3.3 Dynamic Programming Approach

In the previous section, the object-based transcoding framework was presented. We now provide the details of a dynamic programming based approach that jointly considers the spatial and temporal quality of objects.

9.3.3.1 Texture Model for Rate Control

We begin our discussion with a brief overview of the texture models used in the encoder for rate control [9-23],[9-33]. Let R represent the texture bits spent for a video object (VO), Q denote the QP, and (X_1, X_2) denote the first- and second-order model parameters, reflecting the encoding complexity, such as the mean absolute difference. The relation between R and Q is given by Equation (2.21) in Section 2.6.4.2 as

$$R = S \cdot \left(\frac{X_1}{Q} + \frac{X_2}{Q^2} \right)$$

Given the target amount of bits that a VO has been assigned and the current value of S, the value of Q is determined based on the current value of (X_1, X_2). After a VO has been encoded, the actual number of bits that have been spent is known, and the model parameters can be updated. This can be done by linear regression using the results of the past n frames.

9.3.3.2 QP Selections in the Transcoder

The transcoding problem is different in that $\underline{Q}$, the set of original QPs, and the actual number of bits are already given. Also, rather than computing the encoding complexity from the spatial domain, we must define a new DCT-based complexity measure $\tilde{S}$. This measure is defined as

$$\tilde{S} = \frac{1}{M_x} \sum_{m \in M} \sum_{i=1}^{63} \rho(i) \cdot |B_m(i)|^2 \tag{9.27}$$

where the B_m are ac coefficients of a block; m is macroblock index in the set of M coded blocks; M_c is the number of blocks in that set; and $\rho(i)$ are frequency-dependent weighting factors. The proposed complexity measure indicates the energy of the ac coefficients, where the contribution of high-frequency components is lessened by the weighting function. This weighting function can be chosen to mimic that of an MPEG quantization matrix. From the data transmitted in the bit stream, and the data from past VOs, the model parameters can be computed and continually updated. Actually, we can update the model twice for every transcoded VOP; once before transcoding using data in the bit stream, then again after coding the texture with the new set of QPs, $\underline{Q}'$. With this increased number of data points, we can expect the model parameters to be more robust and converge faster. The main objective of the texture analysis is to choose $\underline{Q}'$'s that satisfy the rate constraint and minimum distortion. However, it is important to note that optimality is conditioned on $\underline{Q}$. Therefore, we must take care in how the distortion is quantified. More specifically, we must consider the distortion increase with respect to the operating point defined by the incoming $\underline{Q}$. One way to compute $\underline{Q}'$ is to utilize the same methodology as used in the rate control problem. In this way, we would first estimate a budget for all VOPs at a particular time point, adjust the target to account for the current level of the buffer, and then distribute this sum of bits over the objects. Given these object-based target bit rates, the new set of QPs can be computed with our texture model. The main problem with this approach is that we rely on the distribution of bits to be robust. In general, the distribution is not robust and the ability to control the increase in distortion is lost, since the new QPs have been computed independent of the original ones. To overcome this problem and attempt to solve for $\underline{Q}'$ in some way that is dependent on $\underline{Q}$, we propose an algorithm based on dynamic programming. To maintain as close a quality as possible to the original, it can be argued that the QPs of each object should change as little as possible. Given this, we can express the distortion increase as follows:

$$\Delta D(\underline{Q}' | \underline{Q}) = \sum_{k \in K} a_k [D(Q_k') - D(Q_k)] \tag{9.28}$$

where k denotes a VOP index in the set of VOPs K, and a_k represents the visual significance or priority of object k. This significance may be a function of the object's relative size and complexity, or may even reflect the semantic importance if it is known. It is important to note that $Q_k^{'} \geq Q_k$, for all k. Therefore the solution space is bounded by the incoming QP for each object, Q_k, and the maximum allowable QP, $Q_{\max}$, which is usually 31. Given the above quantification for the increase in distortion, we can solve our problem using dynamic programming. Formally, the problem can be stated as

$$\min \Delta D(\underline{Q}^{'}|\underline{Q}), \qquad \text{s.t.}$$

$$\begin{cases} R_{Total} \leq R_{Budget} \\ Q_k \leq Q_k^{'} \leq Q_{\max,} \end{cases} \tag{9.29}$$

This problem is solved by converting the constrained problem into an unconstrained one, in which the rate and distortion are merged through a Lagrange multiplier, λ. That is, we minimize the cost given by $\Delta D + \lambda R_{\text{Total}}$. Since this cost is additive over all objects, it is possible to solve this problem using dynamic programming. To do this, a 2-D lattice is formed as shown in Figure 9.17. The vertical axis represents the candidate QPs, $Q_k^{'}$, and the horizontal axis denotes the object index k. The incoming set of QPs (Q_k) is highlighted with squares and the valid solution space is the shaded area between these values and Q_{more}. Each node $(k, Q_k^{'})$ is associated

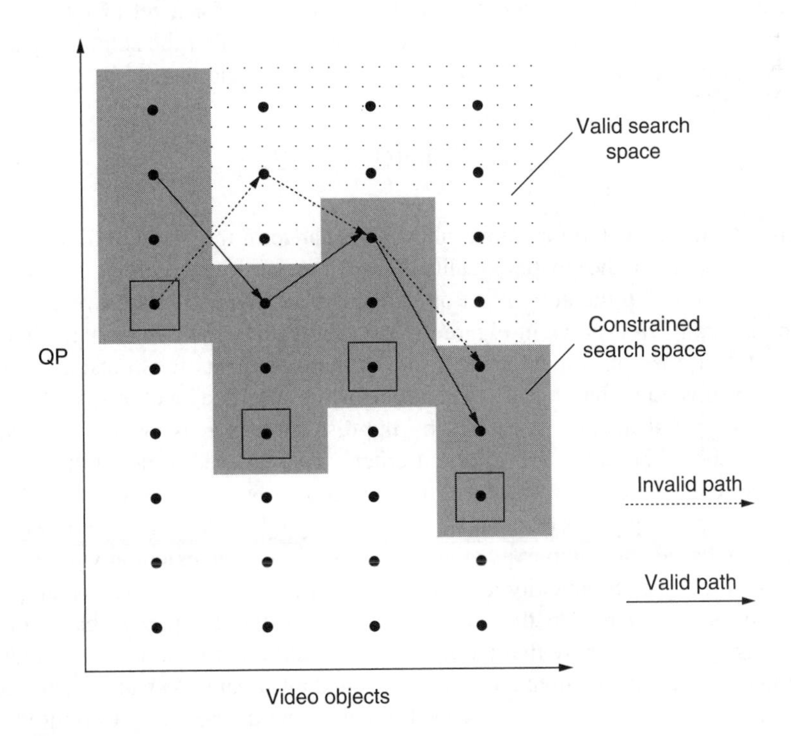

FIGURE 9.17 Illustration of QP search space.

with a cost $\Delta D_{k,a_k} + \lambda R'_{k,a_k}$. It is well known that the distortion with a quantizer step size of Q is proportional to Q^2 instead of Q, especially at low bit rates; therefore the following approximation is used to quantify the increase in distortion: $\Delta D_{k,Q_k^l} = Q_k'^2 - Q_k^2$. The rate R_{k,Q_k^l} for each node is estimated by Equation (2.21) using the modified DCT measure of encoding complexity given in Equation (9.27). This measure of DCT encoding complexity is also used to indicate the visual significance of the VOP, i.e., $a_k = \bar{S}_k$. Given this, the optimal solution can easily be found for any $\lambda \geq 0$. To determine the value of λ that satisfies the constraint on the rate, the well-known bisection algorithm can be used [9-34]. It is important to emphasize that the search space being considered is much less than that for MPEG-2 transcoding algorithms that attempt to find the best set of quantizers for every macroblock. Here, we only consider object-based quantizers; hence this approach is quite feasible.

9.3.3.3 Frameskip Analysis

Generally speaking, the purpose of skipping frames is to reduce the buffer occupancy level so that buffer overflow and ultimately the loss of packets are prevented. Another reason to skip frames is to allow a tradeoff in spatial versus temporal quality. In this way, fewer frames are coded, but they are coded with higher quality. Consequently, if the buffer is not in danger of overflowing, the decision to skip a frame should be incorporated into the QP selection process. Building from the proposed technique for QP selection, which searches a valid solution space for a set of QPs, we can achieve this spatial-versus-temporal trade-off by constraining the solution space. In this way, the quantizer constraint in Equation (9.29) would instead be given by

$$Q_k \leq Q_k^l \leq \min\{Q_k + \Delta Q_k, Q_{\max}\} \tag{9.30}$$

where ΔQ_k indicates the maximum allowable change in the new QP to satisfy this new constraint on the spatial quality. When the constraint cannot be met, the transcoder will skip the current set of VOPs, thereby regulating the overall spatio-temporal quality. As shown in Figure 9.17, a valid path is one in which all elements of Q_k^i fall in the constrained area. If one of these elements falls outside the area, the path is invalid in that it is not maintaining some specified level of spatial quality, where the spatial quality is implied by the distortion increase given in Equation (9.29). It should be noted that different criteria may be used to determine the ΔQ_k for a particular object. For instance, the change can be a function of the object's complexity or simply a percentage of the incoming QP. In the case of choosing the maximum based on complexity, the transcoder is would essentially limit those objects with higher complexity to smaller QPs, since their impact on spatial quality would be most severe. On the other hand, limiting the complexity based on the incoming QP would imply that the transcoder would like to maintain a similar QP distribution as compared to the originally encoded bit stream. Both approaches seem valid; however, more research is needed to determine the best way to limit the QP for each object. Such trade-offs in spatial versus temporal quality have not received significant attention in the literature; some studies have been conducted in [9-35]. Of course, one of the advantages in dealing with object-based data is that the temporal quality of one object can be different from that of another. In this way, bits can be saved by skipping the background object, for example. However, since objects are often disjoint, reducing the temporal resolution of one object may cause holes in the composed scene. This composition problem is discussed in more detail in the next section. To solve this problem, a shape hint that indicates the amount of change in the shape boundary over time is proposed. However, the use of such information has not yet been considered in this framework, but it is surely an interesting topic for further work. Until then, we must impose the constraint that all VOPs have the same temporal resolution.

9.3.4 Meta-Data–Based Approach

In general, metadata include low-level features of the video, which describe color, motion, texture, and shape, as well as high-level features, which may include storyboard information or the semantics of the various objects in a scene. Of course, not all metadata are useful for transcoding, but we identify several forms of metadata that can be used to guide the transcoder. These forms of metadata are referred to as transcoding hints and are specified in the MPEG-7 standard [9-27]. It is the function of the metadata analysis to map these hints into the control parameters of the transcoder. The objectives of the transcoding hints that we propose in the following are the following: (1) to improve the bit allocation among objects; (2) to identify key objects in the scene; and (3) to regulate the variable temporal resolution of objects in the scene. These tasks will be made possible through a difficulty hint, first proposed by Suzuki and Kuhn in [9-36], and a shape change and motion intensity hint, first proposed in [9-29]. The initial motivation of the difficulty hint was to assist the transcoder in performing constant bit rate (CBR) to variable bit rate (VBR) conversions. It would do so by providing the transcoder with information related to the complexity of one segment with respect to the complexity of all other segments being considered. In [9-37], this specific application has been presented in detail; also, several additional motion-based hints have been discussed, including a hint that specifies motion uncompensability and a hint that specifies search range parameters. Here, we examine the use of the difficulty hint for improved bit allocation among multiple video objects. Since encoding difficulty is directly related to the distribution of bits among objects, such a hint is expected to yield better results. Given only the encoding complexity, the task of identifying key objects in the scene would be difficult to do in a reliable way. The reason is that objects that consume a large number of bits, or even a large number of average bits per block, are not necessarily interesting parts of the scene. To enhance the reliability, we also consider the motion intensity of the objects under consideration. In the classic transcoding problem, the selection of control parameters implies that we should determine the spatio-temporal trade-offs that translate into higher PSNR for the scene. However, with object-based transcoding, we allow ourselves to deviate from this traditional notion of quality and begin to consider perceptual quality. In contrast to traditional PSNR measures, this type of measure is more subjective, where the subjective measure is a metric of how well relevant information is being conveyed. Subsequently, it is not only the goal of the metadata analysis to provide a means of improving the bit allocation, but also to identify key objects in the scene for improved perceptual quality.

9.3.4.1 QP Selection

As discussed in the dynamic programming approach, the process of selecting QPs for each object relies on a model that defines the rate-quantizer relationship. We again make use of the model defined by Equation (2.11) and Equation (9.27), where the difficulty hint is used to compute the rate budget for each object. The hint itself is a weight in the range [0, 1] that indicates the relative encoding complexity of each segment. A segment may be defined as a single object or a group of the same object over a specified interval of time. Defining the segment as a group of objects is advantageous to the transcoder in that the future encoding complexity can be accounted for. To compute the weights for each object, the actual number of bits used for encoding each object over the specified interval of time is recorded. The weights are then obtained by normalizing the bits used for each object with respect to the total bits used for all objects (over the specified interval of time). Moreover, given that objects are being considered, the rate for the *i*th object at time *t* is simply given by $R_t(t) = w_i R(t)$, where $R(t)$ is the portion of the total bit rate that has been allocated to the current set of objects at time *t*. Currently, $R(t)$ is equally distributed across all coding times. However, with the difficulty hint, unequal distribution may provide a means to improve the temporal allocation of bits. At this point, we are only interested in the spatial allocation among objects. It should be noted that the difficulty hint is computed based on the actual number of bits spent during encoding. Although the incoming bits to the transcoder may be sufficient, quantization may hide the true encoding complexity. Therefore, when possible, the difficulty hint should be computed from encoding at fine QPs. The semantics of the MPEG-7 description scheme do not make this mandatory, but it should be noted that results may vary.

9.3.4.2 Key Object Identification

In recent years, key frame extraction has been a fundamental tool for research on video summarization (see, for example, [9-38]). The primary objective of summarization is to present the user with a consolidated view of the content. The requirement is simple: give the user a general idea of the content that can be viewed in a short amount of time. In this chapter, we would like to extend such concepts to the transcoder. Time constraints for summarization can be likened to bandwidth constraints for the transcoder. Given this, the transcoder should adapt its delivery so that the content is reduced to the most relevant objects. In this consolidation process, it is important that the transcoder maintain as much information as possible. It can be argued that poorly coded content conveys less information than high-quality content. Consequently, there is a trade-off in the quality of the content versus the content that is delivered. This is a new concept that does not exist in summarization since summarization is mainly focused on high-quality frames only. As one can now imagine, there is scope for a great amount of work in this area centered around the trade-off in delivered content versus the quality of the content, i.e., the choice to transmit more content with lower quality or less content with higher quality. In this trade-off, the transcoder should consider many variations in the spatial and temporal resolution of each object. Also, some measure would need to be devised to handle the decision-making process. As a first step toward such a measure, a measure of "fidelity" has been introduced [9-39]. It is expected that the transcoding framework would need to extend such ideas to incorporate *R-D* considerations. Here, we introduce a simplified perspective to justify these broad concepts. Specifically, we consider a binary decision on each object: whether it is sent or not. To determine the importance of each object, a measure based on the intensity of motion activity σ and bit complexity B is proposed:

$$\eta = (\sigma + c) \cdot \frac{B}{\gamma} \tag{9.31}$$

where $c > 0$ is a constant incorporated to allow zero-motion objects to be decided based on the amount of bits, and γ is a normalizing factor that accounts for object size. Larger values of η indicate objects of greater significance. The intensity of motion activity is defined by MPEG-7 as the standard deviation of motion vector magnitudes [9-26],[9-40].

9.3.4.3 Variable Temporal Resolution

In this section, we illustrate the composition problem encountered in the encoding and transcoding of multiple video objects with different temporal resolutions. We then touch briefly on measures that indicate the amount of change in the shape boundary. Finally, we discuss the proposed shape hint. The uses of this hint are reserved until the next section, in which simulation results are discussed. To motivate our reasons for considering a shape hint, the composition problem is discussed. The Foreman sequence is shown in Figure 9.18. This sequence has two objects, a foreground and background. The left image shows the decoded and composited sequence for the case when both objects are encoded at 30 Hz. The right image shows the decoded and composited sequence for the case when the foreground is encoded at 30 Hz and the background at 15 Hz. When these two objects are encoded at the same temporal resolution, there is no problem with object composition during image reconstruction in the receiver, i.e., all pixels in the reconstructed scene are defined. However, a problem occurs when the objects are encoded at different temporal resolutions. When the objects are being encoded at different rates, movement of the shape of the object in the scene causes undefined pixels (or holes) in the reconstructed frames. These holes are due to the movement of one object, without the updating of adjacent or overlapping objects. The holes are uncovered areas of the scene that cannot be associated with either object and for which no pixels are defined. The holes disappear when the objects are resynchronized, i.e., background and foreground are coded at the same time instant. We now shift our discussion to

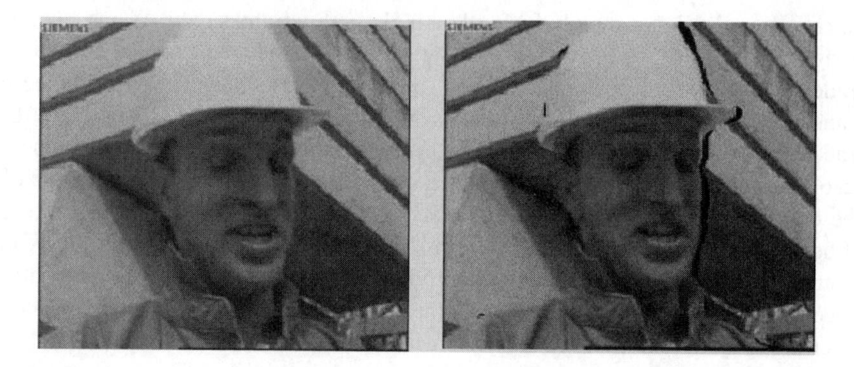

FIGURE 9.18 Illustration of composition problem for encoding multiple objects at different temporal rates. (a) Foreground and background both coded at 30 Hz, no composition problem. (b) Foreground coded at 30 Hz, background coded at 15 Hz, composition problem is shown.

measures that define the shape change or distortion between the shapes of an object over time. This measure could serve as an input to control the temporal resolution of objects. We now review two shape-distortion measures that have been proposed in [9-41] for key frame extraction. The distortion is measured from one frame to the next; however, such shape metrics can also relate to the scene at various other levels, e.g., at the segment level over some defined period of time. The first difference measure that is considered is the well-known Hamming distance, which measures the number of different pixels between two shapes. As defined in [9-41], the Hamming distance is given by

$$d = \sum_{n=0}^{N-1} \sum_{m=0}^{M-1} \left\| \alpha_1(m,n) - \alpha_2(m,n) \right\| \tag{9.32}$$

where $\alpha_1(m, n)$ and $\alpha_2(m, n)$ are corresponding segmentation planes at different time instances. The second shape difference measure is the Hausdorff distance, which is defined as the maximum function between two sets of pixel

$$h(A,B) = \max\{\min\{d(a,b)\}\} \tag{9.33}$$

where a and b are pixels of the sets A and B of two video objects, respectively, and $d(a,b)$ is the Euclidean distance between these points. The above metric indicates the maximum distance of the pixels in set A to the nearest pixel in set B. Because this metric is not symmetric, i.e., $h(A, B)$ may not be equal to $h(B, A)$, a more general definition is given by

$$H(A,B) = \max\{h(A,B), h(B,A)\} \tag{9.34}$$

It should be noted that the above measures are taken between corresponding shapes at different time instants. For a fixed amount of time, small differences indicate that greater variations in the temporal resolution for each object could be used, whereas larger differences indicate that smaller variations are required. If the duration of time is made larger and the differences remain small, then this indicates a potential for savings through variations in the temporal quality. To derive the proposed shape hint from the distortion measure described above, these measures must be normalized in the range [0,1]. For the Hamming distance, we may normalize the distortion by the number of pixels in the object. A value of 0 would indicate no change, and a value of 1 would indicate that the object is moving very fast. If the Hausdorff distance is used, we may normalize with respect to the maximum width or height of the object's bounding box or of the frame. Due to the nonsymmetrical nature of the Hausdorff distance, it may also be useful to distinguish differences between a growing shape boundary and a reducing shape boundary; a shape boundary may grow as the result of a zoom in and reduce as the result of a zoom out. With regard to the composition problem, shape boundaries that grow may be treated differently than those that reduce. We should note that these difference measures are most accurate when computed in the pixel domain, however, as in [9-41], approximated data from the compressed domain can also be used. The pixel-domain data are readily available in the encoder, but for the transcoder, it may not be computationally feasible (or of interest) to decode the shape data. Rather, the macroblock coding modes (e.g., transparent, opaque, border) could be used. In this case, a macroblock-level shape boundary could be formed and the distortion between shapes can be computed. In summary, there are many possibilities for using the proposed shape hint. The following are among the uses:

1. The transcoder can determine time instances when it is feasible to vary the temporal resolution among objects.
2. The shape hint can assist in computing the temporal rate for individual objects, i.e., large difference between a 30- and 10-Hz shape hint could indicate limitations in temporal rate reduction.
3. The shape hint can allow for better bit allocation, since the transcoder is taking full advantage of spatio-temporal redundancy, i.e., coding some objects at lower temporal resolution gives more bits to other objects.

9.3.5 Transcoding Architecture

The purpose of this section is to relate the conceptual object-based transcoding system of Figure 9.16 with an MPEG-4 transcoding architecture that more closely resembles the flow of data and control information. This architecture is shown in Figure 9.19 and resembles an open-loop scheme, which is used in the simulations. It is straightforward to extend this open-loop scheme to a closed one, where the drift in simulations would be significantly decreased. As with previous transcoding architectures, the syntax of the standard dictates the architecture. One high-level difference is that the bit streams for each object are independent of other bit streams. As a result, each object is associated with a video object layer (VOL) and a VOP header.

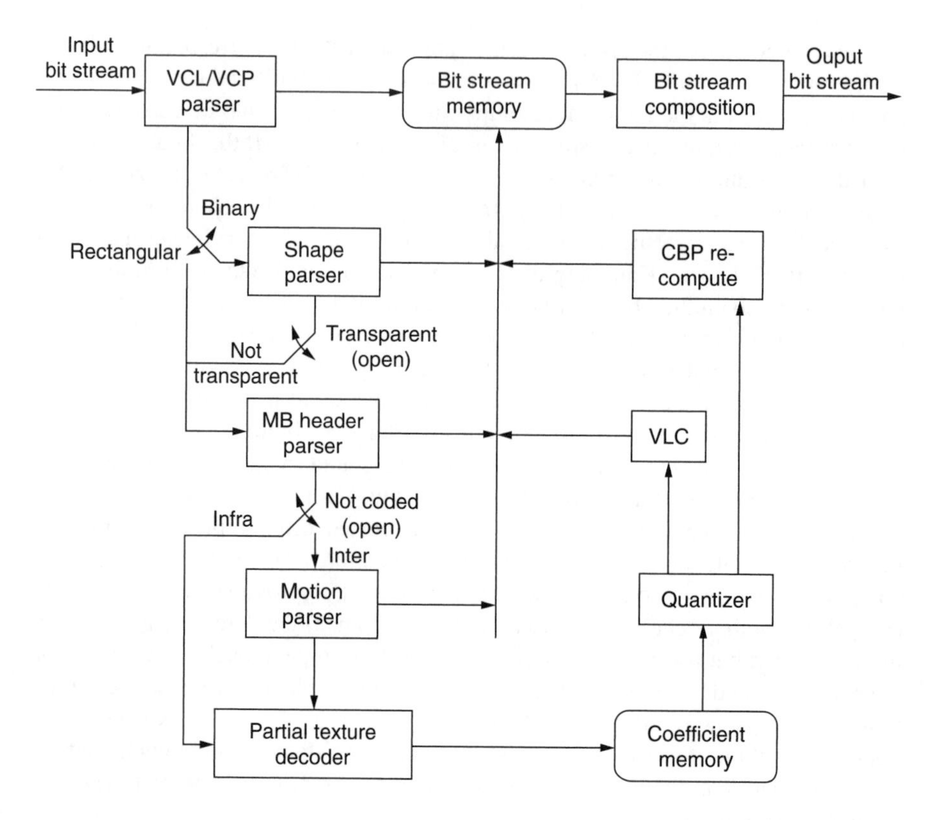

FIGURE 9.19 Architecture to transcode an MPEG-4 object to avoid composition problems, i.e., when all shape hints show low movement.

The VOP header contains the quantization parameter (QP) with which the object was encoded. The QP for each object is later used in the modeling and analysis of the texture information. All other bits are stored until the bit stream is recomposed. During the bit stream recomposition, information about the temporal rate is received. If a VOP should be skipped or the temporal rate should be changed, bits related to the skipped VOP will not be part of the outgoing bit stream. Additionally, fields in the VOL/VOP header that convey information about the timing and referencing may also need to change. The most significant difference from other standards is that MPEG-4 is capable of coding the shape of an object. From the VOP layer, we find out whether the VOP contains shape information (binary) or not (rectangular). If it is a rectangular VOP, then the object is simply a rectangular frame and there is no need to parse shape bits. In the case of binary shape, we need to determine if the macroblock is transparent or not. Transparent blocks are within the bounding box of the object, but are outside the object boundary, so there is no motion or texture information associated with them. The remainder of the MPEG-4 syntax is somewhat similar to that of MPEG-2, with a few exceptions. At the MB layer, there exist bits that contain the coded block pattern (CBP). The CBP is used to signal to the decoder which blocks of a macroblock contain at least one ac coefficient. Not only does the

CBP affect the structure of the bit stream, but it also has an impact on intra ac/dc prediction. The reason that the transcoder must be concerned with this parameter is that the CBP will change with the requantization of DCT blocks. For this reason, we must recompute the CBP.

9.3.6 Simulation Results

The results presented in this section demonstrate the various ways that object-based bit streams can be transcoded. First, we provide experimental results for the bit allocation among objects. Next, we show results using key object identification. Then, we discuss the potential use of shape hints for a variety of test sequences. Finally, experimental results that overcome the composition problem and yield gains in coding efficiency are shown.

9.3.6.1 Bit Allocation Among Objects

To study bit allocation in the transcoder, several results are simulated with the Coastguard and News sequences. Each sequence is CIF resolution (352 × 288), contains four objects, and is 300 frames long. For all simulations, the original bit stream is encoded at a rate of R_{in}. The rate-control algorithm used for encoding is the one reported in [9-33] and implemented in [9-42]. To achieve the reduced rate R_{out}, three methods are compared. The first method is a brute-force approach that simply decodes the original multiple video object (MVO) bit streams and re-encodes the decoded sequence at the outgoing rate. Recall that this method of transcoding is complex due to the added CPU time to perform motion estimation and added frame memory, but it serves as our reference. The next two methods have similar complexity (much lower than the reference) and make use of the same architecture. Actually, they only differ in the algorithm used to select the QP for each object, so the comparison is very fair. One method simulates the dynamic programming approach presented in Section 9.3.3, and the other method simulates the metadata–based approach that uses difficulty hints, as presented in Section 9.3.4. A plot of the PSNR curves comparing the reference method and two transcoding methods is shown in Figure 9.20. The Coastguard sequence is simulated with R_{in} = 512 kbits/s, R_{out} = 384 kbits/s, a frame rate of 10 fps, and GOP parameters N = 20 and M = 1. Similarly, the News sequences was simulated with R_{in} = 256 kbits/s, R_{out} = 128 kbits/s, a frame rate of 10 fps, and GOP parameters N = 20 and M = 1. It is quite evident from the plots that the transcoding method that uses difficulty hints is always better than the DP-based approach. This is expected, since the difficulty hint can provide information about the scene that is difficult (if not impossible) to obtain. For one, the difficulty hint is computed based on the bit usage (complexity) that is obtained from a fine quantizer, e.g., Q = 1. Also, the difficulty hint reflects the bit usage of future frames. In our experiments, the difficulty hint was computed regularly over a period of five frames. We tested the algorithm with different values, e.g., 30 frames or two frames, but the results did not seem to change much. Better results could be expected if the difficulty hint were computed on an irregular basis, where the interval boundaries would be determined in such a way that the difficulty is more consistent

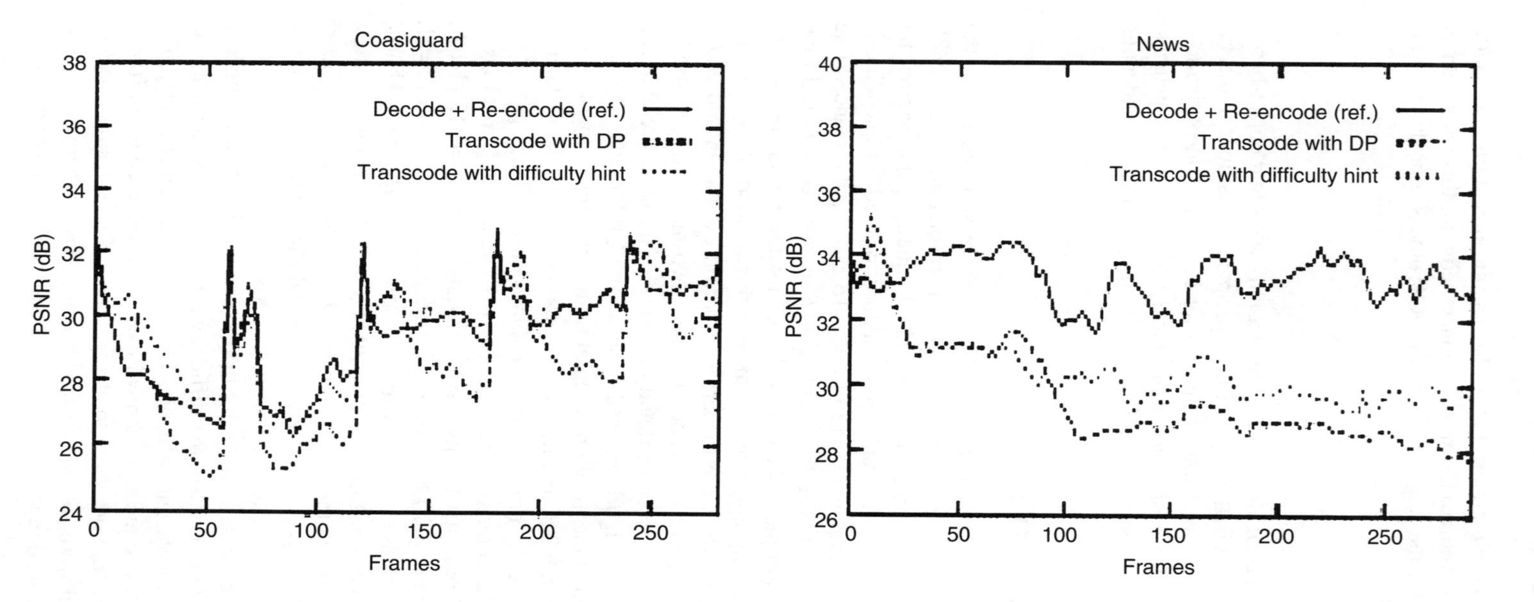

FIGURE 9.20 PSNR comparing reference and transcoded frames using dynamic programming (DP) approach and metadata–based approach with difficulty hints. (a) News, 256 kbits/s → 128 kbits/s, 10 fps, $N = 300$, $M = 1$. (b) Coastguard, 512 → 384 kbits/s, 10 fps, $N = 20$, $M = 1$.

over the segment. With more consistency, the hint would not smooth out highs and lows in the bit usage, resulting in a more accurate description of difficulty for each segment of the program.

From the plots in Figure 9.20, a number of other things are worth discussing. First, for the Coastguard sequence, the results of transcoding using the difficulty hint provide a very close match to the reference method. In fact, for many frames it is even slightly better. This indicates that with the use of noncausal information it is possible to overcome some of the major drawbacks that conventional low-complexity open-loop transcoders suffer from, mainly drift. Speaking of drift, it is clear from both sequences that it is significantly reduced in comparison to the DP-based approach. The News simulation shows an extreme coding situation in which only one I-VOP is used at the beginning of each coded sequence. In this case, the drift can clearly be seen. As reported in earlier works [9-1],[9-21], by using a closed-loop transcoding architecture could also minimize the effects of drift. In [9-43], an additional study regarding the cause of drift is presented. Any of these architectures and techniques could be applied here to achieve further gains. We present our results on the simple open-loop system so that gains are easily seen and compared and insight regarding the bit allocation among objects is not clouded by other factors. Overall, the DP-based approach does demonstrate comparable results to the reference method with a significant reduction in complexity. Furthermore, when the difficulty hint is available, further gains over the DP-based approach can be achieved. As an additional test, we would like to compare object-based transcoding to frame-based transcoding. To do so, we consider a single video object (SVO) simulation of the Coastguard sequence. The SVO reference, i.e., decoded and re-encoded, is compared to the MVO simulations described above, with all simulation parameters equal. However, rather than plotting the PSNR over the entire frame, we only consider the PSNR for the primary object in the scene, the large boat. Both the alpha plane for this large boat and the PSNR plots are shown in Figure 9.21.

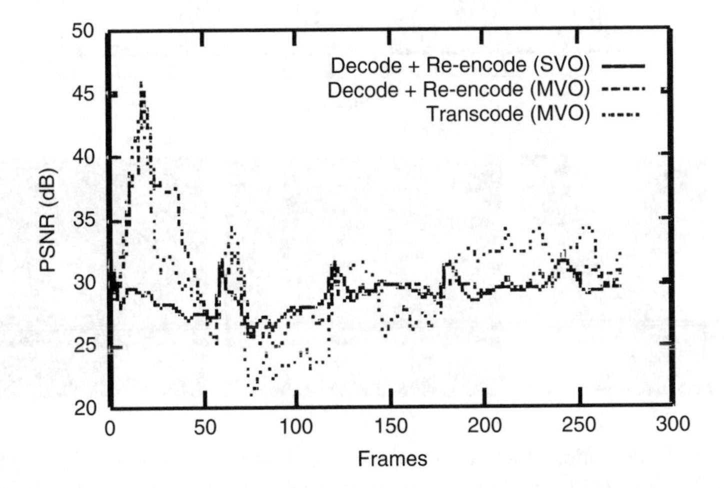

FIGURE 9.21 PSNR comparing SVO and MVO references versus MVO transcoded results for one object in Coastguard. Transcoder results are obtained with the *DP* approach.

It can be seen from these plots that the PSNR of the object as it first enters the scene is dramatically increased with the MVO simulations, and is on average better in the MVO simulations than the SVO result. The MVO transcoded result that is plotted in this figure obtained with the DP-based approach. These results support two main points: (1) it is possible (even with the DP-based approach) to improve the quality in more relevant parts of the scene; and (2) the quality of MVO simulations can outperform SVO simulations. Of course, only one case is illustrated here, but these results encourage further study of this topic.

9.3.6.2 Results with Key Object Identification

To demonstrate how key object identification can be used to adapt content delivery and balance quality, the News sequence is considered at CIF resolution, 10 fps. The News sequence contains four objects of varying complexity and of distinctly varying interest to the user. The sequence was originally encoded at 384 kbits/s, and then transcoded down to 128 kbits/s. Figure 9.22 illustrates several transcoded versions of frame 46 from the News sequence. The versions differ in the number of objects

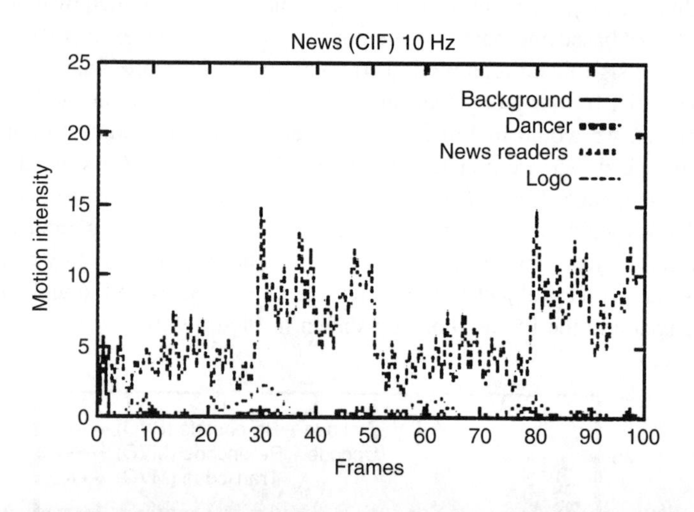

FIGURE 9.22 Illustration of adaptable transcoding of News sequence with key object identification, 384 $\rightarrow$ 128 kbits/s. The motion intensity and difficulty hints are used to order the importance of each object and the difficulty hint alone is used for bit allocation. Fewer objects can be delivered with noticeably higher quality.

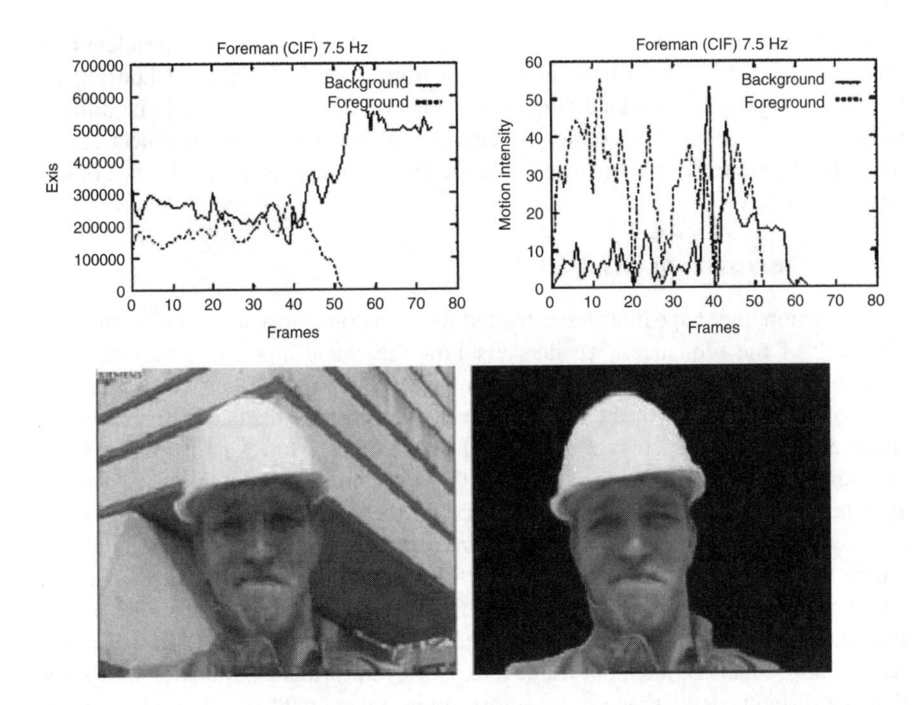

FIGURE 9.23 Illustration of adaptable transcoding of Foreman sequence with key object identification, 256 → 192 kbits/s. The motion intensity and difficulty hints are used to order the importance of each object and the difficulty hint alone is used for bit allocation. Fewer objects can be delivered with noticeably higher quality.

that are actually transmitted. In the leftmost version, all objects are transmitted, with relatively low quality; in the rightmost version, only one object is transmitted, with relatively high quality. The center image lies somewhere between the other two, in that only the background is not sent. It can be considered a balance between the two in terms of quality and content. For all versions, the objects are identified using the computed measure of importance based on motion and difficulty hints, but the bit allocation for objects under consideration were done using the difficulty hint only. Similar results are obtained for the Foreman sequence, as shown in Figure 9.23.

The sequence was originally coded at a rate of 256 kbits/s and transcoded down to 192 kbits/s. Here, since there are only two objects, the choice is much simpler. Again, based on the motion intensity and difficulty hints, the relevant object is identified. It should be observed from the figure that the increase in quality is quite clear, as all the additional bits are used for the foreground. Of course, this method of identifying key objects will not be suitable for all scenes, especially ones in which the nature of the content is not known and meaningful segmentation is difficult to achieve. However, for certain scenes that can be segmented with some confidence, possibly using some *a priori* information about the contents of the scene, this added flexibility of dropping objects from the scene can be very useful for improving the quality of delivered content. A new optimization problem can also be considered.

This optimization can be viewed as an extension of the bit allocation problem over objects, where the objects must be selected from a total of objects. Choosing the best set of objects is the added dimension. However, this problem could be difficult to formulate since there is a subjective element to choosing/excluding objects. Also, bit-to-bit differences from the original scene are meaningless in the absence of one or more objects.

9.3.6.3 Discussion of Shape Hints

In this section, the shape hints are extracted for each object in a number of sequences. For each of the sequences, we interpret how the shape hint could be used by a transcoder. In many cases, the final outcome and transcoder decision is based on the motion or difficulty of the objects as well. The plots of the shape hint are given for every frame and across a number of frame rates. However, in practice and according to the current draft of the MPEG-7 standard, the shape hint will only be specified for one frame rate, which is equal to the frame rate of the content to be encoded or transcoded. Also, the shape hint can be specified over a period of time, where the average of the shape hints at each time may be used. Finally, it should be noted that the differences in the shape hints between different frame rates are for illustrative purposes only, to demonstrate that the shape hints yield fairly consistent results independent of the input frame rate. However, further examination may show that the differences in shape hints across frame rates could be a useful metric to determine the reduced temporal rate for each object.

1) *Shape Hints for Akiyo* (Figure 9.24): The shape hints for this sequence indicate very low movement of objects. This would be a very good candidate for variable temporal rate reduction. Depending on the motion and /or difficulty, a transcoder may reduce the resolution of either object.
2) *Shape Hints for Dancer* (Figure 9.25): The shape hint for the background indicates no change (it is a full rectangular frame), while the foreground object is experiencing large movement in shape. Is it possible to have variable resolutions with this scene, since the full background is coded

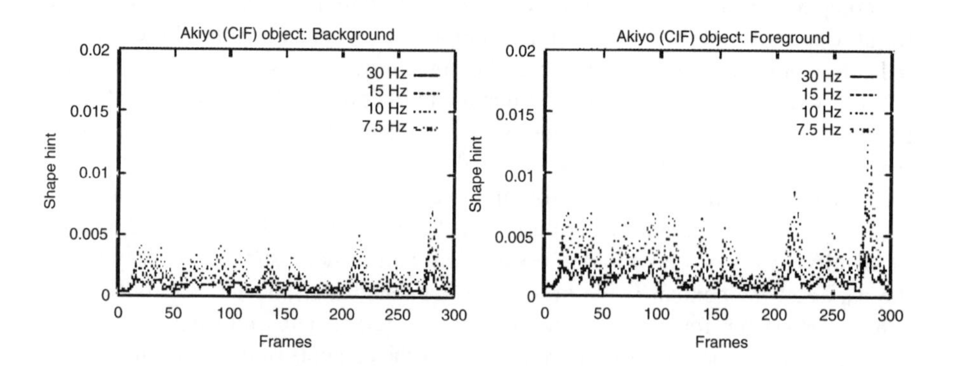

FIGURE 9.24 Shape hints for Akiyo.

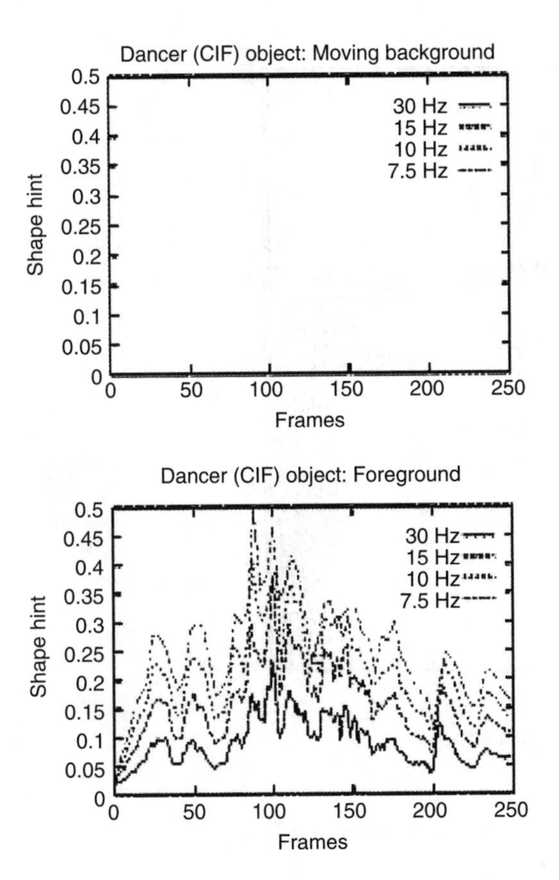

FIGURE 9.25 Shape hints for Dancer.

and it will never cause holes, but it is probably not a good idea since the shape movement of the foreground indicates very large movement.

3) *Shape Hint for Foreman* (Figure 9.26): The movement of the background and foreground agree with each other and show large movement for the first 200 frames. After that, the shape hint indicates that the foreground no longer exists and the background eventually fills the entire frame.
4) *Shape Hint for Children* (Figure 9.27): As with the Dancer sequence, the background takes up the entire rectangular frame and is unchanging. From the shape hint of the logo, we see that it has extremely fast movement (or is not in the scene), then sits still for a while, then moves very quickly again and sits still until the end. This is a very good example where the shape hint can be used to identify key frames to code or identify localized segments in which the temporal resolution can be dramatically reduced. The final object, the children playing with the ball, has very large movements throughout, as indicated by its shape hints.
5) *Shape Hints for News* (Figure 9.28): Overall, the shape hints for this sequence indicate very low movement. There is one place where one of

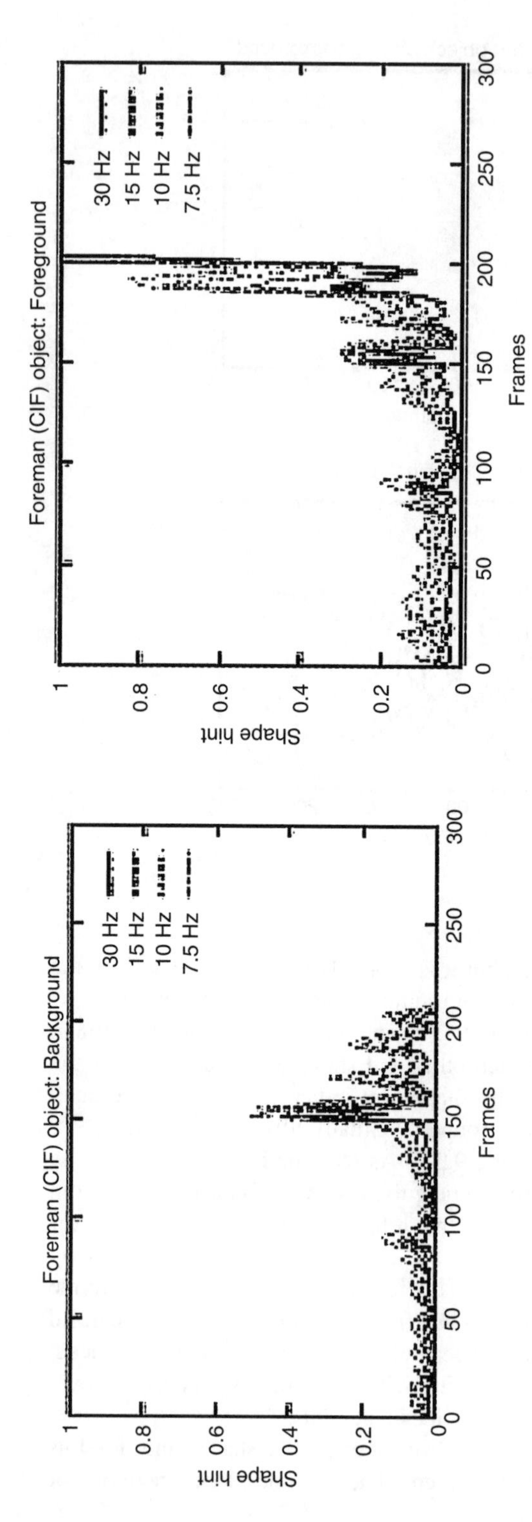

FIGURE 9.26 Shape hints for Foreman.

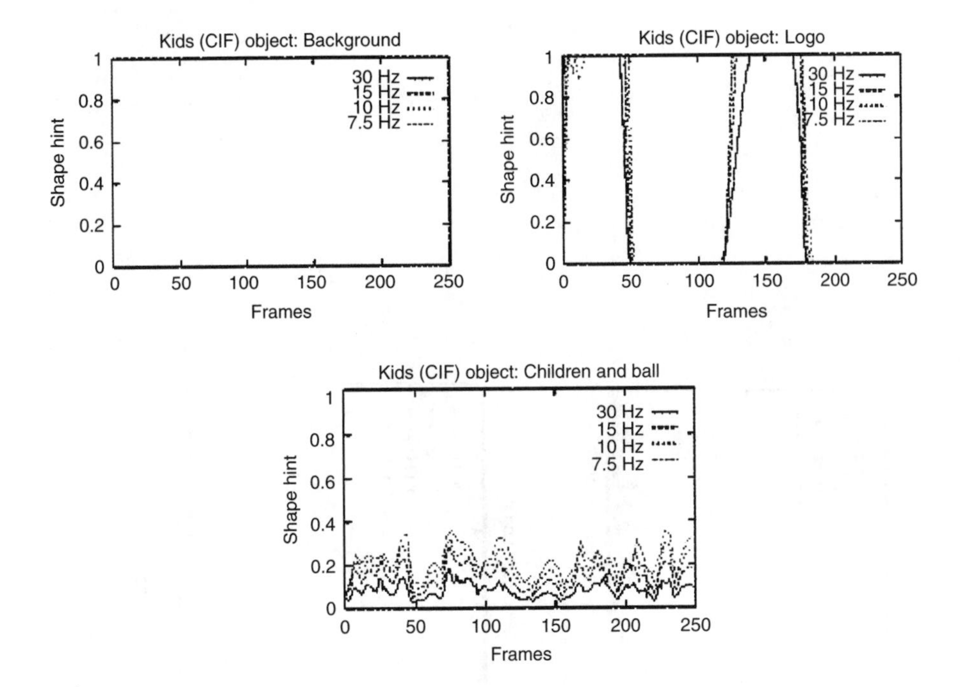

FIGURE 9.27 Shape hints for Children.

the newsreaders nods her head; this is picked up between frames 75 and 100. But, for the most part, variable temporal resolution can be tolerated for this sequence. In fact, the logo object does not move at all and can be coded once at the very start, without interfering with other objects.

9.3.6.4 Results with Varying Temporal Resolution

Given that the shape hint will identify scenes (or segments) that can take advantage of variable temporal resolution, we provide results to demonstrate the gain. In the previous section, the Akiyo sequence has been identified as a scene that can be encoded or transcoded with variable temporal resolution without creating disturbing artifacts due to composition. To examine this further, we generate the *R-D* curves for two cases. The first case is our reference and simply encodes both foreground and background objects at 30 Hz with a constant QP. This is referred to as the fixed case. As a comparison, we maintain the temporal resolution of the foreground at 30 Hz, but reduce the temporal resolution of the background to 7.5 Hz.

This case is referred to as the variable case. A comparison of the fixed and variable *R-D* curves is shown in Figure 9.29. The set of QPs that were used to generate the curves were 9, 15, 21, and 27. The finest QP was 9 since reducing the temporal resolution for high-bandwidth simulations is not of much interest. It should be emphasized that the PSNR for each simulation is computed with respect to the original 30-Hz sequence.

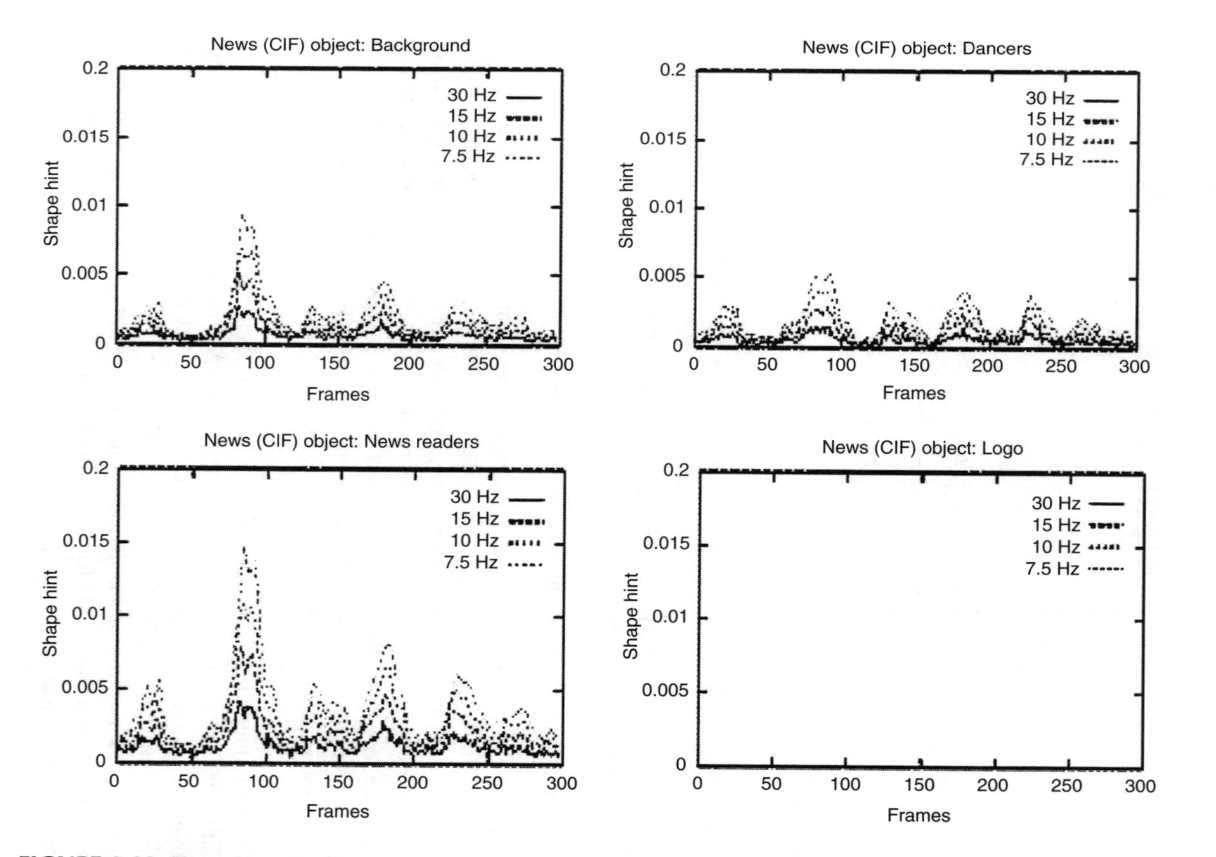

FIGURE 9.28 Shape hints for News.

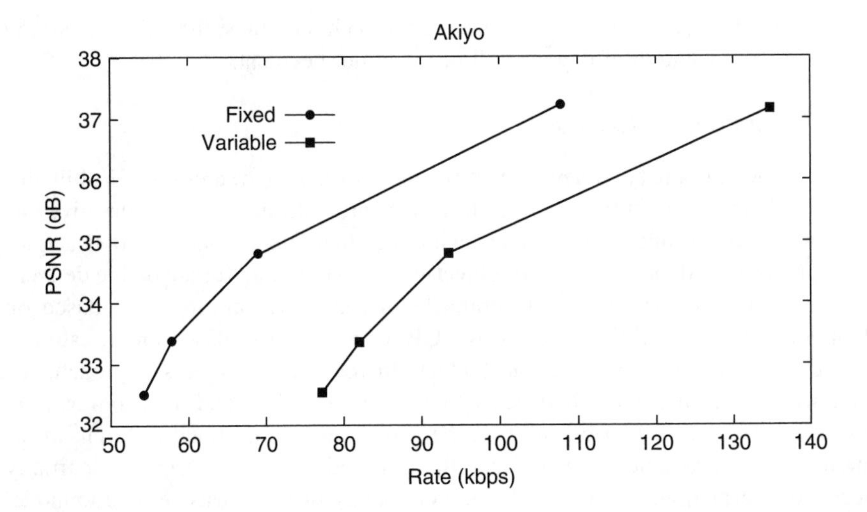

FIGURE 9.29 *R-D* curves comparing the Akiyo sequence coded using fixed and variable temporal resolution. With fixed temporal resolution, both foreground and background are coded at 30 Hz. With variable temporal resolution the foreground is still coded at 30 Hz, but the background is coded at 7.5 Hz.

The curves in this figure show that a 25% reduction in bit rate can be achieved by reducing the resolution of the background only, with no loss of quality. As indicated by the shape hints for this sequence, the movement among shape boundaries is very small. Consequently, any holes in the reconstructed image are negligible and do not impact the PSNR values. Due to limitations in space, it is not feasible to test the various ways that the content can be encoded and transcoded with variable temporal resolution. However, we should mention that the indication of the shape hint to allow variable temporal resolution without holes may not always provide a benefit. For example, the background of the Children sequence does not contain any motion and is a full rectangular frame, so the border of other objects does not affect it. In this case, you can send this background once in the beginning; however, the encoder would not spend any bits on the updates anyway, i.e., it would only use 1 bit per macroblock to signal that a block is skipped. As a general rule, gains will be possible if there is sufficient information that is being coded, but does not need to be. We do however expect gains in other sequences besides Akiyo, such as News and others like it. The gains discussed above use traditional PSNR measures as a major part of the evaluation. Such measures are suitable to test the gains in variable temporal resolution when the movement of the object is low. The reason is that comparisons are made with respect to the original sequence, and so losses between a 30-Hz reference and a 7.5-Hz object rendered at 30 Hz remain small. However, for faster moving objects, PSNR is not a suitable measure to quantify the differences between the same objects coded at different temporal resolutions. The bit-to-bit differences will seem large, though the perceptual loss may only be noticeable to a very trained eye. For this reason, measures to evaluate the quality of objects at different temporal resolutions are needed. With such measures, the gains in varying the temporal resolution of objects will become more evident.

9.3.7 Concluding Remarks

This section introduced a new object-based transcoding framework. Within this framework, two distinct transcoding strategies were introduced: a conservative one and an aggressive one. Transcoding techniques that support these two strategies were proposed and the flexibility of object-based transcoding for adaptable delivery of content was shown. The first approach presented was an approach based on dynamic programming that selected the QP for each object based on an estimate of the distortion increase and rate budget. In the context of this approach, the similarities and differences between object-based encoding and transcoding were discussed. We focused specifically on the texture model used and different information that is available to each. Since the transcoder only has access to partially decoded information, a new DCT-based complexity measure was defined to model the texture. To consider spatial-versus-temporal trade-offs, we discussed how the search space could be constrained. The second approach that we discussed is an approach that is based on the availability of metadata. We identified several transcoding hints that could be used for different purposes. The difficulty hint was shown to be useful for the allocation of bits among multiple objects in a scene. Then, using difficulty and motion, it was shown that key objects could be identified, and coding gains for the most relevant objects can be achieved. Finally, with the shape hint, we discussed several ways to use the shape hint for adaptive content delivery, then gave examples of the gains that can be achieved. As far as future work goes, we feel that there are still many open issues within the object-based framework. Certainly, we have not fully exploited all of the degrees of freedom that are possible, but we hope to at least raise some of the issues here and encourage future work in this area. One area of great potential is the incorporation of the ability to resize objects. For mobile appliances with limited display size, this will surely be a necessary technique, and combining this with some of the approaches presented in this chapter would also be of great interest. However, even for appliances with sufficient display size, it is still of interest to resize objects before transmission, then reconstruct the original resolution on the receiver side. In the past, this has been difficult to do, since the transcoder should conform to a standard coding scheme. By reducing the resolution, e.g., from CIF to QCIF, one would experience problems with the resampling of motion vectors and regaining MB structure, among other things. However, with the adoption of the dynamic resolution conversion (DRC) tool into Version 2 of the MPEG-4 standard [9-44], it becomes possible to further exploit gains that would result in down-sizing a background object, for example. The trade-offs between QP selection, temporal rate, and object resizing would require further investigation. Most likely, a multiple-resolution texture model would be needed to estimate impacts on the rate and distortion. Another avenue of future work is to identify other transcoding hints that improve the bit allocation or provide a new set of transcoding functionality. However, even with the existing hints, there is still scope to optimize the use of them. For example, it would be interesting to examine whether irregular partitions, or intervals of time, for computing and applying the difficulty hint would lead to improved results. Also, we feel that the potential to encode and transcode with variable temporal resolution has not yet been fully exploited. Finally, a very challenging problem is to extend the measure of fidelity that has been defined by those working on summarization to include *R-D* considerations. This is most applicable to transcoders that operate with an aggressive transcoding strategy. Given such a measure, a robust control mechanism can be designed to account for the wide range of transcoding options. Such a measure is consistent with the content-aware framework proposed by Chang and Bocheck [9-45]. In this framework, features of the objects are used to generate utility functions and classify objects into utility classes. Given these functions and classes, various scaling operations can be applied to the content according to current network behavior. Being able to make such predictions allows us to eventually improve upon the problem of dynamic resource allocation, which has been studied in [9-46] within the context of an end-to-end solution.

9.4 SUMMARY

In this chapter, application and implementation issues have been presented. An application example of MPEG-2 to MPEG-4 transcoder has been given. The error resilience coding that is important for wireless or IP networks has been introduced. Finally, the concept of object-based transcoding was presented.

9.5 EXERCISES

9-1. What are the main factors that cause the drift error in the MPEG-2 to MPEG-4 transcoder, including both temporal and spatial resolution reduction? Explain why the closed-loop methods can reduce the drift.

9-2. In the MPEG-2 to MPEG-4 transcoder, if temporal resolution, i.e., frame rate, reduction is required, what kind of GOP structure of the input MPEG-2 compressed bit stream may make the conversion easier? (Hint: For $N = 12$, $M = 2$, it is easy to convert 30 fps to 15 fps by dropping B-picture only.) What problems will be caused if there is no constraint on the GOP structure? (For example, if the original GOP structure is $N = 15$, $M = 3$, how can we convert a video with 30 fps to 15 fps?) Make your proposals to solve these problems.

9-3. Propose an algorithm for arbitrary size down-sampling transcoding in the frequency domain (avoid full decoding and re-encoding). Develop software to simulate your results.

9-4. In the intra refresh architecture of error resilience transcoding, which factors have to be considered when selecting the intra refresh rate? Explain the reasons.

9-5. In object-based transcoding, why do composition problem result from using variable temporal resolution for different objects? How can this problem be overcome?

9-6. In object-based transcoding, using variable temporal resolution may increase the overall video quality with a limited bit rate. What kind of video sequences may obtain more benefits by using variable temporal resolution? What factors are affect the trade-off between spatial quality and temporal resolution?

REFERENCES

[9-1] P. Assuncao and M. Ghanbari, A frequency-domain video transcoder for dynamic bit-rate reduction of MPEG-2 bitstreams, *IEEE Transactions on Circuits and Systems for Video Technology,* 8, 953–967, 1998.

[9-2] R. Mokry and D. Anastassiou, Minimal error drift in frequency scalability for motion-compensated DCT coding, *IEEE Transactions on Circuits and Systems for Video Technology,* 4, 1998.

[9-3] B. Shen, I.K. Sethi, and V. Bhaskaran, Adaptive motion vector resampling for compressed video down-scaling, *Proceedings of the IEEE International Conference on Image Processing,* Santa Barbara, CA, Oct. 1997.

[9-4] P. Yin, M. Wu, and B. Liu, Video transcoding by reducing spatial resolution, *Proceedings of the IEEE International Conference on Image Processing,* Vancouver, BC, Canada, Oct. 2000.

[9-5] S.J. Wee, J.G. Apostolopoulos, and N. Feamster, Field-to-frame transcoding with spatial and temporal downsampling, *Proceedings of the IEEE International Conference on Image Processing,* Kobe, Japan, Oct. 1999.

[9-6] A. Vetro, H. Sun, P. DaGraca, and T. Poon, Minimum drift architectures for three-layer scalable DTV decoding, *IEEE Transactions on Consumer Electronics,* 44; 3, pp. 527–536, 1998.

[9-7] K. Stuhlmuller, N. Farber, M. Link, and B. Girod, Analysis of video transmission over lossy channels, *Journal of Select Areas of Communication,* Vol. 18, pp. 1012–1032, June 2000.

[9-8] N. Bjork and C. Christopoulos, Transcoder architectures for video coding, *IEEE Transactions on Consumer Electronics,* 44, 1, 88–98, 1998.

[9-9] N. Björk and C. Christopoulos, Transcoder architectures for video coding, *Proceedings of the IEEE International Conference on Acoustics, Speech and Signal Processing,* 5, 2813–2816, Seattle, WA, 1998.

[9-10] J. Youn, M.T. Sun, and C.W. Lin. Motion vector refinement for high performance transcoding, *IEEE Transactions on Multimedia,* 1, 1, 30–40, 1999.

[9-11] M.J. Chen, M.C. Chu, and C.W. Pan, Efficient motion estimation algorithm for reduced frame-rate video transcoder, *IEEE Transactions on Circuits and Systems for Video Technology,* 12, 4, 269–275, 2002.

[9-12] J. Xin, M.T. Sun, and K. Chun, Motion-re-estimation for MPEG-2 to MPEG-4 simple profile transcoding, *Proceedings of the International Workshop on Packet Video,* Pittsburgh, PA, 2002.

[9-13] A. Lan and J.N. Hwang, Context dependent reference frame placement for MPEG video coding, *Proceedings of the IEEE International Conference on Acoustics, Speech, and Signal Processing,* 4, 2997–3000, 1997.

[9-14] J.N. Hwang, T.D. Wu, and C.W. Lin, Dynamic frame-skipping in video transcoding, *Proceedings of the IEEE Workshop on Multimedia Signal Processing,* 616–621, Redonda Beach, CA, 1998.

[9-15] A. Vetro, P. Yin, B. Liu, and H. Sun, Reduced spatio-temporal transcoding using an intra-refresh technique, *Proceedings of the IEEE International Symposium on Circuits and Systems,* Scottsdale, AZ, 2002.

[9-16] S.F. Chang and D.G. Messerschmidt, Manipulation and compositing of MC-DCT compressed video, *IEEE Journal of Selected Areas of Communications,* 13, 1, 1–11, 1995.

[9-17] G. de los Reyes, A.R. Reibman, S.-F. Chang, and J.C.-I. Chuang, Error resilience transcoding for video over wireless channels, *IEEE Journal of Selected Areas of Communications,* 18, 6, 1063–1074, 2000.

[9-18] S. Dogan, A. Cellatoglu, M. Uyguroglu, A.H. Sadka, and A.M. Kondoz, Error-resilient video transcoding for robust inter-network communications using GPRS, *IEEE Transactions on Circuits and Systems for Video Technology,* 12, 6, 453–464, 2002.

[9-19] S. Dogan, A. Cellatoglu, M. Uyguroglu, A.H. Sadka, and A.M. Kondoz, Error-resilient video transcoding for robust inter-network communications using GPRS, *IEEE Transactions on Circuits and Systems for Video Technology,* 12, 6, 453–464, 2002.

[9-20] R. Swann and N.G. Kingsbury, Transcoding of MPEG-II for enhanced resilience to transmission errors, *Proceedings of the International Conference on Image Processing,* Switzerland, Sept. 1996.

[9-21] H. Sun, W. Kwok, and J. Zdepski, Architectures for MPEG compressed bitstream scaling, *IEEE Transactions on Circuits and Systems for Video Technology,* Vol. 6, pp. 191–199, 1996.

[9-22] A. Vetro, H. Sun, and Y. Wang, Object-based transcoding for adaptive video content delivery, *IEEE Transactions on Circuits and Systems for Video Technology,* 11, 3, 387–401, 2001.

[9-23] Information Technology—Coding of Audio/Visual Objects, ISO/IEC 14496-2:1999.

[9-24] T. Shanableh and M. Ghanbari, Heterogeneous video transcoding to lower spatio-temporal resolutions and different encoding formats, *IEEE Transactions on Multimedia,* 2, 101–110, 2000.

[9-25] MPEG Requirements Group, Seoul, Korea, Overview of the MPEG-7 Standard, ISO/IEC N2729, La Baule, France, Oct. 2000.

[9-26] Information Technology—Multimedia Content Description Interface, ISO/IEC 15938-3.

[9-27] Multimedia Description Schemes, Information Technology—Multimedia Content Description Interface, ISO/IEC 15938-5.

[9-28] R. Mohan, J.R. Smith, and C.-S. Li, Adapting multimedia internet content for universal access, *IEEE Transactions on Multimedia,* 1, 104–114, 1999.

[9-29] A. Vetro, H. Sun, and A. Divakaran, Adaptive object-based transcoding using shape and motion-based hints, Geneva, Switzerland, ISO/IEC M6088, May 2000.

[9-30] A. Vetro, H. Sun, and Y. Wang, Object-based transcoding for scalable quality of service, in *Proceedings of the IEEE International Symposium on Circuits and Systems,* Geneva, Switzerland, May 2000.

[9-31] H. Sorial, W.E. Lynch, and A. Vincent, Joint transcoding of multiple MPEG video bitstreams, in *Proceedings of the International Symposium on Circuits and Systems,* Orlando, FL, May 1999.

[9-32] S. Gopalakrishnan, D. Reininger, and M. Ott, Realtime MPEG system stream transcoder for heterogeneous networks, in *Proceedings of the Packet Video Workshop,* New York, Apr. 1999.

[9-33] A. Vetro, H. Sun, and Y.Wang, MPEG-4 rate control for multiple video objects, *IEEE Transactions on Circuits and Systems for Video Technology,* 9, 186–199, 1999.

[9-34] K. Ramchandran and M. Vetterli, Best wavelet packet bases in a rate distortion sense, *IEEE Transactions on Image Processing*, 2, 160–175, 1993.

[9-35] F.C. Martins, W. Ding, and E. Feig, Joint control of spatial quantization and temporal sampling for very low bit rate video, in *Proceedings of the IEEE International Conference on Acoustics, Speech, Signal Processing,* Vol. 4, pp. 2072–2075, 1996.

[9-36] T. Suzuki and P. Kuhn, A proposal for segment-based transcoding hints, Noordwijkerhout, Netherlands, ISO/IEC M5847, Mar. 2000.

[9-37] P. Kuhn and T. Suzuki, MPEG-7 metadata for video transcoding: Motion and difficulty hint, in *Proceedings of the SPIE Conference on Storage and Retrieval for Multimedia Databases,* San Jose, CA, Jan. 2001.

[9-38] A. Hanjalic and H. Zhang, An integrated scheme for automated video abstraction based on unsupervised cluster-validity analysis, *IEEE Transactions on Circuits and Systems for Video Technology,* 9, 1280–1289, 1999.

[9-39] H.S. Chang, S. Sull, and S.U. Lee, Efficient video indexing scheme for content-based retrieval, *IEEE Transactions on Circuits and Systems for Video Technology,* 9, 1269–1279, 1999.

[9-40] A. Divakaran and H. Sun, A descriptor for spatial distribution of motion activity, in *Proceedings of the SPIE Conference on Storage and Retrieval from Image and Video Databases,* San Jose, CA, Jan. 2000.

[9-41] B. Erol and F. Kossentini, Automatic key video object plane selection using shape information in the MPEG-4 compressed-domain, *IEEE Transactions on Multimedia,* v2, 129–138, 2000.

[9-42] Information Technology—Coding of Audio/Visual Objects, ISO/IEC 14496-5:2000, 2000.

[9-43] J. Youn, J. Xin, and M.-T. Sun, Fast video transcoding architectures for networked multimedia applications, in *Proceedings of the IEEE International Symposium on Circuits and Systems,* Geneva, Switzerland, May 2000.

[9-44] MPEG Video Group, MPEG-4 Video Verification Model (v14.0), ISO/IEC N2932, Melbourne, Australia, Oct. 1999.

[9-45] S.F. Chang and P. Bocheck, Principles and applications of content aware video communications, in *Proceedings of the IEEE International Symposium on Circuits and Systems,* Geneva, Switzerland, May 2000.

[9-46] Q. Zhang, Y.Q. Zhang, and W. Zhu, Resource allocation for audio and video streaming over the Internet, in *Proceedings of the IEEE International Symposium on Circuits and Systems,* Geneva, Switzerland, May 2000.

[9-47] T. Ebrahimi and C. Christopoulos, Can MPEG-7 be used beyond database applications? ISO/IEC M3861, Atlantic City, NJ, Oct. 1998.

[9-48] H. Sun, A. Vetro, J. Bao, and T. Poon, A new approach for memory efficient ATV decoding, *IEEE Tranactions. on Consumer Electronics,* Vol. 44, pp. 517–525, Aug. 1997.

10 Universal Multimedia Access Using MPEG-21 Digital Item Adaptation

This chapter addresses several transcoding aspects related to the distribution of digital content. The universal multimedia access (UMA) concept is an important topic for the content distribution. The primary function of UMA services is to provide the best quality of service (QoS) or user experience by either selecting/adapting the content format to meet the playback environment, or adapting the content playback environment to accommodate the content. After reviewing the concept of UMA, we describe how the concept of UMA relates to the emerging MPEG standard, digital item adaptation (DIA), which is Part 7 of the MPEG-21 standard. An update on the standards activity in this area is presented. Finally, we discuss the impact that DIA will have on transcoding strategies and analyze some areas of future research.

10.1 INTRODUCTION

During the past two decades, tremendous progress has been achieved in the areas of digital signal processing and communications. In the area of digital signal processing, the techniques for digital video compression, transmission, and storage have progressed at an astounding pace. For example, considering the video-coding standards developed by Moving Picture Expert Group (MPEG) [10-1] and International Telecommunications Union (ITU) [10-2], many successful applications from digital television and to streaming video can be realized. Of course, the major goal of these video coding standards is to greatly reduce the data amount by compression algorithms. However, one very important aspect of the video coding scheme is to standardize the binary data format. In this way, the compressed bit stream can be used by a wide variety of equipment including traditional hardware, such as television set-top boxes, as well as other multimedia devices, such as computers, personal digital assistants (PDAs), and mobile terminals.

With advances in the semiconductor industry, the processing speed of digital signal processors is rapidly increasing; also, the price of storage memory is becoming inexpensive for many applications. This allows many devices, not just TVs and PCs, but also many portable devices such as PDAs or mobile terminals, to handle the compressed video data under certain conditions. Similarly, in the communications area, Internet technologies, network technologies, and mobile communications have also seen tremendous progress recently. These trends have become the major driving force for UMA.

The concept of UMA is to enable access to any multimedia content over any type of network, such as Internet, wireless LAN, or others, from any type of terminal with varying capabilities, such as mobile phones, personal computers, and television sets [10-3]. The primary function of UMA services is to provide the best QoS or user experience by either selecting appropriate content formats, or adapting the content format directly, to meet the playback environment, or to adapt the content playback environment to accommodate the content.

Toward the above goal, Part 7 of MPEG-21, referred to as *digital item adaptation* (DIA) is defining a set of tools to enable transparent and augmented use of multimedia resources across a wide range of networks and devices [10-4]. In the context of MPEG-21, a *digital item* is defined as a structured digital object with a standard representation, identification, and description. This entity is also the fundamental unit of distribution and transaction within the MPEG-21 framework [10-5]. Although DIA will not specify the adaptation engines themselves, there are a variety of interesting factors to consider with respect to the resource adaptation engine given complete knowledge of a user's environment.

The rest of this chapter is organized as follows. In Section 10.2, we first give an overview of UMA. Then in Section 10.3, an overview of MPEG-21 is introduced, the objectives of DIA are described, and the current status of this part of the standards is given. Finally, we describe the resource adaptation engine for a DIA system in Section 10.4 and description adaptation engine in Section 10.5, respectively.

10.2 OVERVIEW OF UNIVERSAL MULTIMEDIA ACCESS

The history of information technology shows that during the recent two decades a tremendous number of new technologies have been developed to meet the growing needs for obtaining any information from anywhere and anytime. This is the major motivation to propose the concept and carry out projects for UMA by many researchers. The concept of UMA is illustrated in Figure 10.1.

The concept of the UMA has two aspects. From the user side, UMA allows users access to a rich set of multimedia content through various connections such as Internet, optical ethernet, DSL, wireless LAN, cable, satellite, terrestrial broadcasting, and others, with different terminal devices. From the content or service provider side, UMA promises to deliver timely multimedia contents with various formats to a wide range receivers that have different capabilities and are connected through various access networks.

Now let us discuss what are today's driving forces for addressing the problems of UMA. There are several facts that provide the evidence of the growing need for UMA. From the viewpoint of contents, one fact is that content is available everywhere. This situation is created by the recent revolution of content format representation. The digitization of content format provides the possibility of letting different devices access the content. The advances of compression algorithms, especially digital video compression standards, greatly reduce the amount of data. This provides strong tools to reduce the difficulty for content delivery. Another fact is the capability growth of the communication networks. The Internet provides a very convenient and powerful tool for content transmission, but its big limitation is the narrow

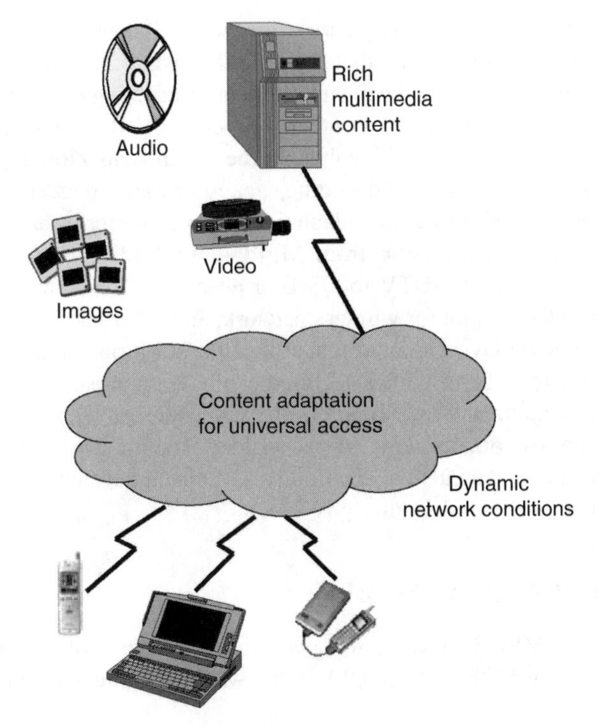

FIGURE 10.1 The Concept of UMA.

bandwidth at present. However, with advances in wireless communication, video transmission becomes possible through the 3G and beyond-3G networks. In summary, mature content representation formats, improving communication technology, and an increasing capability of terminals support the possibilities for UMA.

Given the above, what are the problems that have to be addressed today to achieve UMA? The major problem for UMA is to fix the mismatch between the content formats, the conditions of transmission networks, and the capability of receiving terminals. A mechanism for adaptation has to be created for this purpose and it can be considered into the following two ways. One way is to adapt the content to fit the playback environment and other is to adapt playback environment to accommodate the existing input contents. In either of these scenarios, it should be noted that MPEG-7 plays a key role in providing a description of the content [10-6].

Currently, multimedia content is encoded with various schemes, including MPEG-1, MPEG-2, MPEG-4, wavelets, JPEG., AAC, AC-3, etc.; all these formats encode at various spatial resolutions, frame rates, and bit rates. The communication networks also have different characteristics such as bandwidth, bit error rate, latency, and packet loss rate depending on the network infrastructure and load. Likewise, the receiving terminals have different content playback capabilities and different user preferences that affect the type of content that can be played on the terminals. The mismatch between the rich multimedia content and the content playback environment is the primary barrier for the fulfillment of the UMA promise. The adaptation engine is the entity that bridges this mismatch by either adapting the content to fit to the content playback environment or adapting the content playback environment to accommodate the content. For content adaptation, the nature of content determines the operations involved in the actual adaptation. For example, if the mobile terminal with MPEG-4 decoding capability wants to receive DTV signals that are encoded with MPEG-2, the adaptation needs to perform the conversion for several things: syntax conversion from MPEG-2 to MPEG-4, spatial resolution conversion from SDTV or HDTV to QSIF or even lower resolution, bit rate conversion to reduce the bit rate for wireless network, and other necessary conversions. Adapting the content playback environment involves acquiring additional resources to handle the content. The resources acquired can be session bandwidth, computing resources at the sending and receiving terminals, decoders in the receiver terminals, or improving the network latency and packet loss. The playback environment can change dynamically and content adaptation should match the changing environment to deliver content at the best quality possible.

10.3 OVERVIEW OF MPEG-21

Before discussing MPEG-21 DIA, we give a brief overview of MPEG-21, which may help to understand the role of DIA in the whole multimedia framework.

10.3.1 WHAT IS MPEG-21

MPEG-21 is a new working item within MPEG and is formally referred to as ISO/IEC-21000 Multimedia Framework. Why do we need MPEG-21? In the information age, more and more people are eager for getting and consuming information anytime and anywhere. This is the concept of UMA, which is to enable access to any multimedia content over any type of network from any type of terminal. Currently, there exists a large set of access devices with different capabilities and networks with varying conditions. The users do not have tools to efficiently deal with all the intricacies of this new multimedia usage context. Additionally, with advances of multimedia technology the content creation and re-creation become much easier than before. Individuals are able to produce more and more content with digitized format, not only for professional use but also for their personal use. Based upon the above observation, there are several problems we have to address for achieving the goal of UMA. The first is the growing mismatch between rich content and a lager number of access devices with different capabilities and network conditions. The second is how to handle the intellectual property management and protection (IPMP) during the process of retrieving, transmitting, and consuming the content. The major problems include management of content, protection of rights, protection from unauthorized access and modification, and protection of privacy of providers and consumers. Therefore, we need to provide solutions that satisfy the requirements for accessing, delivering, managing, and protecting the different content types in an integrated and harmonized way, to be implemented in a manner that is entirely transparent to the many different users of multimedia services. This kind of need has motivated the MPEG-21 Multimedia Framework.

MPEG-21 is based on two essential concepts. The first is the definition of a fundamental unit of distribution and transaction, which is referred to as a *digital item* (DI). The second is the concept of users interacting with digital items. The digital items can be considered as the "what" of the multimedia framework and the users can be considered as the "who" of the multimedia framework. The goal of MPEG-21 is to seek solutions to meet the increasing challenges from the growing mismatch between rich contents, varying network conditions, and devices with different capacities. For this purpose, MPEG-21 focuses on defining the technology needed to support users to exchange, access, consume, trade and manipulate digital items in an efficient, transparent, and interoperable way. The scope of MPEG-21 standard is to define a normative open framework for multimedia delivery and consumption for use by all the players in the delivery and consumption chain. The standard will allow the content consumer access to a large variety of content in an interoperable manner and will provide equal opportunities in the MPEG-21 enabled market for content creators, producers, distributors, and services.

Currently, MPEG-21 consists of the following parts, but other parts may be added in future when needed:

Part 1: Technical Report
Part 2: Digital Item Declaration
Part 3: Digital Item Identification
Part 4: Intellectual Property Management Tool Representation and Communication System
Part 5: Rights Expression Language
Part 6: Rights Data Dictionary
Part 7: Digital Item Adaptation
Part 8: Reference Software
Part 9: File Format
Part 10: Digital Item Processing
Part 11: Evaluation of Persistent Association Tools
Part 12: Resource Delivery Test Bed
Part 13: Scalable Video Coding
Part 14: Conformance

Before we give a brief introduction to each part, we elaborate on two important terms in MPEG-21 mentioned above: digital item and user.

Digital item (DI) is the fundamental unit of distribution and transaction in the MPEG-21 framework. It is commonly called *content* and is a structured digital object with a standard representation, identification, and metadata. It is a challenge to define a clear definition for a digital item due to the existence of many different kinds of content and probably just as many possible ways of describing it to reflect its context of use. The Digital Item Declaration specification (Part 2 of MPEG-21) provides such flexibility for representing digital items.

The definition of user in the MPEG-21 is given as follows [10-5]: A User is any entity that interacts in the MPEG-21 environment or makes use of a Digital Item. Users are identified specifically by their relationship to another user for a certain interaction. The interactions include many kinds of activities, such as creating, providing, delivering, and archiving content. From a purely technical perspective, MPEG-21 considers both content provider and consumer as users, and all users are equally when they are interacting within MPEG-21 framework. However, a user may assume specific or even unique rights and responsibilities according to their interaction with other users within MPEG-21.

Part 1 of MPEG-21 is a technical report with the title, Vision, Technologies and Strategies [10-7], which has been approved by MPEG in September 2001. This technical report describes the functional requirement of multimedia framework and its architectural elements. The fundamental purposes of this report include three parts as described by its title. First, it defines a vision for a multimedia framework to enable transparent and augmented use of multimedia resources across a wide range of networks and devices to meet the needs of all users. Second, it outlines and recommends the development and standardization of technologies for digital item creation, management, transport, manipulation, distribution, and consumption. The final purpose is to define a strategy for developing specifications and standards for multimedia framework based on well-defined functional requirements through collaboration with other standard bodies.

Part 2 of MPEG-21 is called digital item declaration (DID). The purpose of DID is to develop a specification, which can be used to form a useful model for defining digital items. The DID includes three normative sections. The first section is the model. The model is formed with a set of abstract terms and concepts. Within this model a digital item is considered as a digital representation of a work that is manipulated (managed, described, exchanged, etc.) with the model. The second section is the representation, which is the normative description of the syntax and semantics of the digital items as represented in XML. Some non-normative examples are also included in this section for illustrative purpose. The third section is the normative XML schema, which comprises the entire grammar for digital items declaration represented in XML. The detailed description of DID can be found in [10-8].

Part 3 of MPEG-21 is digital item identification (DII). The DII specification addresses the problems including how to uniquely identify digital items (DIs) and their parts; IP related to DIs and their parts; description schemata; how to use identifiers to link DIs with related information, such as descriptive metadata; and how to identify types of DIs. The DII will use the existing identification and description schemata for some DIs such as ISRC for sound recordings. The DII identifies the DIs and their parts by encapsulating uniform resource identifiers (URI) into the identification DS. A URI is a compact string of characters for identifying an abstract or physical resource. Identifiers of MPEG-21 can also be a specific subset of URI, such as the uniform resource locator (URL) that is extensively used as a pointer to information on the Internet.

Part 4 of MPEG-21 covers intellectual property management and protection (IPMP). issues, which have garnered much attention since MPEG started standardization work on IPMP, after its first call for proposals in 1997. The IPMP solution

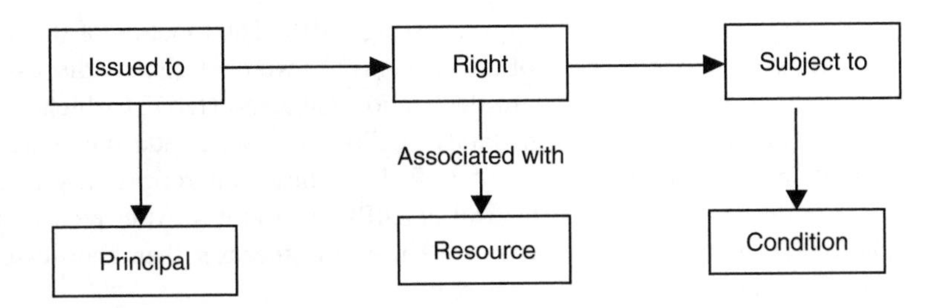

FIGURE 10.2 The REL data model.

is a very important item for the distribution and exchange of digital content between content providers, content servers, and end users. The IPMP provides tools to reach the goals for transparently, seamlessly, and intuitively exchanging and distributing the digital contents. Currently, MPEG-21 focuses on developing the requirements on an IPMP language and an accompanying dictionary. These requirements address the need of MPEG-21 for achieving the goal of interoperability. Actually, MPEG-4 has developed IPMP hooks for the standard but they may not be interoperable for many similar devices and players built by different manufacturers. The IPMP of MPEG-21 will define an interoperable framework for IPMP, which includes defining the standardized ways for retrieving IPMP tools from remote location, and exchanging messages between IPMP tools and between these tools and the terminals.

Part 5 of MPEG-21 is rights expression language (REL), which is considered a machine-readable language that can be used to declare rights and permissions to support the transparent and augmented use of DIs in many applications among different systems and services. The REL data model consists of four basic parts and the relationships among these parts as shown in Figure 10.2.

A principal encapsulates the identification of principals to whom the grant is issued. The grant specifies the rights. The resource is the object to which the right in the grant applies. A condition specifies the terms and obligations under which the rights can be exercised. A simple of example of condition is a time interval within which a right can be exercised.

Part 6 of MPEG-21 is rights data dictionary (RDD). The RDD contains a set of clear, consistent, structured, integrated, and uniquely identified terms to support the MPEG-21 REL.

Part 7 of MPEG-21 is the most related part to the topic of video transcoding. We are going to give it detailed discussion in next subsection.

Part 8 of MPEG-21 is the reference software. The development of the reference software will be based on the requirements that have been specified in the architecture for processing DIs.

Part 9 of MPEG-21 is the file format. Since a DI of MPEG-21 can be a complex collection of information, which may include compressed video, audio data, metadata, layout information, and so on, the data could be textual or binary format. The MPEG-21 file format will use several concepts from MP4 and make multipurpose files possible.

Part 10 of MPEG-21 is digital item processing (DIP). The function of this part is to specify tools for the processing of digital items. As we have already discussed, the DIs are defined by the digital item declaration language (DIDL) which is an XML language and the digital item methods to allow the user to add functionality to a digital item declaration; the role of DIP then allows interoperability at the processing level. The operation mechanism of DIP is as follows. After receiving a DID, a list of DI methods is presented to the user and the user then chooses one method that is executed by the DIP engine.

Part 11 of MPEG-21 is a technical report titled, Evaluation of Persistent Association Tools. The term p*ersistent association* is used to categorize all the techniques for managing identification and description with content and resource. This technical report describes evaluation methodologies for some characteristics of two classes of technologies, watermarks and fingerprints applied to video and audio contents, but it does not attempt to define methodologies for evaluating the resistance of these technologies to deliberate attack on the association.

Part 12 of MPEG-21 is test bed for MPEG-21 resource delivery. This is a new topic for MPEG-21 and led by Professor Tihao Chiang, a co-author of this book. Detailed information is described in the next chapter.

Part 13 of MPEG-21 is scalable video coding. Scalability is an important feature for video coding, which was addressed by the standards, including MPEG-2 and MPEG-4. However, there is no big impact of the existing scalability schemes on the video industry due to the coding performance penalty introduced by adding the capability of scalability on the coding scheme. The purpose of scalable video coding standard will focus on the coding quality improvement with large range of scalability. The detailed requirements can be found in the call for proposals issued in the 66th MPEG meeting [10-24].

Part 14 is the conformance of MPEG-21. The definition of the MPEG-21 DIA conformance is given in [10-25]. It is a set of requirements on a DIA description such that can be utilized by a DIA engine as intended in the DIA specification. The conformance of DIA descriptions is defined for an XML document or fragment.

10.3.2 Overview of Digital Item Adaptation

As we discussed in the Section 10.2, the key problem of UMA is to fix the mismatch between rich multimedia contents, networks, and terminals. To address this problem, the mechanism of adaptation is one of the most important issues for UMA. The current situation is that many standards exist for content representations, such as audio/video coding standards developed by MPEG and ITU, as well as many standards for communications and many protocols for networks. All elements for building an infrastructure of UMA exist and have been standardized or their standards are under development. However, there is no "big picture" of how these elements are related to each other. In other words, the way to fix the mismatch between elements has not been defined. To accomplish the promise of the UMA, we need to develop new standards to meet this need. Now, the emerging standard MPEG-21, especially Part 7 of MPEG-21, digital item adaptation (DIA), aims at fixing these gaps between elements and achieving interoperable transparent access

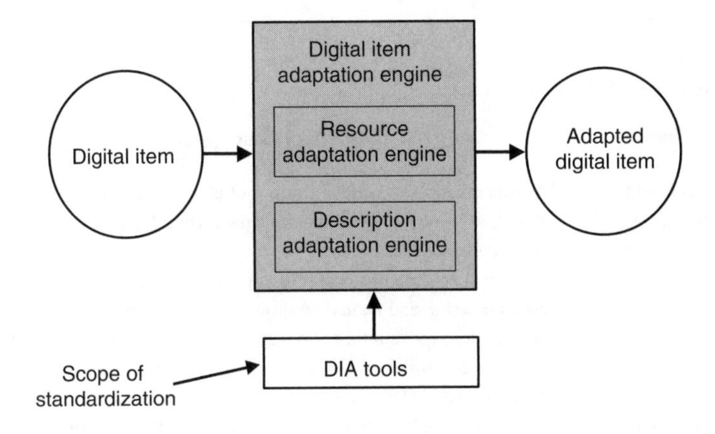

FIGURE 10.3 Concept of Digital Item Adaptation.

to the multimedia content by shielding users from network and terminal installation, management, and implementation issues. To achieve this goal, the adaptation of digital items is required. This concept is illustrated in Figure 10.3.
From this architecture it can be found that the adapted digital item that meets the content playback environment is obtained from the original input digital item through either a descriptor adaptation engine or resource adaptation engine. Digital item adaptation aims at providing the standardized descriptions and tools that can be used by these adaptation engines [10-4]. Therefore, the scope of standardization includes descriptors and format-independent mechanisms that provide the support for DIA in terms of resource adaptation, description adaptation, and quality of service (QoS) management. The adaptation engines are non-normative part of DIA standard.

The DIA tools can be classified into eight major categories according to their functionality and use for adaptation of digital items. The different tools and a brief description of them are summarized in Table 10.1.

The usage environment description tools are an important category of DIA tools. These tools provide descriptive information about these various dimensions of the usage environment, which originate from users, to accommodate, for example, the adaptation of digital items for transmission, storage, and consumption. Four major areas of usage environment descriptions have been considered until now. All of these descriptions may be used to support future UMA services. These areas include user characteristics, terminal capabilities, network characteristics, and natural environment characteristics. User characteristics include five aspects: user type, content preferences, presentation preferences, presentation preferences, accessibility characteristics, and mobility characteristics. The descriptions in the terminal capabilities currently include encoding and decoding capabilities, display and audio output capabilities, as well as power, storage, and input-output characteristics. The items in the network descriptions are network capabilities and network conditions.

TABLE 10.1
Overview of DIA Tools

DIA Tool	Tool Description
Schema Tools and Low-Level Data Types	The schema tools provide uniform root elements for all DIA descriptions as well as some low-level and basic data types that can be used by several DIA tools independently.
Usage Environment Description	This tool specifies User characteristics, terminal capabilities, network characteristics, and natural environment characteristics. These tools provide descriptive information about the various properties of the usage environment, which originate from Users, to accommodate, for example, the adaptation of Digital Items for transmission, storage, and consumption.
Bit Stream Syntax Description Link	Provides the facilities to create a rich variety of adaptation architectures based on tools specified DIA, DID, and MPEG-7 among others.
Bit Stream Syntax Description	Describes the syntax — in most cases, the high level structure — of a binary media resource. Using such a description, a Digital Item resource adaptation engine can transform the bit stream and the corresponding description using editing-style operations such as data truncation and simple modifications.
Terminal and Network Quality of Service	The tools specified in this category describe the relationship between QoS constraints (e.g., on network bandwidth or a terminal's computational capabilities), feasible adaptation operations satisfying these constraints and associated media resource qualities that result from adaptation. This set of tools therefore provides the means to trade-off these parameters with respect to quality so that an adaptation strategy can be formulated and optimal adaptation decisions can be made in constrained environments.
Universal Constraints Description	Enables the possibility to describe limitations and optimization constraints on adaptations.
Metadata Adaptability	This tool specifies hint information that can be used to reduce the complexity of adapting the metadata contained in a Digital Item. On the one hand, they are used for filtering and scaling, and on the other hand, for integrating XML instances.
Session Mobility	The configuration state information that pertains to the consumption of a Digital Item on one device is transferred to a second device. This enables the Digital Item to be consumed on the second device in an adapted way.
DIA Configuration	Provides information required for the configuration of a Digital Item Adaptation Engine.

The descriptions in the natural environment characteristics are currently defined by location, time, and audio-visual environment characteristics.

In addition to the usage environment descriptions, MPEG-21 DIA is also targeting the specification of tools that support format-independent manipulations, tools to help make trade-offs between coding parameters and content characteristics, and tools that support metadata adaptation. Several tools have been specified to enable this functionality: the bit stream syntax description, terminal and network QoS, bit stream syntax description link, and metadata adaptability. The relationships between

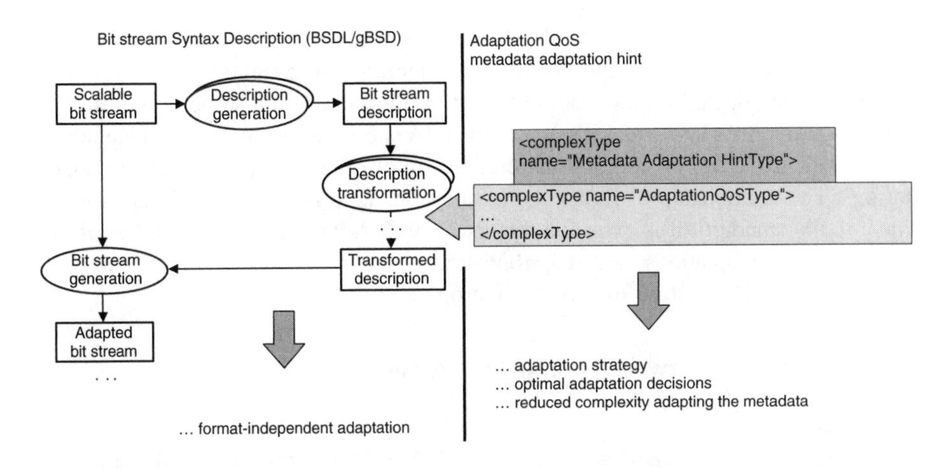

FIGURE 10.4 Example of using several resource adaptation tools.

these tools is shown in Figure 10.4. Tools in the bit stream syntax description can be used to facilitate the bit stream scaling in a format-independent manner. These tools are used for the adaptation of resources contained within a Digital Item at network nodes that are not knowledgeable of the specific resource representation format. In the case of terminal and network quality of service, tools are used to describe the relationship between constraints, feasible adaptation operations satisfying these constraints, and associated qualities. These tools provide solutions, which are able to obtain the compromise between the constraints and quality of services with possible adaptation operations using some parameters. The bit stream syntax description link tool provides the facilities to create a large number of adaptation architectures. The tools of metadata adaptability enable the filtering and scaling of XML instances with reduced complexity.

10.3.3 Relation between DIA and other Parts of MPEG-21

We briefly introduced the parts of the MPEG-21 standard in Subsection 10.3.1. Because we know that the fundamental unit of distribution and transaction in the multimedia framework is the digital item (DI), the purpose of all existing parts of the MPEG-21 together is trying to facility a complete, interoperable multimedia framework, while the individual parts of MPEG-21 deal with different aspects of the fundamental unit of MPEG-21, the digital item. From the overview of MPEG-21 we may easily understand the relationships among different parts of MPEG-21. However, our major focus in this book is on transcoding-related topics. For MPEG-21, the DIA is the most related part of MPEG-21 to the transcoding. In the following, a detailed description about the relation between DIA and the other parts of MPEG-21 is provided.

Part 2 of MPEG-21 (DID) enables the declaration of digital items, where digital item is a packaging of resources, descriptions, and rights expression. A digital item may contain elements that conform to all part of MPEG-21.

In the adaptation framework, a digital item is input to a digital item adaptation engine. The adaptation engine can modify the input digital item by adapting the resources or metadata within the digital item or the declaration of the digital item to the usage environment. Additionally, the identifiers and rights expressions pertaining to the adapted digital item need not be the same as those pertaining to the input digital item. The current version of the DIA specification deals with adaptation, but specifically does not address the relationship of rights and permissions to adaptations. The relationship of rights and permissions is to be addressed in an amendment to the standard. It is expected that users of DIA will register terms describing their specific adaptations with the Registration Authority described in Part 6 of MPEG-21 (RDD) in order to provide interoperability.

10.3.4 Relation between Digital Item Adaptation and MPEG-7

The MPEG-7 is a standard developed by MPEG. It is formally called the *multimedia content description interface.* The task of MPEG-7 is to specify a standard way of describing various types of multimedia content. The scope of MPEG-7 includes descriptors, description schemes, and description definition language. The standardized descriptors and description schemes are used to define a comprehensive set of audiovisual description tools, such as the metadata elements and their structure and relationships. The objective of MPEG-7 is to use these description tools to facilitate the quick and efficient access (search, filtering, and browsing) to an interesting and relevant multimedia content and efficient management of that content. From above summary of MPEG-7, there are several relations between MPEG-7 and MPEG-21.

First, some MPEG-7 descriptions can be used for describing the resource contained in a particular digital item defined in MPEG-21. The description is used to satisfy search and retrieval request of users query for that resource. The search operation would be carried out with a search engine.

Second, several MPEG-7 descriptions can be also used in the digital item adaptation process. Specially, the MPEG-7 tools used for video and audio summarization and media transcoding hints can be used for the purpose of media content adaptation. For example, the transcoding hints description in MPEG-7 can be used to easily adapt audio and video data of interest according to the hints so that the adapted data can fit the transmission bandwidth or terminal capability.

Third, besides serving as an input to the digital item adaptation engine, several MPEG-7 descriptions are also referenced by the DIA specification. These MPEG-7 tools include those indicating user's type and preference, coding formats, and location and time, which are adopted as part of the DIA usage environment description tools. The MPEG-21 is also considering for developing description adaptation engine, which can be used to scale or convert the MPEG-7 metadata to match the need of network and terminal capabilities. The detail of description adaptation engine will be presented later.

Finally, the multimedia description schemes specified by MPEG-7 can be used as description tools for MPEG-21 to specify the decoding and encoding formats in the DIA terminal capability part. With these tools the media resource can be easily adapted to match the terminal capability.

10.4 RESOURCE ADAPTATION ENGINE

Resource adaptation includes a variety of conversions or transformations, including conversion between video coding formats, such as from MPEG-2 to MPEG-4, and coding parameter conversion, such as bit rate, spatial resolution, and frame rate. This section deals primarily with video transcoding-related topics and covers the basic design goals that motivate research on video transcoding, some recent advances in this area, as well as new challenges related to the video transcoding process.

10.4.1 Design Goals and Issues

The key design goals of transcoding include two aspects: 1) to maintain the video quality during the transcoding process, and 2) to keep complexity as low as possible. The most straightforward solution is to simply decode the video signal, perform any postprocessing if necessary, and re-encode the modified signal according to any new constraints. Although the quality is good, the complexity of this approach is relatively high. To avoid such a conversion, some researchers have proposed scalable coding schemes that can be easily scaled down according to the requirements of terminal or network. The MPEG-4 Fine Granular Scalability is one such scalable coding scheme [10-9]. However, in the entertainment industry, many contents exist with a fixed single-layer representation format. For example, the contents of digital television and DVD are encoded with an MPEG-2 format. In order for receivers, such as mobile terminals, to receive this signal, we must convert the bit stream from MPEG-2 to MPEG-4. Therefore, there are instances in which transcoding is certainly needed; the application scenarios are very clear. Such conversion must be done with low complexity and high quality.

In most of video coding standards, the temporal redundancy is removed by predictive coding techniques, where the current field/frame is predicted with the previously coded field/frame. The predictive difference is then quantized and coded with variable run-length codes. During transcoding, the reference used for prediction is typically transcoded to a lower quality or spatial resolution; therefore, the error in this reference frame will accumulate. This accumulation of error is referred to as *drift error.* Minimizing the amount of drift incurred during the transcoding process, while keeping complexity low, is a major goal of the transcoding design.

10.4.2 Transcoding Background

Reducing the drift with a low-complexity architecture is a significant challenge for transcoding and a great deal of research effort has been dedicated to this task. In the context of transcoding to a lower bit rate, several papers have described architectures that avoid full decoding and re-encoding (e.g., [10-10]) and techniques that rely on the on the efficient reuse of macroblock layer information (e.g., [10-11]). More recently, several architectures and techniques for spatial resolution reduction have been presented in [10-12]. Most notable may be the use of intra-block refresh to combat the effects of drift errors incurred by transcoding to a lower spatial resolution. Also, new work in the area of temporal resolution reduction has been presented [10-13]. For all the above methods, motion vector refinement can also be applied for improved quality without a significant increase in complexity [10-14].

10.4.3 Transcoding QoS

Existing transcoding techniques provide a powerful tool for the resource adaptation engine, where changes in bit rate, frame rate, and spatial resolution can easily be controlled. However, controlling these parameters to achieve the best perceptual quality for any user, or group of users, is a challenging problem. In this subsection, we introduce relevant work in this area and discuss possibilities for further work.

The concept of a utility function to measure the user's satisfaction of a coded video bit stream was introduced in [10-15]. Based on features extracted from the video, machine learning and classification techniques were used to estimate the subjective quality of the video coded according to different scaling profiles, e.g., drop B-frames, scale DCT coefficients, drop color components. Based on the quality estimate, some decision regarding the best way to scale the contents could be made.

In [10-16], a video coding algorithm that is considered the trade-off between spatial and temporal quality was presented. This algorithm was based on analytical models that estimated actual MSE and rate for a set of possible frameskip factors and quantization parameters. A similar problem was considered in [10-17], where the authors sought to optimize the coding across bit rate, frame rate, and spatial resolution. In this work, various reconstruction patterns in the receiver were considered. Since the emphasis was on an optimal framework and ways to solve the multidimensional problem, actual rates and distortions were used rather than estimates obtained from a model.

In [10-26], a rate-distortion model for DCT-based video coding incorporating the macroblock intra refreshing rate has been developed. This model can accurately estimate the R-D function for a given intra-refreshing rate before a video frame is coded. Based on this model and the theoretical analysis of channel distortion caused by channel errors and inter-frame propagation, an end-to-end optimization can be reached by adaptive intra mode selection and joint source-channel rate control.

More recently, rate-distortion models to estimate the rate and distortion resulted from specific transcoding operations, such as requantization, spatial scaling, and temporal scaling, have been developed [10-18]. These models allow one to consider the choices between various scaling operations directly from data contained within the compressed bit streams.

The above works indicate that we can optimize the quality of a single transcoded output with some confidence. Given constraints on the bit rate and spatial resolution specified by the capability of a terminal and the current network condition, the transcoder can be designed to automatically skip more frames in the low motion area, thereby using more bits for coded frames, to guarantee the spatial quality. On the other hand, in the fast motion area, more frames will be coded and fewer bits will be assigned to each of coded frame.

Considering the work on utility functions, this kind of control can also be applied according to user preferences instead of objective quality measurements. For example, for surveillance video, the users may not care about overall quality of the transcoded output, but be very interested in a certain time period of video or require a minimum level of frame quality to view certain details in the scene.

Finally, one emerging research area could be on the transcoding of multiple streams; in other words, given a single processing engine that is required to transcode several bit streams, one may consider algorithms to allocate the quality among those streams. Algorithms to allocate the processing resources as well may be considered. The optimization of quality and processing complexity may also be considered in a joint manner.

10.4.4 Comparison between Transcoding and Scalable Coding

In Section 10.4.1, we mentioned that both transcoding and scalable coding are the useful for content adaptation. To make clear the relationship between these two techniques, we would like to give a more detailed comparison. First of all, scalable coding specifies the data format at the encoding stage independently of the transmission requirements, while transcoding converts the existing data format to meet the current transmission requirements. The holy grail of scalable video coding is to encode the video once, and then by simply truncating certain layers or bits from the original stream, lower qualities, spatial resolutions, and/or temporal resolutions could be obtained. Ideally, this scalable representation of the video should be achieved without any impact on the coding efficiency. While current scalable coding schemes fall short of this goal, preliminary results based on exploration activity within MPEG indicate that the possibility for an efficient universally scalable coding scheme is within reach [10-25].

Second, besides the issue of coding efficiency, which seems likely to be solved soon, scalable coding will need to define the application space that it could occupy. For instance, content providers for high-quality mainstream applications, such as DTV and DVD, have already adopted single-layer MPEG-2 video coding as the default format, hence a large number of MPEG-2 coded video content already exists. To access these existing MPEG-2 video contents from various devices with varying terminal and network capabilities, transcoding is needed. For this reason, research on video transcoding of single-layer streams has flourished and is not likely to soon diminish. However, in the near term, scalable coding may satisfy a wide range of video applications outside this space, and, in the long-term, we should not dismiss the fact that scalable coding format could replace existing coding formats. Of course, this could depend more on economic and political factors rather than on technical ones. The main point to all of this is that scalable coding and transcoding should not be viewed as opposing or competing technologies. Instead, they are technologies that meet different needs in a given application space and it is likely that they will coexist.

Finally, we would like to indicate that both scalable coding and transcoding have to consider the issue of digital right management (DRM). For the DRM issue, scalable coding may handle it easier than transcoding. In scalable coding, the content protection information may be directly embedded in the bit stream during encoding whereas transcoding needs to do something to transfer the content protection information during the procedure of transcoding, which may add extra complexity.

10.5 DESCRIPTION ADAPTATION ENGINE

In the previous section, we have presented the resource adaptation engine. In this section we present the principle of description adaptation engine. The description adaptation engine provides mechanism for converting the descriptions or metadata instead of media resource itself.

10.5.1 MOTIVATIONS AND GOALS

In the multimedia framework, the advances of communication environment provide the possibilities for consumers to access multimedia content in any time and from any location. However, due to the large number of multimedia contents, it is very important to solve the problem of how to quickly and easily identify the user-preferred content, such as for the web-based distributed multimedia contents. For this purpose, the MPEG-7 standard defines a large set of descriptions or metadata as multimedia interface for consumers to efficiently find the contents of interests from vast amount of distributed multimedia content resources. Now several proposals have been discussed and adopted by the MPEG-21 digital item adaptation about description or metadata (including MPEG-7 descriptions and metadata, but not limited to) adaptation and scaling [10-19], [10-20], [10-21], [10-22]. The motivations and goals of these proposals are as follows.

First, MPEG-7 provides the metadata and tools for describing the structured multimedia content, such as video, audio, and other data. However, there is a practical problem to apply these descriptions, because such content is usually described from various aspects, which sometimes results in quite complicated, hierarchically annotated instances. The MPEG-7 system tries to alleviate the problem by using a binary compression schema (BiM) or XML texts, which can be compressed by various compression tools such as Zip to reduce the size of description or metadata. However, for some applications the size of metadata is still too large and further scaling or adaptation is needed.

Second, the MPEG-7 standard contains a very large set of descriptors and description schemes, and some are optional. Although MPEG-7 is considering the profile issues and provides some constraints on what kind of descriptions should appear in actual application case, there is still a large set of descriptors or description schemes in the selection pool. This results in difficulties accessing the needed MPEG-7 metadata. Under this sense, the DIA needs not only to perform the media resource adaptation, but also the metadata or description adaptation. The metadata or description adaptation aims at reducing the size of metadata by operations of scaling or filtering to satisfy the actual need for certain applications such as content retrieval and to identify the required subset of metadata to handle the adaptation for actual application environment, such as bandwidth, terminal capability, and user's preference.

For the above reasons, several syntactical elements have been proposed as hint information for description or metadata adaptation in MPEG-21 DIA [10-24].

10.5.2 Metadata Adaptation Hints

In the MPEG-21 DIA, two scenarios have been proposed and under the procedure of core experiment (CE) [10-20]. The purpose of CE is to evaluate the effectiveness of hint information by which metadata are to be adapted.

The first scenario is web contents conversion using hint information. The web content can be considered as widely used digital items. Usually, web contents are presented in personal computers, which are considered as high-performance terminals. With great advances of network technology, more and more appliance terminals can easily be connected to the network through Internet or various wireless networks. Therefore, it is possible to use the low-end performance terminals, such as cellular phones and PDAs, to access the web contents. For these applications the web contents need to be scaled or converted to the formats that the low-end terminals can access. Web contents are usually described by XML for the PC applications and need to be converted to HTML formats, which are applicable to a variety of terminals. Furthermore, as a huge amount of distributed web contents are available, it is also important to allow users with their low-end terminals to search the contents. It is necessary to convert not only the contents into the size and the format suitable for the terminal performance, or to summarize the contents, but also the metadata to satisfy such requirements. The conversions are based on the requirements of the terminal capabilities, the user preference, and the designes of web content creators. This procedure is considered as digital item adaptation. The adaptation process includes two aspects: one is the adaptation of resources and the other is the adaptation of description or metadata. In order to make web content conversion be easier, the content creator can add some hint information or metadata to the content during the processing of content creation; such hint information can be the number and location of the important parts of the resource. An illustration of a description or metadata adaptation engine is shown in Figure 10.5.

In Figure 10.5, the software implementation that performs the adaptation of XML metadata using an adaptation hint descriptor is described. In this implementation, the function of the description or metadata adaptation engine is to adapt, find, or create suitable target metadata using hint information from the metadata used to describe the web-based contents. The resulted metadata would be suitable for various terminals. For example, when a user wants to find the information about a restaurant in a big city, with a cellular phone or PC, the user can search the target web-based contents using metadata that describes the information such as the location, topic, or other keywords associated with the restaurant. If the user uses a cellular phone with limited capability, it is necessary to perform "filtering" or "scaling" operations on the metadata, thus converting the original metadata to a smaller amount of data but still satisfying the minimum requirements for the search. Of course, content adaptation is also necessary at the same time for displaying the result on the screen of cellular phone.

The second scenario is the integration of distributed metadata. In some applications, two or more persons add the metadata to long video sequences in order to efficiently generate it. In this case, some redundancy or overlapping about the metadata may result since those metadata are used to describe the same video

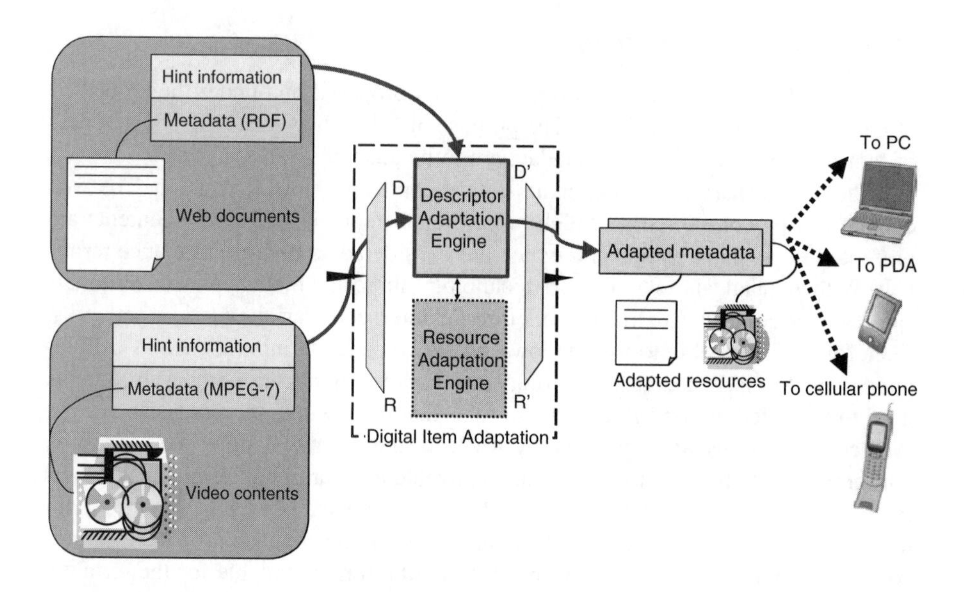

FIGURE 10.5 Illustration of description adaptation engine [10-23].

content. If the redundancy or overlapping can be identified, it is possible to integrate or compact the metadata. Furthermore, it is also possible to automatically generate the information about number and location included in the metadata in the process of distributed metadata integration.

10.6 SUMMARY

In this chapter, an overview of UMA has been presented. The purpose of UMA is to deliver or to use rich multimedia contents with a range of network and receiver capabilities. The emerging MPEG-21 standard, Digital Item Adaptation, which aims to solve the mismatch between components in the end-to-end delivery chain, has also been reviewed. Additionally, we have described the current status of description tools that have been standardized as part of DIA. Finally, in this chapter, the key technical challenges for transcoding have been reviewed, and an important research direction, transcoding QoS, has been introduced.

10.7 EXERCISES

10-1. Give an example of UMA application and explain the key components of UMA.

10-2. Describe the role of transcoding in UMA. Is the transcoding a normative part of MPEG-21? Why?

10-3. According to your knowledge, explain the function of adaptation engines. What is the difference between content adaptation engine and description adaptation engine?

10-4. Describe the relationships between digital item adaptation and other parts of MPEG-21.
10-5. Describe the relationships between digital item adaptation and MPEG-7. According to your knowledge, give an example application scenario.
10-6. According to your understanding, explain why those tools are adopted by MPEG-21 for digital item resource adaptation.
10-7. Based on your knowledge, describe what technical issues must be considered in the multiple streams transcoding processing with one transcoder.

REFERENCES

[10-1] ISO/IEC 14496-2:2001, Coding of Audio-Visual Objects – Part 2: Visual, 2nd Edition, 2001.
[10-2] ITU-T Recommendation H.263++, Video Coding for Low-Bit-Rate Communication, ITU-T Standardization Sector (Geneva), 3, 2000.
[10-3] R. Mohan, J.R. Smith, and C.S. Li, Adapting Multimedia Content for Universal Access, *IEEE Transactions on Multimedia,* 1, 104, 1999.
[10-4] ISO/IEC 21000-7 FDIS, Information Technology – Multimedia Framework – Part 7: Digital Item Adaptation, December 2003.
[10-5] MPEG-21 Overview v.5, ISO/IEC JTC1/SC29/WG11/N5231, October 2002.
[10-6] MPEG-7 Overview v.8, ISO/MPEG N4980, Klagenfurt, Austria, July 2002.
[10-7] MPEG-21 Part 1: Technical report Vision, Technology and Strategy, ISO/IEC MPEG/N6042, Brisbane, October 2003.
[10-8] MPEG-21 Part 2: Digital Item Declaration.
[10-9] W. Li, Overview of Fine Granularity Scalability in MPEG-4 video standard, i.e., EE Trans. Circuits and Systems for Video Technology, Vol. 11, No. 3, March 2001, pp. 301–317.
[10-10] P. Assuncao and M. Ghanbari. Post-processing of MPEG-2 coded video for transmission at lower bit-rates, *Proceedings of the IEEE International Conference on Acoustics, Speech and Signal Processing,* Atlanta, GA, May 1996.
[10-11] H. Sun, W. Kwok, and J. Zdepski, Architectures for MPEG compressed bitstream scaling, *IEEE Transactions on Circuits and Systems for Video Technology,* Vol. 6, pp. 191–199, April 1996.
[10-12] P. Yin, A. Vetro, B. Liu, and H. Sun, Drift compensation for reduced resolution transcoding, *IEEE Transactions on Circuits and Systems for Video Technology,* 12, 1009, 2002.
[10-13] K.T. Fung, Y.L. Chan, and W.C. Siu, New architecture for dynamic frame-skipping transcoder, *IEEE Transactions on Image Processing,* 11, 886, 2002.
[10-14] J. Youn, M.T. Sun, and C.W. Lin, Motion vector refinement for high performance transcoding, *IEEE Transactions on Multimedia,* 1, 30, 1999.
[10-15] R. Liao, P. Bocheck, A. Campbell, and S.F. Chang, Content-aware Network Adaptation for MPEG-4, Proceedings of NOSSDAV, NJ, June 1999.
[10-16] A. Vetro, Y. Wang, and H. Sun, Rate-distortion optimized video coding considering frameskip, *Proceedings if the IEEE International Conference on Image Processing,* Thessaloniki, Greece, Oct. 2001.
[10-17] E.C. Reed and J.S. Lim, Optimal multidimensional bit-rate control for video communications, *IEEE Transactions on Image Processing,* 11, 873, 2002.

[10-18] P. Yin, A. Vetro, and B. Liu, Rate-distortion models for video transcoding, Proceedings of the SPIE Conference Image and Video Communication Processing, San Jose, CA, Jan. 2003.

[10-19] H. Nishikawa, Y. Isu, S. Sekiguchi, K. Asai: Description for Metadata Adaptation Hint, ISO/IEC JTC1/SC29/WG11/m8324, Fairfax, VA, May 2002.

[10-20] K. Kazui: Result of CE on Metadata Adaptation, ISO/IEC JTC1/SC29/WG11 M8576, Klagenfurt, July 2002.

[10-21] M. Sasaki: Report and Study of CE on Metadata Adaptation, ISO/IEC JTC1/SC29/WG11 M8560, Klagenfurt, July 2002.

[10-22] N. Adami and R. Leonardi: Content and description scalability issues in a home network environment, ISO/IEC JTC1/SC29/WG11 M8693, Klagenfurt, July 2002.

[10-23] H. Nishikawa, Y. Isu, K. Asai: Report of CE Metadata Adaptation for MPEG-21 DIA, ISO/IEC JTC1/SC29/WG11/M8962, Shanghai, October 2002.

[10-24] P. Kuhn, T. Suzuki, and A. Vetro, MPEG-7 Transcoding Hints for Reduced Complexity and Improved Quality, International Packet Video Workshop, Kyongju, Korea, April 2001.

[10-25] ISO/IEC JTC1/SC29/WG11 MPEG2003/N5958, Call for proposals on scalable video coding technology, Brisbane, October 2003.

[10-26] ISO/IEC JTC1/SC29/WG11 MPEG/N6069, MPEG-21 DIA conformance WD 0.1, Brisbane, October 2003.

[10-27] Z. He, J. Cai, and C. Chen, Joint source channel rate-distortion analysis for adaptive mode selection and rate control in wireless video coding, *IEEE Transactions on Circuits and Systems for Video Technology,* 12, 511, 2002.

11 End-to-End Video Streaming and Transcoding Systems

In this chapter, we introduce the concept of real-time transport protocol and the carriage of multimedia contents over IP networks. The first section covers the various elements of the video streaming and transcoding system protocol. The second section describes how contents can be carried over IP networks. Details of the real-time transport protocol will be described. An implementation example of such a system is described in Section 11.3 and 11.4. The last section describes some simulation results and conclusions.

11.1 ELEMENTS OF VIDEO STREAMING AND TRANSCODING SYSTEM

There are three major components in a video streaming and transcoding system: the server system, network interface, and client terminal.

The first part is the server system, which includes several components, such as multimedia database, multimedia indexing system, and interface to networks. The multimedia database can be implemented with either a real-time encoding mechanism, such as a web camera, or archive of previously compressed bit streams, such as a DVD archive. In the case of real-time encoding, the system is more expensive since each client needs to be served with a real-time encoder. In the of case of precompressed information, the source should be precompressed with several bit rates and resolutions in order to adapt to the device capability of the client terminal or the available bandwidth when connected. In case unavailable content of certain formats or bit rates are requested, a transcoder is necessary to adapt from the precompressed material to match the needs of the client device. Such a transcoding system is more cost-effective than a real-time encoder system. An example is to use a scalable format to store the compressed contents and transcode it upon request. The advantage is that the transcoder can adapt to the statistics of the channel in real-time similar to the performance using real-time encoder. However, the additional cost of adding a transcoder is much smaller as compared to using a real-time encoder. Such a system is depicted in Figure 11.1. The client side can be either Fine Granularity Scalability bit stream format or MPEG-4 Simple Profile bit stream format depending on the terminal capability. The detailed block diagram is described in Figure 11.2.

The second part is the network interface. Depending on the specific implementation, the transmitting side should include one transport protocol to carry the multimedia content with a best-efforts approach and another transport protocol to reliably

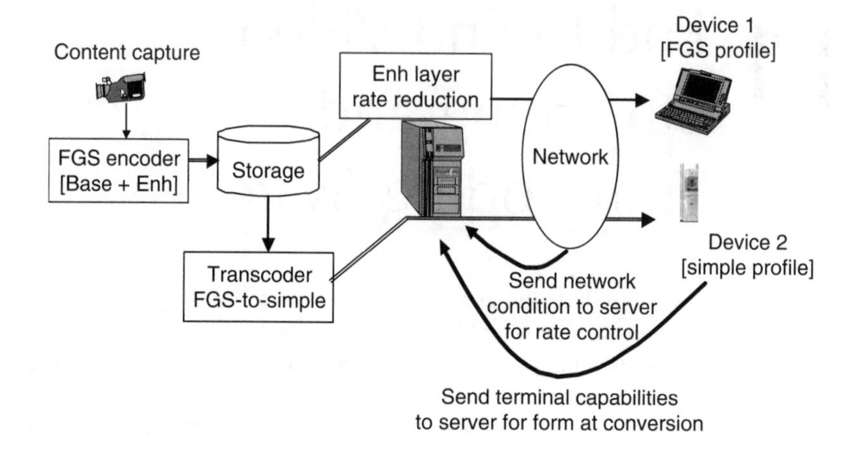

FIGURE 11.1 Real-time transcoder with the content precompressed with MPEG-4 fine granularity scalability (FGS) format.

carry the entire control signal associated with the transmitted multimedia information. For example, the Internet can use a real time protocol (RTP) such as UDP to transport the multimedia information [11-1]. As for the control signal, a more reliable protocol real-time streaming protocol, or RTSP, can be used to carry the associated control signal [11-2]. The RTSP can be implemented with TCP protocol.

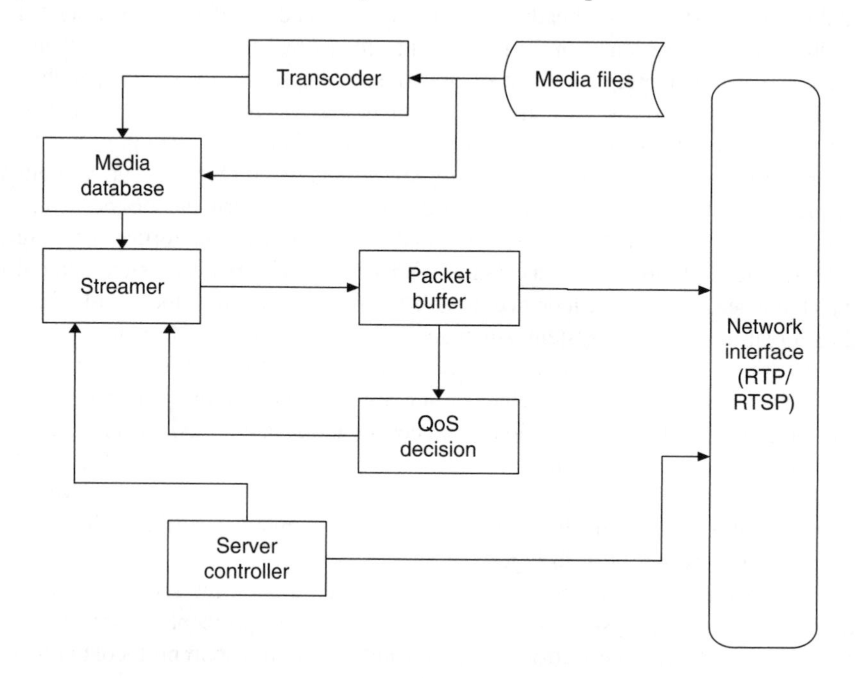

FIGURE 11.2 Block diagram of a real-time server containing a transcoder with the precompressed content.

The third part is the client side, which should contain a robust or error resilient decoder and a user interface. The decoder needs to be robust enough so that it can survive the transmission errors caused by packet loss or random errors. The client should send the relevant packet loss information through a back channel to the server so that it can adapt the content for a one-to-one communications scenario.

11.2 MPEG-4 OVER IP NETWORKS

In this section, we will describe an example system protocol that allows transport of MPEG-4 multimedia content over IP networks. The system is currently recommended by the MPEG committee as part 8 of the MPEG-4 specifications [11-3]. This section provides a abridged summary of the system specification detailed in [11-3].

11.2.1 MPEG-4 PROTOCOL LAYERS

The transport of MPEG-4 multimedia data over Internet protocol (IP) networks is a challenging task. The protocol layers include IP, UDP, Real Time Transport Protocol (RTP), MPEG-4 FlexMux, SyncLayer (SL) headers. Various RTP packetization schemes can be used for the MPEG-4 multimedia contents. The packetization process should adapt to the type of medium, such as video, speech or audio with various selections of standards. For example, the video packets should be packetized with only fraction of one frame but not across multiple frame boundaries. Similarly, the audio information should be packetized with a few number of frames that preferably has no dependencies among packets. This will facilitate error-resilient decoding and possibly random access. The MPEG-4 Part 8 specifies only a generic framework that accommodates any payload formats adequately. However, this specification requires that there are a number of common key features and such features depend on the fact that each RTP session should contain either a single elementary stream or a FlexMux-packetized stream.

If an RTP session contains a single elementary stream, the following protocols should be f protocol are recommended. The RTP time stamp corresponds to the presentation time, for example, composition time stamp (CTS) as defined in the MPEG-4 Part 1, of the earliest access unit (AU) within the packet. The RTP packets should embedded sequence numbers in the order of transmission. At the MPEG-4 synchronization layer, the payloads are logically or physically arranged by the SyncLayers(SL) sequence numbers, which should be in decoding order, for each elementary stream. As for the time scale, the MPEG-4 time scale (clock ticks per second), which is the time Stamp Resolution in the system specification of MPEG-4, will be used as the RTP timescale. For example, such a time scale should be declared in SDP for an RTP stream. The SDP is an Internet protocol used to describe a multimedia session to perform the functions of session announcement, session invitation and session initation. It is required to send at least one control packet within 5 seconds.

In order to achieve minimal interoperability, and to ensure that any MPEG-4 stream can be carried, all the decoders should implement the generic payload format as defined in "draft-ietf-avt-mpeg4-multisl-04.txt" as a default RTP payload mapping scheme [11-4]. Any new payload format needs to be a reconfigurable subset of the generic payload format. During transmission, the streams can be synchronized

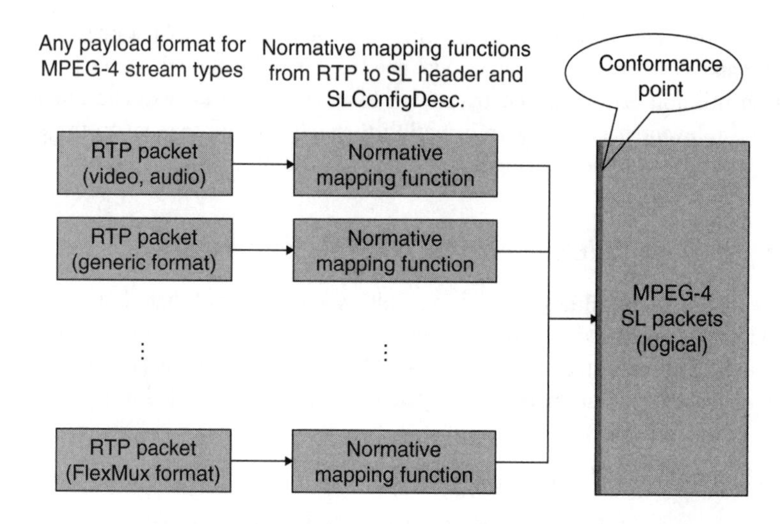

FIGURE 11.3 RTP packet to SL packet mapping [11-3].

using typical RTP techniques as described in the Real Time Control Protocol (RTCP) sender reports. The RTCP provides QoS feedback through this sender report at the source and the receiver report at the client terminal. Typically, the RTCP keeps the 5% of the control packets of the total session bandwidth for sending these reports. A quarter of the reserved bandwidth is used for the sender report and the remaining three quarters bandwidth is used for the receiver report. When the MPEG-4 object clock reference (OCR) is used, it is logically mapped to the NTP time axis used in RTCP.

The RTP packetization schemes can be used for the MPEG-4 elementary streams such as audio and video streams without the MPEG-4 system elementary stream, they can also be used in the context of an overall MPEG-4 presentation and the object descriptor framework [11-3]. In the second case, a SLConfig Descriptor is sent to describe the stream as shown in Figure 11.3. Each RTP stream is passed through a mapping function that is specific to the payload format used. This mapping function yields an SL packetized stream. The SL Config Descriptor describes this logical stream, not the actual bits in the RTP payload. For example, the RTP sequence number can be used to make the SLPacketHeader sequence number, while the other SL fields may be similarly, dynamically, or from static values in the payload specification.

For example, as all RTP packets carry a composition time stamp, the flag in the SL header indicating its presence can normally be statically defined as "true." Each payload format for MPEG-4 content specifies the mapping function for the formation of the SLConfigDescriptor and the SLPacketHeader.

The other payload formats are signaled as dynamic payload IDs, defined by a suitable name specified in SDP RTPMAP attribute. In particular, the development of specialized RTP payloads for video with respect to video packets and audio that provides interleave is expected. These schemes can be compatible with the default scheme required here. The RTP payload format of a bit stream can be a video elementary stream, audio elementary stream, or an SL-packetized stream using FlexMux using the MPEG-4 TransMux Layer protocol.

Most terminals implementing a given subsystem (for example, video) can accept at least an elementary stream and the default SL format of that stream. This means the terminal should accept the RFC 3016 and also the generic payload format for MPEG-4 visual part [11-5]. It is also recommended that terminals implementing a given payload format should accept any stream over that format for which they have a decoder, even if that packetization is not the best packetization scheme for that stream.

For multimedia streams such as video and audio, a certain quality of service is particularly important for an acceptable presentation of content. Since the current Internet only offers best-efforts service, it is critical the server and the client actively investigate possible solutions, such as forward error correction code (FEC), retransmission, or repetition. Another alternative is to perform feedback control so that the server can adjust its transmitted bandwidth using either a real time encoder or transcoder. Also, it is preferable that efficient packetization mechanisms and low overhead are considered. The efficient packetization mechanism means the ability to send in a single RTP packet that contains multiple consecutive access units (AU) each with its own SL information.

We assume that any session described by SDP protocol has at most one MPEG-4 session. The senders should alert the client that an MPEG-4 session is included, by means of an SDP attribute [11-3] as defined with the following syntax.

a = mpeg4-iod [<location>]

The attribute "location" is optional in an RTSP session. If such an attribute is not supplied, the IOD is retrieved over the RTSP session by using DESCRIBE with an acceptance of type application/mpeg4-iod or application/mpeg4-iod-xmt [11-3]. If the SDP information is supplied by some other means, such as a file or in SAP, the location information is mandatory. The location should be a universal resource locator (URL) enclosed in double-quotes, which will supply the IOD. The small ones may be encoded using "data:," otherwise with "http:" or other suitable file-access URL. When the application/mpeg4-iod-xmt type is used, the IOD in XMT format shall be supplied. The InitialObjectDescriptor is defined in subclause 8.6.3.1 of the MPEG-4 system specification and its XMT format is defined in subclause 15.8.3. of the second amendment of MPEG-4 system specification. Any terminals using IOD needs to understand the binary IOD and the textual IOD.

The new encoding names for the a = rtpmap attribute is defined as

a = rtpmap:<payload> <name>/<time scale>/<parameters>

The attribute "payload" is the dynamic payload number. The attribute "name" is defined in the IETF specification for the payload format. Independent of the payload format, each media stream is placed in an appropriate media section. For example, a payload format that can carry both video and audio streams may be used in sections of SDP starting both with "m = video" and "m = audio." The MIME name for the payload format is thus registered under all applicable branches [11-3].

The mapping of RTP streams to elementary streams needs to cover the FlexMux case and the single stream case. Within the SDP information, a stream-specific attribute is presented for each MPEG-4 stream. It takes one of two forms, depending on whether a single elementary stream or a FlexMux format stream is carried. In the case of a single elementary stream, the attribute is defined as

a = mpeg4-esid : a

where the symbol "a" is the ESID. Other SDP attributes should carry values consistent with those carried in MPEG-4 system specification.

11.2.2 Multipurpose Internet Mail Extensions (MIME) Types

As described in [11-3], for the top-level MIME types, such as the attributes including audio, video, and graphics. The attribute "video" is used for MPEG-4 visual streams. The video is defined in MPEG-4 visual specifications as the Streamtype equal to 4 and the graphics is defined in MPEG-4 system specifications as the StreamType equal to 3. Another approach is to use the MPEG-4 systems streams to convey information needed for an audio/visual presentation. Similarly, the attribute "audio" is used for MPEG-4 audio streams defined as the Streamtype equal to 5. Similarly, the MPEG-4 system streams can convey information needed for an audio-only presentation. The attribute "application" is used for the MPEG-4 systems streams with the StreamType values set to all other values for other purposes than audio/visual presentation. For example, MPEG-J streams can be transmitted accordingly.

For the elementary streams, lower-level MIME types are used. When a visual MPEG-4 elementary stream is served (e.g., over HTTP) and is identified by a MIME type, the type "video/MPEG4-visual" is used. This MIME type may require optional parameters to carry all necessary information to configure a receiver. Therefore, no further meta-information defined by the MP4 file format or by the MPEG-4 object descriptor framework needs be provided in the data. The data itself merely represents the media content. The format of the bit stream including timing as defined in the MPEG-4 visual specifications.

The payload names used in an RTPMAP attribute within SDP, to specify the mapping of payload number to its definition, also come from the MIME name space. Each of the RTP payload mappings defined above has a distinct name. The visual streams are identified under "video" and audio streams are identified under "audio," and otherwise "application" is used. In some cases, the initial object descriptor needs to be identified with a MIME type. In this case, the type "applications/mpeg4-iod" is supported, and the type "application/mpeg4-iod-xmt" may be supported. In the latter case, the IOD will be described in an XMT textual format.

When an MP4 file is served (e.g., over HTTP) or otherwise is identified by a MIME type, the type "video/mp4" should be used. The types "audio/mp4" may be used when the MPEG-4 presentation contained within the MP4 file has no visual presentation and refers to a pure audio presentation.

11.2.3 Real Time Streaming Protocol (RTSP)

As described in [11-3], the real-time streaming protocol (RTSP) may be used as a session control protocol for sessions that carry MPEG-4 information. When RTSP is used as a session-control protocol, the RTP should be used as the transport protocol. The initial DESCRIBE format should be SDP. If the SDP information reveals that an IOD is needed, and the terminal does not have it, then a second DESCRIBE accepting an IOD is performed. If all ISO/IEC 14496 streams are closed (TEARDOWN), then the RTSP session ID will be lost. The next (re-)opened stream will supply a new session ID. Note that the target of the URL has not changed in the interval but new DESCRIBEs may be needed.

As described in [11-3], when it is necessary to reference an RTP stream directly from an elementary stream descriptor (ES_Descriptor), the URL field of the descriptor can be used. For example, the URL could contain the SDP description of the stream using the "data:application/sdp" scheme. When it is necessary to embed stream data directly inside an ES_Descriptor, the URL field of the descriptor can be used. For example, the URL could contain the data using the correct MIME type. In this case, the data consists of one SL packet that contains one access unit.

When using IP Multicast, the SDP information describing the MPEG-4 session should be made available to the terminal. In addition, elementary stream descriptors may use URLs to directly address ESs. The goal of such URL would be to convey information to enable the terminal to directly connect to the RTP channel carrying the ES. The URL may contain the SDP information required to access the stream as described in Section 11.3.1.

11.3 MPEG-4 OVER IP TEST BED

It is a challenging problem to stream real-time video over heterogeneous networks for a wide range of consumer electronic applications under universal multimedia access (UMA) [11-6]. These media suffer from bandwidth fluctuation and several types of channel degradations, such as random errors, burst errors, and packet losses [11-7]. Thus, the MPEG-4 committee has adopted various techniques to address the issue of error-resilient delivery of video information for point-to-point multimedia communications. The MPEG-4 committee further developed the MPEG-4 streaming video profile (SVP) and fine granularity scalability (FGS) profile [11-8] that provide a scalable approach specifically for steaming video applications.

In a heterogeneous network, the receiving devices may have limited display and processing power, or may only be able to handle a particular compression format. For devices having FGS decoding feature, the server can truncate the enhancement layer bit stream to fit the variable channel bandwidth. For other devices that are only capable of decoding MPEG-4 Advanced Simple Profile decoder require only the base layer for display. Moreover, the consumers may have different preferences in terms of terminal resolution, transmission bandwidth and picture quality. Thus, MPEG-21 further develops digital item adaptation techniques [11-6] to interact with the streaming server about the receiver's capabilities and user characteristics. However, it becomes more difficult to develop the minimum set of digital item description

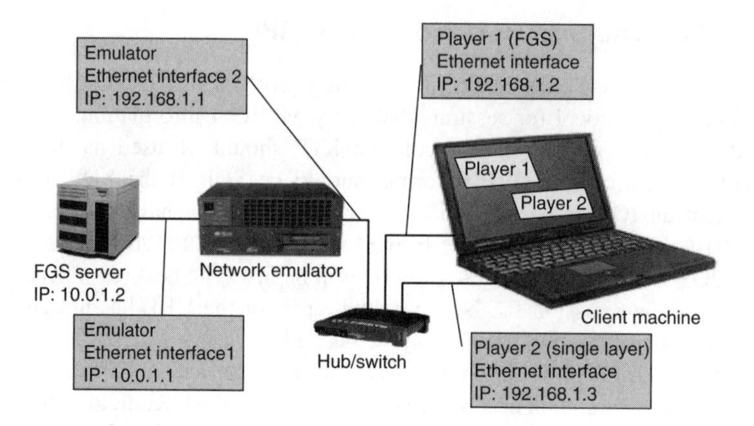

FIGURE 11.4 A simulation system for an end-to-end transmission of multimedia content.

schemes since there is no reference software that supports DIA scheme so far for functionality emulation and advanced analysis. This platform provides such a solution and platform in order for the MPEG-21 committee to experiment with various user scenarios.

Some schemes [11-9]-[11-10] have been proposed to simultaneously stream or multicast video over Internet or wireless channels to a wide variety of devices using MPEG-4 FGS. These schemes show promising results of scalable coding techniques. However, it is difficult to evaluate the results without common test conditions. Thus, the MPEG committee has drafted the testing procedures for evidences on scalable coding [11-11] and applications and requirements for scalable coding [11-12] in July 2002. Moreover, a practical network environment behaves differently for each experiment, which makes it difficult to test the streaming systems with results that can be studied and duplicated. In Figure 11.4 a simulation system is shown that contains a network simulator, namely, NISTnet [11-13] . The NISTnet provides easy controls to create duplicable network conditions, such as packet loss ratio, jitter, bandwidth variation, and delay.

11.3.1 FGS-Based Streaming Test Bed

The goal is to support the MPEG-21 DIA scheme with a more strict evaluation methodology according to the specified common conditions for scalable coding. Figure 11.5 shows the proposed test bed system architecture, which covers four key modules: the FGS-based Video Content Server, Video Clients, Network Interface, and Network Simulator.

11.3.1.1 FGS-Based Video Content Server

The content server covers seven submodules, including FGS encoder, video database, stream buffer, streamer, packet buffer, IP protection, and sending controller. The graphical user interface is shown in Figure 11.6. The FGS encoder offline compresses

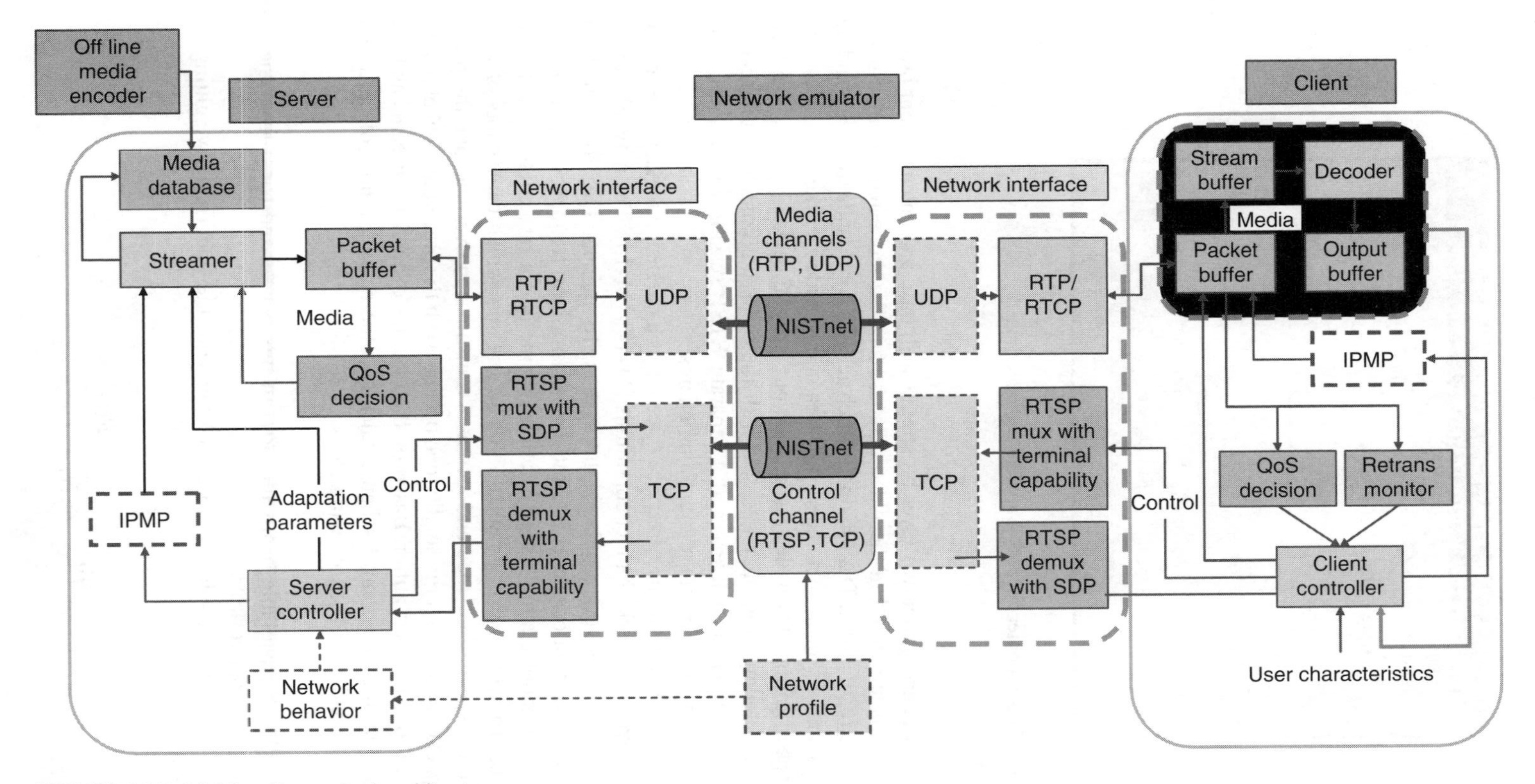

FIGURE 11.5 Multimedia test bed architecture.

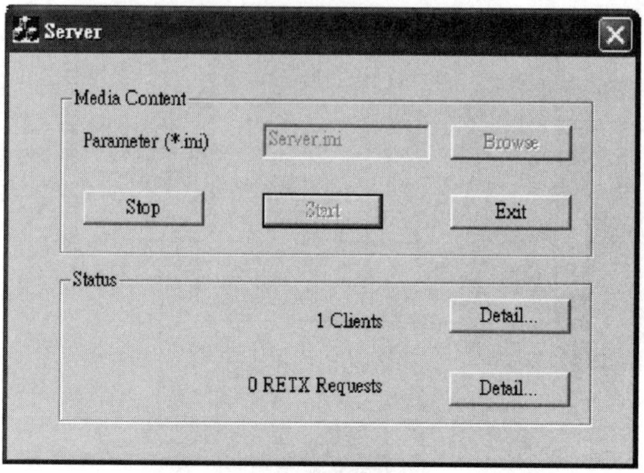

FIGURE 11.6 Video content server graphical user interface.

each video sequence into a two-layer bit stream with base and enhancement layers. Both bit streams are stored in the video database and the requested bit streams are moved to the stream buffer. The streamer, which accepts commands from the sending controller, segments each demanded bit stream into video packets according to MPEG-4 specification. The video packets are put into the packet buffer as the RTP payload. The sending controller interacts with the receiving controller to create a media session for video delivery and a separate RTSP session for accepting the retransmission requests from the individual clients.

To provide the best quality of service (QoS), it is a challenge to adapt the source rate to the current network channel conditions [11-14]-[11-15]. The network conditions are defined as network profiles in this system. To maintain QoS for each client under the consideration of packet loss ratio, retransmission frequency, and effective bandwidth, we adopt a simple segmentation scheme and a rate control scheme in the streamer. To avoid overfragmentation of bit streams, which causes lots of RTP packets with very small payload and increases transmission overhead and the probability of packet loss, we merge small video packets with the preceding packet before storing them to the packet buffer. Based on the bit rates of FGS bit streams and the network profile, the rate control scheme calculates the actual number of delivered packets per second with the following steps.

Step 1: Get the available bandwidth $R(t)$ in bits per second (bps) from the prespecified network profile.

Step 2: Allocate available bits to each VOP with the underlying weighting function

$$R_I(t) = \frac{R(t)}{w_I N_I + w_P P_I + w_B B_I}$$

where the weights are w_I, w_p, and w_B for I-, P-, and B-VOPs, respectively. The values $w_I = w_p = 1$ and $w_B = 0.6$ are found empirically.

Step 3: With the allocated bits for each type of VOP, send all of the packets carrying the base layer bit stream.

Step 4: With the remaining bandwidth available in a 1-s window, send the maximal number of packets covering the enhancement layer bit stream. If the remaining bandwidth is not sufficient for a full packet, the bit stream will be truncated before a FGS resynchronization marker or bit-plane start code. The bits actually used are bounded by the allocated budget.

A simple time scheduling approach is used to spread the packets. Since the packet size is specified in the encoding processing, all video packets in the encoded bit stream have almost identical sizes. We use the time interval that is taken to recognize and retrieve every packet data from the bit stream to put each packet into the channels. Thus, we stream out the packets in evenly spaced time intervals to prevent burst packet loss at the receivers.

11.3.1.2 Video Clients

As illustrated in Figure 11.5 the client has seven submodules: the FGS decoder, stream buffer, packet buffer, video display, QoS Monitor, and receiving controller.

At the client side, we use the packet buffer and QoS monitor to check the packet reception status and request packet retransmission when any packet loss occurs. To prevent the retransmitted packets from occupying a large percentage of the effective bandwidth, the QoS monitor adopts a simple approach in managing the occurrence of retransmission requests. When the streaming session between the server and client is established, the client will fill the packet buffer with 3 s of packets and then move the media data into the stream buffer. The buffer fullness is monitored by the QoS

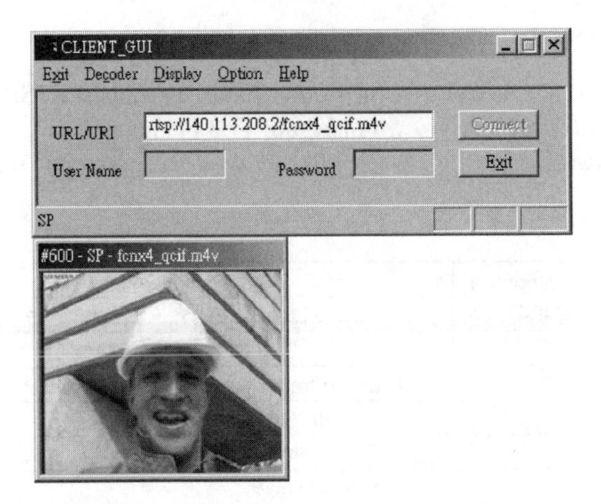

FIGURE 11.7 Video client graphical user interface.

monitor with the following steps for error protection of FGS base-layer bit stream using retransmission.

Step 1: *Packet collection.* Each packet is identified with its "sequence number" from the RTP header. Since the arrival time of each packet may not be in the order of sequence number, we use the first second of the 3 s to wait for the packet arrival.

Step 2: *Packet loss detection.* Within the packet buffer, we have flags to register the successful arrival of every packet. The loss detection procedure is activated periodically each second to balance the overhead and QoS status. As the flag of a packet is triggered as "OFF" in this period, the packet loss is detected.

Step 3: *Retransmission.* When packet loss is detected, the receiving controller sends a retransmission request with the sequence number of the packet. We use the remaining 2 s to wait for the requested packet to arrive at the packet buffer. As the 2 s nearly expire, the retransmitted packets that have not arrived at the packet buffer in time are declared as permanent loss. To avoid the retransmitted packets from occupying the effective bandwidth of the normal media delivery, any lost packet may have N chances for retransmission, where N is set as 3 empirically.

To decode and display the bit stream corrupted by lost packets, a crash-proof decoder with error resilience and concealment is implemented in our proposed system. At the enhancement layer, we utilized the error resiliency tools to verify the robustness of decoding process [11-16]. At the base layer, the bit stream errors are detected with a prior knowledge of the bit stream syntax and its semantics. For the syntactic errors, we will detect errors that are caused by invalid codewords or stuffing bits for the decoding frame or structure. In particular, we will check the cases in which there are more than 64 coefficients in a block, the MB number exceeds the VOP's MB number, or the codeword not in the VLC table [11-17].

11.3.2 Network Interface

To link the server, client, and transmission channel, a standard RTSP/RTP-based network interface is adopted. As shown in Figure 11.8 the network interface adopts three categories of network protocols covering the network-layer protocol, transport protocol,

<table>
<tr><td rowspan="2">Application control commands</td><td colspan="2">Layered video data</td></tr>
<tr><td>Base layer</td><td>Enhancement layer</td></tr>
<tr><td>RTSP</td><td colspan="2">RTP</td></tr>
<tr><td>TCP</td><td>UDP with retransmission</td><td>UDP</td></tr>
<tr><td colspan="3">IP</td></tr>
<tr><td colspan="3">Data link</td></tr>
<tr><td colspan="3">Physical layer</td></tr>
</table>

FIGURE 11.8 Network protocol stack.

and session control protocol. Similar protocol stack can be found in [11-14]. The network-layer protocol using IP networks serves the basic network support, such as the network addressing. The transport protocols, including UDP, TCP, and real-time transport protocol (RTP) [11-1], are used to provide an end-to-end network transport for video streaming. The session control protocol uses real-time streaming protocol (RTSP) [11-1] that specifies the messages and procedures to control the media delivery during an established session. Since there is no standard RFCs or Internet drafts that specify ways to map FGS bit stream to RTP payload and to support MPEG-21 DIA via RTSP, we create our own RTP payload mapping and RTSP content messages for streaming.

For simplicity, basic RTSP methods and a nonstandard RTSP retransmission request are employed for session control. There are four basic client-server RTSP messages, including DESCRIBE, SETUP, PLAY, and TEARDOWN. For an instance in Figure 11.9 the message DESCRIBE is used to describe the terminal capabilities and user characteristics. Within the DESCRIBE message, a new content type as application/mpeg21_dia is declared to support MPEG-21 DIA scheme. The MPEG-21 DIA descriptions are transmitted through a RTSP packet when a client wants to subscribe to a server. After successfully subscribing to the server, the client uses the SETUP message to create the media delivery session with specified terminal capabilities, transport protocols, and port numbers. PLAY starts to transport the media under the built session and TEARDOWN ends the transport and the underlying session. Due to the tradeoff between the best QoS and usage of the effective network

```
DESCRIBE rtsp://140.113.24.154/lasgow.mp4 RTSP/1.0
Cseq: 0
User-Agent: NCTU FGS Player
Accept: application/mpeg21_dia
Content-Type: application/mpeg21_dia
Content-Length: 457
<xml version="1.0" encoding="UTF-8?">
<DIDL xmlns="urn:mpeg:mpeg21:2002:01-DIDL-NS">
   <UserCharacteristics>
      <User>
<Account>johndoe</Account>
          <Password>@%$FHG%^^^&SS</Password>
</User>
   </UserCharacteristics>
   <TerminalCapabilities>
      <Decoding>
       <Format>FGS</Format>
      </Decoding>
      <Hardware>
               <ScreenSize>
            <Width>352</Width>
            <Height>288</Height>
               </ScreenSize>
      </Hardware>
          </TerminalCapabilities>
</DIDL>
```

FIGURE 11.9 Illustration of the client-server RTSP DESCRIBE message with MPEG-21 DIA description.

bandwidth, the retransmission is employed only for the base layer bit stream. Whenever there are packet losses, the client can send (nonstandard) RTSP GET_ PARAMETER requests to the server for retransmission of the missing packets from the base layer bit stream.

To support both the network protocol stack and MPEG-21 DIA scheme, the proposed network interface has six submodules. The network interface connecting the server and network includes RTP Mux, UDP transmitter, RTSP Mux with DIA descriptor, RTSP DeMux with DIA parser, TCP transmitter, and TCP receiver. The network interface connecting the client and network has the same submodules except for RTP DeMux and UDP receiver.

The media data is packetized into RTP packets in the RTP Mux module prior to the transport. The RTP packets are then transmitted using UDP, and received and demultiplexed by RTP DeMux into the packet buffer at the clients for playback. The RTSP stream of the control messages is transported via TCP. In addition, for the DIA descriptions, an XML parser and an XML compositor are included under RTSP DeMux and Mux modules, respectively.

11.3.3 Network Simulator

In the test bed, an IP network emulator, NISTnet [11-13], is used to provide repeatable network environments that are close to practical wide-area heterogeneous network environments. NISTnet is a public domain Linux-based IP network emulator developed by the National Institute of Standards and Technology. It provides a simulated IP environment with many controllable channel parameters, such as packet loss ratio, jitter, bandwidth variation, and delay. A Graphical User Interface (GUI) module that interacts with the NISTnet kernel module is developed for this test bed project. The time-varying network conditions are recorded in a network profile text file as illustrated in Figure 11.10. In addition to controlling the NISTnet module to simulate the prespecified network profile, the GUI module also displays real-time channel usage plots and statistics of both the up-link and down-link connections.

Network profile text file
A simple network profile
#

# Time (sec)	PLR (%)	BW (bps)	MeanDelay (ms)	SDD	Buffer
0	0	100000	0	0	150
20	0	50000	0	0	150
40	0	10000	0	0	150
60	0	50000	0	0	150

FIGURE 11.10 The network profile used to control NISTnet; PLR = the packet loss ratio; SDD = standard deviation of the delay.

TABLE 11.1
PSNR of the Y Component for Two Sequences at Target Bit Rates

	Bit Rate	PSNR (dB)	Bit Rate	PSNR (dB)	Bit Rate	PSNR (dB)
Foreman	64 kbps	27.9	128 kbps	26.8	256 kbps	25.7
News	64 kbps	30.0	128 kbps	32.5	256 kbps	33.7

11.3.4 Experimental Results

According to the requirements from the MPEG committee [11-11], [11-12], the test bed is set up as in [11-13]. Based on this setup, scalable and single-layer bit streams can be transported simultaneously to the clients via an identical network environment, which is controlled by the NISTnet, to achieve a fair comparison of their performance and error robustness.

To demonstrate the streaming performance of this test bed, we adopt two test sequences, "foreman" and "news," which are in YCbCr 4:2:0 QCIF format. Each sequence has 300 frames, and the source frame rate is 30 frames per second (fps). For each sequence, the first frame is encoded as I-VOP and remainders are encoded as P-VOPs. The output frame rate is 10 fps. The base layer bit stream is encoded at 32 kbps. The network profiles cover three average network bandwidths, 64 kbps, 128 kbps, and 256 kbps. In addition, to show the quality variation over time, a novel sequence in CIF and with 2000 frames is used. The sequence is encoded in 30 fps into a structure of IBBPBBP… The base layer is encoded at 512 kbps. In the network profile, the time-varying bandwidth in a range of [512 kbps, 1,024 kbps] and three packet loss ratios—0%, 5%, and 10%—are adopted for simulations. In all simulations, the TM5 rate control scheme is used to fit with the target bit rate.

Table 11.1 illustrates the PSNR of the Y components of both sequences for streaming without retransmission, and Figure 11.11 shows two bandwidth curves of network conditions. It is interesting that the PSNR of foreman sequence is decreased as the average bit rate is increased. The PSNR degradation may be caused by the following two reasons. First, since the increased bit rate may increase packet loss rate when the channel bandwidth is insufficient. Second, the retransmission is not enabled. In addition, Figure 11.12 gives the PSNR of the Y component of novel sequence under the specified network conditions. The results show that the proposed test bed system can provide better QoS for the client under large network bandwidth v ariation.

11.4 MPEG-4 TRANSCODING ON THE TEST BED

In this section, we described a transcoding system that is built on top of the multimedia resource delivery test bed. The concept is to address the design of a multimedia server. There are several possible approaches to design such a system. The feasible approaches include real-time encoding, bit stream switching, and transcoding.

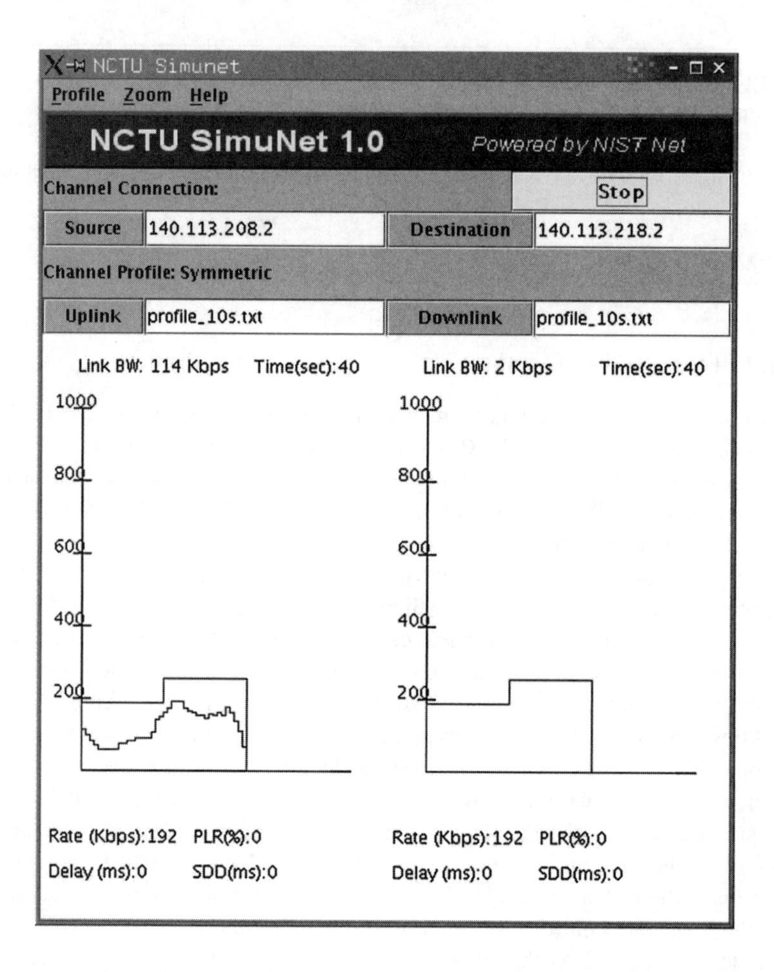

FIGURE 11.11 NISTnet network profiles and real channel conditions over time for streaming the foreman sequence. The straight lines are the specified bandwidth in the network profiles and the curved lines are the real bandwidth.

The real-time encoding is straightforward by building a stream service with a farm of real-time encoders that communicates with the clients in a one-to-one manner. The advantages are that the content only needs to be stored in several uncompressed resolutions. The real-time encoder can adapt to the bandwidth fluctuations and delay jitter. The obvious disadvantage is the cost of such implementation.

The second approach is the bit stream switching that utilizes only precompressed bit streams with prescribed bit rates and resolutions. The client selects a resolution suitable for his or her device and the server will adapt to the actual channel bandwidth using bit streams of different bit rates. The server will receive feedback back information to determine which bit streams it should use. The adaptation is done at a predetermined random access point. The advantages are that the real-time encoder is not used and the storage is much less. This is the most commonly used approach for Internet streaming. The disadvantages are that the bit rates are predetermined so

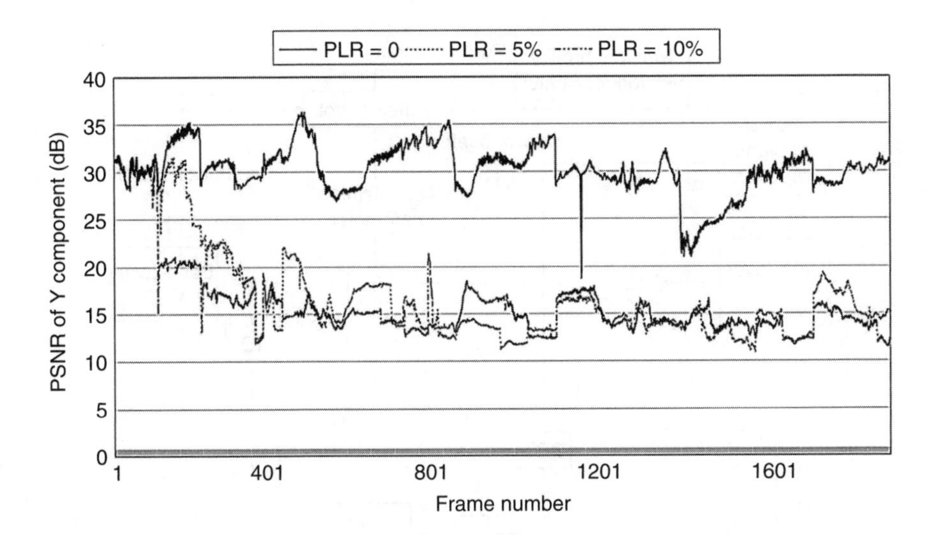

FIGURE 11.12 PSNR of the Y component of the novel sequence for various packet loss rates.

the server has to take a conservative bandwidth utilization control so that the networks are not congested. The discrete levels of bit rates will result in underutilization of the channel capacity. For example, if the server can only serve 128 kbps or 256 kbps, a channel that has 200 kbps can only be served with 128 kbps. The remaining bandwidth is not fully utilized.

Another approach to address this issue is to use scalable coding technique. The scalable approach transmits a scalable bit stream that allows the network truncates the bit stream upon congestion. The client can also decide, on the fly, how to truncated its bit stream to match the bandwidth, computation power, and device display capability. Therefore, it is obviously the most convenient approach. However, the scalable bit stream suffers from less coding efficiency as compared with the nonscalable approach and the scalable decoder is not as widely deployed as the nonscalable decoder. Therefore, the transcoding can be considered as a viable and efficient solution for a streaming service.

11.4.1 Rate Control for a Transcoder

As studied in [11-18], the ρ-domain source model based rate control presents a very accurate mathematical description of the rate-distortion behavior of transformed-based coding. In [11-19], a transcoder based on archived FGS bit streams is proposed. In that design, the truncation of the enhancement layer bit stream needs a novel rate control strategy.

The issue is to allocate the bits for different types of VOPs. To solve this problem, it is well known that TM5 rate control can meet the goal adequately but not very accurately. In [11-19], Lin proposed the following approach to perform rate control in a transcoder precisely. For a given budget $R(t)$, and a VOP type x, the bit rate for each frame is computed with Equation 11-1. The weightings for different

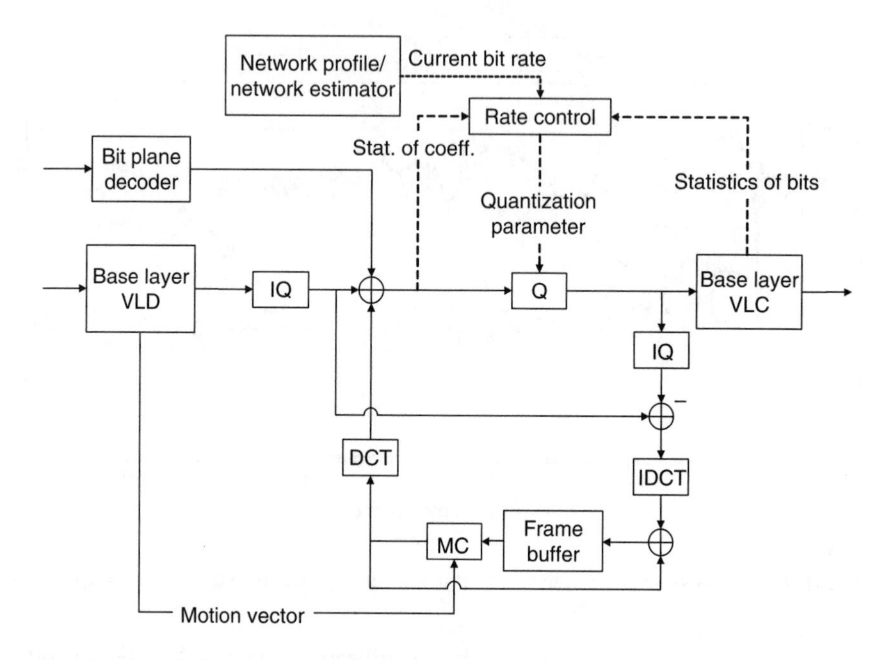

FIGURE 11.13 A transcoder with a rate control module for video streaming.

types of VOPs are found empirically. For the following experiments, the values $W_I = W_P = 1$ and $W_B = 0.6$ for I-VOPs, P-VOPs, and B-VOPS are used, respectively.

$$Rx = \frac{R(t)}{W_I N_I + W_P N_P + W_B\ N_B} \times W_x \tag{11.1}$$

where N_x is the number of x-VOPs; x = one of I, P, or B.

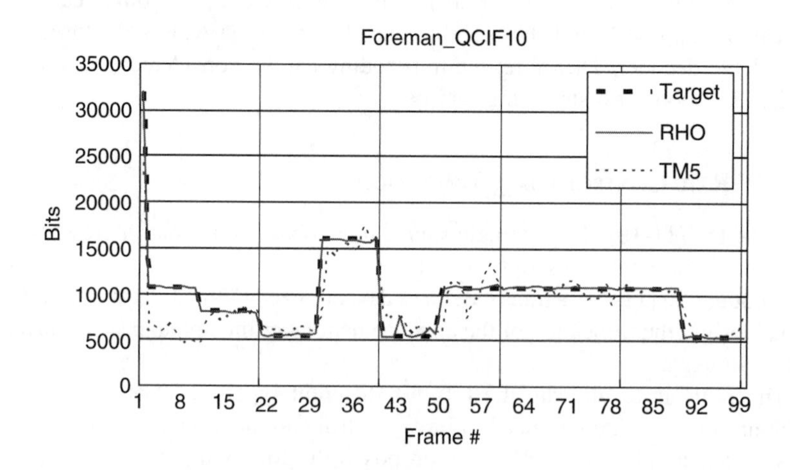

FIGURE 11.14 Actual bit rate encoded by the transcoder.

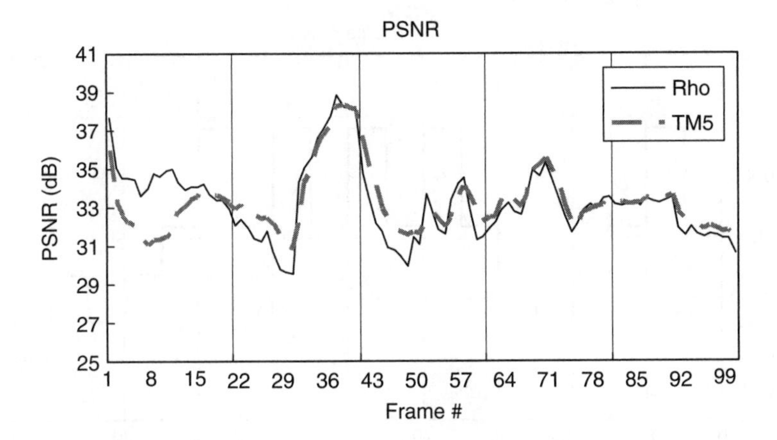

FIGURE 11.15 PSNR plot encoded by the transcoder.

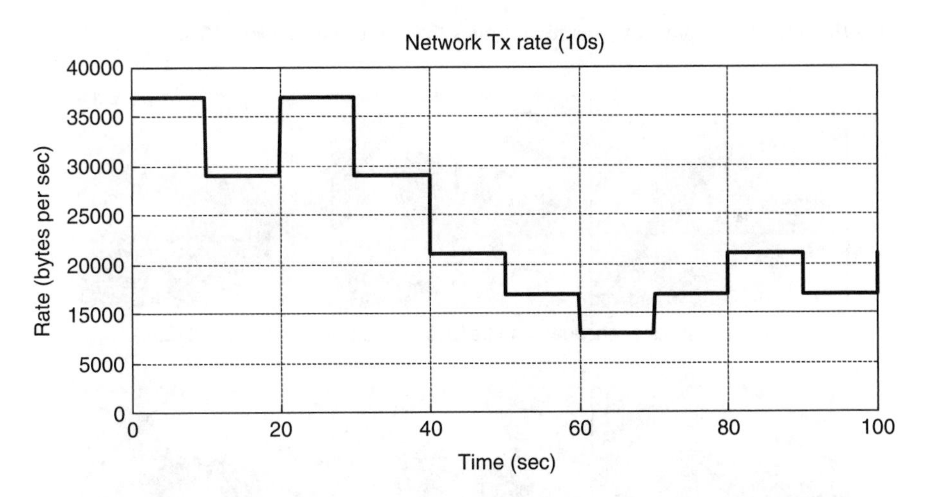

FIGURE 11.16 The plots of dynamic bit rates that are changed every 5 s.

Figure 11.13 illustrates the rate control module of the proposed transcoder. The rate control module in the transcoder computes the quantization parameter based on either TM5 approach or the linear rate-distortion model. As shown in Figure 11.14 the transcoder can perform more accurate rate control with ρ-domain source model while the PSNR results are comparable, as shown in Figure 11.15.

11.4.2 Dynamic Test for a Transcoder

To demonstrate the advantages of such a real-time transcoder, the multimedia resource delivery test bed is combined with a transcoder. The ρ-domain transcoder is also used in the experiment. As shown in Figure 11.16 and Figure 11.17 the bit rate is dynamically

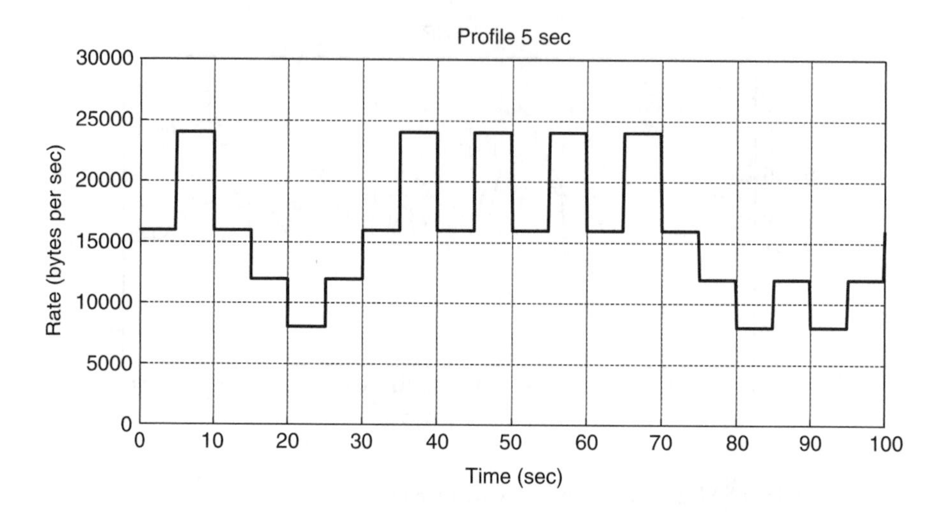

FIGURE 11.17 The plots of dynamic bit rates that are changed every 10 s.

(A) Snap shots of encoded and transcoded sequences at high bit rate.

(B) Snap shots of encoded and transcoded sequences at low bit rate.

FIGURE 11.18 Snap shots of encoded and transcoded video sequences. Left: the reconstructed image by encoding the original sequences. Middle: the reconstructed image from the cascaded transcoder. Right: the reconstructed image from the FGS-to-simple transcoder. The set of images in panel (A) is for high bit rates and the set of images in panel (B) is for low bit rates.

changed from 3 Mbps to 700 kbps at every 5- or 10-s intervals. This is about a factor of 4 in dynamic range swing. The resultant picture is shown in Figure 11.18.

In this set of experiments, four types of encoders are compared, namely FGS, transcoder, pseudo-real-time encoder, and FGS base layer. The FGS encoder is a

TABLE 11.2
Average Bandwidth Utilization for Different Streaming Methods under Various Network Profiles

5-second Profile		10-second Profile	
Method	**Bandwidth Utilization**	**Method**	**Bandwidth Utilization**
FGS	87.87%	FGS	91.25%
Transcoding	93.41%	Transcoding	90.38%
Pseudo RT Enc.	97.65%	Pseudo RT Enc.	94.18%
FGS Base Layer	17.92%	FGS Base Layer	14.96%

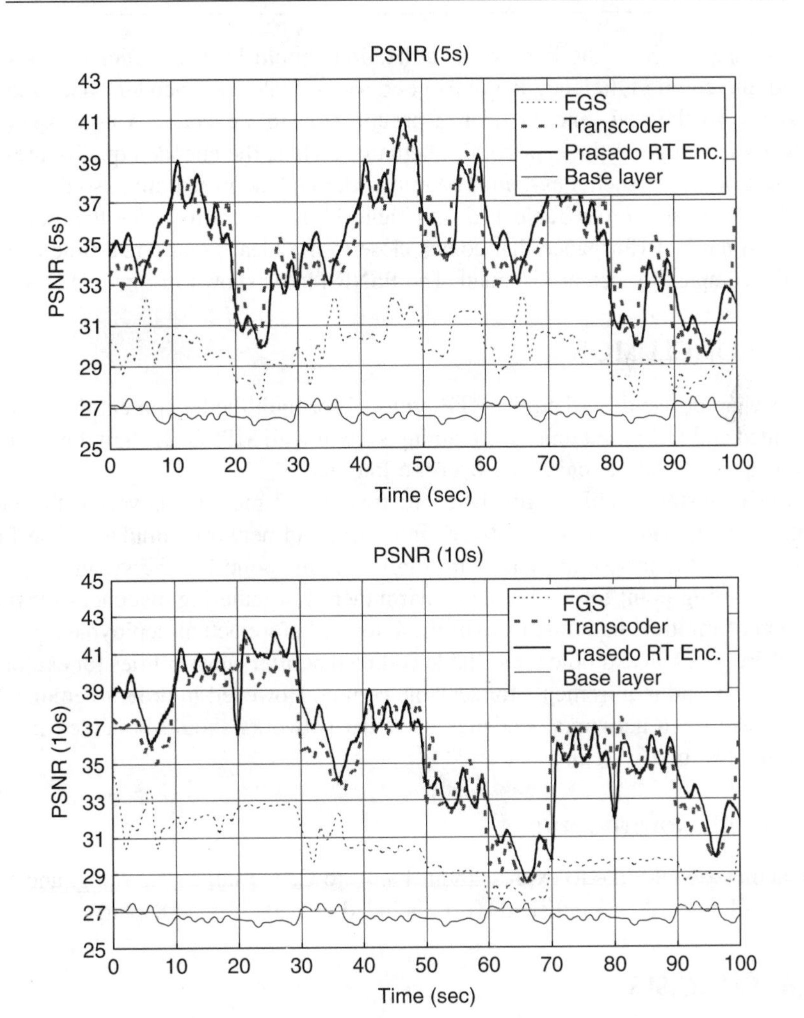

FIGURE 11.19 PSNR plots corresponding to the bandwidth plots of Figure 11.16 and Figure 11.17.

TABLE 11.3
Averaged PSNR Values of Different Streaming Methods under Various Network Profiles

Methods	5-second Profile			10-second Profile		
	Y	U	V	Y	U	V
FGS	29.71	36.10	36.17	30.28	36.54	36.56
Transcoding	34.56	39.68	39.78	35.47	40.34	40.45
Pseudo RT Enc.	34.86	39.74	39.86	35.98	40.54	40.67
FGS Base Layer	26.67	34.19	34.21	26.67	34.19	34.21

scalable approach so the bandwidth utilization should be very efficient. As mentioned, the coding efficiency is not as good as the real-time encoder. Since there is no access to the real-time encoder, a pseudo-real-time encoder is used assuming perfect knowledge of the bandwidth fluctuation. Thus, the encoded quality presents an upper bound of such a system. The transcoder is done in real time, so the experiments are realistic. From Table 11.2 and Table 11.3, it is obvious that the transcoder approach has a performance that comes close to the ideal pseudo-real-time encoder and the complexity increase is small. The PSNR plot is shown in Figure 11.19.

11.5 CONCLUSIONS

In this chapter, we described the RTP protocol for multimedia content delivery and presented an FGS-based unicast streaming system with MPEG-21 digital item adaptation as a test bed of scalability over the Internet.

In the system architecture, there are four major modules covering the video content server, video clients, network interface, and network simulator. The functionalities of the proposed system are limited at this point but the system provides a good starting point for a complete environment for gathering evidences for transmission of multimedia contents over the Internet before actual deployment.

It also gives an example for scalable coding that offers possibilities for exploiting applications and requirements for scalable coding. However, in order to emulate the real transmission networks, statistics of actual networks should be added into the proposed test bed system.

11.5.1 ACKNOWLEDGMENT

The authors would like to express their thanks to C.-J. Tsai, C.-N. Wang, and Y.-C. Lin for allowing their published work included as part of this chapter.

11.6 EXERCISES

11-1. According to your understanding, give several reasons to explain why the transcoder in Section 11.5 has a performance that is close to the pseudo-real-time encoder.

11-2. Explain why digital item adaptation is necessary for heterogeneous networks.
11-3. Explain the difference between scalable coding and transcoding in the context of end-to-end transmission system.

REFERENCES

[11-1] H. Schulzrinne, S., Casner, R. Frederick, and V. Jacobson, RTP: A transport protocol for real-time application, Internet Engineering Task Force, RFC 1889, Jan. 1996.
[11-2] H. Schulzrinne, A. Rao, And R, Lanphier, Real time streaming protocol (RTSP), Internet Engineering Task Force, RFC 2326, Apr. 1998.
[11-3] MPEG Video Group, Part 8: Carriage of ISO/IEC 14496 contents over IP networks, ISO/IEC JTC1/SC 29/WG 11 N4712, March 2002.
[11-4] P. Gentric et al. RTP Payload Format for MPEG-4 Streams, *IETF Draft, draf, draft-ietf-avt-mpeg4-multisl-04.txt,* Feburary 2002.
[11-5] Yoshihiro Kikuchi et al. RTP Payload Format for MPEG-4 Audio/Visual Streams, IETF RFC 3016, November 2000.
[11-6] A. Vetro, A. Perkis, and S. Devillers, MPEG-21 Digital Item Adaptation WD (v2.0), ISO/IEC JTC1/SC29/WG11 N4944, Klagenfurt, July 2002.
[11-7] W. Li, Overview of fine granularity scalability in MPEG-4 video standard, *IEEE Transactionson Circuits and Systems for Video Technology,* 11, 301, 2001.
[11-8] Streaming video profile—Final Draft Amendment (FDAM 4), MPEG01/N3904, 2001.
[11-9] H. M. Radha, M. van der Schhar, and Y. Chen, The MPEG-4 Fine Grained Scalable video coding method for multimedia streaming over IP, *IEEE Transactions on Circuits and Systems for Video Technology,* 11, 53, 2001.
[11-10] W. Zhu, Q. Zhang, and Y.-Q. Zhang, Network-adaptive rate control with unequal loss protection for scalable video over internet, *Proceedings of ISCAS 2001*, 2001.
[11-11] MPEG Video Group, Draft testing procedures for evidence on scalable coding technology, ISO/IEC JTC1/SC 29/WG 11 N4927, Klagenfurt, July 2002.
[11-12] MPEG Requirements Group, Draft Applications and Requirements for Scalable Coding, ISO/IEC JTC1/SC 29/WG 11 N4984, Klagenfurt, July 2002.
[11-13] NISTnet: http://snad.ncsl.nist.gov/itg/nistnet/.
[11-14] D. Wu, Y. T. Hou, W. Zhu, Y.-Q. Zhang, and J. M. Peha, Streaming video over the Internet: approaches and directions, *IEEE Transactions on Circuits and Systems for Video Technology,* 11, 282, 2001.
[11-15] Q. Zhang, W. Zhu, and Y.-Q. Zhang, Resource allocation for multimedia streaming over the internet, *IEEE Transactions on Circuits and Systems for Video Technology,* 3, 339, 2001.
[11-16] H.-C. Huang, C.-N. Wang, and T. Chiang, Streaming video in error-prone environments, Proceedings of ISCOM2001, Oct., 2001, Tainan.
[11-17] R. Talluri, Error resilient video coding in the ISO MPEG-4 standard, *IEEE Communications Magazine,* 1998.
[11-18] Z. He, and S.K. Mitra, A linear source model and a unified rate control algorithm for DCT video coding, *IEEE Transactions on Circuits and Systems for Video Technology,* 12, 970, 2002.
[11-19] Y.-C Lin, et al. Efficient FGS to Single Layer Transcoding, ICCE 2002, pp. 465–468 June 2002.
[11-20] Multimedia and Expo, 2002. ICME '02. Proceedings. 2002 IEEE International Conference on, Volume: 2, 26–29 Aug. 2002. Pages 465–468, vol. 2.

Index

A

B

D

E

F

G

H

I

Q

R

S

T

U

W

Z